S0-DUU-831

Advances in Computer Methods for Systematic Biology

Advances in Computer Methods for Systematic Biology

Artificial Intelligence, Databases, Computer Vision

EDITED BY

RENAUD FORTUNER

THE JOHNS HOPKINS UNIVERSITY PRESS

BALTIMORE AND LONDON

This book has been brought to publication
with the generous assistance of
the National Science Foundation.

Printed in the United States of America on acid-free paper

The Johns Hopkins University Press
2715 North Charles St.
Baltimore, Maryland 21218-4319
The Johns Hopkins Press Ltd., London

Library of Congress Cataloging-in-Publication Data

Advances in computer methods for systematic biology :
artificial intelligence, databases, computer vision /
edited by Renaud Fortuner.
p. cm.
Includes bibliographical references and indexes.
ISBN 0-8018-4492-4 (alk. paper)
1. Biology—Classification—Data processing—Congresses.
2. Information storage and retrieval systems—Biology—
Congresses. 3. Expert systems (Computer science)—Congresses.
I. Fortuner, Renaud.
QH83.A38 1993 574'.012—dc20 92-24949

A catalog record for this book is available from the British Library.

Contents

V
Computer Vision and Feature Extraction

VI
Conclusions

Preface

This volume originates from the ARTISYST workshop, an interdisciplinary conference on the application of modern computer methods to research in systematic biology. Organized by the University of California, Davis (UCD), the workshop was held on September 9–14, 1990, at Napa, California. Funding was provided by the National Science Foundation (NSF) under grant BSR-8815706 from the Systematic Biology Panel.

Researchers in both computer science and systematic biology deal with the acquisition, management, and dissemination of computerized information, but for different ends, established by different research backgrounds. The ARTISYST workshop provided a forum in which members of each group could discuss their problems, requirements, and recent research findings with members of the other group, and in so doing, move toward a common language with which to study the natural world. As is so often the case when researchers from different disciplines meet, the workshop participants examined the extent to which one group's research could complement the other's, once both were stripped of their specialized terminology.

A steering committee composed of computer scientists Mike Walker (co-chair), Jim Diederich, Jack Milton, Eric Horvitz, and Peter Cheeseman and systematic biologists Julian Humphries, George Lauder, Jim Rohlf, Jim Woolley, and Renaud Fortuner (co-chair) chose five areas of research in systematic biology that show particular promise of benefiting from modern computer techniques: expert workstations for systematics, identification, phylogenetic trees, database and geographical information systems, and machine vision. The committee selected 43 participants from 123 applications. Selection was governed by the need to have an appropriate distribution of computer scientists and biologists over the five discussion topics.

The goals set for the workshop were: first, to review the state of the art in these areas, to identify the points that remain unresolved, to specify important needs, and to begin to develop solutions; second, to make recommendations to NSF on the directions for research that offer the best perspective of success and that should be funded in priority over the next few years; and third, to get computer scientists and systematists together to promote understanding of each other's field in the hope that new collaborations would be established that may eventually result in more requests for funding being sent to NSF. Although the need to keep the workshop to a manageable size and focus necessitated the exclusion of research in molecular systematics, the reader will note that existing computer methods for managing molecular information are used by a number of the contributors as examples to what can be accomplished. The reader will also note that comparisons were made to computer systems developed for larger, more supportive markets in business and administration. Both comparisons raise important points about the structure of various kinds of data.

The present volume is offered as a practical guide to the proceedings of the ARTISYST workshop. The workshop presentations and discussions form the core of thirty chapters that seek to provide both an introduction to the areas of research considered and a report on the latest developments. Care has been taken to provide definitions of specialized terms, both in the text and in a glossary. An annotated bibliography lists publications that will further the understanding of the subjects treated and familiarize members of the different scientific disciplines with each other's research literature. A subject index is included, as is a comprehensive list of cited references. In addition to the present volume, the workshop produced a report to NSF with forty-four recommendations, as well as six new collaborative research teams. The NSF report is reproduced as an appendix to help people preparing collaborative grant requests in the area.

I close by extending my gratitude to all the people who made the ARTISYST workshop and the present volume possible. I thank everyone who attended the workshop, participated in the discussions, and presented their work. I thank NSF and David Schindel, whose program in Systematic Biology funded the workshop; Mike Walker (Stanford), the co-chair and workshop program organizer; the Steering Committee, whose members set the workshop objectives and reviewed the applications; the California Department of Food and Agriculture for its support of my involvement in this project; CeCe Price in the UCD Dean's office, who helped with administrative matters; Phyllis Leavitt and the UCD Nematology Department, who administered the grant and helped with the local arrangements for the workshop; Kambiz Aghaiepour, our assistant; the personnel of the Embassy Suites Hotel in Napa, where the workshop was held; and Capitol Secretarial Services in Sacramento, where the recordings made at the workshop were transcribed. And, last but not least, I thank Jack Milton and Jim Diederich, without whom the workshop and the present book would not have been possible at all.

Renaud Fortuner

Contributors

BOB ALLKIN, Computing Section, Royal Botanic Gardens, Kew, Richmond TW9 3AB, U.K., *vxxloadr@cms.am.rdg.cc.ac.uk*. Management of systematic information, database design, software for biologists, identification, and biological data standards.

JAMES W. ARCHIE, Department of Biology, California State University, Long Beach, CA 90840, *archie@csulb.edu*. Optimality criteria, phylogenetic analysis of diverse sources of data, lizard systematics.

JAMES H. BEACH, Museum of Comparative Zoology and Herbaria, Harvard University, 22 Divinity Avenue, Cambridge, MA 02138, *beach@huh.harvard.edu*. Networked information systems and information models for biodiversity/collections data.

JOHN H. BEAMAN, Department of Botany and Plant Pathology, Michigan State University, East Lansing, MI 48824–1312, *beaman@msu* (bitnet). Botanical systematics, floristics, and automated techniques for managing botanical specimen data.

PAULO A. BUCKUP, Division of Fishes, Field Museum of Natural History, Roosevelt Road at Lake Shore Drive, Chicago, IL 60605, *lrhx@midway.uchicago.edu;* currently at: Department of Ichthyology, Academy of Natural Sciences of Philadelphia, 1900 Ben Franklin Parkway, Philadelphia, PA 19103, *mvelho@pennsas.upenn.edu*. Fish systematics, biogeography, morphometrics, cladistic methodology, neotropical fish fauna, computer mapping and geographic information systems, systematic databases.

PETER CHEESEMAN, RIACS, NASA Ames Research Center, Mail Stop 269–2, Moffett Field, CA 94035, *cheeseman@pluto.arc.nasa.gov*. Bayesian statistical methods, evolutionary tree reconstruction, automatic classification, inductive learning, computer search methods.

MICHAEL J. DALLWITZ, CSIRO, Division of Entomology, GPO Box 1700, Canberra ACT 2601, Australia, *miked@ento.csiro.au*. Development of computer programs for taxonomic description, identification, and information retrieval.

JAMES DIEDERICH, Department of Mathematics, University of California, Davis, CA 95616, *dieder@ucdmath.ucdavis.edu*. Biological databases, object-oriented systems, expert workstations, user interfaces.

STANLEY M. DUNN, Department of Biomedical Engineering, Rutgers University, Piscataway, NJ 08855–0909, *smd@occlusal.rutgers.edu*. Computer vision, image analysis in systematics, and natural and biological structure.

Renaud Fortuner, Analysis & Identification, Room 340, State of California, Department of Food and Agriculture, 1220 N Street, P.O. Box 942871, Sacramento, CA 94271-0001, *fortuner@ucdmath.ucdavis.edu*. Identification, systematics, morphometrics, and intraspecific variability of plant-parasitic nematodes.

Nicole Gautier, Laboratoire de Biométrie, URA CNRS 243, Université C. Bernard Lyon 1, 43 Boulevard du 11 novembre 1918, F-69622 Villeurbanne Cedex, France, *nicole@biomac.univ-lyon1.fr*. Knowledge organization, object-oriented models, expert-systems, identification, and modeling.

Michael D. Hendy, Department of Mathematics, Massey University, Private Bag, Palmerston North, New Zealand, *M.Hendy@massey.ac.nz*. Combinatorics, discrete mathematics, and evolutionary tree-building algorithms.

Eric J. Horvitz, Stanford University School of Medicine, OSA-M105, Stanford, CA 94305, *horvitz@sumex-aim.stanford.edu*. Principles of intelligent problem solving, metareasoning and the control of computation, rationality under bounded computational resources, automation of decision analytic reasoning, and computational decision support and automated discovery in science and medicine.

H. Joel Jeffrey, Computer Science Department, Northern Illinois University, DeKalb, IL 60115, *t90hjjl@niu* (bitnet). Artificial intelligence, expert systems, knowledge engineering, design languages, applications of descriptive psychology.

Bob Kanefsky, Sterling Software, NASA Ames Research Center, Mail Stop 269-2, Moffett Field, CA 94035, *kanef@charon.arc.nasa.gov*. Interface design, evolutionary tree reconstruction, algorithm design.

Peter D. Karp, Artificial Intelligence Center, SRI International, EJ229, 333 Ravenswood Avenue, Menlo Park, CA 94025, *pkarp@ai.sri.com*. Applications of artificial intelligence to biology, biological knowledge bases, integrating database and knowledge base techniques, knowledge representation, machine learning.

Bryce Kendrick, Mycology Lab, Biology Department, University of Waterloo, Waterloo, Ontario, N2L 3GI, Canada, *bryce@mycologue.waterloo.edu*. Ecology, development, and systematics of fungi, fungal toxicity, computer modeling of fungal growth, image analysis, and molecular genetics.

Matthew McGranaghan, National Center for Geographic Information and Analysis, University of Maine, Orono, ME 04469; currently at: Geography Department, University of Hawaii, Honolulu, HI 96922, *matt@uhunix.uhcc.hawaii.edu*. Computer cartography and geographic information systems, design of graphic displays and user interfaces.

LESLIE F. MARCUS, Department of Biology, Queens College of the City University of New York; Mailing address: American Museum of Natural History New York, Department Invertebrates, CPW at 79th, New York, NY 10024, *lamqc@cunyvm.cuny.edu*. Applied multivariate statistics in systematics, evolution, paleontology, and geology; multivariate morphometrics; automatic data acquisition; and museum data management.

CHRISTOPHER A. MEACHAM, Department of Integrative Biology, University of California, Berkeley, CA 94720, *meacham@violet.berkeley.edu*. Phylogenetic theory, morphometric data acquisition, computer software.

JACK MILTON, Department of Mathematics, University of California, Davis, CA 95616 *milton@ucdmath.ucdavis.edu*. Databases, object-oriented systems, expert workstations, interactive systems.

GLEN NEWTON, Mycology Lab, Biology Department, University of Waterloo, Waterloo, Ontario, N2L 3GI, Canada, *knewton@mycologue.waterloo.edu*. Image analysis, search techniques, artificial intelligence, and modeling of biological systems.

RICHARD J. PANKHURST, Royal Botanic Garden, Inverleith Row, Edinburgh, EH3 5LR, U.K., *rjp@castle.ed.ac.uk*. Development of computer programs for taxonomic description and identification and databases for taxonomy, especially with images.

ALAIN PAVÉ, Laboratoire de Biométrie, URA CNRS 243, Université C. Bernard Lyon 1, 43 Boulevard du 11 novembre 1918, F-69622 Villeurbanne Cedex, France, *pave@frcism51* (bitnet). Knowledge organization, object-oriented models, expert systems, identification, and modeling.

DAVID PENNY, Molecular Genetics Unit, Massey University, Private Bag, Palmerston North, New Zealand, *D.Penny@massey.ac.nz*. Evolutionary trees, molecular evolution, science policy, scientific computing, and theoretical biology.

DAVID A. PORTYRATA, ARC Professional Services Group, Information Systems Division, 1301 Piccard Drive, 2d Floor, Rockville, MD 20850, *dzp@helix.nih.gov*. Integration of data collection and analysis tools with database systems, design of database systems for biomedical applications, artificial intelligence, and tools for identification.

PATRICK A. D. POWELL, Department of Electrical and Computer Engineering, San Diego State University, San Diego, CA 92182–0190, *papowell@sdsu.edu*. Real-time distributed systems, vision, and pattern recognition.

SAKTI PRAMANIK, Department of Computer Science, Michigan State University, East Lansing, MI 48824, *pramanik@cpswh.cps.msu.edu*. Parallel processing, distributed databases, software engineering, optical pattern recognition.

FRANÇOIS RECHENMANN, IRIMAG-LIFIA, Batiment HITELLA, 46, avenue Felix Viallet, 38031 Grenoble Cedex, France, *frechen@imag.imag.fr*.

Knowledge-based systems, knowledge representation, object-oriented models, frames, and taxonomic reasoning.

F. JAMES ROHLF, Department of Ecology and Evolution, State University of New York, Stony Brook, NY 11794–5245, *rohlf@sbbiovm* (bitnet). Morphometrics, numerical taxonomy, biometry, and computer applications in biology.

RAMIN SAMADANI, Star Laboratory, 202 Durand Building, EECS, Stanford, CA 94305-4055, *ramin@nova.stanford.edu*. Image processing, image analysis, and computer vision techniques applied to remote-sensing and space physics image databases.

RICHARD E. STRAUSS, Department of Ecology and Evolutionary Biology, University of Arizona, Tucson, AZ 85721, *strauss@arizrvax* (bitnet). Morphometrics and phylogenetic systematics, vertebrate development, morphogenesis, and evolution.

RICHARD J. WHITE, Biology Department, Southampton University, Southampton SO9 3TU U.K., *byi012@ibm.soton.ac.uk*. Design and implementation of biological software for database management, image handling and measurements, population analysis, biological data standards, and variation in populations.

PETER J. WINFIELD, Agricultural Scientific Services, Department of Agriculture and Fisheries for Scotland, East Craigs, Edinburgh EH12 8NJ U.K. Biological database design and implementation, biological data standards, and cultivar databases.

MARIANNE WINSLETT, Computer Science Department, University of Illinois, Urbana, IL 61801, *winslett@cs.uiuc.edu*. Databases, automated reasoning, and computer-aided design.

ROBERT ZERWEKH, Department of Computer Science, Northern Illinois University, DeKalb, IL 60115, *t90raz1@niu* (bitnet). Intelligent tutoring systems and neural networks.

I

Artificial Intelligence

1

Automated Reasoning for Biology and Medicine

ERIC J. HORVITZ

During the last decade, computer scientists have made significant progress in developing techniques for storing and retrieving information and for solving difficult inferential problems with computer-based reasoners. The growth in the power of computer processors, and the parallel decline of the cost of computer memory, have catalyzed the development of innovative software for problem solving. In particular, there have been promising advances in computational methods for acquiring, representing, and manipulating biological and medical information. I will present key concepts of automated-reasoning investigated in the computer science subdiscipline called *artificial intelligence* (AI). I will frame my discussion in terms of the genesis and maturation of AI and related subdisciplines that were spawned during the development of electronic computers and will focus on key themes that have dominated research over the last three decades.

In Pursuit of a Crisp Definition

What is artificial intelligence? It is often difficult to construct a definition that is satisfying to all of the practitioners of this discipline. AI research encompasses a spectrum of related topics. Broadly, AI is the *computer-based* exploration of methods for solving challenging tasks that have traditionally depended on people for solution. Such tasks include complex logical inference, diagnosis, visual recognition, comprehension of natural language, game playing, explanation, and planning.

I have emphasized "computer-based" to stress the centrality in AI of the use of computers to experiment with implementations of abstract theories of rea-

---

Advances in Computer Methods for Systematic Biology: Artificial Intelligence, Databases, Computer Vision, ed. Renaud Fortuner (Baltimore: Johns Hopkins University Press).

soning. Experiments with computational models play an important role in distinguishing AI research from other inquiries into reasoning and problem solving, including philosophy and theoretical psychology. The sculpting of complex computational models gives AI researchers an experimental basis for testing theories of problem solving and for comparing alternative solution strategies. Implementing real systems also provides researchers with opportunities to explore unanticipated or counterintuitive behaviors of complex models.

Several AI endeavors have led to the development of distinct approaches for performing well-defined tasks such as medical diagnosis or species identification from observations. These methods are sometimes referred to as *AI techniques*, especially when they are available as software packages for use by nonspecialists. Such AI reasoning techniques provide a means of representing knowledge and a set of inference procedures for manipulating or drawing conclusions from that knowledge.

Formal versus Descriptive AI Research

There are different methodological approaches to AI research. Two large, coexisting phyla of investigators in the AI world are the *formal* and *heuristic* groups. The scientific methodology of the formalists is analogous to that of early aerodynamics scientists attempting to construct theoretical foundations for building a flying machine. In contrast, researchers in the heuristic group are like early airmen, enthusiastically attempting to take to the air with flimsy assemblies of wood and fabric engineered to reflect intuitions drawn from their observations of birds in flight.

The formalists study systems based on logic or other persuasive sets of axioms, such as probability theory. Such axiomatic systems define consistent theories of reasoning. Formalists are committed to the view that reasoning systems based on axiomatic principles can explain and generate complex intelligent behaviors, much as the set of astronomical laws developed by Newton, Tycho, and Kepler provides a satisfactory explanation of the complex motions of heavenly bodies.

Investigators in the heuristic group reject the requirement for using reasoning techniques that follow from a set of mathematical principles, often noting that scaling up the formal methods to problems of realistic size can lead to computationally intractable problems (Simon 1969; Szolovits 1982; Buchanan and Shortliffe 1984). These scientists have also argued that the parsimonious axiom-based methods lack the richness or *expressiveness* needed for intelligent behavior (Gorry 1973; Davis 1982).

Instead of relying on mathematical principles, investigators in the heuristic group apply intuitively attractive methods they hope can provide a more direct and tractable approach. These intuitive, and often ill-characterized, procedures are referred to as *heuristic methods*. They are based typically on observations and introspection about the ways in which people seem to solve problems. Heuristic approaches that explicitly attempt to describe human problem solving are called *descriptive*. Descriptive theories of problem solving are analogous to the epicycle machines used to describe the motions of the planets before modern astronomy offered satisfying theoretical explanations for observations.

Although there have been debate and tension between people in the formal and heuristic groups, both approaches have advantages and disadvantages. Formal models can help in developing a clear understanding of reasoning methods. A successful formal theory for solving a class of problems can be useful for creating computational methods that are easy to modify and to extend for application in diverse areas. Heuristic approaches are especially useful where no computationally tractable formal method is available for solving a problem. Indeed, heuristic models and solutions are often a precursory step in the development of new theories. Investigators in the heuristic group have also suggested that the heuristic approaches are more natural, and thus make it easier to acquire knowledge from experts for use in computer-based reasoning systems and to explain the conclusions of machine reasoning to people.

The Genesis of AI

Several factors coincided to create the research environment that spawned a set of information-oriented disciplines in the mid-1940s. Key forces that created and shaped the new disciplines were the pressure for solving complex tactical problems during World War II, the development of electronic computers, and the development of the conceptual framework of subjective expected utility.

Under the pressures of war, blueprints for general electronic computing systems were implemented and rapidly refined. The need to solve critical war-fighting problems and the availability of computers stimulated research on *operations analysis*. Operations analysis focused on the design of complex systems, processes, and deployments, through optimization and control. Briefly, the solutions to operations-analysis optimization problems are optimal values for variables that define the behavior of a system. These optimal values maximize some measure of benefit or minimize some measure of cost, such as dollars, time, distance, or lives.

Operations analysis matured into the modern discipline of *operations research* (OR). Following World War II, OR methods were used to analyze a broad spectrum of civilian problems, such as determining an ideal schedule for maintenance of a fleet of airplanes, purchasing an optimal number of ambulances for a county, or minimizing the number of oil tankers to meet a fixed schedule (Dantzig and Fulkerson 1954). Investigators in the OR community developed a set of numerically intensive methods, often referred to as *OR techniques*, for modeling and solving optimization and scheduling problems (Dantzig 1963). These OR techniques included linear programming, nonlinear programming, integer programming, dynamic programming, queuing theory, and control theory.

In 1947, von Neumann and Morgenstern developed an axiomatic framework called *utility theory*, founded on the preexisting axioms of probability theory. Briefly, utility theory provides a formal definition of preference and of rational decisions under uncertainty, that is, decisions taken in light of uncertain knowledge about a situation (von Neumann and Morgenstern 1947; Raiffa and Schlaifer 1961). The axioms of utility theory define a measure of preference called *utility*. Utility theory dictates that people should make decisions that have optimal average, or *expected*, utility.

Under uncertainty, we do not know exactly how an action will affect the world. The fundamental procedure in determining a course of action that has optimal expected utility is to consider the probabilities of alternative outcomes associated with a possible action. The expected value of each action is computed by summing the value attributed to each possible outcome multiplied by the probability of that outcome.

Expected value, teamed with the concept of personal or *subjective probability*, provides decision analysts with a powerful framework for maximizing expected utility. Many people who have been trained in classical statistics are not familiar with the interpretation of probability as a subjective measure of belief. That is, the most widespread interpretation of probability, the *frequentist* interpretation, is that of a measurable frequency of events determined from repeated experiments. For example, probability is viewed typically as the number of times a coin (fair or biased) lands on one side, given some large number of coin flips.

A different perspective on probabilities, known as the *subjectivist*, or *Bayesian*, interpretation, is that a probability is a measure of a person's degree of belief in an event, given the information available (Savage 1972; Hacking 1975). According to this conception, probabilities refer to a *state of knowledge* held by an individual, rather than to the properties of a sequence of events. Subjective probabilities abide by the same set of axioms as do classical probabilities or frequencies. The subjectivist approach is a generalization of the more popular notion of a probability as a long-run frequency of a repeatable event. The use of subjective probability in calculations of the expected value of actions is called *subjective expected utility* (SEU). SEU allows personal opinions and hard data to be integrated in a theoretically sound manner.

The development of SEU contributed to the maturation of OR and to modern economics and psychology. In addition, it provided the basis for the emergence, in the early 1950s, of management science and decision analysis, disciplines that are conceptually and academically related to OR. Decision analysts study techniques for determining ideal personal decisions under uncertainty (Raiffa and Schlaifer 1961; Howard 1966, 1968; Keeney and Raiffa 1976). Investigators in management science develop and test mathematical models of production and organizational decision making.

The 1950s: Early Days of the Discipline

In the early 1950s, AI research evolved separately from OR and from the related disciplines of management science and decision analysis. Rather than relying on numerical analyses, early AI investigators turned their attention to the automated processing of abstract symbols (Simon 1972). Herbert Simon, one of the founding fathers of AI, has described how the concept of manipulating sophisticated symbols with computers was stimulated by work with computer systems developed for performing complex numerical analyses (Simon 1987). In the early electronic computers, simple mathematical functions such as add and multiply were encoded in the form of binary numbers. That is, these mathematical *concepts* were represented in the same way as the numbers they operated on. If simple mathematical notions could be manipulated by computers, perhaps, by analogy, more sophisticated and abstract concepts could be processed.

AI research was born with a great enthusiasm for the possibilities of harnessing computer-based manipulation of abstract symbols to perform "intelligent" reasoning tasks. Early on, the field became dominated by studies of computational inference using principles of logic. AI researchers sought to encode knowledge *about* problem solving in the form of explicit, or *declarative*, representations of objects and relationships in the world, as opposed to the traditional *procedural* approach as typified by OR solution methodologies (Barr and Feigenbaum 1982; Cohen and Feigenbaum 1982).

AI researchers also dismissed numerical methods as relatively unimportant to decisions made by agents with the ability to manipulate abstract symbols. In particular, methods for optimizing the expected value of action under uncertainty seemed inadequate for explaining cognition and intelligent problem solving. Early investigators noted that people successfully address challenges in the world without using numerical optimization procedures. In particular, number-intensive decision analysis or procedures seemed to require inappropriate detail and analytic complexity.

Early investigators pointed out that, in contrast to the crisply defined problems amenable to OR or decision-analytic techniques, almost all real-world problems faced by people in daily life are intrinsically *ill defined*. Thus, rather than pursue an optimal solution to well-defined numerical problems, many AI researchers have worked to develop techniques for finding suboptimal yet satisfactory behaviors and decisions, known as *satisficing* solutions (Simon 1969). Many AI investigators believed that, beyond being inappropriate, attempting to apply numerical techniques to real-world problem solving would require extraordinary amounts of computation (Feigenbaum 1964).

In fundamental work in the mid-1950s, Newell and Simon built systems that could perform logical reasoning. The first logical theorem-proving programs were created to derive simple mathematical proofs. However, the primary goal of the investigators was to identify general principles of intelligent problem solving. Newell and Simon constructed a program called the Logic Theorist and later built a descendant system called the General Problem Solver (GPS) (Newell 1958; Newell and Simon 1963). GPS employed a computer representation of logical axioms. The system derived new logical relationships by repeatedly applying one of several rules of logical implication, or *operators*, to the axioms or to other logical statements stored in the system. The system could generate long chains of inference by applying logical operators to statements stored initially or generated later by the system.

GPS's structure and capabilities highlighted several important concepts. The system set a precedent for separating inference machinery and capabilities from domain knowledge. That is, domain-specific knowledge was stored in a separable, or *modular*, *knowledge base*. Research on GPS also pioneered the use of *means-ends analysis* to guide the selection of operators.

Means-ends analysis is an approach for repeated selection of operators that transform the current version, or *state*, of the solution to a new subproblem that is closer to a final solution or *goal* (Nilsson 1980). Operators are selected and applied until a solution is reached. With GPS, the goal was to prove a logical theorem from the initial axioms and logical statements. As an analogy, a sequence of operators for traveling to Stanford University from the Massachusetts Institute of Technology (M.I.T.) is "walk to Technology Square," "automobile to Logan airport," "plane to San Francisco," "walk to taxi," "taxi to Stanford." In this example, subproblems are different positions in space that are progressively closer to a goal. In logical theorem proving, as in real-world problems, there are typically many possible choices, or sequences of operators, that might be applied to any stage of a solution. Thus, we need reasoning techniques that can identify the best next operator. In the work on logical reasoning, operators were selected by heuristics that considered the relative ability of operators to minimize the conceptual distance between the current state and a goal or a subgoal.

In 1959, the term *artificial intelligence*, as the name of a discipline charged with the study of computer-based problem solving, was first used at a Dartmouth conference by John McCarthy, a computer scientist and mathematician interested in the philosophical foundations of computational models of intelligence. The name stuck, and the conference participants formed the core of an evolving field, distinct from other areas of computer science, dedicated to furthering our understanding of intelligence as the processing of symbols.

The 1960s: Maturation of Logical Reasoning Methods

In the late 1950s and early 1960s, investigation of techniques for reasoning with logical methods continued to mature. Theorem proving was applied to solve mathematical problems beyond logic. For example, Gelernter built a system for performing geometrical proofs (Gelernter 1959). *Search* evolved as a central notion in intelligent problem solving. Systems must often rely on the generation and evaluation of large sets of alternatives to solve logical reasoning problems. In GPS, for example, there are many different ways to generate subgoals and goals from an initial problem statement and set of axioms.

To visualize the large *search trees* that a reasoning system must contend with in solving an inference problem, consider each subgoal to be a point in a conceptual space that separates an initial state from a final answer or goal. We can represent the fact that one subgoal can be transformed to another subgoal with an operator by drawing an arc between the *parent* and *descendant* subgoals. Problem solving can be viewed as choosing a path through a large branching tree created by the following procedure, or *algorithm*: (1) Generate a set of subproblems by first applying all valid operators to an initial problem and (2) perform the same subproblem expansion on each subproblem recursively, until we reach goal states or cannot apply any more operators. The subgoals in such a search tree define the *search space* for a theorem-proving problem.

Consider, as another application, the classic search problem, called the traveling salesperson problem (TSP) (Dantzig et al. 1954; Lawler et al. 1985). The TSP requires us to identify the shortest path connecting a set of cities that each must be visited once by the salesperson before he returns to a home base. In generating the search tree for the TSP, we apply a MOVE operator to generate all possible trips from the current location to another city. We apply the MOVE operator again to each new location, making sure each time that we move only to unvisited cities. The process continues for each path through the tree of cities until we have visited all cities and returned to the starting position. Alternative orderings of cities compose a set of tours. Search algorithms for TSP identify different partial sequences of subgoals on the path to complete tours. The shortest circuit, or tour, is the optimal answer to a TSP problem instance.

Finding the optimal solution to a TSP problem is difficult; TSP and a variety of other difficult problems constitute a family of problems called *NP-complete*, in a taxonomy of problem complexity developed by theoretical computer scientists to classify the fundamental difficulty of problems (Garey and Johnson 1979; Aho et al. 1983). With high likelihood, the time required to solve NP-complete problems exactly grows exponentially with the size of the problem (for example, the number of cities). That is, the amount of time required to solve these problems grows geometrically with linear increases in problem size. A great number of problems that are of interest both to AI researchers and to biologists are in the NP-complete class. For example, many taxonomists are probably familiar with the NP-complete *Steiner Graph Reduction* problem, used to build plausible taxonomic trees from data. Because NP-complete problems are difficult to solve exactly, numerous algorithms have been developed to generate approximate solutions with far less effort than the work required for a naive search of all possibilities (Papadimitriou and Steiglitz 1982).

Intelligent versus Brute-Force Search

Without forethought, we might attempt simply to generate all possible sequences of operators until we find an answer to a search problem. Such *brute-force* searches require a great amount of computation. However, there are many ways to navigate through a large tree. AI investigators have developed search techniques for guiding search more intelligently. Frequently, these procedures can be used to solve a problem more quickly than the worst-case situation described by NP-complete brute-force methodologies.

The design of intelligent search procedures often hinges on the identification or development of an *evaluation function* that is used to assign a score to each intermediate state along the path to an answer (Nilsson 1980). Some search techniques rely on logical analyses of the dominance of one or more states to identify irrelevant paths through a tree. Such dominance information comes from the use of deterministic evaluation functions for identifying the value of a subgoal. For example, in one approach, called *branch and bound*, paths through a search tree are pruned from consideration if subgoals generated by those paths will be less valuable than the best approach discovered so far.

Often, branch and bound is not applicable because we do not have a deterministic measure of value for each subgoal. Instead, for many problems, researchers have relied on heuristic estimates of the value of the "goodness" of an answer. For example, the GPS means-ends analysis depends on the use of a heuristic evaluation function to control the selection of a best operator at each step of an analysis.

Sometimes we can evaluate the behavior of a heuristic search method in terms of an optimal analysis or answer. For example, for some search techniques, investigators have developed formal proofs that show how we can characterize the error in an approximate answer in terms of an exact answer (e.g., the shortest tour possible through a set of cities) or characterize the amount of computation required by an approximate analysis as some fraction of the work required by an exact solution.

Search research spans several types of problem solving beyond theorem proving and tour identification. An active area of search research is called game playing. Several research teams have studied computer-based reasoning for competing in games such as checkers and chess (Samuel 1959, 1967; Good 1977; Barr and Feigenbaum 1982; Berliner 1980). In the case of chess, a search space is generated by applying the rules for moving chess pieces from an initial legal position. AI investigators have designed chess programs that assess the promise of the legal moves that can be made from a current board position by evaluating paths of moves in large trees of successive moves that are made possible by each candidate move. The large trees of moves are generated by considering a volley of moves and countermoves to some depth of player interaction, or *lookahead*.

Production Systems for Capturing Expertise

In the late 1960s, many AI investigators continued to study the application of logical techniques similar to the methods employed by Simon and Newell in GPS. However, instead of remaining within the confines of logical theorem proving, investigators began to examine the manipulation of symbols and knowledge in specialty areas, with the goal of building computer systems to assist people with decision making in scientific and professional endeavors. Work began on the construction of knowledge bases that could be used to bring expertise to decision makers in such professions as chemistry and medicine.

Systems based on the use of sets of logical rules of the form IF A THEN B, and on the use of logical-chaining techniques for deriving conclusions from the rules, came to be called *production systems* and, later, *rule-based systems* (Davis et al. 1977). One of the earliest knowledge-intensive production systems was a program for performing difficult symbolic mathematical inference for engineering applications. In this work, mathematicians constructed knowledge bases of mathematical identities that were used for the symbolic integration and differentiation of equations. This work led to one of the most widely used AI programs in the world today, called MACSYMA. MACSYMA is used by scientists and engineers for performing complex symbolic integration.

Another early application of production systems to real-world problems was the DENDRAL project at Stanford (Lindsay et al. 1980). For this project, expert rules describing chemical structure and decomposition were used to reason about mass-spectroscopy data. The large knowledge base was composed of rules that described how molecules, and portions of molecules (as subgoals), might decompose in a mass spectrometer. Experts in organic chemistry collaborated with computer scientists to build DENDRAL's knowledge base and to evaluate the functioning of the system.

Planning Research

The late 1960s and early 1970s saw a marked growth of interest in an area of research called *planning*. To this day, planning research continues to be a hotbed of AI research. In contrast to the study of production systems to automate expert reasoning and decision making within the confines of a specialty area, planning research has dwelled on *common sense reasoning*, that is, reasoning whose goals are routinely achieved by people in the course of daily life. Investigators interested in planning have developed logical-reasoning systems similar to the older theorem-proving systems. However, instead of storing abstract logical operators and statements in the system's knowledge base, the investigators encode operators and logical rules that capture basic physical and mechanical properties of a domain, such as the information that two objects cannot be in the same place at the same time. The rules also capture the preconditions and effects of *actions*. Also in contrast to theorem proving, planning problems require investigators to consider the effects of actions on the world.

Planning research has demonstrated that the simplest goal-oriented behaviors involve complex chains of reasoning about goals and subgoals. In an early planning project, called STRIPS, investigators tackled tasks in a Lilliputian world of blocks on a surface (Fikes and Nilsson 1971). In this world, which came to be known as *Blocksworld*, the goals were simple tasks such as changing one configuration of stacked blocks to another. The STRIPS reasoner demonstrated the complexity of using computers to perform tasks that appear quite natural and simple to people. For example, at first blush, it might appear relatively straightforward for a computer to stack a block, say Block B, on Block C, starting from an initial state of affairs in which Block D is on Block C and Block B is beneath Block A. Such tasks exposed problems such as costly interactions that occur in making moves to attain subgoals. That is, a careless or shortsighted action, seemingly on the way to a goal, might make it difficult or impossible to reach that goal.

Vision Research

During the 1960s and 1970s, great strides were made in the development of algorithms for identifying shapes and objects in a visual field. Like planning research, automated vision research highlighted the difference between tasks that are easy for people and tasks that are easy for computers. It quickly became clear that computer programs have great difficulty recognizing patterns and objects that people recognize with ease (Winston 1975).

Researchers studying computer-based vision sought early on to extract features through *global processing*, the homogeneous application of simple mathematical transformations to an entire scene. Such processing was applied to generate information about classes of features, including the presence of edges and the shading and textures of surfaces (Marr 1975; Marr and Hildreth 1980; Ullman 1979). The earliest vision-identification techniques were supplemented by increasingly sophisticated methods for the *segmentation* of an image into its chief components (Marr 1982). Search-based methods were developed for building objects out of sets of segmented features. In one set of approaches, a search is applied to identify a feasible *labeling* of edges in a scene as different possible boundaries of an object and to build up sets of edges into three-dimensional objects. In some of this work, identification procedures draw on a database of objects for information about what might be present in a visual field.

A trend in vision research from the 1960s to the present has been to use progressively richer models of objects in the world to reduce the computational complexity of visual identification (Binford et al. 1989). Toward the late 1960s, techniques were developed for modeling three-dimensional objects. In one such approach, structures are identified by mapping a set of cylindroid structures to a visual scene. In many current vision projects, relatively detailed three-dimensional models are used to choose among alternative interpretations of a scene. For example, three-dimensional anatomical models have been used to assist in the automated interpretation of medical ultrasound and radiographic images (Brinkley 1985).

The 1970s: Knowledge Representation and Expert Systems

Many areas of reasoning, including the inferential principles and techniques developed in the 1950s and 1960s, received continuing and intensive attention in the 1970s. The 1970s were marked by growing interest in the representation of knowledge and by a surge of enthusiasm for the development of production systems to assist professionals and scientists. Other areas matured in the 1970s, including the study of methods for parsing and understanding natural language, for controlling the focus of attention of computer-based reasoning systems, and for learning from experience. I will discuss work on representations of knowledge and the use of these representations in expert systems.

Knowledge Representation and Problem-Solving Efficiency

A key theme of AI research in the 1970s was the importance of identifying expressive and efficient representations of knowledge. Early in the decade, investigators studied and discussed the best ways to codify knowledge for computational inference (Amarel 1968; Cohen 1977; Korf 1980). In an interesting piece of research, Saul Amarel (1968), of the AI group at Rutgers, investigated the sensitivity of the complexity of solving a problem to details of the representation of that problem. He used examples from a classic computer science problem called the *Towers of Hanoi*. To solve this problem, a sequence of moves must be generated to transfer a set of disks of different sizes (resembling a Buddhist temple) from one pole to another, given a set of rules about legal moves. Amarel showed how changing the initial formulation of the Towers of Hanoi problem to a new representation could increase the efficiency of computational problem solving.

The studies of Amarel and others on the efficiency of representation came at a time of growing interest in reasoning systems that could perform inference on very large knowledge bases. There was a proliferation of alternative representations of knowledge in reasoning systems.

Expert Systems

Projects such as MACSYMA and DENDRAL were developed to explore how computer-based reasoning might assist people with difficult inference problems. However, it was not until the early 1970s that the expression *expert systems* was first used to refer to computer programs that could draw conclusions by performing logical inference on a large knowledge base of information acquired from experts (Buchanan 1982). The expression was made popular in the medical domain.

In the 1970s, a great amount of work in AI was performed on automated medical diagnosis. A large community of investigators with a primary interest in AI in medicine (AIM) coalesced as a bona fide subdiscipline of AI. Many AIM researchers endeavored to build systems to assist with diagnosis, given information about a patient's symptoms. Expert-systems studies included the MYCIN project at Stanford University, the Present Illness Program (PIP) at M.I.T. (Pauker et al. 1976), CASNET at Rutgers University, and INTERNIST-1 at the University of Pittsburgh (which continues today as the Quick Medical Reference [QMR] project at the University of Pittsburgh).

Attempts to build expert systems for medical reasoning highlighted the inadequacy of straightforward applications of logic for reasoning about large and complex domains. Logical-reasoning systems assign values of complete truth or complete falsity to possible states of the world and to the effects of actions. However, physicians are rarely certain about the state of a patient's physiology or about the effects of alternative therapies.

As work on expert systems in medicine progressed, investigators found that the deterministic nature of these systems did not allow experts to express uncertain relationships. A theme of 1970s work on representation was the modification of logical techniques, developed for automating logical theorem proving, to handle the more general situation of uncertainty. In one aspect of this work, investigators extended the logic-based production systems of the late 1960s with heuristic techniques for capturing uncertainty.

Popular representations of knowledge for systems that considered large quantities of expert knowledge about uncertain relationships included production rules (Davis et al. 1977), *semantic nets, causal networks* (Weiss et al. 1978), *frames* (Minsky 1974; Lindberg et al. 1980), and several mixtures of these representations. I will briefly describe these knowledge representations in the context of several medical expert systems developed in the 1970s.

MYCIN. MYCIN is a production system that was developed at Stanford to study the representation of medical knowledge (Shortliffe 1976; Buchanan and Shortliffe 1984). MYCIN reasons about the diagnosis and treatment of bacterial infections. The system uses a set of logical rules. Logical inference in the system is designed to follow a specific search strategy called *backward chaining*. With backward chaining, a system chains together a set of rules that creates an inference path, or *chain*, from causative agents to observed symptoms. *Forward chaining* refers to chaining from observations in the world to possible causes.

MYCIN investigators explored the extension of logical reasoning to address uncertainty by developing a heuristic calculus, the *certainty factor model* (Shortliffe and Buchanan 1975). With this approach, measures of the certainty or *strength* of a rule, called *certainty factors*, were assessed from experts. For example, consider the following rule from MYCIN:

> If the infection is a primary bacteremia, and the site of the culture is one of the sterile sites, and the suspected port of entry of the organism is the gastrointestinal tract, then there is suggestive evidence (certainty factor = 0.7) that the identity of the organism is Bacteroides.

MYCIN contains a knowledge base including hundreds of such rules. When rules are chained together logically, a certainty factor for the logical result is computed with a certainty-factor combination scheme. This numerical calculus propagates uncertainty to logical conclusions by considering the certainty factors associated with the rules that compose an inference chain.

CASNET. Investigators at Rutgers studied the use of semantic networks with a system for the diagnosis of glaucoma, called CASNET (Weiss et al. 1978). Semantic networks are graphs that consist of a set of directed arcs between concepts. In CASNET, the directed arcs describe relationships at three levels: observations, pathophysiological states, and diseases. Arcs called *associational links* are used to connect observations to pathophysiological states; arcs called *classification links* connect pathophysiological states to disease categories. Observations reported to CASNET activate a subset of pathophysiologic states; these, in turn, activate disease categories. The CASNET team experimented with the use of probabilistic information that describes the strength of association among interdependent concepts.

Present Illness Program. The Present Illness Program (PIP) for reasoning about renal disease was constructed in the 1970s by a team at M.I.T. (Pauker et al. 1976). The system uses expert knowledge represented in the form of files of structured records, called *frames.* A separate frame was created for each disease considered by PIP. Each frame contains several classes of knowledge that experts have identified as being useful in considering the presence of a disease, given a set of observations. The classes of knowledge include typical findings or symptoms associated with a disease, criteria for making decisions, and relationships between the disease under consideration and the symptoms and diseases described in other frames.

The structure, or *architecture*, of PIP is based on descriptive intuitions about human cognition. The heuristic machinery used in PIP to draw conclusions from the knowledge stored in its frames includes a *supervisory program*, a *working memory*, and a *long-term memory*. In reaction to a set of symptoms observed in a patient, the PIP supervisory program activates a set of diseases by introducing these diseases from long-term memory to the active, working memory. Diseases linked to the activated diseases via relationships specified in the frames are added to the working memory. Finally, questions about symptoms of the activated diseases are asked in order to gather additional information about the patient. The process of refinement continues until a single disease, or a small set of diseases, remains in active memory.

INTERNIST-1. The INTERNIST-1 project was initiated at the University of Pittsburgh, in the 1970s, to construct a system for assisting physicians with diagnosis in the broad domain of internal medicine (Miller et al. 1982; Pople 1982). The INTERNIST-1 reasoning system relies on a numerical scoring approach for assigning belief to alternate disease hypotheses, given a patient's symptoms. It uses several classes of numbers obtained from experts. For example, one set of numbers is used to represent the degree of belief in the presence of different individual diseases, given a symptom. Another class of numbers characterize how often, for each disease, a symptom appears.

INTERNIST-1 applies heuristic scoring rules to combine the numbers associated with different symptoms into comprehensive measures of belief in alternative entities. These quantities are used to generate a list of diseases, each with a heuristic measure of likelihood. Such a list of diseases ordered by likelihood, or *differential diagnosis*, is analyzed to select questions about the best next tests to perform to narrow the list.

INTERNIST-1 uses an iterative reasoning cycle called *hypothetico-deductive* reasoning. With this approach, a few salient observations are input to the system, and a measure of likelihood is assigned to each possible disease. Based on this list of hypotheses, new questions and tests are recommended. After new information is input to the system, the system performs the entire cycle again, rescoring all of the diseases and recommending new questions based on the revised disease list. Answers to the new questions are considered in conjunction with the initial salient features to generate a revised set of hypotheses, with revised associated likelihoods. This cycling process continues until only one disease remains or until the cost of gathering additional information outweighs the benefits of further refinement.

INTERNIST-1 continues to this day as the QMR project at the University of Pittsburgh (Miller et al. 1986). Its knowledge base has continued to grow since the inception of the INTERNIST-1 project in 1973, and it is now one of the largest expert systems in the world.

The evolution of INTERNIST-1 to QMR illustrates how, for some applications, the power of commonly available computer hardware has grown more quickly than the sophistication, and concomitant thirst for memory and computation, of our software. INTERNIST-1 was implemented on a large mainframe computer at Stanford, accessed by Pittsburgh researchers via a computer network. Today, QMR operates comfortably on inexpensive MS-DOS–based personal computers.

The 1980s: Maturation of Reasoning under Uncertainty

In the 1980s, several rule-based expert-system architectures and representations matured to the point of commercialization. Several companies were founded with the goal of providing *expert system shells*: software packages that could be used by investigators in specialty areas, outside of AI and computer science, to build their own expert systems. However, although the popularity and use of rule-based systems grew throughout the 1980s, there was a concurrent growth of interest in alternatives, especially in sound methods for addressing uncertainty.

Some AI investigators began to reexamine concepts and reasoning methodologies developed in decision analysis and operations research in order to develop more accurate reasoning methods. These investigators began to borrow ideas that would allow them to define the concept of a theoretically correct answer to a problem, or best action to take, given uncertainties about actions and outcomes. In particular, interest began to grow in the use of SEU to tackle difficult AI reasoning problems (Horvitz et al. 1988). In some of this work, probability and utility were applied at the metalevel to address decisions *about* the best ways to build a reasoning system, or to control reasoning to optimally solve a problem, given limited computational resources (Horvitz 1988; Russell and Wefald 1989).

I will focus on the development, in the 1980s, of SEU-related methods for using probability and utility to represent knowledge and reason under uncertainty; these techniques are particularly relevant to the subject of this volume.

Sufficiency of Heuristics for Reasoning under Uncertainty

Throughout the 1980s, increasing attention was given to the integration of probabilistic and decision-theoretic reasoning methods with AI approaches. Some of this interest was stimulated by attempts to apply computer-based reasoning methods to progressively more difficult decision-making tasks. The importance of uncertainty, and of developing formal reasoning techniques, was underscored by the complexity and the high stakes of application areas such as medicine and aerospace.

Research had never halted on the use of probability and utility in the realm of expert systems and AI. In fact, several groups, working to the side of mainstream AI, had continued to explore the use of simplified forms of probabilistic and decision-theoretic reasoning for medical diagnosis (Ledley and Lusted 1959; Warner et al. 1961; Gorry and Barnett 1968; Gorry et al. 1973; Gorry 1973; Patrick 1977). Typically, assumptions were imposed on these systems to make the representation and reasoning with uncertain information tractable. For example, most researchers investigating the use of probability and utility in the 1960s and 1970s assumed conditional independence among observations and assumed that there was an exhaustive set of mutually exclusive diseases. By assuming conditional independence among observations, the researchers considered the probability of each observation to be unaffected by the presence or absence of other observations. Although the early probabilistic systems performed well in small subdomains, AI investigators showed little interest in them, beyond pointing out their limitations.

Most AI investigators understood that probability and decision theories served as axiomatic bases for ideal reasoning and decision making under uncertainty; however, they were not impressed by the simple probabilistic systems for several reasons. Many were convinced that it is difficult to obtain probabilistic information and to reason efficiently with probability and utility. Also, AI investigators studying heuristic techniques for reasoning under uncertainty in expert systems justified the need for heuristic techniques by focusing on the invalidity of the independence assumptions made by the probabilists to render their systems tractable (Buchanan and Shortliffe 1984). The researchers suggested that attempts to relax such assumptions would lead to unmanageable combinatorial increases in the computation time and in the numbers of probabilities required by the systems.

Interest in probabilistic inference in the mid- to late 1980s was invigorated in part by theoretical work that demonstrated clear parallels between the shortcomings of several heuristic calculi and the simplified probabilistic inference. Several investigators demonstrated that the heuristic methods did not provide a means for escaping from the invalidity of the independence assumptions made in the simple probabilistic systems. In fact, for some of the heuristic methods, the assumptions were shown to be even more restrictive and, moreover, dangerous because they were implicit. For example, the MYCIN certainty factor model and the INTERNIST-1 scoring scheme were shown to be equivalent to the use of highly simplified and constrained probabilistic reasoning (Heckerman 1986; Horvitz and Heckerman 1986; Heckerman and Miller 1986; Horvitz et al. 1988).

AI scientists were also stimulated to reinvestigate the applicability of probabilistic reasoning by the development of analyses of the epistemological inadequacy of rule-based methodologies for capturing uncertain knowledge (Heckerman 1986; Pearl 1988). One study showed that most classes of uncertain knowledge cannot be represented efficiently in rule-based systems (Heckerman and Horvitz 1987). Other discussions demonstrated that probability theory uniquely satisfies a set of essential desirable properties of uncertain belief (Cox 1946; Tribus 1969; Horvitz et al. 1986).

Belief Networks and Influence Diagrams

Beyond theoretical analyses, the most important stimulus to renewed work on probability theory and SEU in AI was the development of efficient and expressive representations of probabilistic knowledge called *belief networks* (Pearl 1988; Howard 1989) and *influence diagrams*. Belief networks and influence diagrams allow people to express qualitative, in addition to quantitative, knowledge about beliefs, preferences, and decisions.

Belief Networks. A belief network is a graphical representation that allows a computer scientist, working with an expert, to efficiently encode expert knowledge about probabilistic dependencies among important distinctions in a domain. More important, belief networks allow an expert to specify independence in a domain. After the construction and assessment of a belief network, algorithms can be applied to the network to assign probabilities to alternative hypotheses, given a set of observations.

In the language of computer science, a belief network is a directed acyclic graph (DAG) that contains nodes representing propositions (for example, hypotheses and observations) and arcs representing probabilistic dependencies among nodes. Each node representing a proposition is associated with an exhaustive set of mutually exclusive values that represent alternative possible states or events.

To construct a belief network, an expert defines nodes that represent important distinctions about the world. These distinctions include hypotheses, important states of interest (e.g., diseases in a patient or disorders in a jet engine) that may not be confirmed directly, and observations that are useful for discriminating among the alternative hypotheses. The expert also provides information about the probabilistic dependencies among nodes.

Figure 1.1 displays a simple belief network, showing a dependency between coronary artery disease and chest pain. A directed arc from the node CORONARY ARTERY DISEASE to the node CHEST PAIN, means that an expert asserts that coronary artery disease affects the probability distribution over the possible values of chest pain. As indicated in Figure 1.2, beyond constructing the high-level dependency graph, an expert specifies the possible values of each node. For example, the expert might specify that coronary artery disease is either NONE, MINOR, MODERATE, OR SEVERE, and that chest pain on exertion may be either ABSENT, MINOR THROBBING, SEVERE BURNING, or SEVERE KNIFELIKE. An arc between the two nodes means that the expert believes that the probabilities of having the different kinds of chest pain depend on the severity of the artery disease.

As part of building a belief network, we would need to assess the *conditional probability* that we would see the different types of chest pain when coronary artery disease was at each of its possible values. Conditional probabilities have the form: "The probability of observation A, given the presence of hypothesis H, is x." We say that the probability of A is *conditioned* on H. For the cardiac belief network, we need to assess the probability that chest pain is knifelike given that coronary artery disease is minor. The convention for writing this conditional probability statement is, p(CHEST PAIN = KNIFELIKE|CORONARY ARTERY DISEASE = MINOR, ξ). The portion of the statement that follows the vertical line is called the *conditioning clause*. The symbol ξ in the clause refers

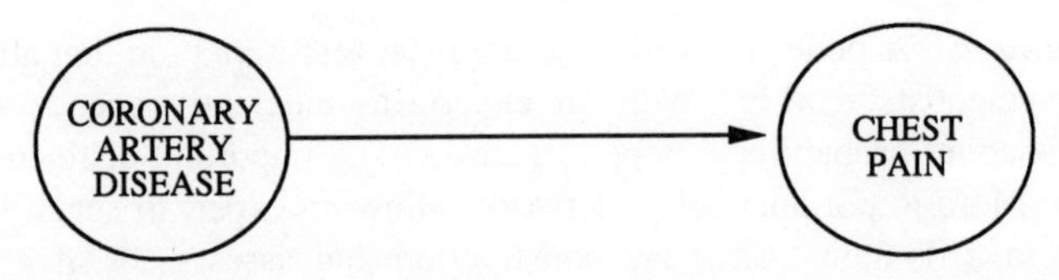

Figure 1.1. A small belief network representing the dependency between chest pain and coronary artery disease (from Horvitz 1990).

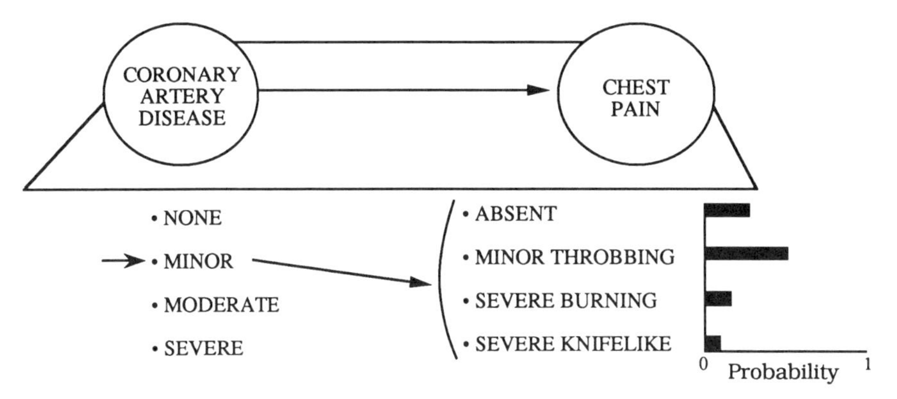

Figure 1.2. Values of propositions and a probability distribution for chest pain, given minor coronary artery disease (from Horvitz 1990).

to the general context, or *background state of information*, that is not explicitly listed in the conditioning clause. Figure 1.2 uses a bar graph to denote the probability distribution over chest pain, given minor coronary artery disease. Figure 1.3 depicts the different probability distribution over different values of chest pain, given a patient has severe coronary artery disease.

Of course, belief networks can be more complicated. Figure 1.4 displays how we can represent multiple causal dependencies with a belief network. In this case, an expert diagnostician has added a node and arc representing his knowledge that gastric ulcer can lead to reports of chest pain. To assess the probability distributions over alternative degrees of chest pain, we must now consider all combinations of the values of gastric ulcer and coronary artery

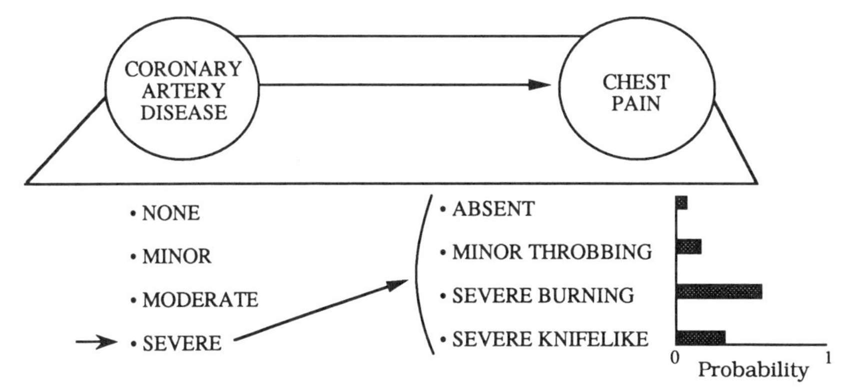

Figure 1.3. Probability distribution for chest pain, given severe coronary artery disease (from Horvitz 1990).

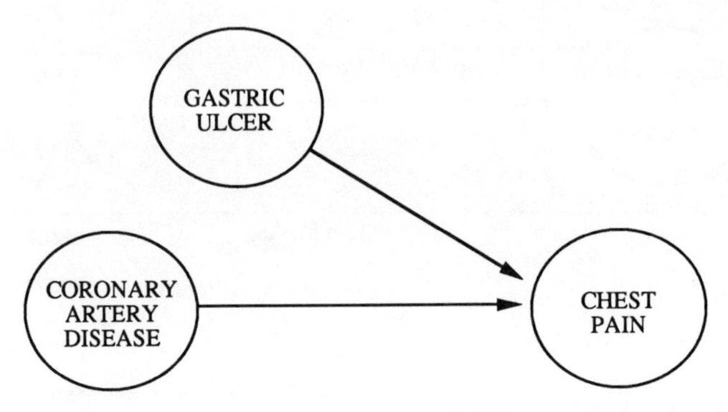

Figure 1.4. Representation of multiple causes with a belief network (from Horvitz 1990).

disease. In general, we can add many more manifestation and hypothesis nodes to a belief network. Decision-support systems have been constructed which rely on belief networks consisting of hundreds of interconnected nodes.

Typically, we assess belief networks in the *causal* direction. That is, we draw arrows and assess conditional probabilities in terms of the likelihood of an observation, given a causal hypothesis. However, we often wish to use this information by performing inference in the *diagnostic* direction; that is, we would like to determine the probabilities of alternative hypotheses, given the observation (or assumption of the truth) of one or more symptoms. To do this, we apply belief-network inference algorithms. For example, we can apply an inference algorithm to determine the probability of various levels of severity of coronary artery disease when we hear the age of a patient and also hear a particular complaint about chest pain on exertion. A number of exact and approximate algorithms have been developed for revising the assignment of likelihoods to alternative hypotheses in a belief network, given observations of some evidence (see reviews in Pearl 1988 and Horvitz et al. 1988).

Influence Diagrams. Belief networks capture knowledge only about probabilistic dependencies among hypotheses and observations. An influence diagram (Howard 1968; Howard and Matheson 1981; Olmsted 1983) is an extension of belief networks which, in addition to probabilistic-dependency information, represents alternative actions and outcomes, as well as information about a decision maker's preferences for different outcomes. The influence diagram is in many ways a more compact representation of the more familiar *decision-tree* representation of actions, events, and outcomes.

Figure 1.5 displays an influence diagram that represents the problem of deciding whether to perform an angiogram test on a patient with severe chest pain. In determining the expected value of taking, versus not taking, the costly and potentially dangerous test, we consider how different outcomes of

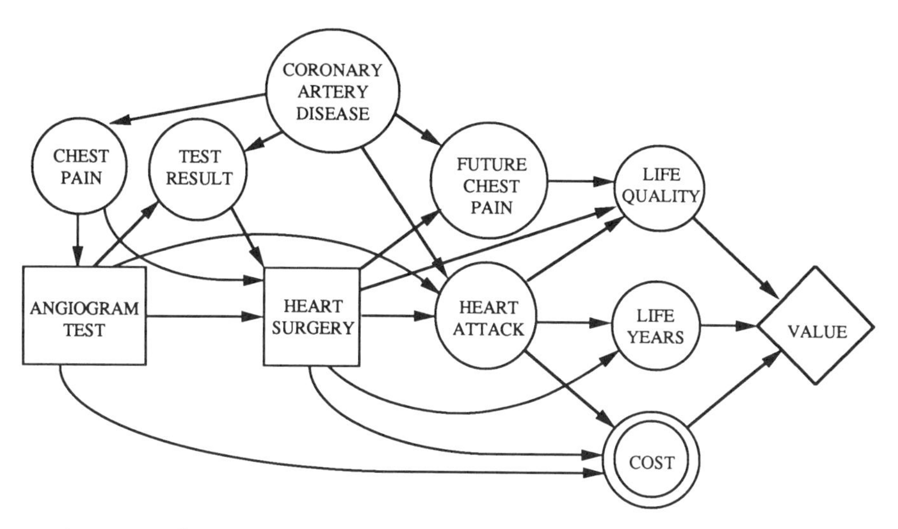

Figure 1.5. Influence diagram for a cardiac decision problem (from Horvitz et al. 1988).

the test will affect a decision to have heart surgery. We also consider such factors as how the test, and future surgery, will affect future chest pain, and the possibility of future myocardial infarction, or heart attack. All arcs point into a set of nodes that represent fundamental attributes of value to the patient. The value node represents a function that combines the value of the basic attributes into a scaler utility. This influence diagram also depicts another type of node, called a deterministic node (double-circled). The value of a deterministic node is a deterministic function of the inputs to the node. In this case, a physician has specified that the monetary cost is a deterministic function of the values of its predecessors (whether the angiogram is given, whether surgery is performed, and whether the patient has a myocardial infarction).

Algorithms have been developed that perform inference directly on influence diagrams through direct manipulation of the graphical structure of the representation (Shachter 1986). The output of such decision-theoretic inference is the best decision to make, given a decision maker's preferences, the uncertainties about the world, and the information that is available.

Acquiring Probabilities from Experts and from Data

Typically, the qualitative dependency relationships defined by the structure of a belief network or influence diagram require hundreds, or even thousands, of conditional probabilities to be gathered, or assessed, from experts. Some investigators have been concerned about the difficulty of acquiring these conditional probabilities and a specific class of probabilities, called the *prior probabilities* of alternate hypotheses. A prior probability is the prevalence of a state of the world, conditioned only on the background state of information. For example, the prior probability, p(coronary artery disease = severe|ξ), is the prevalence of severe coronary artery disease in a population.

Conditional and prior probabilities can be gathered through new statistical studies or through extraction of data from preexisting studies. Most often, however, we find that the distinctions used in a belief network have not been defined and analyzed before. Thus there is little statistical information available on dependencies in the network. Moreover, we generally do not have the resources to gather information about the frequencies of thousands of events.

The subjective probability methodology has proved extremely successful for gathering the probabilities needed by an expert system. Our work on the construction of belief networks with expert diagnosticians from the fields of medicine, aviation, and electric-power generation has demonstrated that many experts can assess prior probabilities and conditional probabilities with confidence and ease.

One of the systems we have constructed, called PATHFINDER, performs pathology diagnosis given information about several hundred features that might be seen under a microscope (Heckerman et al. 1989; Heckerman et al. 1990; Horvitz et al. 1989b). PATHFINDER uses a belief network containing about 30,000 probabilities assessed from an expert pathologist with many years of experience. To construct a large knowledge base for the system's domain of lymph-node pathology, we assessed numbers from experts such as the probabilities that MONOCYTOID CELLS are ABSENT, MONOCYTOID CELLS are PRESENT, and MONOCYTOID CELLS are PROMINENT, given NODULAR-SCLEROSING HODGKIN'S DISEASE. Such assessments were made for all diseases in the domain. The assessment procedures begin only after such distinctions as ABSENT, PRESENT, AND PROMINENT for monocytoid cells are defined precisely. Assessments can be more complex: In cases where an expert has identified dependencies among features, probabilities must be conditioned on the values of other features, in addition to the presence of a disease.

Studies with the chief expert on the PATHFINDER project showed that probability assessments are quite stable and repeatable. Perhaps more important, we found that the diagnostic accuracy of the system is not very sensitive to small variations in individual conditional or prior probabilities. Typically, it is the *order of magnitude* of a probability that is most important. Regarding error in prior probabilities, we have found that, for diagnosing disorders within the pathology application area, the influence of a prior probability is often dominated quickly by probabilistic updates with evidence. That is, small errors in prior probabilities are quickly washed out by the updates based on observations. PATHFINDER has been shown in formal trials to perform at the level of expert pathologists. More detailed clinical trials are under way to ascertain the value of PATHFINDER as a diagnostic aid for nonexperts.

Well-defined techniques have been developed for helping experts with the assessment of probabilities for use in belief networks. For example, methods have been developed (Spetzler and Stael von Holstein 1975) to help experts avoid biases in probability assessment that have been identified by cognitive psychologists (Tversky and Kahneman 1974). Other methods test an expert's consistency in probability assessment. The expert's confidence can be measured in terms of a *second-order probability distribution*, a probability distribution that describes the probability of alternative probabilities. Techniques have been developed to simplify the assessment of second-order distributions and for ideally updating probabilities given new data. Also, software tools have been developed that allow an expert to avoid redundancies in the assessment of similar probabilities. These tools enable a user to combine diseases into groups and to perform assessment at greater levels of abstraction. For example, Heckerman recently developed a graphical software tool that streamlines the probability-assessment process through abstraction and pairwise comparisons (Heckerman 1990).

Because subjectivist and frequentist probability are defined by the same system of axioms, we can refine an expert's initial assessment by performing statistical studies. There are well-characterized techniques for updating an expert's beliefs with statistical data, based on the assessment of the expert's confidence in probabilities, in addition to the probabilities themselves. Furthermore, we can apply formal cost-benefit analyses to direct the collection of better subjective assessments or to supplement the subjective probabilities with statistical information, depending on the importance of the probabilities and on the confidence of the expert. Such analyses draw on techniques in decision science known as *sensitivity analyses*. Sensitivity analyses are used to identify the probabilities that have the greatest impact on the accuracy of a system's conclusions.

Belief Network Tools and Applications

Belief networks hold promise as a representation for helping biologists to acquire and reason with knowledge about the likely identity of organisms given incomplete sets of observed features. Belief networks also can provide a coherent means of integrating diverse classes of knowledge to solve taxonomic problems. For example, investigators might use belief networks to reason about the relative likelihood of alternative phylogenetic trees by considering genetic, ontogenetic, and paleontologic data.

Belief network programming shells (Andreassen et al. 1987), as well as stable expert systems, have been produced commercially, ranging in application from sleep disorders to jet engine repair (Henrion et al. 1991). Perhaps the most widespread application has been in the area of pathology, stemming from the work on Pathfinder. Pathologists at over 300 sites throughout the world are using multimedia expert systems that are based on belief networks integrated with videodisc systems. These systems allow pathologists to employ one of several belief networks, each representing the leading expertise on the histological analysis of tissue from a major human organ system.

Interest in belief networks has been growing in a variety of areas of AI, beyond diagnostic problem solving. For example, there has been recent work in the field of machine learning on the automatic construction of belief networks from data. Belief networks have also been used recently in systems for performing visual identification (Binford et al. 1989) and for understanding natural language and identifying plans (Charniak and Goldman 1989).

Other Developments

The last decade has seen a broad spectrum of advances in automated reasoning methods, beyond techniques for reasoning under uncertainty. I will briefly mention a few areas of research with application to systematic biology.

Given the explosion of biomedical literature and information, it is becoming difficult and time consuming for investigators to access and integrate information that is relevant to their problems. Thus, tools for managing this information and helping biologists, especially systematists, to search through a number of large knowledge bases will become increasingly useful. Techniques have recently been developed to assist researchers in the design of databases (Barsalou 1989). Work has also increased on methods for performing intelligent searches in databases (Frisse 1988). In this realm, tools have been developed that allow a user to search more efficiently through disparate knowledge bases and to generate query results that take into account the preferences and interests of that user.

Another area of research with relevance to the study of biological processes is work on *qualitative reasoning*. In this area, investigators have worked to capture the essential aspects of physical systems with abstract qualitative models. These models are used to predict behaviors of a system without using numerical information, by considering the orders of magnitude of effects and modeling the competition and synergy between basic influences in a system. Several projects have focused on the construction of qualitative models for reasoning about biological systems (Karp 1988). Related research has also been done on analyses of qualitative probabilistic reasoning with belief networks. This work centers on abstracting the networks to qualitative networks that consider only positive and negative influences on probabilities (Wellman 1988).

Exciting research is under way on computer-based methods for understanding and assisting scientific investigation. Studies of computers in scientific reasoning include work on tools for assisting with the discovery of new theories and with validation, or *confirmation*, of competing theories of scientific phenomena. For example, in recent work, computer models and programs have been developed to perform probabilistic *meta-analyses* (Lehmann 1991). Such analyses supplement standard statistical analyses to determine the effects of an experimental outcome on the probability of alternative hypotheses. In other work, planning techniques have been applied to the task of designing experiments in molecular biology (Friedland 1979; Stefik 1981). That is, given the experimental goals, a computer-based planner determines a sequence of steps that will perform the analysis.

Another area of growing interest is the control of trade-offs in reasoning systems. The goal of this work is to optimize the value of computing systems, given limitations in the time available for computation imposed by a scientific investigator, a physician, or an industrial process. Several studies have examined methods for trading off the precision or accuracy of a computational result in exchange for the timeliness of that result (Horvitz 1987; Dean and Boddy 1988). By applying decision-analytic techniques to control computation in belief networks and influence diagrams, we can develop a model of rational belief and action under bounded reasoning resources. As an example of this work, we explored a computational model of bounded rationality with the PROTOS system (Horvitz et al. 1989a; Horvitz and Rutledge 1991). PROTOS considers the ideal trade-off between the precision of probabilistic inference and the timeliness of action for medical intensive-care decision problems.

Conclusion

I have dwelled on the advances in automated reasoning that I believe are most relevant to the problems that challenge systematists and biomedical scientists in related disciplines. I organized the presentation of fundamental areas of research in terms of the major research themes that have characterized three decades of AI research. In my attempt to make the best use of the few pages allotted to this chapter, I have unfortunately had to omit broad areas of research on automated reasoning. I trust readers will seek more comprehensive understanding of the topics they find most interesting.

II

Systematic Biology and Phylogenetic Inference

2

The Goals and Methods of Systematic Biology

LESLIE F. MARCUS

Introduction

To attempt a short overview of modern systematics and taxonomy is an awesome task. Here, I present an introduction to some important terminology, some issues concerning the nature of biological diversity, and a brief discussion of how organisms are studied and documented when they are classified. Systematic biologists generally agree that their major goals are to describe biological diversity and to produce natural classifications. There are, however, important differences in terminology and concepts among systematists (or from one group of systematists to another) that lead to differences in the classification of this biological diversity and its description.

It is sometimes difficult to determine the meaning of much of the common vocabulary of systematics. Some of the variation in meaning and usage arises from the adherence of systematists to different schools of thought and differing experiences conditioned by the groups of organisms on which they work. The biological world is very complex, and any one worker has usually sampled only a small part of the vast biodiversity, but is nevertheless tempted to generalize broadly. A misunderstanding of differing points of view may also arise from the difficulty of determining nuances of current practices from the literature. I have attempted to present several definitions for terms when they are used in different ways and to summarize what I consider the important approaches to the description and classification of biological diversity. Strauss (this volume, chap. 3) provides more detailed comments on particular aspects of systematics, especially those pertaining to the reconstruction of evolutionary relationships.

Advances in Computer Methods for Systematic Biology: Artificial Intelligence, Databases, Computer Vision, ed. Renaud Fortuner (Baltimore: Johns Hopkins University Press).

Systematics and Taxonomy

Systematics is the scientific discipline that is concerned with the discovery and identification of the diversity of living and fossil organisms. Its practitioners, systematists, analyze variation among organisms, patterns of shared common ancestry, and the evolutionary processes that gave rise to both diversity and patterns of phylogenetic relationship (Chernoff 1986). Taxonomy, which may be thought of as ancillary to systematics, is concerned with the theory and practice of classifying this biologic diversity (Chernoff 1986).

This definition of systematics is comprehensive and includes material from many of the biological sciences. Some workers use the term *taxonomy* as a synonym for *systematics*, but to most biologists taxonomy is restricted to the theory and practice of classification.

Classification

Classification makes information about organic diversity accessible. Through classification organisms are ordered into groups reflecting their relationships (Sneath and Sokal 1973). The word *classification* is also used for the end product of the process (e.g., a classification of plants).

The term *classification* is used differently in nonbiological fields. In statistics and computer sciences, it is used for what biologists call *identification*, the assignment of unidentified objects to the correct group or class once a classification has been constructed (Sneath and Sokal 1973). In what follows, I will use these terms in the biologists' sense.

Nomenclature

Systematists use a set of ordered ranks—species, genus, family, class and phylum—following the system developed by Carl Linnaeus in the mid-eighteenth century. Additional levels or divisions may be used by adding the prefix *sub* or *super* to any of these words or by using different categories as needed, such as tribe, series, or division, among others (see Beach et al., this volume, chap. 15). There are zoological, botanical, bacterial, and viral international codes of rules about forming names and naming conventions (see Ride [1988] for a general discussion; Greuter [1988] for botany; Anonymous [1985] for zoology; and Lapage et al. [1975] for bacteria). Most of the rules are concerned with lower-level categories, especially genera and species where a binominal or *binomial nomenclature* is used. Except for some viruses, each species has a two-part name. The first part is the genus name, and the second part is the species modifier. This combination need be unique only within plants, within animals, or within bacteria. There is no universal biological code.

The word *taxon* (plural *taxa*) refers to any formal unit in the taxonomic system, at any level or rank. It is used as a substitute for any group of organisms, for example, species, genus, or family, when one does not want to specify the rank. The family Hominidae is a taxon; the genus *Homo* is a taxon, as is the species *Homo sapiens*, the order Primates, or the phylum Chordata.

A system of priority in publication is followed to avoid duplication and proliferation of names. The names reflect concepts of relationship that may change a great deal, and one tries to use a name that has already been used whenever possible when the same group is recognized again. One-time valid, but long-unused, names are suppressed by the international commissions to maintain stability of nomenclature.

Taxonomic Material

Diverse Nature of Taxonomic Material

The materials of systematics are organisms living in natural habitats, organisms in living collections such as zoos and botanical gardens, or the remains of once-living organisms preserved in museums or herbaria, including specimens of contemporary species and fossilized remains. In addition, records of geographic and habitat characteristics of the organisms are essential items of information for classification and systematics. Any information on sex, life stage or maturity, physiology, behavior, and development is also relevant to our understanding of relationships. Associations with other individuals of the same species and other species are also important.

Artifacts (e.g., nests or trails) and behavior (e.g., bird songs) are also studied. Almost any aspect of organisms that is inherited and thus has a genetic basis is studied and compared among organisms. Comparison of organisms is the subject of comparative biology. Some people think of comparative biology as a larger field while others subsume it as a part of systematics.

Species the Basic Unit of Classification

For most biologists, species are the basic units of classification. There has been a long debate on the meaning of the term *species*, and once again we are in the midst of a reanalysis of what is meant by this word (see papers by Templeton, Cracraft, Nelson, Endler, and others in Otte and Endler 1989; Heywood 1988; Patterson 1988; Szalay and Bock 1991).

Biological Species. The biological species has been widely discussed as a desirable concept or definition to be used for sexually reproducing organisms.

Biological species are defined as groups of organisms that are reproductively isolated from other such groups (Mayr 1969). This means that a species consists of populations of individuals in nature capable of breeding with other individuals of the opposite sex, or at least are capable of being linked genetically through chains of other breeding individuals (Wilson 1988). Biological species thereby exchange genes only with members of the same species and do not exchange genes with other species. However, details on reproduction potentialities are poorly known for most species of organisms and are not known at all for fossils. This definition is inapplicable for non–sexually reproducing organisms. The name *biological species* does not imply that other definitions are not biological.

Phenetic Species. A more descriptive or phenetic definition based "on morphologically similar populations in a definite geographic area and morphologically distinct from other populations assigned to different species" is preferred by some workers (Sneath and Sokal 1973), although the definition should not be limited to morphology.

Evolutionary or Phylogenetic Species. One definition of the evolutionary, or phylogenetic species, concept is that of Cracraft (1989), which considers that members of a species share a history and that the species can be diagnosed or differentiated from other such species. These are the units of phylogenetic analysis.

These three definitions, the phenetic definition, the biological definition, and the phylogenetic definition, reflect differences in the three main schools of systematics (discussed below).

Number of Species

How Many Species Are There? The species definition used can affect the number of species recognized. About 1.4 million species have been described in the biological literature (Wilson 1988), but nobody knows the exact number described. There are two reasons for this. The first is that there is no comprehensive list of all of the species that have been described. The second is that the numbers change constantly as new collections are studied and older collections are restudied. Systematists may recognize additional species in material already assigned to one species, or they may decide through more detailed study to combine species. As new biological specimens frequently cannot be assigned to known species, new ones must be described, named, and classified into the appropriate higher taxon.

How many species are there? Nobody knows. There is not even agreement to an order of magnitude (Wilson 1988; May 1988). The numbers may be in the millions, in tens of millions, or even higher. There have been some thought-provoking experiments conducted recently in estimating diversity among insects. Insects include a large majority of described species, and the most diverse group among the insects is the beetles.

Experiments to Estimate the Number of Species. Erwin (1988) estimated the diversity of species of beetles from the high canopy in tropical forests, a habitat where the fauna, including the beetle fauna, has not been studied much because of difficulty of access. The first experiments were done in Panama, by fogging the forest canopy with insecticide bombs. These bombs kill most insects in the canopy and cause them to fall onto ground-sampling collectors where they can be recovered and identified. Erwin found that the majority of species he recovered were new. From the knowledge that most species of beetles are specific to particular types of trees or canopy forests, that there are many types of canopy forests, and that beetles represent a certain proportion of the number of insects, Erwin extrapolated his results to an estimate of thirty million species of insects. About 1.4 million species have been described so far.

Erwin recently repeated the experiment in the Amazon basin of Peru and has increased his estimate to fifty million species. The method is fraught with difficulties, as Erwin himself is the first to recognize, and others are conducting experiments in other habitats. To reiterate, we do not know how many species there are and what the true diversity of life is. A little additional thought suggests that even Erwin's estimates may be too low. For example, it is known that most terrestrial organisms are infested with one or more species of external parasites such as mites, many restricted to one or a few species. Nematodes are also common parasites for most if not all species, and they, too, frequently have restricted host species distributions. These facts have not been entered into the estimation procedure (May 1988).

A beginning has been made in terms of thinking about the allocation of limited systematic resources to this problem in terms of relative diversity and uniqueness of groups (Stiassny 1992). May (1988) has also suggested some ways of thinking about the problem in ecological terms. It is estimated that thousands of unknown, or poorly known, species become extinct each year. We do not know how many become extinct because we do not know how many there are. The International Union for Conservation of Nature and Natural Resources (IUCN) publishes lists of threatened or endangered known species in the form of *Red Data Books* (Anonymous 1988a).

There is renewed political interest in biological diversity, that is, the problem of the numbers of kinds and abundance of species that occur in various habitats. In this country, federal legislation is now in committee (Hoagland 1990) on a bill to provide some funds for the study of biological diversity.

Taxonomic Lists. In a spring 1990 conference at Oxford, reported in the *International Biometrics Bulletin*, Robert May suggested making a database to contain information on each and every species (Anonymous 1990a). He pointed out that this would be an endeavor comparable to the Human Genome Project and the space exploration program. For animals, the *Zoological Record*, founded in 1864, documents published taxonomic names each year. All names published since 1978 are available on a computer dial-up service (Anonymous 1989). Besides the necessity of having an idea of diversity, a list of names is required because there are international rules that forbid duplicate names. Such a published list eases the task of searching for duplicates. The *Zoological Record* is indexed in several ways, and one index is by generic and subgeneric names. Books, monographs, and about 5 500 serials were scanned in 1989 (Anonymous 1990b).

For botanists there are the *Index Kewensis, Kew Record, Bibliography of Systematic Mycology, Index of Fungi*, etc. (Dextre Clarke 1988). For bacteriologists, there is the *International Journal of Systematic Bacteriology.*

Lists of species in selected groups have been compiled for specific disciplines such as ornithology, where there is a checklist of more than 9000 species. This makes birds the best-known higher-level taxon (class) of vertebrates (Peters 1934–1986). There are plans to produce an online database from this list. The published list cannot represent the latest thought because of constant revision, though the discovery of previously unknown species is rare.

In groups with more species, monographs on selected genera, families, or orders often include the then-current list of species in the taxon studied. These lists are quickly outdated. In some monographic revisions there can be drastic changes in the number of species. Entomologists publish catalogs for groups or taxa of insects that include up-to-date lists of the taxa within those groups and bibliographic references to pertinent literature. These taxonomic catalogs are not to be confused with museum and herbarium specimen catalogs.

Bacteriologists have the most stringent rules, in that no new names are recognized after 1975 unless they appear in the *International Journal of Systematic Bacteriology.* An approved list of names was adopted at that time, in which only 2300 of the 30,000 names proposed were accepted (Ride 1988; Hawksworth and Bisby 1988).

Specimens in Collections

Importance of Systematic Collections

At present, there are hundreds of millions of specimens of organisms preserved in the world's herbaria and museums and relatively small living collections in zoos, botanical gardens, and culture collections. Culture collections maintain frozen live cultures of microorganisms that can be thawed and studied.

Collections require large amounts of space. In addition, considerable space is required for the activities of preparation, conservation, and study. The condition of specimens must be monitored. Automated sensors to maintain the collections and their documentation are only now coming into use. Major collections have been lost through fire, poor environmental conditions, war, and natural disasters. Extant collections are threatened more than ever because of lack of funds. New collecting is, in many cases, even more difficult.

It is almost impossible to estimate what it would cost to replace current collections. Some simply cannot be replaced because they represent extinct species or drastically altered habitats. Whitehead (1990) gives a few examples of current collecting costs. There is a pressing need for careful conservation of collections and more collecting efforts because there is no area of the globe where all the living taxa have been described. Exhaustive surveys may have been conducted for some kinds of organisms in specific areas, and there have been some projects launched, for instance, to do a complete survey of smaller regions or even whole countries such as Costa Rica (see Tangley 1990 for a discussion of this project).

Collections Are in Jeopardy. Some major collections are not even staffed at this point or have small or inadequate staffs. One of the premier institutions, the British Museum of Natural History in London, has been recently subjected to severe monetary and staffing cutbacks (Whitehead 1990).

Some botanists at the University of Queensland have suggested that we not keep all of the herbaria specimens now maintained as vouchers for ecological studies, as well as the large numbers or replicate specimens or series. They suggest that we rely instead on the published documentation of species. This would economize the maintenance of collections and increase the resources available for research (Clifford et al. 1990). This suggestion is one extreme response to some of the funding problems systematists face. Rejoinders were published in the September 20, 1990, issue of *Nature* (347: 222–24). The responses dealt with anecdotal and general statements on the value of collections in revisions and study. One point the letters did not make is that museum collections also document historical environmental conditions as the organic material in collections contain an isotopic and chemical record of the conditions in which the organism lived. This has proved useful in tracking DDT and other organochlorine levels in birds, for example (Kiff 1989; Davies and Randall 1989).

Taxonomic Identification. The duty of curators, or keepers of collections, and other scientists who work with museum collections is the maintenance of those collections and the identification or assignment of new specimens to known taxa at the lowest possible category (i.e., closest to the rank of species). If specimens are recognized as not belonging to a known species, then they have to be described as new to science either in a short publication or in larger monographs on taxonomic groups or geographic regions. Naming species and publication of their descriptions are known as alpha taxonomy (Mayr 1969). It is the major occupation of a number of taxonomists. Aids to identification are necessary when one needs to identify specimens from less familiar groups or collections from different, unfamiliar geographic regions. They are absolutely necessary for busy curators in small institutions who are responsible for identification of species in widely differing families, orders, classes, or even phyla.

In taxa with large numbers of species, or where diagnostic characters are difficult to describe or ascertain, for example, in some insects and other invertebrates, identification aids are indispensable. The most widely used aids are dichotomous keys (see Pankhurst, this volume, chap. 8). These are used by curators, scientists needing to identify material for biological study, agricultural agencies, amateurs, and students. There has been considerable success in computerizing aids to identification, and more sophisticated computer-based interactive keys with graphics are becoming more common (Pankhurst 1978a, 1986, 1991; Dallwitz and Paine 1986). Many of the following chapters in this volume discuss these computerized procedures. A worldwide database of taxonomists retrievable by taxon and geographic interest will be useful for identification and recognition of new taxa. Valdecasas et al. (1989) have developed DIRTAX, a database for taxonomists, implemented at present for Spain.

Nature of Specimens for Different Taxa. Traditions within taxonomic disciplines differ with respect to the parts of organisms stored in collections and investigated in systematic studies. Also, the materials studied change with new discoveries and advances in techniques for gaining information about organisms.

Some of the more traditional kinds of materials kept in museums and in herbaria are: stuffed study skins, including feathers, bills, and feet, of birds (earlier, nothing else was kept, but more often now skeletons are removed and preserved, and frequently whole individuals are preserved in alcohol); skins and skulls of mammals and, more rarely, entire skeletons; whole individuals for reptiles, amphibians, and fish in preservative liquids such as alcohol; shells in the majority of cases for clams and other shelled mollusks; dry whole mounts for arthropods, especially insects, but also alcoholic specimens; whole specimens for nematodes and other worms preserved in vials with alcohol, formalin, or other media or mounted in glycerine on slides; slides or vials for single-celled animals with hard parts; and living cultures for soft-bodied single-celled animals, plants, and simpler organisms where possible. Plants are kept mainly as dried parts mounted on sheets of paper, although there are also collections of pollen, seeds, and other kinds of specimens kept in suitable preservative. Bacteria are preserved on slides or in living frozen cultures.

Increasingly, tissue samples of living or recently killed specimens in the field are collected and frozen in liquid nitrogen for subsequent molecular biology or biochemical work.

Documentation of Museum Specimens. The documentation of specimens can also be very different depending on the group of organisms and traditions within zoological and botanical disciplines. A bird skin and mammal skin with skull are usually individually labeled with tags and unique numbers. These numbers are also kept in a card catalog, a series of ledgers, or, more recently, a computer-based catalog. Fish are stored in lots: groups of specimens in a jar or vessel filled with preservative, usually containing all of the specimens for one species from one collecting locality. Each lot is assigned a unique number, and the number of specimens in the lot is recorded.

Specimen catalogs have usually not been kept for plant specimens in herbaria and insects in entomology collections. Each specimen is essentially self-documented. All the information about the specimen is kept with the specimen. In the case of most insects labels are mounted on the same pin as the specimen; and in herbaria, all relevant information is recorded on a label on the sheet of paper to which the specimen is attached. Usually the only additional specimen documents kept are the original field notes of the collectors and some form of accession catalog that documents when the material entered the collection. Sometimes separate catalogs of collecting localities are kept, but these may or may not be cross-indexed with the specimen and taxon data.

There is increasing activity in the use of computers to store taxon and specimen information. Systematic disciplines for various groups of organisms are proposing data standards for the incorporation of more of this information in national and international computer-based catalogs. A dialogue has begun on standards for data recording, data exchange among disciplines, and how to combine the various diverse data and information so that it will be useful for ecologists and other scientists, conservation groups, government planners, and other interested parties.

Approaches to Classification

Systematics and Classification

Information on species would be useless if it were not organized and classified. Besides identification, another major activity of systematists is the interpretation of relationships among species in order to organize them into a hierarchical classification of nested taxa. This is one of the most important activities of systematists or taxonomists. Systematics provides "a general reference system for all life studies" and "provides the unifying structure around which all biological exploration must be built" (Whitehead 1990).

Without a good classification it is impossible to understand geographic patterns of diversity and the origins of adaptations and function in other than the narrowest sense.

Names must be applied to a hierarchy of inclusive groups in a way that reflects the proposed system of relationships. There is not, however, full agreement on the ways to find the natural relationships or how to name the groups. Some of the alternative methods and difficulties in this process will be discussed. Chapters 3 to 6 of this volume provide fuller details.

The formation of larger groups nested in a hierarchy and described in the form of a hierarchical classification is called beta taxonomy. Gamma taxonomy is concerned with the biological aspects of taxa including intraspecific population studies, speciation, and evolutionary rates and trends (Mayr 1969). These intraspecies- and interspecies-level studies can sometimes be supported by laboratory and field experiments with populations, statistics of individual variation, and field studies of population structure, associations, etc. Mayr (1988) has more recently called this activity *microtaxonomy* as distinct from *macrotaxonomy*, which is the study of the organization of species into genera, families, and larger taxa.

Taxonomic Characters

The classification of taxa is based on the shared features or characters of organisms or lower-level taxa. The type of character studied depends on the organisms; in some cases, organisms have to be examined in great detail, with thin sections studied under a microscope or features magnified greatly using an electron microscope. The organisms themselves may have to be dissected or analyzed using biochemical and molecular biological techniques, and so on. The methodology is vast.

Types of Characters and Character States

Characters are recorded in various states for the organisms and taxa being classified. They can be nominal characters recording alternative forms of complex structures, shapes, and patterns; continuous characters (qualitatively observed as *long* or *short*, or measured in some way); or counts of discrete features (e.g., numbers of toes in a hoofed mammal). Characters include colors, physiological states, behaviors, song types, presence or absence of some feature, alternative amino acid in a protein, or nucleotide in a DNA sequence. There is major interest in automating the collection of all sorts of information from specimens, including more efficient measuring using automated calipers, computerized imaging methodology, and automatic sequencing of nucleic acids.

Data Collection and Problems

A first step in constructing a classification is evaluating and recording character states for specimens and species. This is no easy task as some characters may have a complex developmental history. Any single specimen in a collection usually represents only a single stage in the life cycle of the organism, but organisms change constantly throughout their life span. Collections typically document each organism only at the very instant in its life that it was collected. Collections of living specimens offer better opportunities.

In some cases, different stages of the life cycle of one species may have been collected as individual specimens without understanding that they belong to the same species. A caterpillar and a butterfly might be collected and classified separately until someone sees one transform into the other and understands that they are the same species. Slime molds take on an animallike motile existence during one part of their life cycle and a vegetative plantlike existence in another. Some nematodes have a double life cycle, one cycle being plant-parasitic, and the second cycle insect-parasitic. The females of these two cycles are so different that they were once put into different suborders.

Fossils

In the case of fossils, typically only hard parts or impressions of organisms are preserved. The fossil represents only a small fraction of the complete individual—a skeleton, a shell, one bone, a tooth, a leaf impression, or even a footprint or trailway. Even in the latter two cases, an experienced paleontologist may be able to identify the fossil taxon to which the trace fossil belongs or recognize it as representing a new, undescribed species.

Sexual dimorphism is not easily recognized in the fossil record. Some described pairs of species of ammonites, fossil relatives of the chambered nautilus, a mollusc, have since been recognized as most probably males and females of half as many species as had been previously described (Callomon 1981). Polymorphic species may develop several distinct forms, even in some cases from one genome, depending on their developmental and environmental history, as in the case of some castes of social insects: queens, drones, workers, and soldiers.

Almost nothing is known about the life history of many described species, living and extinct. For example, in the Radiolaria (a group of living and fossil single-cell protozoans with silicified skeletons that are very important in determining the age of geological deposits), it is not even known whether most species reproduce sexually.

Taxonomic Schools

In the 1930s and 1940s, as the mathematical theory of population genetics was incorporated into evolutionary and systematic thinking, much attention was paid to the study of closely related groups of species and subspecific groups down to population-level relationships; this is gamma taxonomy (Mayr 1969) or microtaxonomy (Mayr 1988). The biological species concept was emphasized at that time. Population genetic terminology and speciation concepts were incorporated into the training and thinking of most systematists. There was an increased use of statistics to study and describe within- and among-species variation and patterns and causes of speciation. Classification of higher categories, or macrotaxonomy, was not much affected by this development, but it provided a theoretical foundation for an understanding of how the evolutionary hierarchy might have evolved.

Evolutionary Taxonomy

Evolutionary taxonomy, a name coined later to distinguish it from more recent taxonomic schools discussed below, built on ideas widely in use since the publication of Darwin's *Origin of Species* in 1859 that classification should be based on genealogy, that is, the evolutionary history, or phylogeny, of groups.

Since Darwin, it has been common to represent the phylogenies of organisms in the form of treelike diagrams or dendrograms (from *dendron*, tree in Greek). Classifications reflected a compromise between historical relationship or recency of common ancestry as depicted in the phylogenies and the amount of difference or divergence among taxa.

Evolutionary taxonomy holds that, when some (but not all) of the descendants of a particular ancestor have experienced many changes, they should be placed into a separate taxon. A classic example is the classification of birds. These feathered organisms are currently believed to share their most common recent ancestor with one group of dinosaurs; and among living taxa they are considered to have shared most common recent ancestry with the taxon containing alligators and crocodiles. Because birds are very distinctive compared to all living reptiles, the tenets of evolutionary taxonomy hold that they should be recognized as a separate class of vertebrates; not a sister order to the order Crocodilia. According to evolutionary taxonomy, the evolution of birds represents a drastic biological reorganization that should be reflected in its classification. All living birds have feathers, lack teeth, have a separate oxygenation pathway for the blood (as do mammals but very different in form and development), have lost and fused many bones in their skeletons, and have many other distinctive features in common.

Evolutionary taxonomy, therefore, does not and is not intended to clearly represent the phylogenetic relations of the taxa involved. Its classifications cannot be directly translated into a diagram showing the nested hierarchy of a phylogeny. The phylogenetic interpretation and inferences of shared ancestry must be depicted by accompanying graphs and discussion.

Phenetics and Numerical Taxonomy

Phenetics (from the Greek *phaino*, to appear) bases its classification on overall similarity of taxa (Sneath and Sokal 1973). Similar species are combined to form genera, genera to form families, and so on.

Numerical taxonomy was developed in the late 1950s as computers that could rapidly process large numbers of characters became available. The term *numerical taxonomy* logically describes numerical phenetics and numerical cladistics (discussed below), but the term is usually used for numerical phenetics alone. Overall similarity is best determined by a random sample of a large number of characters. The character states for p characters numerically coded for n organisms or taxa are organized into an n by p data matrix. Such numerical codes are available for binary characters, for nominal data, for discrete and ranked data, and continuous measures (Dunn and Everitt 1982; Sneath and Sokal 1973). Similarity or difference measures are computed to characterize the relation between each pair of taxa based on information from all of the coded characters. The association measures are in turn organized into a symmetrical n by n matrix of similarity or distance coefficients between all pairs of taxa. Algorithms cluster the taxa into dichotomous or polytomous hierarchical branching diagrams called phenograms, which are dendrograms in which the taxonomic units are combined into ever more inclusive groups. Large numbers of equally weighted characters are usually analyzed. It was thought that in some cases large numbers of characters could be collected automatically using scanned images (see Rohlf and Sokal 1967 for an interesting experiment). Names are then assigned to clusters; that is, a classification is constructed using names in the literature for the groups that clustered at some level on the scale of similarity (or difference).

Numerical taxonomy produces repeatable results for a given data matrix, and it was expected that as the number of characters increased the classification would converge to a natural classification. However, different measures of similarity and different clustering algorithms lead to different dendrograms and different classifications. A great deal has been written about the properties of the various approaches and kinds of patterns that emerge.

The term *operational taxonomic unit* (OTU) is used in numerical taxonomy studies. An OTU can be an individual organism, a species, or a group of organisms at any taxonomic level usually of the same rank in a given study. Operationally, it is one of the n entities being classified.

The n by p data arrays can be entered into one of a number of computer packages that produce dendrograms, and one can then form a classification from the resulting phenogram. Such clustering packages are widely used in many fields outside of biology, though some such as NTSYS-pc (Rohlf 1991) were especially motivated by biological taxonomy.

While numerical taxonomy has been an important development in helping us to rethink much of systematic methodology, it is not clear what its long-range contribution is going to be. Its workers fostered the development of some of the numerical cladistic procedures. Its greatest success seems to have been in microbiology where characters can be relatively easily scored or coded, in some cases using chemical tests.

It was not a goal of numerical taxonomy to find the true phylogeny or history of the groups studied, but rather to produce a natural classification, one that was predictive of features not included in the original analysis and of character states of taxa not included. It is suggested, however, that the relationships that are found are to be explained by evolution. It is hoped that characters that do not reflect evolution will be swamped by those that do. Debate continues on the naturalness of numerical taxonomic methods and their information content. The algorithms are very fast in contrast to those for numerical cladistics, discussed below.

Cladistics

History

Most attention in systematics is now focused on cladistic methodology (see Strauss, this volume, chap. 3). Several years before numerical taxonomy was flourishing, Willi Hennig (1950), an entomologist, published a book in German on a method he called *phylogenetic systematics*. Its impact was not felt, at least in the English-speaking world outside of entomology, until it was translated into English (Hennig 1966; a newer 1979 edition is available). Since that time much of the methodological literature has been concerned with his methods.

Hennig presented a logical methodology for constructing phylogenies and suggested that if our goal was to determine the true historical pattern of evolution, then we should study our materials in a way designed to do this. The branching diagram used in phylogenetic systematics is called a cladogram. A cladogram (from the Greek *klados*, branch) is a branching, treelike diagram that reflects at each node or branching point the branches or *sister taxa*, that is, those taxa believed to be most closely related by common ancestry. The order of branching from the stem ancestor upward delineates less and less inclusive evolutionary groups. One may or may not attach values to the length of branches. The phenogram used in numerical taxonomy is also a branching tree diagram or dendrogram, but it records at each branching level only the overall numerical value of similarity or difference at that level. The phenogram was not designed to depict hypotheses of shared ancestry.

Cladistic Methods and Terminology

Apomorphy and Plesiomorphy. A major point in cladistic methodology was that only shared derived character states should be used to determine common ancestry. Characters states that are shared among all members of a group at a given level represent an ancestral condition and are of no value in dividing that group into subgroups. For example, in a systematic study of mammals, presence of hair does not tell us anything about how to group certain mammals. But in comparison with other vertebrates where the ancestral state is hairless, presence of hair characterizes all mammals. Hair is a shared derived condition for mammals and presumably was the primitive, or ancestral, condition for the species giving rise to all mammals. The word *ancestral* is to be preferred to *primitive* as it is less value laden. Another character state shared by all living mammals is the presence of a single bony element in each side of the lower jaw, while other jawed vertebrates and presumably their common ancestor had several distinct bony elements in the jaw.

Ancestral states are called plesiomorphies (from the Greek *plesio*, near), and such states that are shared among related taxa are called symplesiomorphies. Derived character states are called apomorphies (from the Greek *apo*, away from), and derived states shared among related taxa are called synapomorphics. Plesiomorphies at one level (e.g., hair for all mammals when we are differentiating mammals from each other), become apomorphies at another level (when we are differentiating all mammals from the other vertebrate classes). Distinctness of taxa is indicated by apomorphies. Synapomorphies, the shared derived character states, allow us to place sister taxa together in higher groups. Symplesiomorphies are of no value for understanding relationship.

It is necessary to establish whether character states are ancestral or derived. This operation is called character polarization. One of the most widely used methods of polarization is that of out-group comparisons. A character state will be found more widely distributed over a broader group of taxa if it is ancestral and more restricted to fewer taxa if it is not. Systematic studies thus extend to include additional taxa more distantly related to the group whose detailed relationships are being analyzed. The out-group is hypothesized to have the ancestral state. For example, in studying the phylogeny of apes and humans, one might want to include some members of sister groups, such as Old World monkeys, to help discover which character states are more generally distributed, for example, presence or absence of a tail or bipedal versus quadrupedal locomotion.

Another approach to character polarization relies on the assumption that those characters that appear more widely or generally in the development or ontogeny of an organism are more likely to be ancestral. Gill slits appear very widely among embryos of vertebrates, suggesting that presence of gills is ancestral. Hecht (1989) has criticized this approach on the grounds that in some species the supposed more general stages of embryonic development are bypassed completely.

Hennig (1979) lists some other criteria for determining character transformation series. Some of these are briefly described in Janvier (1984).

Homology. A critical concept in all schools of systematics is the idea of homology. Evolutionary biologists recognize homology in terms of a character state inherited from a common ancestor. According to Hennig (1966), "Different characters that are to be regarded as transformation stages of the same original character are called homologous," so the definitions appear to be similar. On the other hand, numerical taxonomy uses a more operational definition of homology based on similarity, so that character states can be entered into a data matrix.

The phylogenetic definition poses a methodological problem because the ancestral character condition is not known until some form of classification is established. Many critics have therefore felt that cladistic methodology is circular in this regard. One must test homologies with additional characters. For some (Patterson 1982a), the word *homology* has become synonymous with the idea of synapomorphy or shared character state derived from a common ancestor.

Most of the work in reconstructing phylogenies is in character selection and analysis, that is, identifying the homologies and arranging the character states into a character-state tree, or transition series, that represents the pattern of derivation of one state from another. Neff (1986) has offered a description of the steps in constructing character-state trees prior to cladistic analysis that does not depend on a priori classifications. Neff, together with Szalay and Bock (1991), argues that this is the stage at which one determines homologies, not in conjunction with or after constructing a phylogeny.

Homoplasy. Convergence, or evolution of what appears to be the same character state in distantly related taxa, creates false synapomorphies that are called homoplasies. Warm-bloodedness as a trait for birds and mammals is a homoplastic character that has evolved more than once from presumed cold-blooded reptilian ancestors of birds and mammals, respectively. Other groups such as the flying reptilian pterosaurs and some dinosaurs may also have evolved warm-bloodedness. Salvini-Plawen and Mayr (1977) discuss the multiple origin of eyes and photoreceptors in animals. Parallel evolution is sometimes distinguished from convergence, when the taxa share a more common recent ancestor, but it may be recognized as a special case of convergence.

Another type of homoplasy is character reversal, or the gain, loss, and subsequent reevolution of a character state. The presence of finlike appendages in fish and whales is an example. The presumed common fish ancestor of vertebrates had paired fins; these evolved into supporting limbs with separated toes or digits in the ancestral mammal, and then reevolved as fins in an ancestral whale. Homoplasies, being similarities in character states that are not inherited from a common ancestral state, are not to be confused with the homologies that form synapomorphies.

Distinguishing true synapomorphies from homoplasies is the most difficult part of cladistics. It requires careful analysis of characters to determine the evolutionary relations between character states, so that homoplasies can be identified. Parsimony, discussed later, is considered by many cladists to be the most important criterion for distinguishing true synapomorphies from homoplasy.

If enough carefully selected characters are included, the true synapomorphies should outweigh homoplasy and provide the best estimate of a phylogeny. As additional characters and character systems are studied they can lead to new or different phylogenetic interpretations, in which case what were thought to be synapomorphies may prove to be homoplasies. Or the additional characters may corroborate earlier hypotheses of relationship.

Monophyly and Paraphyly. The concept of monophyly is used somewhat differently by different groups of comparative biologists and systematists. Following Hennig, a monophyletic taxon is one that contains the common ancestor of a group and all of the descendants of that common ancestor. A taxon at any level is monophyletic if it contains all subbranches and terminal twigs. Ashlock (1974) proposed the term *holophyletic* for this restricted definition of monophyly.

The concept of monophyly used by Simpson (1945, 1961) is not necessarily monophyletic in the cladistic sense. The word is used by many evolutionary taxonomists for a group originating from a single ancestor, whether or not the group contains all the descendants of that ancestor or the ancestor itself. Those groups that are not strictly monophyletic in the sense of Hennig are called paraphyletic in cladistic analysis.

Informal groups like *reptiles* are useful for a level of vertebrate organization that is generally recognized as paraphyletic. The word *grade*, as opposed to the strictly monophyletic *clade*, is sometimes used for such informal groups to refer to a level of organization not necessarily characterized by synapomorphies.

Examples of paraphyly abound in textbook classifications. For example, traditional teaching held that the family Pongidae contained the great apes (gorilla, chimpanzee, orangutan, and gibbons), while the related family Hominidae contained *Homo sapiens* and some fossil relatives (Simpson 1945). Later, this paraphyletic classification persisted even when it was agreed that some apes shared a more common ancestor with *Homo sapiens* than others. Many of the latest classifications now place us all in one monophyletic family, the Hominidae (the name with priority), and the new much-debated question is more refined: Is the genus *Homo* a sister taxon with the gorilla (genus *Gorilla*) or with the chimpanzee (genus *Pan*)? The orangutan (genus *Pongo*) is generally recognized as a sister taxon to all three.

It is agreed that polyphyly is always to be avoided, but paraphyletic classes sometimes have the advantage of cutting down the number of names that may be required for each branch of a highly resolved phylogeny. Some rules have been suggested for listing taxa in a phylogenetic classification to convey maximum information and avoid proliferation of names, but there is no widely accepted standard (see Wiley 1981; Janvier 1984; and Schoch 1986 for a review of these ideas). However, paraphyletic taxa may not be that useful in comparative biological studies, where communication of the best understood phylogeny will help in the study and comparison of other features of organisms, such as functional studies and biogeographic distribution.

Strict Dichotomy. Cladists would like to fully resolve relationships among taxa by having all branching dichotomous. In theory, an ancestral taxon does not give birth to a new taxon, but it "dies" by splitting into two new sister taxa. This idea is strongly criticized by many systematists.

One attempts to attain complete dichotomous resolution when constructing a cladogram. The cladogram, however, is not the phylogeny, but only a step in the analysis. More than one phylogenetic interpretation is available for any given cladogram if some of the species are believed to be ancestral species and occupy internal branch segments. In fact each of the three possible strictly dichotomous cladograms for three taxa can generate seven phylogenetic interpretations if internal branch segments are recognized as ancestral species (Schoch 1986).

Pattern Cladistics. This school holds that evolutionary theory is not a prerequisite for construction of cladograms and for understanding the nested hierarchy of living organisms. The arguments for this point of view have been clearly stated by Patterson (1982a) and Janvier (1984) among others, and they have been criticized by Ridley (1986). A broader viewpoint is presented by Rieppel (1988).

Numerical Cladistics

Algorithms and computer software for cladistic analysis or constructing cladograms are available from many sources. Among the most widely used software packages are PHYLLIP, HENNIG86, MacClade, and PAUP. Sources for these packages and an excellent up-to-date discussion of some of the methodology are given in Swofford and Olsen (1990).

Maximum Parsimony. As in numerical taxonomy, one starts with a matrix of

n OTU or EU (Stuessy 1990 for *evolutionary units*; Estabrook 1972) and p characters, usually coded as discrete character states. The program algorithms search for cladograms that are most parsimonious; that is, a tree that depicts the fewest character changes. This is the most popular optimality criterion. Some others are discussed later and in Swofford and Olsen (1990).

If there were no homoplasies, that is, no reversals and no repeated evolution of the same character state, then the length of the tree or cladogram would be just the sum of the number of possible character-state changes over all characters. For example, with twenty binary characters, the tree length would be twenty. In real data homoplasy is common, and true homology is not easy to evaluate. Therefore, algorithms have been designed to produce the least amount of homoplasy in the cladogram.

From a computer science viewpoint, finding the most parsimonious cladogram is an *NP-complete* or *NP-hard* problem. NP or *nondeterministic polynomial* problems (Lewis and Papdimitrion 1978; Graham 1978) are among the most difficult computational problems (Felsenstein 1983) and grow exponentially with the number of taxa in the analysis. In the case of finding the most parsimonious cladogram (i.e., the shortest tree) the only rigorous solution for n taxa is to form all T possible trees where T is the product of the first $n - 2$ odd integers greater than 1 (assuming only dichotomous branching and not allowing any of the n taxa to occupy internal segments). For three taxa there are three possible cladograms, for four there are $3 \times 5 = 15$, for ten OTUs there are $3 \times 5 \times 7 \times 9 \times 11 \times 13 \times 15 \times 17 = 34{,}459{,}425$ different possible trees, and by twenty OTUs the number is truly astronomical, already near or beyond the capacity of the largest computers. Current algorithms examine a subset of possible solutions. Frequently there may be many equally shortest trees; and there is some argument over whether it is necessary to find "the" shortest tree.

Character Compatibility Analysis. Compatibility analysis is another way to form cladograms. It looks for *cliques* of characters, the largest group of characters for a set of taxa so that each character state arises only once. In other words, the method finds a set of characters for which there is no homoplasy over the taxa. Meacham (1981) gives a clear exposition of the method for finding cliques.

Because characters showing homoplasy are discarded, the method has been criticized by users of parsimony methods. However, there are algorithms that, after finding the largest clique over all taxa, then find additional character compatibilities among those earlier discarded characters for smaller monophyletic groups or branches within the cladogram. There is some debate (Felsenstein 1983; Farris 1983) as to whether parsimony methods or compatibility analysis methods are the algorithms that best reflect Hennig's nonnumerical methodology.

Maximum Likelihood Cladistics. Felsenstein (1984a) has employed maximum likelihood methods for estimating trees or cladograms. This method requires a probabilistic model for the evolutionary processes and explicit assumptions about rates of evolution in the branches. It has been worked out only for the neutral or random-walk model. Using standard statistical maximum likelihood theory, one finds the tree-branching pattern and branch lengths that maximize the likelihood function.

The maximum likelihood estimate is defined by Felsenstein (1983) as follows: "*D* is the data we have, *T*, a possible evolutionary tree (phylogeny), and *M*, a probabilistic model of evolutionary change. Given the model, we can compute the probability $P(D|T,M)$, the probability of obtaining the data, given the tree *T* and the model *M*. The maximum likelihood estimate of the tree is simply that tree that maximizes $P(D|T,M)$."

One would like to be able to compute $P(T|D,M)$, the probability of the tree given the data and the model, but we do not have the prior probabilities of the trees to use Bayes' theorem other than in the trivial sense of considering all prior probabilities equal. Thus, the likelihood is computed over all possible trees to find the maximum likelihood estimate. Note that, unlike probabilities, likelihoods do not sum to one over the trees. The computational task is very heavy. Maximum likelihood programs are available in the program package PHYLLIP (Felsenstein 1984a).

The strongest attraction of maximum likelihood is that it takes a statistical approach to estimating phylogenies. In addition, the method is quite general, as it can explicitly yield a parsimony solution and a maximum clique solution as two extreme answers if appropriate statistical assumptions are made about rates of evolution. Some numerical taxonomic algorithms, if they are used to construct phylogenies, may also be viewed as a maximum likelihood method when appropriate assumptions are made. However, maximum parsimony has been the most widely used procedure.

Farris (1983) offers a critique of both compatibility and maximum likelihood techniques. Maximum likelihood methods have been more widely used by those who deal with allelic and molecular data.

Conclusions

Biodiversity

At present, biological diversity is understood in terms of the most common taxa, observed in or collected from relatively more easily sampled habitats. Most major groups of life have probably been discovered, but only a fraction of the world's species have actually been preserved in our collections and described in publications. I am not sure that we should collect and describe all taxa. However, we must set priorities in our attempt to discover the true diversity of life. It is important to focus limited resources on the kinds of information that we will need to develop and store, so that we can focus research on framing the important questions to pursue.

Documentation

The documentation accompanying collections, as well as the published record, must be made more accessible to researchers and interested parties who use collections and collection-based information. Taxonomic lists need to be widely expanded and made accessible online and in other cheap and transportable ways to the international community of potential users. Identifying all of the potential users for this information is not always easy, and we must keep an open mind.

It is important to go beyond keeping track of lists of specimens and taxa and to provide an audit trail of taxonomic identifications and opinions and ideas concerning systematic materials. A recent book by Hughes (1989), *Fossils as Information*, has made some suggestions in this direction.

We need to improve our ability to identify the backlog of already collected specimens and of new specimens and to recognize specimens as belonging to new taxa that need to be described.

Phylogenetic Reconstruction and Classification

Research efforts in understanding phylogeny and production of classifications must be expanded again to disseminate information and serve other biologists. At the present, cladistic methodology seems to be the most popular and best articulated and comes closest to the goals of describing the history of life. On the other hand, computer-aided cladistics has real computational restrictions, as best solutions are not available for more than a small number of taxa, and the task is immense when one considers the amount of diversity to be classified.

Acknowledgments

Our indefatigable editor shepherded this manuscript through very rough drafts to the present form. His effort has helped me to improve its quality. F. James Rohlf stimulated my thinking on the relations of cladistics and numerical taxonomy. Mario C. C. di Pinna offered several helpful criticisms. Antonio Valdecasas, Max Hecht, Kumar Krishna, and Therese Wojtowicz gave many helpful comments and additional references, not all of which were used. Antonio continued to provide references and suggestions. Scott Miller provided an early reading list, and Colin Patterson and Daryl Siebert pointed me to some other useful references. None of these people is responsible for any of the shortcomings of this chapter.

Most of the reading and writing was done while the author was on sabbatical from Queens College of the City University of New York. I thank Claudio and Penelope Vita Finzi for providing a relatively isolated haven in London while I did a great deal of the reading.

Discussion

——: Would you please give a more complete illustration of monophyly and paraphyly.

Marcus: I will use the following classification of Woese et al. (1990) to illustrate paraphyly and monophyly.

- Domain Bacteria
  - Kingdoms ?
- Domain Archaea
  - Kingdom Euryarchaeota
  - Kingdom Crenarchaeota
- Domain Eucarya
  - Kingdoms ?

Note that the word *domain* is used for the highest levels, because the more traditional category, kingdom, is a subgroup of the domains of Woese et al.

Woese et al. (1990) claimed to have discovered the very earliest division of living organisms, based on characteristics of ribosomal RNA fractions. Ribosomes are the cell organelles that translate the genetic code and assemble proteins. An important distinction for living cells is whether a nucleus is present (eucaryote), or absent (procaryote). The primary division of all living organisms is within the procaryotes, which become a grade. One monophyletic major clade consists of the Bacteria with many bacteria, blue-green algae, and their procaryotic relatives; and the unnamed sister clade is divided into another monophyletic group of bacterialike procaryotic organisms (the domain Archaea), and the monophyletic eucaryotes (the domain Eucarya). The latter include all nucleated single-cell life and multicellular organisms, plants, animals, and various groups of fungi. Woese et al. interestingly do not provide names for the two most basic monophyletic sister groups of their most fundamental dichotomy, but instead they define three monophyletic groups. They propose this classification for textbooks, which have traditionally presented a quite different classification.

Mayr (1990) presents an alternative classification based on Woese's results. It is purposely paraphyletic to emphasize the "most drastic change in the whole history of the organic world," that is, the attainment of a nucleus, and places the prokaryotes together in one paraphyletic domain, Prokaryota, and the eucaryotes in a second domain, Eukaryota.

Brown (1990) in a *New Scientist* discussion of the Woese et al. paper presents a broader perspective, noting that results based on other molecular data, for example, that of John Lake of UCLA, splits the Archebacteria between the eucaryotes and other procaryotes. Lynn Margulis (1988) disagrees that molecular data are so important for the fundamental divisions. She recognizes a more traditional five-kingdom classification.

My classification based on the Woese et al. cladogram uses strict monophyly and recognizes their postulated most basic dichotomy. I avoid assigning category names to the ranks, as I do not want to add to the nomenclature here:

Major Group Bacteria
Major Group Unnamed
  Subgroup Archaea
  Subgroup Eucarya

This classification has a one to one relation with the diagram of Woese et al., and it clearly represents their hypothesis of relationship. I feel it is much clearer. Woese et al. in their conclusion say:

> The system we propose here will repair the damage that has been the unavoidable consequence of constructing taxonomic systems in ignorance of the like course of microbial evolution, and on the basis of flawed premises (that life is dichotomously organized; that negative characteristics can define meaningful taxonomies). More specifically, it will (i) provide a system that is natural at the highest levels; (ii) provide a system that allows a fully natural classification of microorganisms (eukaryotic as well as prokaryotic); (iii) recognize that, at least in evolutionary terms, plants and animals do not occupy a position of privileged importance; (iv) recognize the independence of the lineages of the Archaea and Bacteria; and (v) foster understanding of the diversity of ancient microbial lineages (both prokaryotic and eukaryotic).

I applaud the intent of these statements, but feel that Woese et al. have fallen into the group they are criticizing by not naming the fundamental division they have discovered. I would agree with their proposed classification and nomenclature only if they had shown an unresolved trichotomy. Mayr (1990), from my viewpoint, makes things even less clear by presenting a paraphyletic classification.

3

Phylogenetics

RICHARD E. STRAUSS

Phylogenetics, a term often used as a shorthand synonym for phylogenetic inference or phylogeny reconstruction, is the attempt to assess and portray how organisms are related historically in terms of their evolutionary paths of descent and divergence. The basic steps of phylogeny reconstruction would seem to be straightforward: (1) defining the primary biological units of study, called *taxa* or OTU (operational taxonomic units), which are often species but may be units at a lower (e.g., subspecies, populations) or higher (e.g., genera, families, orders, classes, phyla) level of organization; (2) characterizing taxa by the states of their characters, which might be any attributes (genetic, morphological, physiological, etc.) that vary in their expression among taxa; and (3) arranging the taxa by grouping them according to shared character states or degree of similarity. The arrangement is usually hierarchical, with more inclusive groups being characterized by more and more shared characters. The hypothesized hierarchical arrangement (a phylogeny) is usually portrayed graphically as a tree and may be used, for example, as the basis for a taxonomic classification or as an interpretation of temporal evolutionary relationships.

Despite the apparent simplicity of these steps, phylogenetic inference is not straightforward; on the contrary, it is a highly controversial discipline. This is because different characters do not necessarily define the same groups. In fact, different characters may be completely incompatible in their information about hierarchical relationships. For example, possession of a dorsal hollow nerve chord defines the chordates, which are almost universally considered to be an evolutionarily monophyletic group (i.e., derived from a single ancestral taxon). Possession of a complex eye defines a quite different group, on the other hand, partly overlapping with the chordates, that is not generally considered to be monophyletic. The incompatibilities among characters, and the consequent necessity for systematists to choose (i.e., weight) one kind of

---

Advances in Computer Methods for Systematic Biology: Artificial Intelligence, Databases, Computer Vision, ed. Renaud Fortuner (Baltimore: Johns Hopkins University Press).

character over another, provide both the fundamental sources of controversy in phylogenetics and the reason why there is more to systematics than the simple definition of groups.

Model-Free versus Model-Laden Phylogenetics

Characters can be classified in different ways, such as absolute (green) versus relative (forelimb longer than hindlimb); discrete (five toes) versus continuous (wing 8 mm long); developmentally early versus late; evolutionarily stable versus labile; evolutionarily ancestral versus derived. The first three of these contrasts can usually be evaluated objectively using the organisms at hand, while the latter two depend on prior hypotheses about evolutionary patterns and processes.

There is a spectrum of opinion among systematists about the desirability of incorporating process-level evolutionary assumptions into phylogenetic inference, depending to some extent on the intended use of the resulting phylogeny. If the object of a study is to obtain the best (i.e., most realistic) phylogeny possible from the data, then it may be desirable to use a tree-building method that incorporates as much as is known or surmised about evolutionarily processes. The resulting phylogeny will then reflect the biological assumptions built into the methodology. Molecular systematists characteristically take this approach, attempting to incorporate specific models of nucleotide-sequence evolution (Jukes and Cantor 1969; Kimura 1980; Lake 1987; Cheeseman and Kanefsky 1990). However, if the resulting phylogeny is to be used to test independent hypotheses about biological phenomena such as sequences of character evolution, rates of evolution, or biogeographic distributions, then the tree-building method should be as independent as possible of prior biological assumptions about the relevant phenomena. For example, if the objective of a study is to test for heterogeneity of rates of diversification among different lineages, then the phylogeny on which the test is based should have been constructed without invoking an assumption of rate homogeneity. The most extreme position in assumption-independence is that of the so-called transformed or pattern cladists, who have attempted to avoid the evolutionary framework entirely in inferring phylogenetic relationships among organisms (Platnick 1979; Patterson 1980; Ridley 1986). Most systematists consider the complete exclusion of evolutionary assumptions to be untenable (Sober 1989).

Trees, Cladograms, and Phylogenies

Putative historical relationships, and their implied time components, are conventionally portrayed as branching phylogenetic trees. However, because information about nested sets of character states (from which historical relationships among taxa are inferred) is also portrayed in the form of branching trees, several important distinctions must be made (Eldredge and Cracraft 1980).

Although not all systematists agree on terminology, the terms *cladogram, tree*, and *phylogeny* are often used to imply different kinds of information about character states and taxa. A *cladogram* is a branching diagram of species, clustered by their shared derived character states (see Marcus, this volume, chap. 2). The connections among the species indicate only the joint possession of derived characters, and there is no implicit suggestion of ancestral or temporal relationships. A *tree*, on the other hand, is usually considered to be a branching diagram depicting patterns of character-state distributions and transitions among taxa at various hierarchical levels. Hypothetical ancestral species are placed at the branching nodes, rather than at the tips of branches as in cladograms. The branching relationships among taxa then imply or represent evolutionary lineages.

A tree may be either *rooted* or *unrooted*; the root of a tree is the position of the putative ancestral taxon. An unrooted tree is often called a network, although this term is discouraged by some theorists because it conflicts with the standard usage in topology. Many of the quantitative procedures described below produce unrooted trees, for which the position of the root must be judged from other information. Typically one or more related taxa that are assumed to lie outside the group of interest (the in-group) can be included in the analysis. The position at which these so-called out-group taxa join the tree then defines the root with respect to the in-group (Maddison et al. 1984).

The term *phylogeny* or *phylogenetic tree* often implies the addition of a time dimension, particularly when stratigraphic, molecular-clock, or other time-referential data are available to indicate relative or absolute rates of evolution.

Phylogenetic Inference under Uncertainty

The expression *phylogeny reconstruction* implies a deterministic process of reconstructing the past, but phylogenetics is best viewed as an estimation procedure. The objective of the procedure is to produce the best possible estimate of historical relationships based on incomplete and often contradictory information about the past, primarily utilizing contemporary morphologies and molecules (although fossils can sometimes serve as a valuable source of phylogenetic information). Putative historical relationships are then portrayed as branching cladograms or trees, but many such trees are possible, and systematists need a basis for selecting one or more preferred trees from the spectrum of possibilities.

Contemporary phylogenetic methods determine preferred trees in one of two ways (Swofford and Olsen 1990): (1) by defining a specific sequence of steps (an algorithm) by which the preferred tree is produced; or (2) by defining a quantitative criterion by which to compare alternatives and then deciding which of a series of trees is best by this criterion.

The first class of methods combines the definition of the preferred tree and the process of tree building into a single step. All of the various forms of hierarchical pair-group cluster analysis are of this type, as are the additive-tree methods described below. The primary advantage to these methods is their efficiency; because a single tree is generated algorithmically according to some optimality criterion, the current implementations are quite fast and require relatively little computer memory. Their primary disadvantage is that, although the resulting tree can be compared with the original data to assess the degree of fit, it is generally not possible to construct and evaluate alternative suboptimal trees. A second problem is that these methods are heuristic and often fail to address, or even make explicit, the underlying evolutionary assumptions.

The second class of methods divides tree building and tree assessment into two distinct logical steps. (1) An optimality criterion (also called an objective or merit function) is first defined, by which to evaluate individual trees. Optimality criteria are often based on explicit evolutionary assumptions and allow scores to be assigned to trees indicating their relative suitability. (2) Various algorithms are then used to search for the one or more trees having the best value (maximum or minimum) of the criterion. The evolutionary assumptions invoked in the first step, presumably based upon scientific criteria, are thus decoupled from the searching algorithms of the second step, which are based on computer science research and are improving rapidly with time. Because every tree can be assigned a score, the primary advantage to this class of methods is that alternative phylogenies can be ranked in order of preference and assessed in terms of their biological properties. The cost of this is reduced speed, a consequence of having to search for the best tree among an often enormous number of possibilities.

Kinds of Data

There are two broad categories of data used in phylogenetic inference: character data, which provide information about individual taxa, and distance or similarity data, which describe pairwise relationships among taxa.

Character Data

Character data are specified as single numeric values codifying the state of each character for each taxon. There is some disagreement among systematists about the terms *character* and *character state*, but those who employ numerical methods generally use the term *character* to be a synonym for *variable* and consider a character state to be one of the set of possible observed values of the character. A character matrix $X_{(nxp)}$ for n taxa and p characters is a rectangular matrix of character-state values x_{ij}. Characters can be either *qualitative* or *quantitative*, and qualitative characters can be further classified as *binary* (having two states, often indicating the presence or absence of traits) or *multistate* (having three or more states). For tree-building methods requiring binary characters, multistate characters can be recoded as a series of binary characters by additive binary coding.

Multistate characters are often considered to be *unordered* (nominal), such that any state can transform (over evolutionary time) to any other state. This is the usual assumption for DNA-sequence data, for which the four nucleotides (A, C, G, T) are the possible states at a particular sequence position. Alternatively, multistate characters can be ordered into a *transformation series* reflecting a set of assumptions about sequences of character-state evolution. For example, the relative development of a particular bony crest might be coded as an ordered sequence 0–3, for which the states represent crest absent (0), crest present but small (1), crest present and elongated (2), and crest present and bifurcated (3). The corresponding assumption would be that evolution from no crest to a bifurcated crest must proceed through the intermediate states, thus ruling out a punctuated evolution of the bifurcated crest. Whether this assumption is tenable is a biological problem, not a numerical one. Because ordering represents a constraint on character evolution, the tree length for an ordered character must be greater than or equal to that for the corresponding unordered character.

Character polarity is a separate but related concept. Whereas ordering refers to the permitted character-state transformations, polarity refers to hypothesized directions of character evolution through time. Polarity assessment thus presumes the determination of the ancestral state, that of the most recent common ancestor of the group in question. Putative character polarities can be assessed in one of two ways: (1) one character at a time at the beginning of a study (prior polarities), usually by comparison with an out-group condition; or (2) from the results of the study (posterior polarities), after the preferred tree has been rooted. A prior polarity represents a constraint on character evolution beyond the assumption of ordering and thus increases the lengths of preferred trees.

Quantitative character data differ from discrete characters in that they vary incrementally on an interval scale or continuously on a ratio scale with a fixed zero point. Thus they are inherently ordered and have a set of possible states limited only by the resolution of the measuring device. However, the problem of polarity, the assessment of the ancestral state, still holds.

Regardless of whether characters are discrete or continuous, ordered or unordered, there are two assumptions that are fundamental to all character-based phylogenetic studies. The first is that of *character independence*. The assumption that evolutionary character-state transitions are uncorrelated across characters permits phylogenetic problems to be partitioned into simpler, independent subproblems. For example, in parsimony methods corresponding branch lengths can be summed across characters to give a total branch length, and corresponding probabilities can be multiplied across characters in maximum likelihood methods. The assumption also allows us to ignore character covariances, which greatly simplifies computational methods. It should be noted that the assumption of character independence, though necessary, is biologically unrealistic and very unsatisfactory to many systematists (Donoghue 1989).

The second fundamental assumption is that of *homology* (Patterson 1982b; Roth 1988; Smith 1990), usually defined to be the similarity of structure due to descent from a common ancestor. Characters are defined by systematists such that the states among observed taxa are assumed to have been derived via modification from the state of a common ancestor. The critical assumption of homology implies that states among taxa are scientifically as well as numerically comparable and is the methodological basis for studies of character-state transformations and the diagnosis of monophyletic groups. This is not to say that the recognition of homology, or even its operational definition, is a trivial problem, and several different kinds (e.g., operational, taxic) have been proposed and implemented by systematists of various philosophical persuasions (Smith 1990). Note, however, that the evolutionary definitions of homology are not synonymous with mere similarity, as in the term *sequence homology* used by some molecular biologists.

Pairwise Similarity and Distance Measures

Whereas character data describe individual taxa, similarity and distance data describe the relationship between pairs of taxa. Some procedures, such as immunology and nucleic acid hybridization, provide distance data directly; other kinds of character data (e.g., allozymes, nucleotide sequences) are routinely converted to distances for analysis (but see Farris 1981, 1986). A distance matrix is a symmetrical matrix of which the elements are the pairwise distances among taxa.

Phylogenetic methods based on distance data usually assume the distances to have particular mathematical properties, of which two are of special interest: the additivity and ultrametric properties (Swofford and Olsen 1990).

Additive Distances. A set of additive pairwise distances are mutually compatible to the extent that all distances can be fitted exactly by an unrooted tree, such that, for all pairs of taxa, the branch lengths along the paths connecting the taxa sum exactly to the observed distances between them. A number of methods described below (the additive-tree methods) assume that the observed data are additive; if so, then all should provide the same tree. Because real data are seldom exactly additive due to systematic and random variation, the various procedures often produce different results and can be assessed by their performance when the assumption is violated.

Ultrametric Distances. Ultrametric distances are much more highly constrained. They must be additive, but in addition can be fitted exactly by a tree such that all terminal taxa connected to a particular node are equidistant from that node. This in turn implies that the tree must be rooted such that all of the terminal taxa are equidistant from the root. The ultrametric criterion corresponds to the strict evolutionary assumption that all lineages have diverged to equal extent with equal rates of change (the molecular-clock assumption).

It is very unlikely that ultrametric data can be sampled from nature; even if amounts of evolutionary divergence are identical for two or more lineages, statistical sampling fluctuations will lead to deviations from the ultrametric condition. The criterion is important in systematics primarily because cluster analysis techniques (discussed below), which have been widely used to construct phylogenetic trees for more than twenty years, implicitly assume that the distances being clustered are ultrametric.

Kinds of Methods

The methodologies currently used in phylogenetics can be classified into three broad categories (Felsenstein 1982; Swofford and Olsen 1990): distance methods, parsimony methods, and methods invoking explicit models of evolutionary change over time. The best known of the latter methods are the maximum-likelihood procedures, although the category could include a number of specialized methods geared toward particular kinds of data (such as DNA sequence data).

Pairwise-Distance Methods

Cluster Analysis Methods. The cluster analysis algorithms typically used in phylogenetic applications are a family of algorithms known as *hierarchical agglomerative procedures* (Rohlf 1970; Williams 1971), because they produce hierarchical trees by beginning with individual taxa and iteratively fusing them into more and more inclusive clusters. Such algorithms are conceptually straightforward. Given a matrix of pairwise distances among taxa, all of these methods first link together the most similar (i.e., least distant) pair, followed by a series of successive linkages involving more and more distant pairs. At each step the two taxa or groups of taxa being linked are merged into a single cluster. The result is usually called a *dendrogram* because it is a rooted dichotomous treelike structure that portrays clusters of taxa based on their relative similarities.

All hierarchical agglomerative algorithms involve four basic steps: (1) finding the current taxa or clusters C_i and C_j having the minimum distance value d_{ij}; (2) defining the depth of branching between the clusters (the position of the node) to be half the distance d_{ij}, according to the ultrametric criterion; (3) merging the two clusters to form a new cluster C_h; and (4) defining (and thus computing) the distances between the new cluster C_h and all of the other existing taxa or clusters (C_k)to be a linear function of d_{ik} and d_{jk}. For N terminal taxa, the steps are repeated $N - 1$ times, each time with one less taxon or cluster, until all taxa are merged into a single cluster.

Out of a large number of possible clustering algorithms, only a few are typically used in phylogenetics, particularly single linkage or nearest neighbor; complete linkage or furthest neighbor; UPGMA (unweighted pair-group method using arithmetic averages); and WPGMA (weighted pair-group method using arithmetic averages). These differ only in the definition of the distances between new and existing clusters (step 4 above). Indeed, Lance and Williams (1967) showed that these and many other algorithms are special cases of a four-parameter general linear model (the so-called flexible strategy) that is applicable to any measure of intertaxon distance. For a given distance matrix, a series of dendrograms ranging from balanced (nearly symmetrical) to chained (completely asymmetrical) can be produced by altering the parameters of the model. The agreement between the resulting dendrogram and the original data matrix can be assessed by the so-called cophenetic correlation between the elements of the matrix and the pairwise distances implied by the minimal nodes of the dendrogram (Rohlf 1970). Of the commonly used procedures, empirical studies have shown UPGMA to yield rather good fits.

Clustering procedures have historically been very popular, primarily because they are fast and easy to implement and because they invoke few assumptions about the data, other than that the intertaxon distances are ultrametric (Colless 1970). The primarily disadvantages are that there seems to be no objective definition of what constitutes the best tree when the data are not ideal. Furthermore, alternative methods are available that are suitable for data that are merely additive, which is a more moderate assumption than that of ultrametric relationships.

Additive Tree Methods. Additive tree methods comprise a broad class of methods based on the assumption that the lengths of the branches of a tree can be summed to provide a quantity that is proportional to the amount of evolution that has taken place along a lineage. There are three basic kinds of additive tree methods, all of which produce unrooted trees.

1. The Fitch-Margoliash methods (Fitch and Margoliash 1967; Cavalli-Sforza and Edwards 1967) differ conceptually from cluster analysis in that

they involve the minimization of a merit function that measures the summed deviations between the original pairwise intertaxon distances and those predicted or accounted for by the branch lengths along the tree. The deviations to be minimized may be specified in various ways, for example, as absolute differences or squared differences (the least-squares criterion), and can be weighted by various criteria. As with other minimum-length tree methods, finding the tree with the minimum value of the merit function generally requires two considerations: optimizing the branch lengths of a given tree topology to find the best correspondence between observed and predicted and identifying the particular tree topology having the minimum function value of all possible tree topologies. For a particular topology, the merit function can be minimized iteratively by successive refinement or analytically by linear or quadratic programming. The more general problem of finding the shortest of all possible trees is discussed below ("Searching for Globally Optimum Trees").

2. The so-called distance-Wagner methods (Farris 1972; Swofford 1981; Tateno et al. 1982) are similar conceptually to the Fitch-Margoliash methods in that they find a tree having a minimum value for the merit function: in this case, the total of all absolute branch lengths of the tree. Corresponding to this computational difference, however, is an important conceptual distinction. The conceptual basis of the Fitch-Margoliash methods is that the observed intertaxon distances are estimates of the true evolutionary distances, with some estimates being greater than the true evolutionary values and some being smaller. Thus the estimated or predicted distance, as determined from the branch lengths of the resulting tree, may be either greater or less than the observed distances. In contrast, because a significant amount of evolution may have taken place that is not reflected in the single observed distance value for two taxa, the distance-Wagner method presumes that the observed intertaxon distances are actually lower bounds for the true values, uncorrected for superimposed changes. Thus the observed distances are taken to be lower bounds for the fitted branch lengths, so that the length of the path connecting any two taxa is constrained to equal or exceed the corresponding observed distance. A number of algorithms for finding such minimum-length trees have been described. If the observed distances are exactly additive then the optimal solution can always be found, but if this is not the case then the effects on the resultant tree are unpredictable.

3. The neighbor-joining method (Saitou and Nei 1987; Studier and Keppler 1988) is conceptually related to cluster analysis, in that it is iterative and involves a sequential recomputing of linear functions of distances, but it relaxes the assumption that the intertaxon distances are ultrametric. The procedure begins with the standard distance matrix, but keeps track of the positions of nodes rather than clusters. It computes a modified distance matrix in which the amounts of separation between pairs of nodes are adjusted on the basis of their average deviations from all other nodes. The tree is then con-

structed by joining the least distant pair of adjusted nodes into a new node, removing the branches originating from the new node, and successively repeating this process until only two nodes remain, separated by a single branch. The entire tree is reconstructed from the set of accumulated nodal values. As with the distance-Wagner methods, if the observed intertaxon distances are exactly additive then the optimal solution will always be found. The procedure is relatively new and has few applications in the literature.

Parsimony Methods

Parsimony methods, which invoke the criterion of *maximum parsimony*, have been the most widely used tree-building procedures in systematics. The principle of maximum parsimony as it applies to phylogenetic inference equates simplicity of explanation with the explanation of shared attributes of two or more taxa being due to inheritance from a common ancestor (Sober 1989). When character conflicts occur this simplest explanation cannot hold, and additional assumptions of *homoplasy* (convergence, parallelism, or reversal of character states) must be invoked to account for the observed data. The parsimony methods summarized here are all based directly on character data, as coded in a character matrix $X_{(nxp)}$. The result of parsimony analysis is a cladogram depicting nested sets of derived character states, from which inferences about the historical relationships are then made. Such cladograms are often called *most-parsimonious reconstructions* (MPR).

Parsimony algorithms operate by finding, for any particular tree topology, the distribution of character states that minimizes the total tree length, where each branch length is taken to be proportional to the number of required character-state changes (evolutionary steps) along it (Fitch 1975, 1977). Such trees are known in graph theory as *Steiner minimum trees* (Gilbert and Pollak 1968). The tree-construction methods differ primarily in the underlying evolutionary assumptions or constraints that they invoke about the nature of character-state changes along branches, assumptions that must be made to determine the total tree length. However, they share the difficult necessity of searching for the particular tree topology that results in the shortest total length. Thus, parsimony methods in general are distinguished by their optimality criteria rather than by the particular algorithms used to search for optimal trees.

By far the most widely used parsimony methods used have been the Wagner and Fitch procedures. The so-called Wagner method, formulated by Kluge and Farris (1969) and Farris (1970) on the basis of Wagner's (1961, 1969) tree construction protocol, assumes only that each character is measured on an interval scale, which is an appropriate assumption for continuous characters and for binary and ordered multistate characters. Fitch (1971) generalized the method to allow the use of unordered multistate characters, as are common with DNA (nucleotide) and protein (amino acid) data. The Wagner and Fitch procedures both permit the free reversibility of character states along the tree; that is, they assume that the probabilities of changes among contiguous character states are symmetrical. The Wagner and Fitch algorithms as originally formulated find only one tree (MPR), even though others may exist (Swofford and Maddison 1987), but they can be modified to handle multifurcations.

While the Wagner and Fitch methods assume probabilities of character-state changes to be symmetrical along branches, the *Dollo parsimony model* (Farris 1977) imposes an asymmetry on possible directions of change: a derived character state may be independently evolved any number of times, but once lost, it cannot be regained. Such a model seems to be appropriate for molecular restriction-site data, for which the "gaining" of the character state can be interpreted as the loss of the cleavage site, and vice versa (Debry and Slade 1985); however, if a particular restriction site does originate independently in two lineages, the Dollo model can significantly overestimate the actual number of evolutionary changes. The original parsimony method of Camin and Sokal (1965), which was the first discrete-character parsimony procedure to be formalized, makes an asymmetry assumption that is far more strict than the Dollo assumption: namely, that any derived condition cannot reverse to the ancestral character state. This is the strongest assumption about character-state evolution made by any of the parsimony methods.

These various assumptions about ordered versus unordered character states and the symmetries of character-state transitions can be subsumed into a generalized parsimony method (Sankoff 1975; Swofford and Olsen 1990) that assigns a weight or "cost" for the transformation of each character state to every other state. The transformation weights are tabulated in a *cost matrix*, special cases of which can be determined for the Wagner, Fitch, and Dollo parsimony models. An exact dynamic-programming algorithm can be used both to determine the minimum tree length required on a given tree topology for any particular assignments of costs and to obtain one or more of the MPR corresponding to this length (Sankoff and Cedergren 1983). Such an algorithm is very expensive computationally, which is a current if not future limitation. A more general problem is that it may not be clear in any particular case how to assign the costs between character states (Williams and Fitch 1989). Details about conceptual and computational aspects of these parsimony algorithms and particular examples of each are provided by Swofford and Olsen (1990).

Maximum-Likelihood Methods

A final (if arbitrary) category of tree-building methods includes those that invoke explicit probabilistic models of evolutionary change. The most familiar of this category are the maximum-likelihood methods, originally proposed by Edwards and Cavalli-Sforza (1964; Cavalli-Sforza and Edwards 1967) but further developed and described in detail by Felsenstein (1973; 1981a,b; 1982; 1984b; 1985; 1988). Maximum-likelihood methods are common in hypothesis-testing applications in which a clear pair of alternative explanations of the data is presented. They are more difficult to apply when numerous alternative and complex hypotheses are evaluated, which is the case in phylogenetic inference.

However, the rationale behind maximum-likelihood procedures becomes clearer if we view phylogenetic inference as a means of finding the phylogenetic hypothesis (as summarized by a tree having a specific topology and set of branch lengths) that is most consistent with the data. Given an explicit model of the evolutionary process that converts one character state into another, a maximum-likelihood approach would estimate the likelihood that the given evolutionary model will yield the observed set of character-state distributions among taxa. If we could evaluate this likelihood for a collection of possible phylogenies, we could then select the one (or more) that provides the greatest likelihood of "explaining" the data.

The sources of controversy about maximum-likelihood procedures center on the necessity of defining particular models of evolution. Several models have been used, most of which are very simplistic. For morphological data the model typically assumes that all characters are evolving independently, each following a random sequence (Brownian motion) with a mean displacement of zero and a constant variance in displacement per unit time. This model is tractable analytically because, after t units of time, the resulting phenotypes can be considered to change by an amount drawn from a normal distribution with a mean of zero and a variance proportional to t. It is then possible to obtain maximum-likelihood estimates of the best-fitting evolutionary trees. Such trees can be estimated from continuous or discrete data, given such a random-walk model of independent evolution.

For molecular sequence data the evolutionary models can be made more explicit in terms of patterns of nucleotide substitution; however, the procedures in current use are still based on very simple models of sequence evolution. For example, the Jukes and Cantor (1969) model invokes the assumptions that the four nucleotides are equally frequent and that all substitutions are equally likely. The Kimura (1980) model is somewhat less restrictive in that it assumes independent substitution rates for transitions versus transversions. Both models make explicit quantitative assumptions about the nature of the relationship between substitution rate and sequence similarity. Felsenstein (1988) has generalized the Kimura model, replacing the assumption that the nucleotides are equally frequent with a more relaxed assumption that the relative nucleotide frequencies are maintained at equilibrium.

Regardless of the assumptions by which the branch lengths of a maximum-likelihood tree are optimized, the result is a likelihood value for the particular tree topology. Several different methods have been proposed for assessing the statistical significance of the likelihood values and for comparing the likelihoods of two or more alternative trees (Felsenstein 1988; Kishino and Hasegawa 1989).

Searching for Globally Optimum Trees

As noted above, all of the tree-building methods that are associated with specific optimality criteria distinguish the problem of finding the particular optimal combination of branch lengths for a given tree topology from that of finding the one or more trees (from the extremely large number of possibilities) that provide the globally optimum solution. For data sets of up to approximately twenty taxa the exact solutions can usually be found, either by exhaustive search (systematically evaluating all possible trees) or using the so-called branch-and-bound methods (Hendy and Penny 1982; Swofford 1990). The latter are conceptually related to the combinatorial exhaustive-search algorithms but instead systematically search through the possible branch rearrangements in such a way as to localize the subset of possible trees of minimum length.

For data sets of more than about twenty-two taxa (given the current technology) even the most efficient branch-and-bound methods would require a prohibitive amount of computer time, and consequently other methods must be used, methods that are approximate and that cannot be guaranteed to find the globally optimum trees. These heuristic approaches generally operate by *hill-climbing methods*, seeking rearrangements of the tree that result in improved values of the optimality criteria and continuing until no further improvements can be found (Swofford 1990). Like hill-climbing methods in general, such approaches will find local optima that might or might not correspond to global optima, and various heuristic fixes can be applied to perturb the search or to begin it from a different starting position. One particularly useful type of perturbation is *branch-swapping*, a systematic predefined interchanging of branches designed to move the search off of local optima in the hope of finding a better hill to climb.

Other Issues

There are a large number of other important issues that follow from these basic considerations but that cannot be elaborated here. These include such things as: (1) decisions about weighting or not weighting different characters and about how to manage missing data; (2) methods for detecting internal inconsistencies and systematic biases in data sets; (3) problems associated with estimating ancestral character states and identifying appropriate and inappropriate character-state transitions; (4) evaluation of the reliability and robustness of cladograms and trees, based either on asymptotic statistical assumptions or on resampling techniques (bootstrapping and jackknifing); (5) comparisons of trees derived from different kinds of data sets; (6) choice of the most appropriate optimality criterion for continuous characters; (7) justification of the use of parsimony as a universal optimality criterion; (8) mapping of trees onto so-called character spaces and other kinds of geometric manifolds; (9) potential ramifications of character correlations, either among terminal taxa or along evolutionary lineages; (10) methods of dealing with multiple solutions (e.g., consensus trees); and (11) the role of fossils as hypothetical ancestral taxa. Some of these problems are inherently biological and must be tackled by evolutionary biologists, while others are numerical and involve the development or refinement of computational algorithms and more realistic probabilistic evolutionary models. The discipline of phylogenetic inference lies at the heart of contemporary evolutionary biology and hence is likely to remain a dynamic and controversial enterprise for some time to come.

4

Randomness and Levels of Uncertainty in Phylogenetic Inference

JAMES W. ARCHIE

Technological advances are rapidly increasing the ability of systematists to collect comparative nucleotide sequence data for use in phylogenetic analysis. Although there is undoubtedly a tremendous amount of information on phylogenetic history in these data, three major problems confront systematists trying to use this information for phylogenetic inference: (1) computational complexity; (2) sequence alignment; and (3) repetitive or redundant sequence change (homoplasy). The first two problems are being attacked by the use of faster computers and more efficient computer programs. The third problem is related to the others since homoplasy levels inferred from the data will either increase or decrease depending on the particular alignment chosen. However, there is also an underlying biological phenomenon that implies that repetitive or convergent changes at nucleotide sites over evolutionary time are expected to be frequent. In this chapter, I address the problem of the estimation of homoplasy levels in nucleotide sequence data and the importance of estimating homoplasy levels in evaluating the quality of and expected confidence in both the data and the resulting estimates of phylogenetic relationships.

Estimating Levels of Randomness in Nucleotide Sequences

Two different approaches can be used to estimate the levels of randomness in nucleotide sequences: (1) derive a theoretical expectation of the accumulation of repetitive changes based on a model of sequence divergence and observed differences between aligned sequences and (2) scale an estimate of the observed amount of repetitive change to the worst possible case using a *lower*

Advances in Computer Methods for Systematic Biology: Artificial Intelligence, Databases, Computer Vision, ed. Renaud Fortuner (Baltimore: Johns Hopkins University Press).

bound on the amount of change estimated directly from the observed sequence data using the principle of parsimony (Farris et al. 1970). The difficulty with the first approach is that the estimated amount of repetitive change associated with a particular level of divergence depends on intrinsic and unknowable properties of the biological system (initial nucleotide frequencies before divergence and nucleotide substitution probabilities). The parsimony approach is based directly on properties of estimates of relationships derived from the data by weighting each site equally and by treating each site independently. The worst possible case, that is, the maximum amount of repetitive change for the observed sequences, can be obtained by estimating the distribution of lengths of minimum-length (ML) trees for comparable but phylogenetically random nucleotide sequences. The degree of randomness in observed sequences can be estimated simply by comparing the length of the ML tree from the original data to the distribution of tree lengths for the phylogenetically random data. The methodological problem is the computational difficulty of estimating the distribution of tree lengths for phylogenetically random sequence data.

A Randomization Test for Phylogenetic Information

The distribution of steps on minimum-length trees for random phylogenetic data can be obtained by repeating the following operations some large number (r) of times: (1) randomly permute the nucleotide state assignments within each site and (2) determine the length of the minimum length tree from the resulting data. Permutation eliminates all hierarchical correlations between sites but maintains the distributions of character states within each site. The length of the tree calculated from the original (nonrandomized) data can then be compared to this distribution. An hypothesis of no phylogenetic information on relationships in the data can be tested by determining if the length of the original data tree is less than 95 percent or 99 percent (or some other percentage) of the lengths of the randomized data trees (see Archie [1989a] for details). An example of these distributions will be given below.

Indices of Levels of Homoplasy

Three indices have been introduced as measures of the amount of homoplasy (repetitive change): (1) the consistency index (Kluge and Farris 1969); (2) the homoplasy-excess ratio (Archie 1989b); and (3) the retention index (Farris 1989; see Archie 1990). Formulas for these indices are given in Table 4.1. The consistency index (CI) compares the observed amount of homoplasy for the data (actually, the minimum number of changes plus the number of

Table 4.1. Formulas Comparing Three Available Homoplasy Indices

1. Consistency Index (Kluge and Farris 1969)

$$CI = \frac{\sum \text{(number of states in character} - 1)}{\sum \text{character lengths of ML tree}}$$

$$= \frac{\text{Minimum changes}}{\text{Length of minimum length tree}}$$

$$= \frac{\text{Minimum changes}}{\text{Minimum changes} + \text{homoplastic changes}}$$

2. Homoplasy-Excess Ratio (Archie 1989b)

$$HER = 1.0 - \frac{\text{Observed homoplasy on ML tree}}{\text{Maximum possible homoplasy}}$$

$$= \frac{\text{Maximum possible changes} - \text{observed changes}}{\text{Maximum possible changes} - \text{minimum changes}}$$

Minimum changes = Σ (number of states in character − 1)
Maximum possible changes = mean number of steps on minimum length tree for randomized data.

3. Retention Index (Farris 1989)

$$RI = \frac{\text{Maximum possible changes} - \text{observed changes}}{\text{Maximum possible changes} - \text{minimum changes}}$$

Minimum changes = Σ (number of states in character − 1)
Maximum possible changes determined for each character from the distribution of character states among the taxa

homoplastic changes) to the length of a tree obtained if there were no homoplasy, while the homoplasy-excess ratio (HER) and the retention index (RI) compare the observed amount of homoplasy to the maximum possible homoplasy for the data. The determination of the latter is in the only way that the two indices differ.

In the formula for the HER, the maximum possible homoplasy is estimated as the mean number of steps on minimum length trees for randomized data, as described above. In contrast, the RI uses the maximum possible homoplasy calculated from the distribution of character states among the taxa. The maximum number of steps possible for a single character on any tree can be obtained by determining the number of steps required to evolve the character on a *bush* (a multifurcate tree in which all branches originate at the base). The maximum possible number of steps is obtained by choosing the state with the highest frequency among the terminal taxa as the ancestral state. For nucleotide sequence data (four possible states) we subtract the most frequent state at each site from the number of taxa, t. The maximum number of steps for the entire data set used in calculating the retention index is determined by summing across all sites the maximum number of steps for each site. Both the CI and the RI are easier statistics to calculate than the HER because the computationally intensive procedure of character randomization followed by minimum-length tree determination is not required. However, as general and comparative measures of homoplasy, both the CI and RI fail critical tests in their behavior related to (1) the number of taxa in the study; (2) the number of characters in the study; (3) the character-state distributions among the taxa; and (4) the achievable values of these statistics when the data are random with respect to phylogenetic relationships. Each of the three indices achieves a value of 1.0 when no homoplasy is present and decreases toward 0.0 as the amount of homoplasy increases. One would expect that if the data were random with respect to phylogenetic relationships, each index should equal 0.0; otherwise, a particular value would be difficult to interpret. For comparative purposes the indices should be scaled to account for different numbers of taxa and characters as well as character-state distributions. Two examples of the use of these statistics and the randomization procedure are presented.

Examples

An Example Using Protein Sequences on Plant Families

Bremer (1988) analyzed the unambiguous, informative nucleotide sequence sites derived from amino acid sequences for four proteins in an analysis of the relationships of nine plant families. One of the sequences was available for only six of the families, so two separate analyses were performed. Bremer's basic observations from the results of his analyses were that trees that differed by only two or three steps from the minimum-length tree for the data, differed substantially in inferred topological relationships. In fact, for the analysis with nine taxa, if a consensus tree (Rohlf 1982) was made of all trees that differed from the minimum-length tree by at most three steps (151–154 steps), there was no structure to the consensus tree. Similarly, for the six-taxa study, a consensus tree of trees that differed from the minimum-length tree by at most two steps (161–163 steps) had no structure. The consistency indices for the minimum-length trees were 0.583 and 0.689, neither of which appeared alarmingly low. Archie (1989c) presented an analysis of these data using the randomization technique described here to elaborate on the Bremer's results. As shown in Table 4.2, the range of tree lengths (r = 200 and 100, respectively) for nine and six taxa included the length of the minimum-length tree.

Table 4.2. Results of the Randomization Analysis of Plant Families Based on Nucleotide Sequences Derived from Proteins

Number of taxa	9	6
Number of informative sites	64	84
Number of states in data	88	111
Length of minimum-length tree	151	161
Consistency index (CI)	0.583	0.689
Number of minimum-length trees	2	1
Number of trees 1 step longer	3	1
Number of trees 2 steps longer	9	3
Number of trees 3 steps longer	24	—
Mean tree length for randomized data	155.0	160.61
Range of tree lengths	(146–162)	(148–166)
Homoplasy-excess ratio	0.046	−0.008

The hypothesis of no phylogenetic information in these data cannot be rejected. In contrast to the CI, the homoplasy-excess ratios for these data are 0.046 and −0.008, agreeing completely with the conclusion that the data contain no information on phylogenetic relationships.

An Example Using Ribosomal RNA Sequences

Field et al. (1988) presented an analysis of sequences of 18S ribosomal RNA for twenty-four animal taxa from diverse animal phyla plus four nonanimal taxa (Table 4.3). Using the Fitch and Margoliash (1967) distance-based method they derived a phylogenetic tree for these twenty-eight taxa. The relationships inferred from this study have been particularly controversial for two reasons: (1) their analysis implies that the kingdom Animalia and, also, Metazoa are polyphyletic, with the Cnidaria more closely related to the Fungi, Plantae, and a ciliate protozoan than to the rest of the animal kingdom; and (2) their analysis implies that the Arthropoda is a relatively early derivative from the primary animal lineage, having branched off the main lineage prior to other Protostomia, including the Annelida. Traditional interpretations have the Mollusca and Annelida being derived before the Arthropoda within the Protostomia.

Although these data have been reanalyzed extensively (Ghiselin 1988; Lake 1990) with certain of the controversial results reinterpreted, one of the main reasons for the difficulty in the analysis of these data is the large amount of convergent and parallel changes. The amount of residual nonrandom phylogenetic information in these data can be examined using Wagner parsimony and the randomization technique (described here) along with the three homoplasy indices.

Table 4.3. Taxa Used by Field et al. (1988) in Their Phylogenetic Analysis of Animal Phyla

- Protista
  - Ciliate
  - Cellular slime mold
- Fungi-Yeast
- Plantae-Corn
- Animalia
  - Acoelomates
    - Cnidaria
      - Hydra
      - Anemone
    - Platyhelminthes-Planaria (Flatworm)
  - Coelomates
    - Protostomia
      - Annelida
        - Polychaete
        - Oligochaete
      - Mollusca
        - Clam
        - Clam
        - Chiton
        - Nudibranch
      - Arthropoda
        - Fruit fly
        - Brine shrimp
        - Millipede
        - Horseshoe crab
      - Sipuncula-Peanut worm
      - Pogonophora-Vent worm
    - Lophophorata
      - Brachiopoda-Lamp shell
    - Deuterostomia
      - Echinodermata
        - Starfish
        - Brittlestar
        - Urchin
        - Crinoid
      - Chordata
        - Tunicate
        - Amphioxus
        - Human
        - Frog

The results from these analyses are presented in Tables 4.4 through 4.6. In each analysis, I first used either PAUP (Swofford 1985) or HENNIG86 (Farris 1988) to determine the length of the minimum-length tree for the particular subsample of taxa chosen. This provided a value for the consistency index and retention index. I derived a value for the homoplasy-excess ratio using twenty-five or more randomizations of the original data followed by estimation of the length of the minimum-length tree using PAUP. For each repetition, the length of the ML tree was recorded, and the values of the consistency index and the retention index were calculated. The average values from the replicates provide estimates of the minimum possible value achievable for each of these statistics assuming that the data are random with respect to phylogenetic relationships. In all cases the average minimum value for the homoplasy-excess ratio will be identically equal to 0.0.

For all 28 taxa and 346 informative sites, the analysis yielded two minimum-length trees with 1420 steps (CI = 0.44; RI = 0.50). In both trees the Metazoa and Animalia are monophyletic, but the Arthropoda are paraphyletic and the Mollusca are polyphyletic. Using 180 informative sites (due to program limitations), I obtained a tree with 735 steps (CI = 0.437, RI = 0.336, HER = 0.349; Table 4.4). The range of steps on randomized data trees was 943–964. Clearly, the ML tree length is well below that for the randomized data trees, allowing rejection of the hypothesis of phylogenetic randomness. However, the low values of all three indices indicate that there is a substantial amount of homoplasy in these sequence data. An important observation is that the mean values for the CI and RI derived from the randomized data were not 0.0, but 0.336 and 0.358, respectively. The HER for these randomized data is 0.0.

Table 4.4. Results of the Randomization Analysis of Entire Field et al. (1988) Data Set

Number of taxa	28
Number of informative sites	180
Number of states in data	501
Maximum possible number of steps	1156
Mean tree length for randomized data	954.2
Range of tree lengths for 50 replicates	(943,964)
Minimum length tree	735
Consistency index [CI = (501 − 180)/735]	0.437
Mean CI for randomized data	0.336
Retention index [RI = (1156 − 735)/(1156 − 321)]	0.336
Mean RI for randomized data	0.358
Homoplasy-excess ratio [HR = (954.2 − 735)/(954.2 − 321)]	0.349
Mean HER for randomized data	0.00

Table 4.5. Results of the Randomization Analysis of Arthropods (*Limulus*, Millipede, *Artemia, Drosophila*, Oligochaete, Human, Starfish, Planaria)

Number of taxa	8
Number of informative sites	141
Number of states in data	348
Maximum possible number of steps	419
Mean tree length for randomized data	361.6
Range of tree lengths for 150 replicates	(352,376)
Minimum length tree	353
Consistency index [CI = (348 − 141)/353]	0.586
Mean CI for randomized data	0.573
Retention index [RI = (419 − 353)/(419 − 207)]	0.311
Mean RI for randomized data	0.271
Homoplasy-excess ratio [HR = (361.6 − 353)/(361.6 − 207)]	0.055
Mean HER for randomized data	0.00

Table 4.6. Results of the Randomization Analysis of Deepest Relationships (Ciliate, Slime Mold, Yeast, *Zea*, Hydra, Planaria)

Number of taxa	6
Number of informative sites	136
Number of states in data	335
Maximum possible number of steps	361
Mean tree length for randomized data	314.8
Range of tree lengths for 150 replicates	(304,329)
Minimum length tree	304
Consistency index [CI = (335 − 136)/304]	0.655
Mean CI for randomized data	0.633
Retention index [RI = (361 − 304)/(361 − 199)]	0.352
Mean RI for randomized data	0.285
Homoplasy-excess ratio [HR = (314.8 − 304)/(314.8 − 199)]	0.093
Mean HER for randomized data	0.00

Because of the large amount of homoplasy and the controversial results from the analysis of the original data by Field et al. (1988), I performed several subsequent analyses to determine if I could identify particular parts of the data that might be considered problematical or exhibit exceptionally high average levels of homoplasy. Two of these analyses are presented here (Tables 4.5 and 4.6). These two analyses address the problems from the complete data set of (1) the placement of the arthropods and (2) the relationship of the

Cnidaria to the other animal taxa. In the analysis of the relationships of the Arthropoda (four taxa) to several distantly related taxa (Oligochaete, Human, Starfish, and Planaria), the CI is higher in value (0.586) than that from the entire data set (Table 4.4). The RI is comparable in value to that seen in the analysis of the complete data set (RI = 0.311). In contrast to both of these, however, the HER value for this analysis is close to 0.0, HER = 0.055. In addition, the length of the minimum-length tree (L = 353) falls in the range of values observed for lengths of randomized data trees (352–376, r = 100). Both of these results indicate that there is essentially no information in these data on the relationships of the Arthropoda to these other taxa. Although this result is not apparent from the observed values of the CI or RI, note that the observed values are very similar in magnitude to the mean values obtained for these two statistics for the randomized data trees.

The second analysis used six taxa (t = 6) to investigate the deepest relationships addressed by these data—the relationships of representatives of the nonanimal kingdoms plus the Cnidaria and a single representative of the other animal phyla. The results are presented in Table 4.6. The CI for these data is the highest value observed for any of the analyses carried out, CI = 0.655. The retention index is slightly higher than in the analysis of the Arthropoda. Again, however, the HER is close to 0.0, HER = 0.093, and the minimum-length tree (L = 304) was equal to that observed for at least one of the randomized data trees (range = 304–329, r = 150). As with the arthropod analysis, the mean CI and RI for the randomized data differ very little from the observed values, even though the observed CI was higher than in any other analysis. The interpretation of these results is again that because of the large amount of homoplasy these sequence data contain no significant residual information on the deepest relationships addressed in the analysis, including the relationship of the Cnidaria to the other kingdoms.

Conclusions

Knowledge of levels of homoplasy are important in evaluating the data and results from a phylogenetic analysis. High homoplasy levels affect both the stability and reliability of estimated relationships. For phylogenetically informative data with some noise, it is often observed that there may be two to several trees that are either of minimum length or near minimum length. However, the common observation is that these trees resemble one another in tree shape (topology). For sequence data that exhibit little residual phylogenetic information, trees that differ by only one or a few steps can be expected to differ drastically from one another in topology. As longer sequences are gathered, trees of nearly equal length would be expected to differ from each other more and more.

In order to evaluate the level of homoplasy in a set of data, it is convenient to have an appropriate index for its measurement. Both the consistency index, which has been used for the past twenty years as a comparative measure of homoplasy, and the retention index have recently been shown (Archie 1989a,b; 1990; Sanderson and Donoghue 1989) to have a series of undesirable properties related to the number of taxa (CI only), the number of characters, N, and character-state distributions. The minimum achievable values for the CI and RI are directly related to these same factors as well. Archie (1989b) showed that the HER is unaffected by either t, N, or character-state distributions, and its minimum expected value is always 0.0. The HER is directly and appropriately affected, however, by the absolute level of apparent or estimated homoplasy determined by the number of repetitive changes on the minimum-length tree.

Because of the way in which the HER is calculated, although its expected value for phylogenetically random data is 0.0, as was seen in the analysis of the plant family data, the HER can take on negative values. This is to be expected for sequence data of finite length. Rather than using the maximum observed tree length from the randomized data, the formula uses the mean from the r repetitions. The maximum from a series of repetitions is sample-size dependent while the mean is a stable statistic and has a low variance among replicates. For phylogenetically random data, if tree lengths are symmetrically distributed about the mean, approximately 50 percent of all trees should produce negative HER values. Such values indicate that the data have more homoplasy than the average expected due to chance hierarchical correlations alone.

The analyses presented here demonstrate several possible uses of the character randomization procedure and the homoplasy-excess ratio as tools for investigating the amount and variability in homoplasy levels among a series of taxa. The most important point from these results is the necessity that biologists evaluate the level of homoplasy in their data set as a whole and investigate the variability of inferred evolutionary rates and estimated homoplasy in different parts of their data. The randomization test is quite powerful and easy to reject. For example, if the anemone is added to the final data set and reanalyzed, the null hypothesis is easily rejected, and the HER increases to 0.1841 (the CI actually decreases, CI = 0.649). The residual information in the data identified by the test and HER is thus restricted primarily to the two relatively closely related Cnidaria.

There is clearly room for additional development of the randomization technique and, in particular, to more precisely specify the nature of the statistical tests being carried out. Faith and Cranston (1991) have found several other such data sets for which the null hypothesis tested by the randomization procedure could not be rejected. Faith (1992) has used the randomization procedure to develop specific tests for monophyly and polyphyly of particular groups of interest on the tree. The randomization approach is an exciting new direction that exemplifies the need for high-speed computational equipment and advanced algorithms and programs in systematics.

Acknowledgments

This research was partially supported by NSF grant BSR-8918042.

5

Limitations to Accurate Molecular Phylogenies

MICHAEL D. HENDY AND DAVID PENNY

> A more accurate molecular phylogeny of these species and others should be obtained when the sequences of additional proteins and nucleic acids have been determined. The study of informational macromolecules from an evolutionary standpoint is a young science that was founded only about a decade ago. It is a powerful approach that should make increasingly important contributions to our understanding of biological evolution.

This quotation from F. Ayala (1978), which appeared in a special *Scientific American* issue on evolution and showed the early optimism in building evolutionary trees from molecular sequences, suggests that given a little more time and a little more information the science of phylogeny should be solving important problems. The molecular approach was expected to make an increasingly important contribution to our understanding of biological evolution.

We suggest that from what we currently know we really have not yet attained this goal. This chapter will explain some of the barriers that prevent better answers being obtained.

Problems in Phylogeny

Inconsistency

Ayala (1978) predicted that the results of phylogenetic analyses would improve as more sequence data became available. However, gathering more data does not guarantee a better answer! Inconsistency is the situation where the results are expected to get worse as additional and more accurate data become

---

Advances in Computer Methods for Systematic Biology: Artificial Intelligence, Databases, Computer Vision, ed. Renaud Fortuner (Baltimore: Johns Hopkins University Press).

available. Indeed, when the conditions for inconsistency are met, the results are guaranteed to converge to the wrong answer, provided there is enough good information.

This problem was first reported for evolutionary trees by Felsenstein (1978), who analyzed an example with four taxa. In his example he suggested that the conditions for inconsistency (much faster rates of evolution on two nonadjacent lineages) were so extreme that it was unlikely to be a problem in practice. We have since shown that with five or six taxa the problem of inconsistency can arise with quite reasonable assumptions (Hendy and Penny 1989). With five or more taxa the problem of inconsistency can arise even when all lineages (edges of a tree) have equal rates of evolution. A common case where inconsistency may arise is the introduction of out-groups where the variation between lengths on the edges of the tree may be extreme.

Our initial analysis (Hendy and Penny 1989) was restricted to the parsimony method for nucleotide sequences. Steel (1989, 1990) has since extended the analysis to the class of methods we call *linear*, that is, methods where the predicted edge lengths are linear functions of the nucleotide sequence data. Many phylogenetic analysis methods are linear. Steel has shown inconsistency can arise in any linear method, even with very uniform assumptions about edge lengths, provided the number of taxa is sufficiently large. The first problem of reconstructing accurate phylogenetic trees from nucleotide sequences is that popular methods for inferring trees lack consistency.

Computability

Definition of the Computability Problem. The second problem is a computability problem that results from the astronomical number of trees that need to be considered for a complete search.

In Figure 5.1 the solid line shows how the time for a complete search of all binary trees increases with n, where n is the number of taxa. (The number of steps is proportional to n for each tree.) Binary trees are phylogenetic trees with a single bifurcation at each point. As the number of taxa grows, the proportion of phylogenetic trees that are binary becomes vanishingly small. Searches are usually restricted to binary trees because nonbinary trees can be formed from binary by allowing the number of changes on some edges to be zero. The time axis on Figure 5.1 is a logarithmic scale. If the 1.2 million steps required for a complete search of all binary trees for nine taxa required one second, then a complete search for thirteen taxa will take more than one day. Fifteen taxa would require approximately one year, and twenty-one taxa about one solar lifetime. The hope of analyzing increasing numbers of sequences by a complete search is not going to be realized. Even to progress from sixteen to seventeen taxa requires more than a thirtyfold increase in the number of steps required.

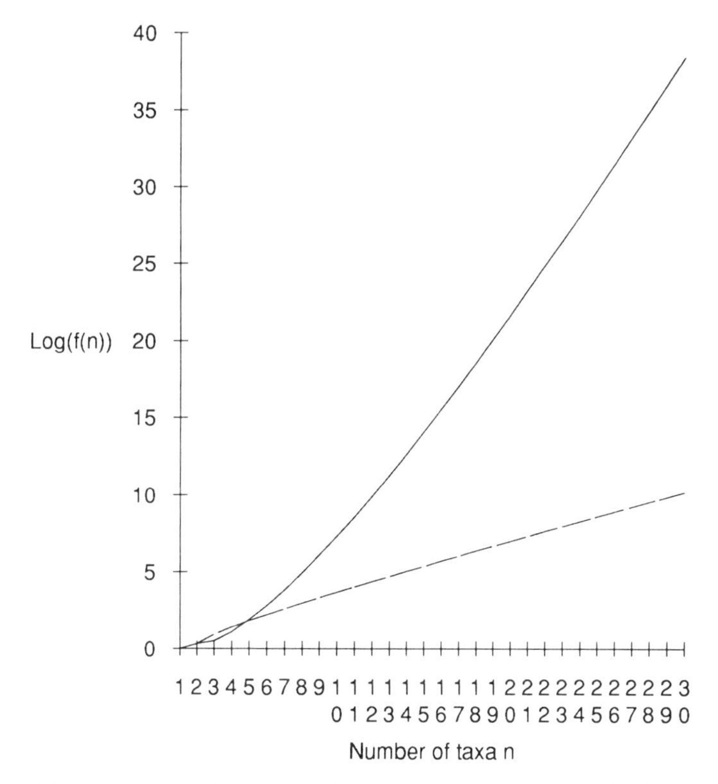

Figure 5.1. Comparison of the number of steps for a complete search of all binary trees for parsimony (solid line) and for the fast Hadamard transform (dotted line).

Restricted Searches. There are a number of ways of trying to bypass the computability problem and to avoid computing for many years in order to obtain a single result. A pragmatic approach is to restrict the search to a small, affordable subset of trees. There are two dangers with this approach. Most of the searches are local searches. You start the search at a particular region, you look around, you see a "hill" that appears to be better (higher) than any other hill, and you climb to the top of it and claim that it is the highest peak. However, there is no guarantee that it will be the global optimum.

A second danger is that it can lead to wild extrapolation. Restricted search methods may have been tested on examples with small numbers of taxa, but their behavior for larger numbers of taxa is uncertain. Problems often get worse as the number of taxa increases. Sometimes it appears that, if things go right with smaller problems, it is almost guaranteed that they will go wrong when the problem gets bigger! The branch-and-bound approach (Hendy and Penny 1982) helps a little, extending the search to about twenty taxa.

Branch and Bound. The branch-and-bound approach is a search procedure that begins with the calculation of the length (number of mutations) of an initial tree. This length serves as an upper bound: the length of the minimal tree cannot be longer than this bound. In principle, any tree could be used, but the lower this initial bound the faster the process. Branch-and-bound methods eliminate many branches of the search tree; every time a branch of the search tree is eliminated a whole category of trees is eliminated from the search. At each step, we record any phylogenetic tree found whose length is less than the current bound, and the bound is reset to this length. When the process is complete, the global minimum is guaranteed. The net result is that branch-and-bound methods allow optimal trees to be found for larger numbers of taxa. It is not possible to predict the time required for a branch-and-bound search because the number of steps taken is dependent on the data. The branch-and-bound search works best when the minimal-length tree is considerably shorter than all other trees.

The histograms of Figure 5.2 show contrasting examples of the distributions of lengths of all binary trees from two sets of sequences on seven taxa. In both cases the distribution appears approximately normal, but in Figure 5.2a a single tree of length 10 is considerably shorter than all others, with the next tree of length 14. In this case the branch and bound works well as all the longer trees can be eliminated early in the search. In contrast, with a distribution of lengths as in Figure 5.2b very few of the trees would be eliminated early, so the branch-and-bound search would have little advantage over a complete search. Also, we would not find a unique minimal-length tree, there being ten trees of minimal length. In this case a set of minimal trees would be presented.

However, it is perhaps more common to find a distribution such as that of Figure 5.2b, where there are a number of trees close to the shortest. In this example there are ten shortest trees, and their length is not very much shorter than those of many other trees. Branch-and-bound searches do not work as well when the distribution of lengths is like that of Figure 5.2b. The time required for a branch-and-bound search is thus dependent on the distribution of tree lengths, and this is a function of the data, as well as the number of taxa.

An Alternative Approach

Many methods for obtaining phylogenetic trees from sequence data, such as parsimony or maximal cliques, consider the set of all possible trees, select one tree at a time, and measure the goodness of the fit of the data to that tree. This computation is repeated for every tree. Finally the single tree, or the subset of trees that best fits the data, is selected as representing the correct phylogeny.

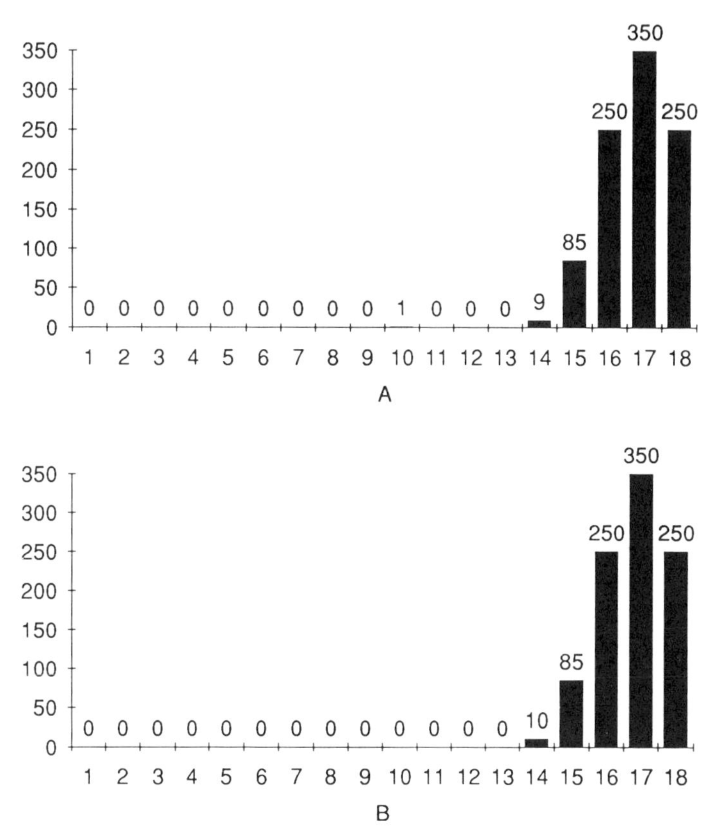

Figure 5.2. Histograms of two theoretical distributions of the lengths of all 945 binary trees for seven taxa.

Of course, we cannot know if it is the true phylogeny, but it is the best interpretation of the data for the method of analysis used. In some methods, weights are put on the edges (branches). These weights represent time, or numbers of events, or some other measure.

Our New Approach

What we have taken now is an approach that avoids the problem of inconsistency and can be applied to larger sets of taxa. It reverses the order of computation, with the weights of the edges being computed first. The optimal tree is then selected by identifying the set of edges that best fits the data.

This approach is based on the simple model formalized by Cavender (1978) that incorporates a Poisson process of character changes of sequences on the phylogeny T. In this model the characters are restricted to two values. We have shown that the expected numbers of character changes on the edges of T (the edge "lengths") are related by a simple invertible relationship to the expected distribution of characters in the resultant sequences the Hadamard transform, similar to that which has been used in image analysis (Andrews 1970).

The Hadamard Transform

The expected distribution of the characters in the sequences is described by a vector s and the set of edge weights is described by a vector γ. Then we found that:

$$s = 1/2^{n-1} \, [H \ln(H\gamma)] \ ,$$

where H is a Hadamard matrix, that is, an orthogonal matrix whose entries are 1 and -1, and ln is the natural logarithm function applied to each component of the vector separately. This is easily inverted to obtain:

$$\gamma = 1/2^{n-1} \, [H \exp(Hs)] \ ,$$

where exp is the exponential function applied to each component separately. With observed sequence data, the frequencies of character patterns give an estimate of the vector s, and the resultant γ vector is an estimate of the edge weights. We refer to this γ vector as the spectrum of the data. We then use a least-squares selection called the closest-tree procedure to estimate T. This analysis is based on sequences with only two characters. For nucleotide values with four characters we have an extension where different mappings of four characters onto two are applied. Then the resulting data spectra are combined to give a phylogeny based on the four characters.

Consistency. The closest tree selection procedure from the spectrum is a consistent procedure, avoiding the problem of converging to the wrong tree. Note that γ is not a linear function of the sequence data s. The Poisson distribution of changes on each edge of T gives estimates for the number of multiple changes (reversals, etc.). Hence the estimated edge lengths are already corrected for unobserved changes. It is also possible to incorporate alternative models, although the Poisson model appears to be the simplest. It also has the property that, when the edge lengths are not too small, the least-squares value for the closest tree to the spectrum tends to be much smaller than that for other trees, so that a branch-and-bound algorithm can select T more efficiently.

Computability. However, in avoiding one computational barrier we have introduced another. For computing the transform for n taxa, the Hadamard matrix H has 2^{n-1} rows and columns, and it increases in size by a factor of 4 for each additional taxon. H is orthogonal, so inversion is simple. The transform has a fast form, like the fast Fourier transform, that gives the product in $n \cdot 2^n$ steps rather than the 4^n steps required for a matrix multiplication. In Figure 5.1, the number of steps for the fast Hadamard transform (dotted line) is compared to the number of steps for a complete search (solid line) for the same number of taxa. We see that this procedure has a computational advantage over the parsimony complete search. Further, with data where the rates of change are not too small, the branch-and-bound selection of the closest tree is also more efficient.

We have found that the transform can be computed in about two and a half minutes for seventeen taxa on a PC. Using the parallel computing facilities of an eight-transputer system we expect to be able to calculate the Hadamard transform for up to thirty taxa.

Example of Application

We have also developed a similar analysis of changes giving a data spectrum from difference data (Hendy and Penny 1992). We conclude with an example using a set of eight mammals (Sarich 1969; Farris 1972) given in Table 5.1. The upper triangle of the table contains the original data. We applied the Hadamard transform to get the γ values that are shown in Figure 5.3. This diagram is the spectrum of the data.

Each line in the spectrum represents a potential edge of a tree. The selection of the closest tree is a matter of selecting which thirteen of these spectral lines can be combined in a tree, assuming that all the other spectral lines have zero value.

Table 5.1. Immunological Differences between Pairs of Mammals

	Dog	Bear	Raccoon	Mink	Seal	Sealion	Cat	Monkey
Dog	—	32	48	51	50	48	98	148
Bear	33.15	—	26	34	29	33	84	136
Raccoon	46.80	25.69	—	42	44	44	92	152
Mink	49.13	31.93	45.59	—	44	38	86	142
Seal	48.95	31.75	45.40	41.57	—	24	89	142
Sealion	47.35	30.15	43.81	39.98	24.13	—	90	142
Cat	97.50	80.29	92.93	86.77	89.93	88.34	—	148
Monkey	151.40	134.20	137.86	140.68	143.84	142.25	148.13	—

Note: Derived by Sarich (1969) and Farris (1972), upper righthand triangle. The corresponding distances obtained from the phylogeny of Figure 5.4, lower lefthand triangle.

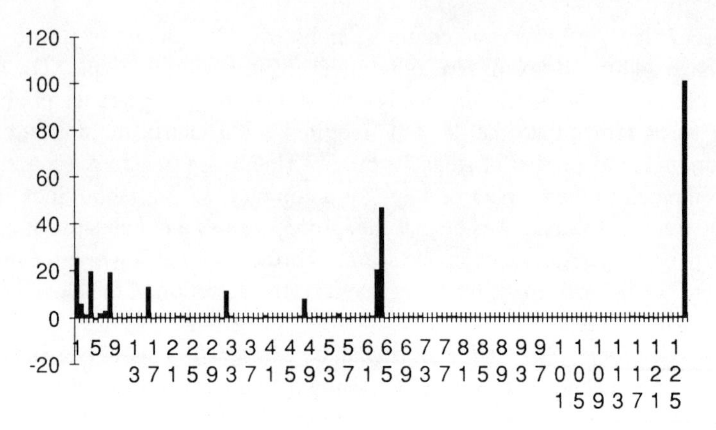

Figure 5.3. The spectrum of potential edge lengths derived from the difference data of Table 5.1.

In this case the thirteen largest spectral values correspond to the edges of a tree, and the other values are all almost zero. This choice immediately returns the closest tree (Fig. 5.4) that has weights on the edges. The outcome of the transformation followed by the closest tree selection always gives a weighted phylogeny with numbers associated with each edge. Because we are using difference data, these numbers count the expected number of differences for each edge of the tree that best fits the data. We can then calculate the expected numbers of differences between the pairs of taxa from these values. For comparison with the original data, these values are shown in the lower triangle of the matrix (Table 5.1).

Discussion

——: When you do the Hadamard transform, are you actually doing a matrix multiplication or are you doing a fast transform?

Hendy: For the example requiring two and a half minutes on the PC we were using the fast transform. In our development of the parallel algorithms on transputers we have only coded for matrix multiplication at this time. For twenty taxa, a single transputer required about two minutes. We estimate that using the fast transform the parallelization will be at least 99 percent efficient. Using eight transputers, for example, we expect to be able to compute the fast transform and find the closest tree for up to about thirty taxa. We're still in the process of developing this.

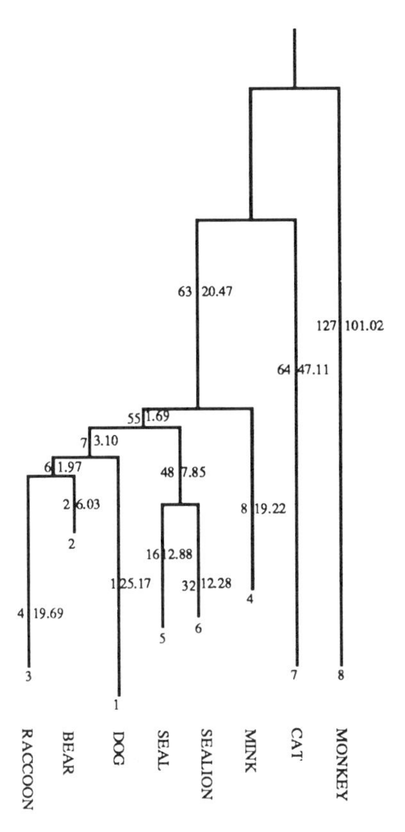

Figure 5.4. The phylogeny for the eight mammalian taxa of Table 5.1 derived using the Hadamard transform and the closest tree selection procedure.

——: Is the method still a parsimony method—does it find the shortest tree to the original data?

Hendy: Our method does not always produce the same tree as the parsimony method. The transform changes the data, correcting for unobserved changes to produce the data spectrum. To find a phylogeny from the spectrum requires a selection procedure. We use a least-squares fit of the spectrum to the subspace corresponding to a particular tree. The tree whose subspace is closest is called the closest tree. The closest point in this subspace then gives us the edge lengths of this tree. In that sense we are still using a minimization in our selection of the phylogeny, but it is a different minimization to that of parsimony.

——: Have you tried this with an ultrametric, like genetic distance, or different genetic distance calculations?

Hendy: We have not yet incorporated other metrics in our method. For the example above with distance data we did introduce corrections for unobserved changes, but we found that this did not lead to an improvement in the fit of the data to a tree.

6

The Reconstruction of Evolutionary Trees Using Minimal Description Length

PETER CHEESEMAN AND BOB KANEFSKY

A major challenge for the 1990s will be making sense of the flood of data generated by the Human Genome Project. If we know the evolutionary history of genes, we can often make strong predictions about their function, by comparison with other, more distantly related genes whose function we do know. Such sequence comparisons are routinely performed for newly sequenced genes using existing gene databases (e.g., GenBank), and many interesting, unexpected relationships have been discovered. In addition to using evolutionary relationships to predict the function of genes, evolutionary tree reconstruction is of interest in its own right, as it often shows relationships between species that cannot be reconstructed from the fossil record alone.

We present a minimum description length (MDL) approach to evolutionary tree reconstruction. This approach is closest to parsimony methods in that both methods try to find the tree with the fewest unjustified assumptions. However, the MDL approach has a probabilistic definition that allows probabilistic models of evolutionary change to be translated into equivalent description lengths. The foundations of MDL in Bayesian probability theory remove much of the ad hoc flavor of the more intuitive parsimony method, so that, for example, different mutation rates of one type of DNA base into another can be modeled or different rates on different branches can be included. The basic modeling assumptions we used are explained in the following, but it is worth remembering that the MDL approach can be extended to fit any evolutionary modeling assumptions.

Advances in Computer Methods for Systematic Biology: Artificial Intelligence, Databases, Computer Vision, ed. Renaud Fortuner (Baltimore: Johns Hopkins University Press, 1993).

Basic Theory

Definition of the Problem

The problem addressed in this chapter is: Given a set of known DNA sequences, find the most probable evolutionary tree (or trees) that relates these sequences. A simple example of an evolutionary tree is shown in Figure 6.1, where the known sequences are shown on the bottom row and all other sequences are reconstructions of the sequences of hypothetical ancestors. The length of a branch from parent to child sequence is proportional to time since divergence. The mutations that change a parent sequence into the corresponding child sequence are shown on the connecting branches. A tree, such as the one shown in Figure 6.1, is an evolutionary model that offers a possible explanation of how the known sequences (on the bottom row) were created. However, there are many possible trees, all having the same terminal sequences; so which tree (or set of trees) should be preferred over another? Intuitively, the simplest tree seems the most plausible, and this intuition has lead to the parsimony approach to evolutionary tree reconstruction (Felsenstein 1988). The usual definition of parsimony is unable to account for more complex evolutionary models, such as unequal rates of mutation on parallel branches. In extreme cases, parsimony tends to give the wrong answer as more data are provided (Felsenstein 1988). The MDL approach described below captures the basic intuition behind the parsimony approach, but defines

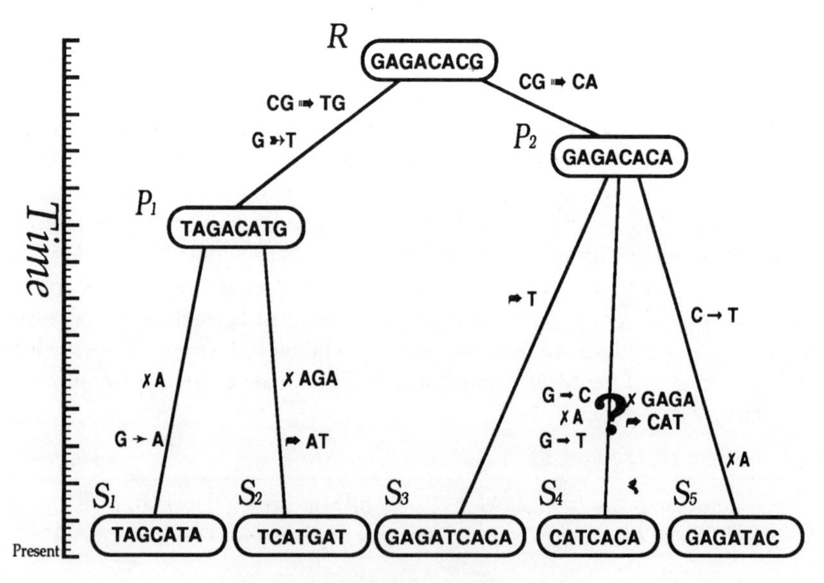

Figure 6.1. A simple evolutionary tree.

parsimony in terms of probabilities, and so avoids the limitations of the standard parsimony approach.

Although Figure 6.1 relates all the known sequences through a single tree, we consider the general problem to include reconstructing multiple trees that jointly *explain* the sequence data. That is, we are not assuming that all sequences had a common ancestor at some time in the past, although our example will be developed under this assumption. For a set of related proteins, a graph (instead of a tree) is sometimes the correct representation of the evolutionary events. For example, a protein may have evolved from a combination of parts of existing proteins (domains), and thus have multiple parents.

MDL Formulation of Problem

By Bayes' theorem, the relative posterior probability ratio of two different trees T_i and T_j given a set of known sequences S is:

$$\frac{P(T_i|S)}{P(T_j|S)} = \frac{P(T_i)P(S|T_i)}{P(T_j)P(S|T_j)}. \tag{1}$$

By taking logarithms of this equation and negating we get:

$$\begin{aligned}-\log P(T_i|S)] - [-\log P(T_j|S) &= -\log P(T_i) - [-\log P(T_j)]\\ &\quad -\log P(S|T_i) - [-\log P(S|T_j)].\end{aligned} \tag{2}$$

From information theory, $-\log P_i$ is the minimum possible message length to encode the ith outcome, if this outcome had probability P_i. An *outcome* in this case would be a result such as "There is a C (cytosine) in the 10th position in the third sequence." We will use base 2 for the logarithm; then the message length is in bits. Note that $-\log P_i$ (bits) is typically a real number, not an integer. It is clear from Equation (2) that the particular tree T_i with the maximum posterior probability relative to any other tree T_j is also the tree with the shortest relative encoding (MDL). For each tree T_i, the MDL consists of two parts: a part to describe the model selected (T_i) and a part to describe the data given the model. To apply Equation (2) we need the prior probabilities of trees, $P(T_k)$, and the likelihood functions $P(S|T_k)$, that is, the probability of sequence S given the particular tree. Note that, in the following, a tree T refers to the hypothetical construct only and does not include the known sequences S. Also, we only calculate the length of the theoretical minimum message; we do not actually do an encoding.

Tree Prior Probabilities

If the user has prior information about the target evolutionary tree, such as from fossil evidence, then this can be used directly in Equation (2), but this information is rarely available. Common evolutionary assumptions are that the probability of a mutational event and a branching event occurring per unit time is independent of the absolute time. These independence assumptions imply that the probability of a subtree is independent of the events in the supertree and depends only on the immediate parent and the time since divergence from the parent. Symbolically, these independences can be expressed as:

$$P(\text{Tree}) = P(\text{Root})P(\#\text{branches})P(\text{Subtree}_1|\text{Root}) \ldots P(\text{Subtree}_N|\text{Root}) ,$$

where the probabilities $P(\text{Subtree}_1|\text{Root})$ are recursively defined by a similar decomposition in which a subtree depends only on its immediate parent. This recursion stops when a subtree has only known terminal sequences for children.

Taking logarithms of this equation, including the recursive expansion of the subtrees, leads to a simple additive form for $-\log P(T_i)$. This additive form corresponds to a recursive coding scheme, where all immediate children are described as the result of a particular set of mutations of the parent. Since the root does not have a parent, it must be described separately. The leaf nodes are not regarded as part of the tree T_i. However, the definition of the conditional probability of a child given its parent is the same everywhere in the tree, including the leaf nodes. Choosing a coding scheme is equivalent to accepting particular prior probabilities (and vice versa).

Sequence Probabilities

Except for the root, which is coded directly, the information required to describe all other nodes in the tree is the information needed to describe each child given its parent. The conditional probability of a child sequence requires describing (1) its parent; (2) a set of mutations that transform the parent into the child; and (3) a time difference between parent and child, that is,

$$P(S_{\text{child}}|S_{\text{parent}}, \text{mutations, time difference}) \qquad (3)$$

This description requires finding the most probable set of mutational events that could have transformed the parent into the child, or at least a very probable set. For example, in Figure 6.1, two alternative sets of mutation between $P2$ and $S4$ are given, and many more are possible. We note that the tree-building procedure described by Felsenstein (1988) finds the maximum likelihood tree topology and branch lengths. It does not infer particular ancestral sequences, but averages over all possible ancestral sequences. This approach is answering a different question than the one addressed here, and it has difficulty taking into account insertions and deletions. Methods we use for finding the most probable mutation events and time differences are described below, but in this section we assume they are known.

There are three types of mutations (with time dependent probabilities) that transform the parent sequence into the corresponding child sequence: point mutations, insertions, and deletions. All three types are illustrated in Figure 6.1. Our coding scheme for a sequence transformation is as follows.

1. *Deletions*. At each letter (either nucleic acid or amino acid, as appropriate) in the parent string, we state whether it is the beginning of a deleted string or not. Thus, if there are 300 letters in the parent, we encode up to 300 *Yes* or *No* messages. Since deletions are rare, each *No* message is typically a small fraction of a bit; its length is given by $-\log\,[1 - P_{del}(t)]$, while a longer *Yes* message has length $-\log P_{del}(t)$ bits. The *Yes* and *No* messages resume from the next letter that was not deleted. When a deletion event occurs, the length of the deletion (beginning at the current letter) is described, using $-\log P(n)$ bits, where n is the length of the deletion. The t parameter is the number of time units between parent and child. Note that we are assuming that deletion length probabilities are independent of time.

2. *Point mutations*. For each letter we give a message describing the fate of that letter of length $-\log P(\text{new}|\text{old})$, that is, $-\log$ of the probability of the new letter given that we know the old letter in the same position. For example, a purine base is about twice as likely to turn into another purine (e.g., A → G) than it is to a pyrimidine (e.g., A → T or G). An example of a *change probability matrix* is shown in Table 6.1. This coding scheme makes the simplifying assumption that the probability of a point mutation does not depend on its neighboring bases or residues or on its location in the sequence. Both these assumptions are biologically incorrect. For example, a C followed by a G is much more likely to be transformed into a T than a C followed by A, C, or T (CpG decay). Similarly, in coding DNA, silent third-position bases have a much higher point mutation rate than nonredundant positions. A more complex probabilistic model could be developed within the MDL framework

Table 6.1. Point Mutation Probability Matrix

	A	C	G	T
A	0.99	0.0025	0.005	0.0025
C	0.0025	0.99	0.0025	0.005
G	0.005	0.0025	0.99	0.0025
T	0.0025	0.005	0.0025	0.99

to include these known effects, but we have not made these extensions here.

3. *Insertions*. These are encoded much the same way as we encode deletions. If the parent is 300 letters long, we encode exactly 301 *Yes* and *No* messages. This time, in addition to its position and length, we must describe the letters in each inserted string. Each inserted letter described uses $-\log P_i$ bits.

The encoding method described above is similar to describing a set of *edit commands* that transform the parent into the child sequence. However, instead of commands such as "Skip the next 20 letters," we prefer to use the probabilities of different mutations at each position in the sequence. The use of biologically meaningful mutation probabilities allows learning to take place during tree reconstruction. This is done by adjusting the probabilities of the different mutations based on information gained during the tree-building process, as described below.

The above coding procedure allows the message length $[-\log P(T_i)]$ of a tree to be calculated provided all the probabilities mentioned are known. The known sequences at the leaf nodes are not encoded in the tree description, since they are implied by it.

Time-Dependent Probabilities

For a particular letter, the probability that it will undergo a point mutation is a function of the time between the parent and the child. For a given number of time units, these point mutation probabilities can be deduced from the unit-time transition matrix. An example of a unit time matrix for DNA is seen in Table 6.1. where A,G,C,T refer to the different bases, and the numbers represent the probability of the row letter (from the parent) transforming into the column letter (the child). Note that in this matrix a child base is very likely to be the same as the parent base. Also, mutations that turn a base into another of the same type (e.g., purine $\rightarrow$ purine) are more likely than a cross mutation (e.g., purine $\rightarrow$ pyrimidine).

To obtain the point mutation probability for n time units, this matrix is raised to the power n. The unit time matrix is chosen so that the probability of a point mutation in one time unit is 0.01. After twenty-five time units, the probability of a base remaining unaltered is 0.778, so that multiple mutation events at the same location are unlikely even for this large time difference: Note that we have quantized time, so that the information required to describe the time between a child and its parent is small, $-\log P(t)$. The message length clearly depends on the quantization level selected, so for a given tree there is an optimal quantization level. For simplicity, we assume that the time in which 1 PAM (1 percent point mutation rate) occurs is adequate. An improved version would dynamically optimize this parameter during tree building.

Probability of Known Sequences

The probability of all the known sequences, $P(S|T_i)$ in Equation (1), is given by:

$$P(S|T_i) = \prod_j P(S_j|\text{Immediate parent of } S_j) \tag{4}$$

where S_j is a particular known sequence. The individual probabilities in Equation (4) are coded the same way as any other child given its parent, as described above.

Dynamic Probability Learning

If all the component probabilities required above are known from experience, then these can be used directly, with no need for learning. However, experience usually only provides rough prior probabilities that can be used as initial values. As the tree is built, further information becomes available from the frequencies of types of mutation events in the current tree. To exploit this additional information, we compute our tree MDL serially, and update the probabilities for the rest of the tree, using information from the tree encoded so far. We use the standard Bayesian probability update formula, illustrated here for the probability of a deletion:

$$P_{del} = \frac{n_{del} + r_{del}}{N + R} \tag{5}$$

where n_{del} is the observed number of deletions encountered so far, N is the total number of letters (deletion or not), r_{del} is a prior weight for deletions, and R is the total prior weight. The situation for updating point mutation probabilities is not as simple, because the mutations typically occur after a number of time steps, while the transition probability matrix is defined for a single time step. We solve this problem by arbitrarily assigning a given point mutation to a single time step, and all the other time steps count as no mutation. This approximation depends on the probability of alternative multiple mutations with the same end result being very unlikely. We have now described how to compute the component probabilities in Equation (2), so that relative MDL can be found. We use this relative MDL to search for the lowest MDL tree, as described below.

Sequence Alignment

The problem of finding the most probable mutation list, given a parent and a child sequence, is the standard pairwise sequence alignment problem. An alignment algorithm typically uses a penalty function that assigns a penalty to any proposed mutational events. The goal is to find the alignment with the minimum penalty. In the MDL approach, these penalties turn out to be mutation probabilities in disguise. There are many alignment algorithms in use. Generally, their performance depends on how well the user adjusts the penalties. We have implemented an MDL-based alignment algorithm in Lisp that uses given probabilities. This allows it to use knowledge of the estimated time difference between a parent and child and to adjust itself to the dynamically changing posterior probabilities for different types of mutations. It is very similar in theory and practice to the procedure conceived independently by Allison et al. (1990).

The Tree-building Procedure

The previous sections describe a MDL measure for deciding which of two alternative trees is more probable. This suggests an iterative improvement procedure for finding an optimal tree: start with a tree that is approximately right, then look for local improvements of the MDL measure. The obvious way of constructing a good initial tree is to build an initial $(n \times n)$ distance matrix based on the MDL alignment values, then build the tree bottom-up.

Unfortunately, the cost of constructing the distance matrix can be prohibitive, especially as the alignment algorithm will spend a lot of time producing alignments of very distantly related sequences that will never be used by the tree-building procedure. To reduce this cost, we have implemented a heuristic initial tree-building procedure. Instead of using an alignment algorithm to produce distances between sequences, we use a combination of approximate measures that correlate with evolutionary distance. The best measures between sequences we found, in order of increasing accuracy, are:

1. *Length ratio*. Sequences that are close evolutionarily tend to be about the same length, because insertions and deletions are rare.
2. *Longest common subsequence*. Since close sequences have fewer mutations, this estimator correlates negatively with distance.
3. *Squared difference of uncommon hexamer densities*, where a hexamer is a string of six adjacent letters. This estimator works because in closely related sequences the probability that short uncommon substrings are unaltered is high.

All these heuristic estimators are cheap to compute compared to a full alignment. The initial tree-building procedure builds a binary tree from the bottom up by adding each new node into the tree at the place that minimizes the total squared error of the combined estimators for that node, that is, it finds the height h and the place in the current tree to insert a new parent of the current sequence so that the following measure is minimized:

$$M(h) = \sum_{k=1}^{3} \sum_{i} \frac{(m_i^k - h)^2}{(\sigma_k)^2} \tag{6}$$

where k is the measure index, i is an index ranging over all sequences in the current tree, m_i^k *is the* kth heuristic measure between the current sequence and the ith sequence, and σ_k is the standard deviation of the kth measure. Because the positioning of the early members of the heuristic tree did not have the benefit of the constraining influence of the later nodes, existing nodes are subsequently re-added, until the tree does not change. This heuristic approach builds remarkably accurate initial trees on our test data. Part of the reason for this accuracy is that the tree is the result of forcing many independent pieces of evidence into a consistent tree, so statistical averaging compensates for the crudeness of the heuristics. The initial tree is used to guide which sequences to align and where to put new ancestors in a bottom-up MDL tree construction.

Unfortunately, this bottom-up tree-building procedure does not construct optimal trees by the MDL criterion. The reason is that in bottom-up tree construction there is a degree of arbitrariness in how to assign insertions and deletions to a new parent. However, once a tree is constructed, we use a set of optimization operations to improve it. For example, whenever common deletions or insertions are detected on neighboring branches, we can change the parent to eliminate them. In all these local optimization steps, the criterion is always Does the proposed change lower the MDL? This opportunistic local optimization procedure runs until it is unable to find further improvements. The resulting tree is not guaranteed to be the globally optimal MDL, but on artificial data where we know the correct answer, we found that this procedure gets close. Typically, our current implementation gets to within 20 percent of the MDL of the true tree and does a fairly good job of placing related sequences together. Preliminary results on a real DNA dataset, containing 128 human *alu* sequences, yield a slightly smaller description length than a multiple alignment based on a consensus sequence. Our description is still somewhat worse than the four-level hierarchy proposed by Jerzy and Smith (1988), because our optimization operations are still too simplistic.

Extended Theory

We made a number of simplifying assumptions in the tree-building procedure described above in order to produce a working program. The theory can be extended by removing many of these assumptions, although the effect of such changes on the search procedure may not be simple. The simplest extension is to allow the probabilities of point mutations to depend on such additional factors as neighboring letters, position in sequence, whether the sequence is a

RNA coding sequence, different point mutation probability matrices (e.g., Table 6.1) on different branches, etc. In the MDL approach, the question to be answered is whether the additional information required to describe these extra probabilities is paid for in improved predictive power. To answer this question, we should observe whether additional probabilities produce lower MDL.

Extending the tree building to allow for multiple parents (a graph) is more challenging because the combinatorial searches are more numerous. Constructing a MDL that reflects the fact that some proteins evolved by combining domains from very different parent proteins is simple in principle. The code must state which parts came from which parents and how they are ordered, in addition to the usual mutation events. A successful graph-building program must rely on efficient methods for identifying potential building blocks (domains). Proteins provide a more interesting possibility for MDL—predicting secondary or higher structure from sequence information and information from proteins whose structure has been determined. The basic idea is that for proteins, instead of just hypothesizing particular ancestor sequences, these sequences are segmented into typed regions, such as α-helix or β-turn. From a MDL perspective, the question is, Is the information required to describe these typed regions paid for by the information required to describe the sequence data given the regions? For example, the statistics of particular amino acids at particular locations in a type of β-turn could provide a reduced encoding. If some proteins in a particular tree have known structure, then this greatly enhances the certainty of inferred ancestral structure, since secondary structure is strongly conserved.

Finally, we note that the goal of trying to find "the" evolutionary tree for the given data is not ideal. Generally in science, the results of an investigation are reported by an estimate *and its error bars*. Without this additional information it is difficult to assess the accuracy of the given result. Current methods for finding the evolutionary tree from data, including this one, return only a single answer. As part of the search procedure, the MDL method generates many evolutionary trees that are close in terms of MDL but differ in specifics. Reporting a set of close MDL evolutionary trees would give the user an indication of how well the tree or parts of the tree are pinned down by the data.

III

Expert Systems, Expert Workstations, and Other Identification Tools

7

Expert Workstations: A Tool-based Approach

JIM DIEDERICH AND JACK MILTON

We wish to explore an alternative approach to building computer-based applications for the support of scientific research in biological domains. In our conclusion, we recommend criteria and guidelines for determining if such an approach is appropriate for each of the four areas of systematic biology under consideration in this volume. Although there are many approaches to building applications, we compare and contrast three here: generic tools, expert systems, and expert workstations.

Comparison of Some Approaches

The Generic Tool Approach

By generic tools we mean off-the-shelf software systems such as word processors, spreadsheets, microbased database systems, and the like. One example is using Hypercard to store and manage information about biological strains in genetic research, a very important activity according to geneticists.

The advantages of such an approach are that the software is inexpensive, no programming is required, little or no collaboration with computer scientists is required, there is a direct fit with current manual practices, and the solution facilitates routine but important activities. In essence, the generic tool fits very well with its intended use. A disadvantage is that it may be difficult to add functionality, for example, to process data in the Hypercard system. This, of course, reduces generality. The data input may be expensive, but this is true for most real systems anyway.

Advances in Computer Methods for Systematic Biology: Artificial Intelligence, Databases, Computer Vision, ed. Renaud Fortuner (Baltimore: Johns Hopkins University Press).

Expert Systems

The second approach is expert systems (ES), or what we call deep problem solvers. An example might be an expert system to aid in phylogenetic inference, based on rules and heuristics used by systematists in their efforts to create phylogenetic trees. A major advantage is that this approach allows us to attack problems that otherwise might be very difficult to solve using conventional programming methods. In some cases the development effort can be reduced by using existing ES shells. The use of ES shells minimizes the need for programming and collaborating with computer scientists.

On the other hand, large rule bases are difficult to manage and update, an expert system tends to be essentially closed to adding various kinds of additional functionality, and it might take some doing to find an appropriate shell relative to a given problem. Knowledge acquisition to extract the knowledge needed to guide the system, often via rules, may take a considerable amount of effort and require the services of a knowledge engineer. Also, the system may be relatively inflexible in its use, with the user still essentially confined to a *do what I say* type of interaction, though ES interfaces are now much more flexible than they were previously.

Expert Workstations

In our work, we are using a third approach, that of an expert workstation (EWS), which can be characterized as a tool-based system. We will examine this approach in detail. An example of an EWS is a computer-aided design (CAD) system where various tools support a variety of steps in the design process and in which inference capability tends to be more local, or tool based. A strong advantage of this approach is that we may be able to create tools that have a long lifetime, that support a multitude of activities, and that may help with unanticipated as well as anticipated problems. Furthermore, the amount of knowledge representation and acquisition may be reduced, as we point out below. On the other hand, this approach is likely to require close collaboration with computer scientists over an extended period, and it may require a programming environment that supports efficient prototyping. Also, as we will elaborate, the state of the art in commercial database systems for handling biological data is primitive, and, in addition, the technology is just emerging to support complex interface development. While rich database access and natural interfaces are important in any system, their importance is underscored in an EWS because the user has more control and flexibility by design in such a system.

Tools

Unfortunately, the word *tool* conjures up different images with different people. To be more specific, consider a set of carpenter's tools: saw, hammer, chisel, square, etc. They illustrate the paradigm we consider when we use the word, and it may be helpful to keep them in mind during the remainder of this discussion.

Tool Characteristics

Our conception of a tool in an EWS context embodies several primary characteristics. First, a tool performs one or more functions, and something significant within the domain can be accomplished with it; otherwise it is a subtool. For example, an editor may be a tool in the context of a programming system in that it allows us to construct and modify programs. But its facilities for cutting and pasting are only subtools within the editor. Second, a tool should be so fundamental to the domain that it enjoys a long life and retains its essential character as changes occur. For example, while there may be changes in hammers, saws, and chisels, over time they retain their essential character. Third, tools can be used with other tools, jointly or sequentially. Fourth, their use should not be confined to a precise regimen. As we see from our metaphor of carpenter tools, sometimes it is appropriate to use tools together like a hammer and chisel, and at other times to use them individually. Their use should be determined by the context, which we define as the current system state plus the user state, that is, what the user would like to do in light of what has already been done.

Another important property of tools as we define them is that tools allow for unanticipated uses. In fact, a good tool will probably be used in ways not imagined by the tool creators. For example, if you have ever done any remodeling you have probably used a hammer in ways that were not anticipated by its creator.

Furthermore, a tool requires an agent to exploit its power. The user is not a passive observer of the tool in action. For example, a compiler would not fit our definition of a tool if one could use it only to compile programs without user intervention. We would call it a subtool rather than a tool.

As is the case with the carpenter's tools, effective use of the set should grow with experience; nevertheless, use of the tools is not confined to experts alone. Other users, even apprentices, should be able to employ them in meaningful activities. This elaborates the notion of an EWS, a possible misnomer in our context, in the sense that we do not intend it for exclusive use by experts; rather we feel it should be useful to nonexperts as well. Also, using a variation of a tool should not involve a significant adjustment; for example, encountering different kinds of hammers and saws should not necessitate dramatic readjustments.

Finally, the tool set of a system that supports scientific exploration will not be fully mastered by all of its users. There may be some kind of uncertainty principle here where building a system that is particularly effective results in one that not everybody will be able to use effectively, because the degree of effectiveness relies on the expertise of the user in the domain. Nevertheless, the value of the tool set should not be restricted entirely to expert users. Examples of tools are illustrated later in this chapter and in Diederich and Milton (this volume, chap. 10).

What a Tool Is Not

To further delineate the concept of a tool, it is important to mention characteristics that true tools should not have. First, a tool is not a complete application. If your initial reaction to encountering a tool is that it does not seem to do everything, that is probably good. Many find this strange at first, and until the client for whom the EWS is being built understands this, it will make matters difficult for mutual understanding between the collaborating computer scientists and biologists designing the system. A tool should not do it all. It should facilitate important steps within various activities. It is the set of tools that should allow complete activities, depending on the user's intentions. If a tool does everything, then it probably binds the user too much in terms of the flexibility one wants in working with a set of tools and thereby reduces the overall power of the EWS.

Second, a tool does not replace an expert. For example, carpenter's tools do not replace a carpenter. Yet there may be some embedded expertise within a given tool, as clearly there is in a saw. In this we include the understanding of wood grains that dictate the shape of the teeth for crosscuts and rip cuts. Third, a tool should not be a form of *training wheels*. Note that the apprentice carpenter would use the same tools as the carpenter, but usually with limited effectiveness. No special training tools should be needed.

Expert Workstation Characterization

An EWS has been characterized by Faught (1986) in terms of representation, presentation, and rules, in that order. You first analyze data and knowledge needed to support the functionality of the system. You then look at what needs to be presented and how to present it. You try to postpone as much as possible capturing the expertise in the form of rules. This should reduce the number of rules required and the amount of knowledge acquisition effort involved, which is very important because we are not trying to replace the expert. So by analogy we would not try to build a carpenter, which may be quite difficult if not impossible to do reasonably well, but only the tools needed by one. We also feel that tools in the hands of an expert can be more effective in the biological domain in which we work than a system that tries to replace the expert.

On the other hand, rule-based expert systems typically follow a paradigm that uses a permutation of these three characteristics, with rules and representation at the top and presentation trailing in importance. We do not follow exactly Faught's paradigm of the EWS either; we move presentation up with representation in priority, if not even higher.

In the remainder of this chapter we examine the representation and presentation (interface aspects) that need further study to support an EWS in biological domains. We do not discuss rules, partially because of the reduced importance of this aspect in our conception and partially because rules typically receive top priority in discussions of ES. First we will look at the data and knowledge concerns and then examine interface, or presentation, concepts. Throughout we will use illustrations from our prototype system, NEMISYS, NEMatode Identification SYStem, for the identification of plant-parasitic nematodes. NEMISYS is being written in Smalltalk-80 (Goldberg and Robson 1983) and will eventually be linked to a large data and knowledge base. Further presentations of NEMISYS are found in Fortuner (this volume, chap. 9) and in Diederich and Milton (this volume, chap. 10).

Representation

In order to build an EWS in biological domains you need to have good interface and database management system (DBMS) support tools. This applies both to the tools that compose an EWS for application by the user and to developmental tools for the system builders. Perhaps these latter tools could be viewed as metatools, or tools to construct tools. Without good generic metatools, there is going to be a large duplication of effort with many people working on their own to solve the same problems. In addition, system construction is going to be much more difficult without a good set of such tools. To illustrate the DBMS support required we need to look at some of the biological database concerns that are not supported in current commercial DBMSs. One of the things that we propose is an effort to build, in conjunction with a commercial DBMS, what we call a Bio-DBMS.

Biological Data Types. To illustrate just a few things not supported in current DBMSs, consider the basic data types used by biologists. Biological data types are continuous, discrete, and binary, and in addition there are data that cannot be compared other than for equality (nominal), data that can be ordered (ordinal), data that can be ordered and for which differences can be measured (interval), and data that have these properties and for which a ratio is also meaningful (real number). In NEMISYS, we have seen the need for several of these types, but they are not supported in commercial DBMSs. In a typical DBMS one can only state whether the data are continuous (real), discrete (integers or strings), or binary (Boolean), in addition to data types used in business settings. With new kinds of database systems being developed, such as extended relational and object-oriented DBMSs (Kim and Lochovsky 1988; Stonebraker 1986; Maier et al. 1986; Winslett, this volume, chap. 16), it is possible to create these data types as user data types, which is certainly a step forward. But user-defined data types alone will not be sufficient to support all of the activities that rely on the types, since each group would be required to build its own types and support their use, again duplicating effort.

Qualitative Characters. A further illustration of requirements for biological data is that qualitative characters are not well supported in database systems. In plant-parasitic nematodes, tail shapes can be roughly arranged into general categories (or general character states), such as filiform, conoid, cylindroid, dorsally convex, and spiked. Then, specific states can be defined within each of these general states. For example, a filiform tail can be elongate, attenuated, or tapering. Similarly the general shape of another organ such as the spermatheca can be defined as round, with specific states circular or elliptical. During the identification process, if the user enters *tail filiform* or *spermatheca round* this can mean any of the specific states mentioned.

In a typical DBMS there is no simple way of saying that a state is indeed a general state or that a state is specific for some general state. Even if it were possible to do so, there are additional implications and complications. For example, consider the query, *Find all taxa with the tail shape filiform.* The system will automatically have to find all of the taxa with its specific states elongate, attenuated, and tapering, in addition to those explicitly tagged as filiform, because they are all valid filiform tail shapes. The problem is different if you ask, *Find all taxa with tail shape elongate*, and it is not properly answered by simply returning all taxa that have been tagged as elongate. Keep in mind that in some descriptions the author may just have indicated that the tail shape is filiform and may not have indicated whether it is elongate or one of the others. So for those taxa only filiform is stored in the database. Thus in response to this query the system should return all taxa with the value elongate but also must return the taxa with value filiform as *maybe* results and handle them as such. Current commercial systems do not support this at all. This kind of support is needed since we are building a database that faithfully represents the data found in the literature, where each piece of data written in published papers bears the stamp of its author. Also one needs to support the idiosyncrasies of the user in the formulation of input, especially when a natural language style of input is allowed as in NEMISYS.

Integer Characters. The treatment of integers in biological domains also requires special care. For example, discrete values may be found in descriptions, as those in the set {*2, 4, 6*}, or as ranges like *2–5*, or as open ranges like *6 or more*, or as approximate values like *half a dozen*. Again, keep in mind that if the author of a description indicates an approximate value of *12 or so* the database needs to store it that way to be as accurate as possible. If someone then poses the query, *Find all taxa with about a dozen lateral field lines*, a typical DBMS will not be able to answer correctly, since neither the storage nor the retrieval mechanisms are designed to handle integers in this way.

Metadata. Another important representation issue in NEMISYS is that of domain-based metadata. In a typical DBMS you build the schema to show what your records look like. You can state that the social security number of the employee cannot be null, its value cannot change once it is input, it is single valued, and it has a particular format. These are data about the data, sometimes called *metadata*. In a biological setting metadata would be viewed as characteristics of characters. We will give only a few highlights about metadata here. A detailed discussion can be found in Diederich and Milton (1991).

In our system we are looking at metadata needed to support the process of identification. We have to specify characteristics of our characters so that we can determine which characters are best suited for identifying each taxon. Suitability is determined by several metadata characteristics such as *conspicuity of the organ* and *ambiguity of the character*. Conspicuity indicates how easy it is to observe an organ, or organ part; ambiguity indicates how easy it is to make a mistake when recording a character. For example, in a taxon it can be said that "the tail is highly conspicuous but its shape may be a little bit ambiguous." That means that there should be no trouble for the identifier to see the tail, but she may have difficulty interpreting exactly its shape and deciding which of the several states of the character *tail shape* it fits best. In addition some taxa may be highly variable for some characters, in the sense that different individuals in the taxon exhibit different states of the character. This would make the character unreliable for identification, particularly for identification of a single specimen.

What we have found to date is that the metadata for identification is not typical metadata. By typical we mean metadata that is global in that it applies to every instance in the database. For example, every employee's social security number will be of a particular form, *ddd-dd-dddd*, and that form does not change from one employee to the next. Within the world of nematology a particular character for one taxon may be highly ambiguous, but for another taxon it may not, so in this case the metadata are local. Also, the metadata change from instance to instance (i.e., taxon to taxon), and this presents interesting problems in database management that have yet to be solved.

These are just a few examples of aspects of biological data that are not supported in DBMSs, and the existence of a significant amount of such data makes systemic support very important for developing EWSs in biology.

Presentation

While the word *presentation* is typically taken to mean user interface, we like to think of the representation, rules, and user interface together as an interactive system, specifically to emphasize that they should all work together. While the decomposition into the three components can be helpful for system analysis, people often maintain a strict separation during system construction.

A typical approach to building user interfaces has been to code an application in its entirety and then tack on an interface. The farther back you look the more typical this was, and we really do mean *tack on* in the sense that often the interface represented a very small part of the code that was added at the last minute. The application code was not created with the interface in mind, and the two largely separate code modules did not work together very well.

An excellent example of coding the application first is furnished by an expert system built in the Netherlands called FAD, a crop disease diagnosis system by Wieringa and Curweil (1986). Their final report reveals a strong awareness of the importance of proper interface design, as it is discussed at some length. They want "extensive analysis of the user information and requirements and interface ergonomics." They then note, however, that traditional expert system techniques dictated that they look at the development of the inference mechanism first and the architecture of the interface last. Their conclusion, noting that they operated under severe time constraints, is that their resulting interface was "a clumsy dialogue which would probably annoy any user." So they clearly recognized the inadequacy of this traditional approach, but seemed tied to it nevertheless.

We believe, on the other hand, that it is very important to progress to the point that interface design and construction are an integral part of system development from start to finish. For the construction of interactive systems in general the representation and the interface should be developed together, and the need for rules may be minimized by embedding expertise in the tools. It is difficult to imagine that all need for rules would disappear, particularly if one goal is to make a system usable by the nonexpert, but it seems clear that better, integrated interfaces can minimize such requirements to some extent. As we mentioned earlier, interface development in NEMISYS has been aided by object orientation, but it has still been difficult in spite of a well-integrated set of user interface building blocks. Winslett (this volume, chap. 16) explained some of the advantages of object orientation, and we will add to her comments after outlining some important questions with respect to interactive systems in general. Better developmental and support tools, or *metatools*, are badly needed for interface development. A recent article in the *New York Times* (Pollack 1991) on ten critical technologies for the decade lists software development as one of them, with object-oriented concepts potentially playing a key role in conquering some of the problems caused by the enormous complexity being built into systems.

After working in biological systems for several years, we have not yet seen reasons to suspect that the types of interaction supported in other areas are much different from what must be supported in biological domains. If this assessment turns out to be correct this is good news, because we can thus make use of knowledge in other fields. It is important to take a look at the general state of the art.

Modern Interactive Systems. On the computer side of the human-machine equation almost all modern interfaces are centered around a collection of windows and an event loop, that is, some way of managing different types of *events* in different ways. We very loosely use *window* to mean panes, lists, icons, buttons, etc.

From the perspective of the user, there are five different basic types of user interaction (Schneiderman 1987): command language, direct manipulation, form fill-in, menus, and natural language. We discuss each type briefly:

1. With *command language*, the user interacts with the system by issuing specialized commands. One example of this type of interaction is the UNIX operating system. For example,

```
$ rm myDataFile
```

is a typical user action, using the command *rm* for *remove* and supplying the name of a file that is to be deleted, in response to the system prompt, $. Advantages of this type of interaction are power and flexibility (actions can be taken at many levels, even at the system level) and performance (there is little command overhead with the very specific and precise commands). Disadvantages are the necessity to become very familiar with a complex and perhaps low-level system, much of which is peripheral to a given application; the difficulty of remembering the specialized interaction in absence of regular use of the system; and the difficulty of dealing with complexity because users are typically working at a level closer to the machine than in other types of interaction.

2. With *direct manipulation* users manipulate icons that represent system entities. Certainly the Macintosh gives a popular example of this interaction style. Icons represent files, folders, applications, etc., and these are moved, opened, closed, and subjected to a variety of other actions with a pointing device, usually a mouse. With this style of interaction the icons can be

tailored to represent the underlying entities as well as actions (such as in a palette). Often, in addition, certain basic actions (moving, duplicating, etc.) are made standard across applications and are thus typically easy to remember. Among the important disadvantages are the somewhat restrictive nature of the representing icons and "canned actions." The tool builder must anticipate quite well the basic interaction that will be desired by users. While a fixed set of commands can be a disadvantage in any interaction style, this is more restrictive with direct manipulation than with command language, for example, because the icons typically represent much higher-level entities and actions. If something has not been built into the system the user cannot make use of it, whereas with the lower-level command language the user can piece commands together to build higher-level actions (assuming sufficient knowledge of the system).

3. With *form fill-in* the user supplies information, usually from the keyboard, in specified places on prebuilt forms. You can easily follow the forms, and within the forms you often have some leeway in terms of the order in which the data are entered. Advantages include clear guidance in entering elementary and well-structured data, but this can be a major disadvantage in some situations with the attendant restriction of flexibility. The forms can also take up much of the screen and obscure other aspects of the application and interaction.

4. With *menus* the user selects an action from a list of possible actions, and there are several different types of menus. Some older systems made heavy use of *full-screen menus*, which were hierarchically organized, and the user could proceed down through the menu tree, making selections and entering data. The Macintosh has popularized *pull-down menus*, accessible via a menu bar at the top of the screen. Hierarchical organization is then not required because each menu is available at a given place in the bar, though some of the individual pull-down menus are indeed hierarchical. Yet more powerful are *pop-up menus*, which appear at the cursor position when the user presses a mouse button. The greater power derives from at least two things: the menus are completely hidden before being activated, and sometimes different mouse buttons (on a multibuttoned mouse) activate different menus. This typically means that many more menus can be made available, tailored to function by position of the cursor and selection of a specific mouse button. Several systems at this workshop make extensive use of pop-up menus, notably Intellipath and NEMISYS. The power of pop-up menus is illustrated by the language Smalltalk in that as a user works a very large portion of interaction is typically taken via pop-up menus. These menus greatly reduce the need to enter symbols at the keyboard that are already in the system. Some disadvantages are that menus are more restrictive than command language, again because their actions are typically at higher levels of granularity, and at the other end of the spectrum they can be too powerful for the casual user,

resulting in such a user taking unwanted actions by accident, with unintended results. This latter problem applies to all types of menus, but it is exacerbated by the hidden pop-up menus. Also, beginners sometimes complain that they cannot remember where to find pop-up menus, but this complaint disappears quickly, even after only limited familiarity with the application has been developed.

5. The final interaction style is *natural language*, in which the user interacts with the system in unrestricted English. Unfortunately, the state of the art is not to the point that completely unrestricted English is a viable option. Systems that do use natural language rely on an appropriate subset of English, or what is called a sublanguage (restricted to the domain). A major advantage is that there is no need to learn a specialized command system. Disadvantages include lack of precision in the restricted English and resultant lack of power, as well as ambiguities that can easily be introduced, of which the user is often completely unaware. Underlying these difficulties seems to be a basic mismatch of human thinking and machine reasoning.

Each interaction style has further advantages and disadvantages. It turns out that combinations can be very powerful, and in NEMISYS we use all five types, each in appropriate places.

A Tool Illustration. When people think in terms of tools, they often mean very different things. It has not been uncommon, therefore, during demonstrations of some of our tools to be told that until the tools were viewed on the screen people had no idea what we meant. Thus the presentation of the tool can be a powerful communications device. So let us demonstrate a specific tool from NEMISYS. We will note requirements and functions later on, but the layout and the coordination of a particular tool should give a much better sense of what a tool is. Figure 7.1 shows the display of one tool in our identification system, which we call the Basic ID tool. Let us deemphasize its total functionality here and concentrate primarily on layout.

The window is composed of several panes. Panes 1–5, 8, 11, and 13 are so-called list panes. Each contains a list, and selection of a specific item in one list causes something to appear in one or more other panes. For example, pane 1 contains a list of *systems*, which are the top-level entities in the hierarchical organization of nematode morphology. Selection of a system in pane 1 causes a list of organs from that system to be displayed in pane 2, and so on. Pane 6 is for free-form entry, which is the example of natural language mentioned above. Although one cannot use completely unrestricted English in that pane and expect to get accurate results, it turns out that the terminology used in nematode descriptions is a rich and descriptive sublanguage, but one which is not standardized. There is usually enough information in what is entered in pane 6 for a search routine in the system to find the characters and

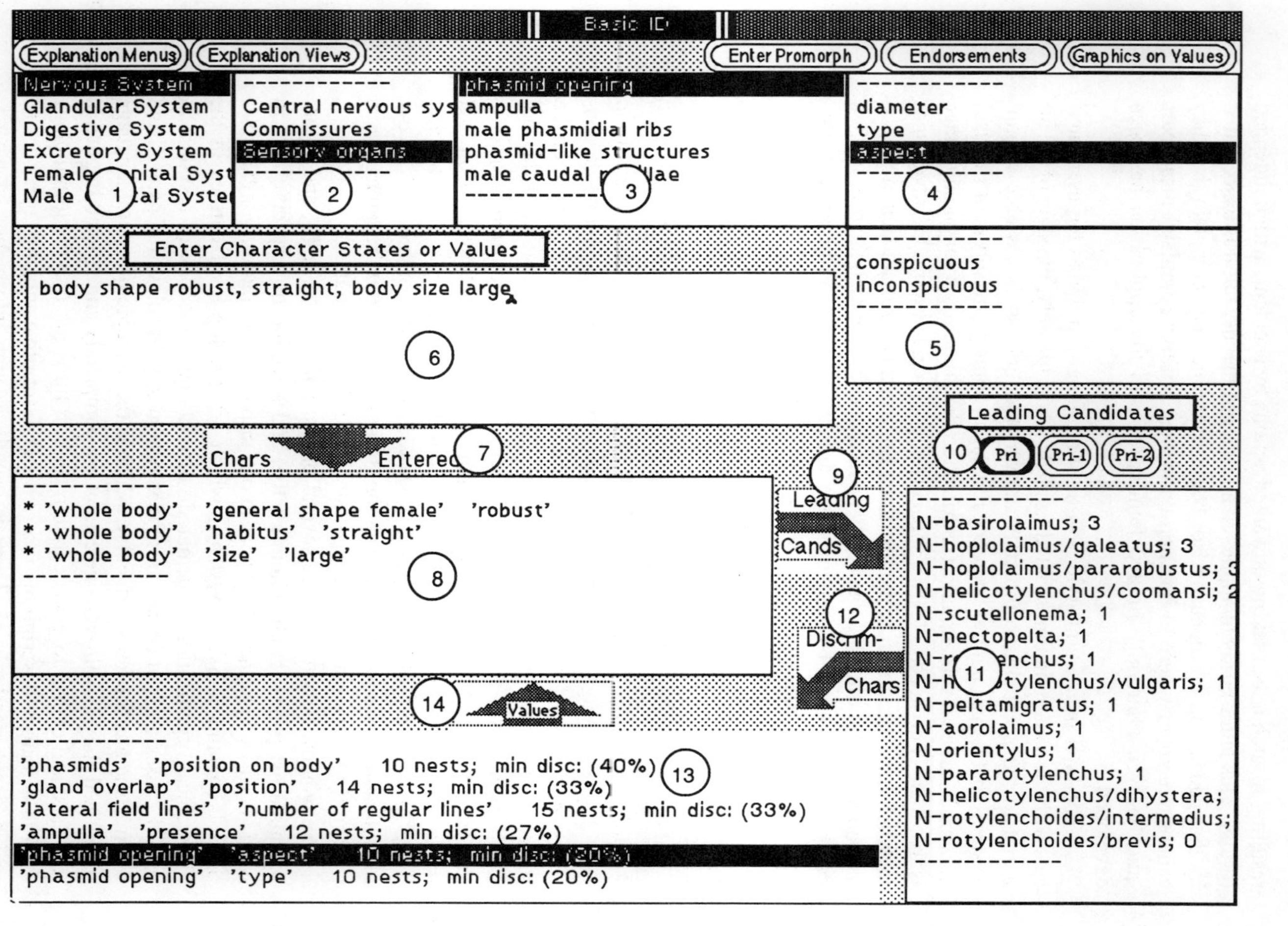

Figure 7.1. The Basic ID tool from NEMISYS. Focus: Hoplolaimids.

states corresponding to the user's observations. When the user has finished entering observations he pushes button 7, and a list of preferred characters will appear in pane 8, formulated in terms of the contents of the underlying data and knowledge bases. These suggestions can be modified before entry into the database if the user intended something else.

Panes 8, 11, and 13, along with buttons 9, 12, and 14, form the basic *identification loop*. The user confirms the characters in pane 8, pushes button 9 to get a list of leading candidates based on those characters to appear in pane 11, and pushes button 12 to get a set of characters that would discriminate between the leading candidates; this set of characters appears in pane 13, etc. This loop can be continued until a definite identification is reached, or the user can move to other appropriate tools at any time.

A typical expert system integrates the knowledge and the plan of attack for the user, which can be an excellent means of interaction for some users but which comes at a large cost in flexibility for others. In an EWS, no plan is normally provided in the interface itself, though documentation and demonstrations can be an aid in that direction. Our approach is to use what we call the *Visual Plan* within the individual tool interfaces.

The Visual Plan is a plan of attack that a user with some expertise would often take within the domain. It is represented visually in the tool interface layout, it is closely related to the normal functionality of the tool, and it provides options to allow for flexibility. Since using a plan is an integral part of the activities for beginners, advanced beginners, and competent users (Dreyfus and Dreyfus 1986), we feel that an explicit representation of a major plan is critical for providing guidance. Options through tools and activities selection allow for varying the plan and avoid constraining the expert. In Figure 7.1 it is clear that the windows in the basic identification loop are laid out in a roughly circular way, to suggest the order in which they should be used. In addition, the user can move from one pane to the next by pushing arrow-shaped buttons 7, 9, 12, and 14, or can alternatively use menu items on the appropriate panes. These arrows form a more explicit indication of the visual plan than the layout of the tool alone.

While it is beyond the scope of this chapter to discuss the Basic ID tool in more detail, we note that other major features of the tool are the use of graphical as well as textual input and extensive help for panes and menus (which remain hidden if the user so chooses). See Diederich and Milton (this volume, chap. 10) for more details. We trust that this illustration suffices as a good example of the sort of tools we have in mind. What can be noted from this brief discussion, however, is that different users will use this loop in different ways, depending on their expertise and their perception of the need to move to other tools at different stages of the identification, which is easily done by pop-up menus on some of the panes.

Building an EWS Prototype

A NEMISYS Retrospective

From the beginning of the NEMISYS project a watchword has been to let user requirements drive the interface, which in turn has driven the underlying architecture. This is perhaps the mirror image of the more standard methodology of attaching a user interface after the application logic and code have been developed. In retrospect this developmental methodology has been highly successful for several critical reasons, and its success is coupled with the use of another developmental methodology—that of rapid prototyping. This latter methodology consists basically of building prototypes of parts of the system, testing the prototypes alone and in conjunction with other prototypes, and eventually building the final system by piecing together the mature prototypes in the appropriate way. Specifically, you can identify important portions of the application, build each portion and its objects, test each independently of the rest of the application, and then piece them together into a bigger portion of the final application. Rapid and sometimes dramatic changes are thereby possible, and this possibility has been particularly helpful. In some cases the prototype is scrapped altogether, and the system is rebuilt. In others the prototype is the basis from which the system is built or extended.

We found, as was to be expected, that the computer scientists did not understand some important aspects of the domain, and the biologists did not understand some important aspects of the computing methodologies. Several significant problems arose, and the ability to prototype and to interact with a consistent facade of the system (often tool interfaces would be developed before their functionality was implemented) led to quick discovery of such situations and allowed us to change the system design to accommodate them. Since user requirements typically had immediate implications for the interface, tool interfaces often were developed first, and thus this methodology could be used from the beginning. In effect, the overall system, including interface, could be exercised as it grew.

For example, the computer scientists did not initially understand that identification to a higher classification level, such as the genus level, often was not a critical identification landmark for the biologist, but it was considered only as an *intermediate* step in the identification process. Identification to species level was the ultimate goal in most cases. This consideration is aside from any focus such intermediate identification might provide. This misconception became readily apparent when the computer scientists began to design and present tools to do that job alone. The biologists, on the other hand, had some trouble understanding the *set of tools* concept, and the confusion revolved around viewing a tool as accomplishing too much of the task.

At various times the computer scientists were asked to add far too much functionality to given tools, and it was not fully appreciated that the tools worked together without the user being forced to take explicit actions to link them. For example, if another tool is evoked from a pop-up menu in one of the panes of the Basic ID tool, all current information in the tool is available to the new tool without the identifier having to cut and paste, etc. This became apparent to the biologists when they began to request special facilities for transferring information between tools, as was customary in some of the microcomputer applications they had used.

What we did not anticipate, however, were situations in which the biologists had not fully thought through the implications of the new ways of viewing identification, or in which the computer scientists were not completely aware of what the full implications of certain computing principles would be in this application and domain. Such examples are much more subtle and are beyond the scope of this chapter, however. Suffice it to say that since requirements were driving our interface development, we believe that basic conceptual flaws in the requirements were more apparent from the beginning and were thus resolved sooner, more easily, more satisfactorily, and in a better integrated way, all of these leading to increased effectiveness of our system. This is generally cited as one of the main advantages of rapid prototyping.

While there is no question that good user interfaces in interactive systems will cost time and resources, our experience suggests that amortizing these costs over the entire project not only makes the costs more palatable but produces a superior interactive system.

On the other hand, the growing complexity of the system has begun to take its toll on development, even in the presence of the powerful environment of rapid prototyping, etc. Certain aspects of our interface will help illustrate this point, and we will indicate some directions in which we think research needs to progress.

The Object-oriented Paradigm

With the introduction to object orientation by Winslett (this volume, chap. 16), we refrain from making more general such comments and concentrate on advantages of the object-oriented language in which we have written our application, Smalltalk-80. Smalltalk-80 is in many respects the main object-oriented language, though C++ is used more widely. Smalltalk-80 runs on several microcomputers, including the Macintosh and the IBM 386 PC and clones. It also runs on a variety of workstations, including Sun 3 and 4, the Hewlett-Packard 9000 300 series, and DEC workstations.

Significant advantages to object orientation are touted in the literature, and we have encountered many of these in our several years of experience. Object orientation provides a different kind of paradigm for program development by focusing on the objects that appear in an application and on the operations that are performed on or by those objects. This is opposed to a more traditional approach of decomposing the functionality of an application in a top-down fashion. Through this paradigm, object orientation facilitates use and reuse of objects in different settings, where functionality is often not as easily carried over from one application to another in traditional methodologies, even if the applications are closely related. This makes tool-based systems more feasible since tools can be developed somewhat independently and then integrated, though some coordination is always necessary. You can write code and use it over and over, and there are a variety of ways to do that, many of which are not supported in non-object-oriented languages. A major payoff is usually realized in terms of how little code has to be modified when changes are made, even major changes.

Object orientation also allows for rapid prototyping, a key element of EWS development, as mentioned above. While prototyping can be done in any developmental methodology, the particularly helpful aspect of object orientation is that the decomposition of a system is by the objects that participate in the application. This is opposed to a functional decomposition of an application that may change radically when a prototype is studied. Also of critical importance is the facility for building user-defined objects, which provides for matching data structures and control structures more closely to the applications. The general thrust of high-level languages is to get the computer language closer to the application rather than being forced to make compromises in an application in order to get a better fit with the computer language, and object orientation has taken a large step in that direction.

While object orientation can make the job of the computer scientists easier, a fundamental question is, What is in it for biologists? It goes without saying that if your computer scientists can develop systems more smoothly and cheaply and they are happy, that is very good for you. That advantage is somewhat indirect, however, and we think there is a more direct advantage. That is, we found that object orientation facilitates building a prototype that provides the basis for coming to a mutual understanding of requirements, and we think it has been very important.

The interface-building tools in Smalltalk-80 have been extremely helpful. The object-oriented facilities and the methodology of rapid prototyping have made certain things possible that we would not have attempted in other languages. While this is all well and good, there remain distinct problems. What is now needed is the development of high-level interactive system-building tools.

Current State of User Interface Technology

It is well known that sophisticated user interface code typically takes 50 to 80 percent of the total code (Lee 1990), and that is a large amount by any standards. Numerous studies have verified this, and we believe that this is often interpreted as a formidable barrier to the potential development of highly interactive applications. There seems to be the attitude that, if you want to develop a highly interactive application, you have to pay a huge price. While it may be that better methodologies would make the interface less painful to develop, there is always going to be that huge price, so the viewpoint goes. Some who face this dilemma then opt to purchase what they can in terms of system functionality and tack on their interface at the end, which is doomed to be inferior. We take a little different view of this, however, and arrive at a different conclusion.

It is an unfortunate fact that good interactive system programmers for interfaces are scarce and in high demand. If one is building a commercial system, one has little choice but to find them, pay whatever it takes to hire them, and get the job done right. Typically this means building a new system from scratch. With so few good programmers to go around, and with many research budgets already strained to the limit, we believe that a different approach must be taken in the research community. In particular, it is clear that much better developmental tools are needed. To review the state of the art, there are three primary approaches to providing better user interface architecture (see Lee 1990).

First is the *toolkit*, in which there is a library of components, typically menus, commands, scroll bars, etc. A toolkit is accessed through the operating system language or library-calling process. Some examples are Microsoft windows, which many people have heard of, and Motif from the Open Software Foundation. There are obvious advantages to being able to forge a specialized interface by connecting pieces from a toolkit rather than handcrafting code. There are some disadvantages, however. In particular, use of toolkits tends to be tedious. Also, user interface code becomes tightly connected to nonuser interface code, which leads to at least one major problem in that it makes the code hard to port (make run on different systems).

Second is the *user interface management system*, or UIMS, whose use is a little different from that of toolkits. It is a software architecture in which the implementation of the user interface is more clearly separated from that of an application. The functionality is essentially the same as with a toolkit, but the access is different. A UIMS makes use of a user interface language rather than pulling things out of the library. An example is Apollo's Open Dialogue. Disadvantages here are somewhat similar in that a UIMS is tedious and time consuming to use. Looking back at the systems that have been developed reveals that this approach has not been as successful as the toolkit approach, and the reason is largely because current architectures restrict the interface/application code communication. This has made programming highly interactive applications difficult.

The third approach, *interactive design tools*, is related to the first two in that it is software that helps you use the toolkit or UIMS more productively. Basically, its job is to give access to the interface through direct manipulation rather than through a library or a language. In this paradigm you can influence functionality and layout by icon manipulation, and this approach is relatively new. Interactive design tools generally do a good job of defining static layout, an example of which is given above in Figure 7.1. However, in addition to the layout of the interface representation of a tool (the tool *window*), a tool will have functionality independent of the interface and having to do mostly with the application, as well as functionality made necessary by its interface layout. These two types of functionality can become very difficult to coordinate as the application (or the tool) becomes more complex.

An example of this latter point is given by list selection in NEMISYS. Pane 13 in Figure 7.1 is a *list pane* (only lists can appear in it), and it is used for three different reasons: (1) it can hold a list of alternative characters to those listed in pane 8; (2) it can display the result of a database search, evoked from pane 6; or (3) it can display a list of characters that will help narrow the identification, as represented by the list in pane 11, if the user can then enter the appropriate state. Even with only three uses for that pane, maintaining its correct behavior through development has become very troublesome, as small, seemingly innocuous changes to other parts of the tool can have unanticipated effects on that pane. We see a strong need for specifying pane interactions in a more declarative way, and we feel that this would be an important feature of a good interactive design tool that can be used for *dynamic design* of tools as well as *static design*.

Based on our experience with building interactive systems in a language that has a rich set of basic primitives or building blocks, Smalltalk-80, we make two primary observations: (1) this basic functionality of Smalltalk-80 made it possible for us to develop interactive systems of a far greater complexity than we would have ever attempted to develop in standard languages such as C or Pascal; and (2) in order to move ahead with yet more complex interactive systems and to maintain them over the years we feel a strong need for good static and dynamic interactive design tools.

Criteria for Expert Workstation Development

With this background on expert workstations, we would like to issue an informal challenge to those who are beginning to develop systems. As you consider modern computing methods and weigh different approaches, consider in which of the three categories your needs best fit: generic tools, expert systems, or expert workstations.

If your needs fit the generic category, which a lot do, it seems that you have a reasonable chance to put together and coordinate off-the-shelf components yourself. Programmers are typically not needed at this level, as is the case, for example, with spreadsheets, word processors, and some microcomputer DBMSs. If it is category two, ES, much outstanding work has been done to date, and we would encourage you to take advantage of the work in other domains, such as in medical informatics. If your needs fit category three, expert workstations, then you will need to identify similar needs to those we present. As the systematic community proceeds to build better EWSs the identification of generic needs will be critical. Certainly some needs are implicit in what has been presented above, but let us make them more explicit with a list:

1. We need to identify tools. Are there tools that are needed to serve data needs that naturally exist within the domain? Can you characterize fairly specifically what the need is and how the tool should work? We feel it important not to worry too much about interface implementation initially. Just, what is the tool and what do you want it to do? We obviously will want to spend a great deal of time on design of tool interfaces before any code is written, but first let yourselves dream about the functionality you could use without being hampered by real or imaginary interface restrictions. In particular, as you see clever interface implementations in use, imagine what functionality such implementations could liberate in your application.
2. As we pointed out, a single tool is not apt to do it all; it only has to do something significant on the way toward a solution. It is thus important to think about sets of tools in addition to single ones.
3. What knowledge is needed to support a tool? What does it need to know?
4. Is there existing DBMS support for the particular knowledge? And, if there is not, and a very important function is under consideration, should we be pushing the commercial database community to get the appropriate type of support built into commercial systems? Our feeling is that a Bio-DBMS needs to be built, but much will need to coalesce before commercial vendors can be attracted. Biologists will need to coordinate requirements across many biological domains and will need to reach a critical mass before such a venture could be worthwhile for vendors. In the meantime it is important that institutions like the National Science Foundation support basic database research in this area.
5. From the interface point of view, we believe that at least part of the answer to developmental difficulties is more declarative implementations of interfaces. In a procedural approach you must indicate exactly how things are to be carried out, while the essence of a declarative approach is that you must indicate what to do rather than specifying the details of accomplishing it. We

do have some ideas, some of our work is on how to do this better, and we think the emphasis on making tools and their interfaces declarative is very important.

6. Are there some primary generic types of activities to go on in the biological workstation interface, for which basic tools could be built?

7. Finally, we do not think that the need for good interfaces is unique to systematics. One approach for this scientific community might be to sit back and let others develop interfaces, in hopes of using them when they come along. However, we think that this would not be appropriate in that such development is a shared responsibility, as similar needs will manifest themselves in most scientific areas. On the other hand, some interests might be unique for systematics, and these might indeed require special attention. For example, there clearly seem to be some with respect to data types for biological workstations as we illustrated above. The data we have found in this biological application have typically not been seen by database researchers.

Acknowledgments

This research was partially supported by NSF grant IRI 88-07475.

Discussion

Pankhurst: I was very interested in your shopping list for a DBMS, the sort of database you need, in particular for coping with different data types. I think we also need variable field lengths. Fixed field lengths are unnecessary restrictions and a nuisance.

Diederich: Some relational databases and object-oriented databases offer this, but you have to look at the particular requirements for the kinds of variable-length fields that you want to manage. Sybase, which is a relational database system, has a particular type of field where you can put anything as large as you want. It is currently used at the Lawrence Berkeley Lab for the Human Genome Project because the data from the project can be stuffed into that field, but you have to write your own procedures to manipulate it.

Pankhurst: The second thing I very much want is a proper way to implement relationships between the data fields, and I mean different types of fields to different tables between the same type of data and the same fields—for example, characters and descriptions related to one another.

Diederich: When you create a schema, most DBMSs have means for saying that a given entity has specific attributes. What they do not provide is something for indicating that between these attributes you have peculiar relationships, which may be based on their values. That is something we have seen on the biological side that needs to be supported.

Pankhurst: And the third thing I wanted to say is that I want a database system on the hamburger model. What I mean is that for the top slice of bread I want my code. For the meaty layer, I want the DBMS to do whatever it does, preferably by subroutine calls that can easily be built into my code. The bottom layer of bread is my specialized subroutines that are manipulating the data, checking data type, and enforcing relationships between them.

Diederich: I think reality would dictate that you are going to have bun, bun, and meat. If we want a DBMS to manage biological data, we will probably have to build on top of the meat, that is, on top of some existing commercial DBMS. Then we will not have to duplicate the effort in building a specialized DBMS, and we can build on the capabilities that are coming with the new generation of DBMS such as user-defined types. So I guess we will have to get our hands dirty eating this kind of hamburger, but maybe someone will do it for us so we do not all have to build our own Bio-DBMS on top of a commercial one.

8

Principles and Problems of Identification

RICHARD J. PANKHURST

Identification Methods

The emphasis in this chapter is to explain the methods used by biologists for the identification of specimens of plants and animals, with an attempt to put these into the context of computer science. One definition is needed: the word *taxon* (plural *taxa*) means any taxonomic category (e.g., a species, a genus, a family) without necessarily stating which. For a general account of identification methods in biology, see the textbook by Pankhurst (1991).

No identification procedure can begin without a prior *classification*. A classification is a division of objects into groups, where the groups have been given names and their distinctive properties stated. An identification is the process of assigning a new object to one of these groups. This distinction is made because in common parlance, and also in the literature of statistics, these two distinct processes are often confused. It is of little or no consequence to the identification process as to how the given classification is created, provided only that the groups are adequately defined.

How Do I Identify This Specimen?

As an illustration of the situation, a specimen of *Senecio vulgaris*, a common weedy plant of the Asteraceae (Compositae) was brought in from a nearby flower bed. In order to find the name of a specimen, the methods are, in order of preference:

1. To know what it is already. This is not an attempt to be facetious; an expert opinion can be given in a few seconds and is far faster than any other method.

Advances in Computer Methods for Systematic Biology: Artificial Intelligence, Databases, Computer Vision, ed. Renaud Fortuner (Baltimore: Johns Hopkins University Press, 1993).

2. To ask somebody else who knows. This works well if an expert is on hand. Unfortunately, such expertise is not often available or easily accessible and is probably getting scarcer.

3. Failing either of the above, go to a collection of preserved and named material and compare the unknown specimen with other specimens in the collection. In the case of a plant, this would mean looking in a herbarium, which is why this is sometimes called the *herbarium crawl* method. Suitable museum collections are few and often accessible only to a few professional biologists. Using a popular guidebook with colored illustrations and searching through the pictures is really just another form of the same method. This is not really such a good technique as it is generally believed to be, as it is ineffective for any group of organisms that are numerous and relatively nondescript (e.g., grasses). Actual specimens, even allowing for the information loss caused by methods of preservation, are generally superior to pictures for this purpose. Either way, these are *comparison methods*.

4. Use a dichotomous (diagnostic) key. This is by far the most commonly used method. Computer scientists might find it convenient to think of a diagnostic key as a decision tree that is written down on paper in a formal way and traced out manually by a human user. Keys are often printed in books and manuals and are therefore very practical for use in the field. There are two forms of keys, corresponding to whether the decision tree is traced out first from top to bottom and then left to right (parallel type), or else in the opposite order (bracketed type). An example of a parallel key appears in Figure 8.1. The user begins by examining the unknown specimen and asking the first question (lead) in the key. This asks whether *sterile rosettes* are *present*, assuming that the user will know what this means. If *yes*, then the next question to ask is number 2, and if *no* (rosettes *absent*) then continue with question 10. Repeat this procedure until the name of a plant appears at the end of the line. Suppose that at question 2 the *involucral bracts* of the specimen were *patent*, and that at question 3 the *cauline leaves* were *auriculate*, then the result is the species *Jurinea polyclonos*. This identification was made by assuming, first of all, that the unknown specimen was a kind of *Jurinea* and that no errors were made in the question-answering process. The proper thing to do at this point is to check the identification by some independent means (e.g., look in a collection or compare with a picture). The important point is that the key method is a procedure of step-by-step deterministic *elimination*.

5. Last, and most important, there are of course computer methods for biological identification, and these will be discussed below. Some methods are fairly direct implementations of existing techniques, whereas others have no historical counterpart.

1 Sterile rosettes present. 2
2 Outer involucral bracts patent, or recurved. 3
3 Cauline leaves auriculate, pappus 0.8 times achene. 11.J.polyclonos
3 Cauline leaves without auricles, pappus 1.1 to 1.2 times achene. 4
4 Cauline leaves amplexicaul. 12.J.ledebourii
4 Cauline leaves not amplexicaul. 5
5 Achene 1.0 to 2.0 mm. 14.J.glycacantha
5 Achene 3.0 to 5.0 mm. 10.J.mollis
2 Outer involucral bracts erect. 6
6 Upper surface of leaves white, or grey, basal leaves arachnoid-tomentose above. 7
7 Basal leaves entire, leaf margins revolute, capitula hemispherical, corona of achene inconspicuous, pappus 1.2 to 1.4 times achene. 7.J.kirghisorum
7 Basal leaves pinnatifid, leaf margins plane, capitula obconical, corona of achene conspicuous, pappus 3.0 to 4.0 times achene. 4.J.pinnata
6 Upper surface of leaves green, basal leaves subglabrous above, or setose above. 8
8 Basal leaves setose above, capitula obconical, corona of achene conspicuous, pappus 3.0 to 4.0 times achene. 3.J.tzar-ferdinandii
8 Basal leaves subglabrous above, capitula subglobose, or hemispherical, corona of achene absent, or inconspicuous, pappus 1.1 to 2.0 times achene. 9
9 Capitula subglobose, corona of achene absent. 17.J.fontqueri
9 Capitula hemispherical, corona of achene inconspicuous. 13.J.consanguinea
1 Sterile rosettes absent. 10
10 Leaf margins revolute. 11
11 Basal leaves setose above. 12
12 Pappus 1.5 to 2.0 times achene. 2.J.stoechadifolia
12 Pappus 3.0 to 4.0 times achene. 3.J.tzar-ferdinandii
11 Basal leaves arachnoid-tomentose above. 13
13 Capitula cylindrical, achene glabrous. 1.J.linearifolia
13 Capitula obconical, achene hairy. 14
14 Outer involucral bracts erect, distal part of bracts purple. 16.J.taygetea
14 Outer involucral bracts patent, or recurved, distal part of bracts green. 15.J.humilis
10 Leaf margins plane. 15
15 Stem woody at base. 16
16 Achene 6.0 to 7.0 mm, corona of achene inconspicuous, pappus 1.5 times achene. 6.J.albicaulis
16 Achene 3.0 to 4.5 mm, corona of achene conspicuous, pappus 2.5 to 4.0 times achene. 17
17 Basal leaves entire, capitula cylindrical. 1.J.linearifolia
17 Basal leaves pinnatifid, capitula obconical. 4.J.pinnata
15 Stem herbaceous. 18
18 Upper surface of leaves white, or grey, basal leaves arachnoid-tomentose above, capitula obconical, outer involucral bracts coriaceous. 5.J.tanaitica
18 Upper surface of leaves green, basal leaves subglabrous above, capitula subglobose, outer involucral bracts herbaceous. 19
19 Cauline leaves without auricles, outer involucral bracts erect, achene 3.0 to 4.0 mm, achenes obpyramidal. 8.J.cyanoides
19 Cauline leaves auriculate, outer involucral bracts recurved, achene 5.0 to 6.0 mm, achenes subcylindrical. 9.J.ewersmanii

Figure 8.1. Key to *Jurinea.*

Characters and States

It is necessary to explain at this point that the data used in identification are expressed in terms of characters and their states. In the above example, *presence of sterile rosettes* was a character, to which there were two alternative states, *absent* and *present* attached. The character is the abstract feature or property, such as *color of petals* or *shape of leaf*, and the states are the values that these characters can take, such as *pink* or *elliptic*. Characters may be qualitative (as above) or quantitative. Quantitative characters do not have states but values. Quantitative characters can be integers (e.g., *number of petals*) or real numbers (e.g., *height of stem*).

In the great majority of cases, there is no information available (nor is there likely to ever be) about the probabilities of occurrence of species or of character states, so that identification methods that use probabilities are not appropriate. There are exceptions to this, such as for medical bacteria, where much information on probabilities has accumulated over the years.

Brief History of Identification Keys

Identification keys have certainly been in use since the eighteenth century and possibly for a century before that. Linnaeus himself published diagrams that look like diagnostic keys, but he did not clearly state whether these were for identification or whether the diagram was just a summary of his classification. Oddly, he published a key for naming botanists, rather than for plants! It was a means of finding out who was the specialist for a particular plant group.

There were variations on the basic scheme but no real innovations until the twentieth century, when keys on punched cards (polyclaves) first appeared in the 1930s. Each card represented a taxon and had holes punched round the edges to represent characters. These cards could be sorted with a needle and had the important advantage that any selection of characters could be used in any order. This is still an elimination method, but it is now *multiaccess*. Another version of the polyclave used center-punched cards, which later on were often IBM eighty-column data cards. The key to families of flowering plants (Hansen and Rahn 1969) is an important example of this.

Around 1970 there was a flurry of activity as the first programs for constructing diagnostic keys were published, for example, Pankhurst (1970, 1971), Dallwitz (1974), Morse (1974), and Hall (1973). Programs of a similar nature, called rule induction systems, have also appeared in the context of expert systems, for example, Quinlan (1979). All programs mentioned above operate in a batch mode, meaning that the program reads a data file that describes taxa by their characters and makes a key from it, without user intervention.

An interactive key construction program, which builds the leads of the key line by line with the guidance of an expert user, has also been published (Pankhurst 1988b). A screen from this is shown in Figure 8.2. This corresponds to the building of the first lead of the key in Figure 8.1. The use of such a program has the great advantage that the key is guaranteed to be as accurate as the data on which it is based; errors in copying facts, or in the structure of the key, are not possible, although it is still possible to produce a

KEY TO JURINEA

BEST CHAR DELE DIAG DIFF EXAM EXPA FINI HELP RENU SAVE SCOP TAXA VIEW
Next Taxa Quit Recombine Delete Order Skip

1 (6)Sterile rosettes (=1)absent.
TAXA 1 2 3 4 5 6 8 9 15 16

2 (6)Sterile rosettes (=1)present.
TAXA 3 4 7 10 11 12 13 14 17

Type D,N,O,Q,R,S or T Tax 17 Cha 18 Sco 1 Leads 0

Figure 8.2. Interactive key construction program with example of *Jurinea*, display of EXAM command, with initial version of lead 1.

badly designed key! This program offers alternative strategies for building keys. At each lead, it is possible to switch between trying to make the shortest key with the most informative characters (data-directed search) or trying to distinguish a particular taxon or group of taxa from the rest (goal-directed search) or just making trials with characters you favor from experience.

When a lead has been selected and is being examined before deciding to keep or reject it, a range of editing functions is provided so that the lead can be cast in many different forms. It is always desirable when writing keys to provide auxiliary characters (i.e., more than one character per question). There is then a confirmatory or alternative character you can try if the first one puzzles you. This is easily done when only a few taxa are being distinguished, but hard to do with larger groups, since conveniently correlated characters are not that common. The program automatically looks for all auxiliary characters, and if it suggests none, then this means that there are none to be found.

Expert-Style Identification Programs

The first interactive or online program, for multiaccess entry of characters at the keyboard and identification by elimination, was published by Goodall (1968). Another program of this type was first published by Pankhurst and Aitchison (1975) and has been transformed in numerous versions as part of the PANKEY package (Pankhurst 1986). Version 3 was distributed along with the CONFOR programs of Dallwitz (1980b), and version 6, with color graphics images in an application to the identification of orchids, is described in Pankhurst (1989). The orchid application was demonstrated at the meeting. Mike Dallwitz has now developed his own interactive identification program called INTKEY. There has been a considerable number of programs of this type published since the beginning of the 1980s. The more notable of these include XPER by Lebbe (1984), which was designed independently but turns out to be remarkably similar to the ONLINE programs. XPER was used via the MINITEL system available to telephone subscribers in France to provide an interactive means of identifying mushrooms and toadstools, perhaps motivated by the national interest in food (Lebbe 1986). Also worthy of mention is the MEKA program (Duncan and Meacham 1986), which was first applied to the identification of world plant families, using the data published by Hansen and Rahn (1969).

In the context of computer science, such programs might be described as expert systems for diagnosis, but their development has been largely independent, and there are some striking differences of emphasis. These are:

1. The knowledge representation has always been a *taxonomic data matrix*, that is, a rectangular table of the taxa by their characters, with the states filled in. The representation of knowledge as rules is not used, possibly because it is unnecessarily general, and because the more specific matrix approach leads to more effective algorithms. Rules do occur in taxonomic identification systems, but as output rather than as input! The leads in keys produced by key-constructing programs are similar to rules. Also, there are programs (e.g., Pankhurst [1983a]) for finding diagnostic sets of characters, that is, sets of characters that are guaranteed to distinguish a given taxon from all others. Such diagnostic characters are equivalent to rules that say, "If the specimen has this and that characters, then it must belong to taxon *X*."

2. In the great majority of identification problems with plants and animals, probability information is not available. What is more, in many cases, such information may never become available, since many species are known only from a handful of specimens. Also, probabilities of the occurrence of taxa and of the states of their characters are often variable, depending on environmental factors, such as climate. It is also worth saying that most systematists have no use for specimens that are doubtfully identified. The usual practice is to either ignore such material or put it back into the collection again until someone does venture to put an authoritative name on it. There are some special groups of organisms, such as medical bacteria, where probabilistic data have been collected, and in such cases, algorithms using Bayes' theorem (or maximum likelihood) or other measures of degree of belief are routinely used (e.g., Willcox et al. [1973]). Probabilistic methods are of course very often applied to medical diagnosis, on the presumption that they are more accurate than deterministic methods. This presumption should perhaps be reexamined, as there seems to be little evidence for it.

3. Very few of the biological identification programs have any control system. This is probably because botanical and zoological users would not be inclined to accept a system that took away their decision making. Skills in

identification have traditionally been (and often still are) regarded as personal and not capable of being expressed in an algorithm carried out by a machine. On the other hand, it may be argued that if this is a scientific process that is being carried out, it must be capable of expression in an objective form and that any stages of subjectivity should be located and removed forthwith.

Apart from the above, other standard features of expert systems are definitely present in identification algorithms. These are:

1. Data-directed or goal-directed search. These appear as commands to find characters that are good for separating any taxon from any other (data-directed), or for recognizing particular taxa (goal-directed). These are often based on a separation number (for a character, the number of pairs of taxa that are distinguished by the character) or on an information coefficient.
2. Graceful degradation. This first appeared (Morse 1974) in the form of the *variability limit*, which is the number of characters in which a specimen may differ from a taxon before it is considered to not belong to the taxon. In fact, if the limit is zero, you have a method that functions exactly by elimination, and as the limit is increased toward the total number of characters, the method is then completely of the comparison type, with intermediate methods between.
3. Explanatory capability. One way in which this occurs in the context of a specimen whose identity is indicated as being something different to what was expected. The question then is, If so, how does it differ from *X*? where *X* is what the correct identity was thought to be.

Matching Programs

The bulk of the discussion so far has been on elimination methods. Comparison methods have also been programmed (e.g., Pankhurst [1975]). In this case, the unknown specimen needs to be as fully described as possible, that is, all its characters need to be scored and coded as input data to the program. The output is basically a measure of the similarity (similarity coefficient) between the specimen and a selection of those taxa that most resemble it.

An example of program output is shown in Figure 8.3 for the genus *Jurinea*. The special characters were chosen by the user as characters that are thought to be of special significance for that particular unknown specimen. In the table of similarities that appears in Figure 8.3, the five taxa with the highest scores (similarities) are shown. The plus marks (+) show where the special characters agree for that taxon. The asterisks mark taxa that belong to the *right* subgroup of taxa. In this particular example, the genus was divided into subgenera that were defined in the input data, and the unknown specimen was compared with the average of each of these groups. The asterisks then mark any taxa that fall in the group which, on average, is the most similar.

```
Jurinea <pinnata?>
 SPECIAL CHARACTERS ARE -
   Stem <shrubbiness>
   Basal leaves <shape>
   Upper surface of leaves <color>
   Cauline leaves <auricles>

   SEQ  SIM.  COUNT        SPECIES

   1    54.8  18  *++      6.J.albicaulis
   2    50.7  19  * + +    2.J.stoechadifolia
   3    49.1  19  *+ ++    4.J.pinnata
   4    45.5  19  *+  +    1.J.linearifolii
   5    42.9  18  *   +    3.J.tzar-ferdinandii

 RESEMBLES GROUP   1

 SPECIAL TAXA COMPARED
    3   49.1 *             4.J.pinnata
```

Figure 8.3. Example of output from a matching program.

More effort is required to make the complete specimen description than with the elimination methods, but the effect of erroneous characters is much reduced.

Other matching programs have also been published, such as a system for diseases of soybeans (Michalski and Chilausky 1980) in which the similarity coefficient is weighted with data based on probabilities, and NEMAID, a system for nematodes (Fortuner and Wong 1984). Matching methods seem to have been much less used than elimination methods, perhaps because they may require more data per specimen.

Taxonomic Databases

By about 1980, it may be said that the major algorithms for biological identification had been discovered. That is not to say that the programs had been perfected, and with the coming of the demand for user friendliness there is no doubt that much can still be done to improve user interfaces. This is the motivation for the author's current development work on a DELTA editor (DEDIT) and a version of the expert program (ONLIN7) with windows, menus, and help screens. However, it began to be clear that the major deficiency in computerized identification lay in the direction of the data (Pankhurst 1983b). There were not, and still are not, sufficient databases with enough good data for the algorithms to act upon. Problems of handling taxonomic descriptions in standard databases have also become apparent. These questions are further addressed by Pankhurst (this volume, chap. 14).

The Relation of Databases to Identification

The two main packages of programs for identification, CONFOR and PANKEY, both utilize the DEscription Language for TAxonomy, or DELTA for short (Dallwitz 1980b). DELTA summarizes the descriptions (morphology) of a group of organisms and contains a list of characters and states and sets of character states for a set of named taxa. These packages are usually run on IBM-compatible PCs, but both originated in mainframes (IBM, CDC), and versions exist for other machines such as VAX and Macintosh (MacPankey).

DELTA is not a database system; it is simply a data format. It has been adopted as an international standard by the International Union of Biological Societies (IUBS) Commission on Taxonomic Databases, an organization previously known as the Taxonomic Databases Working Group (TDWG). Although used with FORTRAN programs, DELTA data are in free format. Efforts are being made to handle such data within database management systems, as, for example, in ALICE (Allkin and Winfield 1989), in the HyperTaxonomy package based on HyperCard for Macintosh microcomputers (Skov 1989), and in the PANDORA database (an early version described by Pankhurst [1988a]; see also this volume, chap. 14).

Brief Description of the DELTA Format

Figure 8.4 shows an example of a file in DELTA format. The data have a TITLE, and then follow definitions of all the characters after the line CHARACTER DESCRIPTIONS. The characters and their states are numbered in a readable fashion. Every time a string of letters occurs as part of the data (e.g., *absent*) this is terminated with a solidus (/). The first character shown is a qualitative character with two states, the presence of the stem. Comments are permitted and are enclosed with sharp brackets (e.g., <presence>). Character 2 is a real quantitative character and does not need a list of states, but it does have units (cm). Character 14 is an integer quantitative character.

The DEPENDENT CHARACTERS are very important because the computer could not otherwise know what are the logical rules that connect characters; in the example, *1,1:2–4* means that if character 1 (stem presence) is equal to state 1 (absent) then characters 2 through 4, which are various properties of the stem, are impossible.

After ITEM DESCRIPTIONS a description of *J. linearifolia* is given. Each

```
*HEADING  JURINEA TEST/
..
..

*CHARACTER DESCRIPTIONS
#1. Stem <presence>/
 1. absent/
 2. present/

#2. Stem <height>/ cm/
..
..

#14. Capitula <no.>/
..
..

*DEPENDENT CHARACTERS  1,1:2-4:12-13

*ITEM DESCRIPTIONS

#1.J.linearifolia/
 1,2 2,12-40 3,3 4,2 5,1 6,1 7,2/3 8,1 10,2 11,V 12,1 13,1 15,0.5-1.8
 16,1 17,1 18,1 19,5/6/7 20,3.5-4.5 21,2 22,1 23,3 24,2.5-3.5

..
..

*END
```

Figure 8.4. Brief example of DELTA data file for *Jurinea*.

group of numbers represents a character (followed by a comma) and the states that go with it. For example, *3,3* means that character 3 (stem leaf distribution) has state 3 (leafy throughout). Similarly, *19,5/6/7* means that character 19 (color of external bracts) is variable and has states 5, 6, or 7 (various shades of pink through purple), and *20,3.5–4.5* means that the fruit (strictly, the achene) is 3.5 to 4.5 mm long.

Given that the data are written in this formal manner with character strings and numeric codes, programs are available in CONFOR and PANKEY (Pankhurst 1978b) for producing the equivalent as written text, for checking, and for publication. For more information, see the DELTA manual (Dallwitz 1980b). There is also a free newsletter for DELTA users that is distributed twice a year.

Attitudes to the Use of Computers in Systematics

Computer scientists need not be told that it is a good idea to apply computers to scientific data processing and will find it perfectly natural to work in this way. Systematists, on the other hand, exhibit a wide range of attitudes to the use of computers, running from those whose point of view could be expressed as *Over my dead body!* at one extreme to computer addicts at the other.

From the author's experience over many years of encouraging systematists to use computer methods, some consistent themes do emerge, which may be worth recording.

1. The advent of cheap desk-top personal computers with user friendly software has tempted many systematists to use computers who would not have had and do not have any intention of going anywhere near a mini or a mainframe. The author has distributed hundreds of copies of identification software over the years and has not had a request for anything other than MS-DOS and Macintosh PC versions for quite a long time. It looks as though many systematists are seduced into computing via word processing, by small and simple databases, and by cladistics programs, but many are failing as yet to see any further potential than that.

2. Many systematists react with horror at what they see as the *complexity* of taxonomic databases, of DELTA and of algorithms for systematic processes, such as the interactive key-constructing program. Computer scientists might find this kind of reaction surprising. The problem seems to be that systematists, although unconsciously using complex data relationships and algorithms, are unaccustomed and even at times unwilling to make them explicit. Can these complexities be hidden in suitable interfaces, or are systematists going to have to become better acquainted with their underlying thought processes?

3. Many systematists privately admit that data processing with numbers and text seems very dull and unappetizing, but are startled and intrigued as soon as they see a system that handles images of plants or animals, especially when these are in color. Computer scientists, on the other hand, probably take for granted the need to handle data, are accustomed to make the necessary effort in collecting it, and have long been well aware of the hardware capabilities of color graphics systems.

Discussion

Horvitz: First, to make a program that is easy to use, about 80 percent of the code will be interface, no matter how fancy the program is. Second, in the early 1960s people were doing Bayesian systems for medical diagnosis. These came to a head in the late 1960s. The word *expert systems* was coined later, but such things existed in the early 1960s.

Pankhurst: The medical diagnosis is a fascinating field, and I wonder why medical people do not seem to ever consider deterministic identification or diagnosis. Why do they insist on using probability in diagnosis? Medical data are uncertain, but so are botanical data.

Horvitz: All the symptoms are not observable that you need for medical diagnosis. Internal symptoms are hidden. It may be costly and painful to gather the evidence. Usually external signs only give you a probability; they do not pin down a disease. Several diseases will compete for the same set of external physical features.

Pankhurst: We have the same problem in biology.

Horvitz: An expert systems person would probably say that what you have is not an expert system, although it has expert knowledge in it, because you do not have a theory of graceful degradation. There is no notion of how you take an incomplete set of characters and use it to establish the likelihood that this flower belongs to a certain species. I am not sure how often in botanical taxonomy you have only a partial set of information, but I cannot imagine you not having probability theory in there.

9

The NEMISYS Solution to Problems in Nematode Identification

RENAUD FORTUNER

This chapter gives an example of an identification problem and its solution by modern computing techniques. This example, using NEMISYS (NEMatode Identification SYStem), illustrates and expands previous discussion on this subject (Pankhurst, this volume, chap. 8). NEMISYS is described from the computer science point of view by Diederich and Milton (this volume, chap. 10).

The Identification of Nematodes

There is a need to identify all plant-parasitic nematodes, not just a few common species, because even a rare species may be destroying cultures. This is also true for animal- and man-parasitic nematodes. Predatory nematodes and insect-parasitic nematodes must be identified for possible applications in biological control. Even the so-called free-living nematodes (mycophagous, saprophagous forms) that have no obvious economical effect must be identified because they are the most numerous animals on Earth, and any study of biological diversity that does not address the identity of nematodes would be incomplete.

Identifying nematodes, particularly the minute plant-parasitic species, is a difficult prospect, and few people have the required expertise. Even these few experts need identification aids, but the difficulty of the subject makes the design of such aids a formidable task. Some characteristics of nematode identification will now be reviewed to define the requirements for a successful identification aid.

Advances in Computer Methods for Systematic Biology: Artificial Intelligence, Databases, Computer Vision, ed. Renaud Fortuner (Baltimore: Johns Hopkins University Press, 1993).

Circumstances of Identification

The design of a system will depend on its projected use by nematologists identifying nematodes. From my experience, I have characterized several general situations for nematode identification.

Field Survey. Typically, a general field survey is conducted on plants or geographical areas where the nematode fauna is unknown. Field surveys are often conducted for the practical purpose of identifying existing nematode attacks on cultivated plants. Here, only plant-parasitic nematodes need to be identified, and often identification is made only for the species in genera that are dangerous parasites. More general surveys attempt to identify all species present, including forms that are not plant parasitic.

A priori selection of the most likely species is impossible if nothing is known of the circumstances (host/location) where the survey is conducted. Sometimes the surveyor has some idea on the species likely to be present (based on experience with similar plants/areas), but surprises can and do occur. Focusing the identification process on the most likely species can be used with prudence if a previous survey has been conducted under similar circumstances.

Consultation done by a systematist who identifies slides received from colleagues is a special case within this first category. Depending on the origin of the slide and the experience of the systematist, focusing on likely species may or may not be used.

Field Test Follow-Up. Nematological field tests are used for the study of chemical control, varietal resistance, population dynamics, etc. The field where a test is established must be checked for the species it harbors. After this initial survey, the nematode population is well known. Focusing on these species can offer valuable identification shortcuts for the following checks of the nematode populations.

Regulatory. Evaluation of plant samples sent to a regulatory agency presents special circumstances. Typically, the samples have been treated with nematicides, and they should be free of any nematodes. Specimens are sometimes observed because of unsuccessful treatment, and they must be identified. The samples come from many different countries and plants. Focusing on likely species is possible when the observer knows what nematodes are usually found on the particular host and origin of the sample. The regulatory agency may have a list of species under quarantine. If a plant-parasitic nematode shows up in the sample despite treatment, identification should make certain that it does not belong to any of the species in that list.

Nematode Identifiers

Across-the-Board Experts. There are several types of persons who identify nematodes. Across-the-board experts are able, or claim that they are able, to identify any nematode they observe. Such expertise may have existed fifty or a hundred years ago when only a few nematode species had been described, but nobody today can truthfully claim to belong in this category.

Experts with Limited Expertise. Most expert identifiers have limited expertise. They can identify species easily, rapidly, and accurately in certain groups of nematodes, for example, from some genera or some families. For example, experts in plant-parasitic nematodes can identify common species in the genera that are most damaging to cultivated plants. They can identify on sight maybe one-half to two-thirds of the genera from a total of one to two hundred genera and perhaps a hundred out of the 3700 nominal species in this category. Those experts are doing most of the total of plant-parasitic nematode identifications, but they are fast disappearing, as traditional systematists are not being replaced as rapidly as they are retiring, or they are being replaced by so-called molecular systematists. Molecular identification is a promising technique for selected species, but it is questionable whether molecular probes can be defined or used for thousands of species.

General Practitioners in Nematology. I call general practitioners in nematology people who study nematode control, nematode biology, population dynamics, etc. They have a broad knowledge of nematodes, including their gross morphology, and they can recognize the most important characters. Some work on only one or a few species, others (ecologists, for example) are interested in more species. Most do not do their identifications. They can recognize the few species they see frequently during their work, but they cannot identify a form they have never encountered before. They rely entirely on experts for any identification problem. With the rapid disappearance of the experts, general nematologists are forced to start doing identification.

Students. Students are a category apart. At the beginning they have no knowledge of nematodes, but as they work toward their Ph.D. they learn the state of the art in identification techniques. By the time they obtain their degree, most can be classified as expert identifiers. Yet, as they enter the professional field, they quickly become general practitioners in nematology. They cannot maintain and update their skills and soon are unable to identify.

Users of an Identification Aid. Nematode identifications traditionally are made by a small group of experts, knowledgeable in only a fraction of existing nematode species. These people do not really need identification aids but just a quick reminder of the diagnostic characters of particular taxa. For decades, they have been using dichotomous keys and notebooks with copies of published species descriptions.

As opposed to traditional identifiers, nonexperts are not interested in identification per se, as they are not taxonomists. Rather, their interest is only in finding a name for the forms they observe. They do not want to waste any more time than is absolutely necessary for an activity they regard as secondary, and they insist in getting an answer in a few seconds.

Traditional experts also want speed from an identification aid. Most of their identifications are done by instant recognition. When they must identify a form out of the area of their expertise using an identification aid, they want to reach the answer as quickly as they do on their own.

Speed is an important requirement for the identification process. This means that the system hardware and software response must be quick, but also that the identification procedures must not require the identifier to follow laborious and detailed operations. An important point is the amount of data required by the system to reach an answer. An expert accustomed to using half a dozen characters to identify species will resent having to enter twenty characters to do the same job. Casual users also want as few requirements as possible for data entry. This means using few characters and easy data entry. For example, while frequent users may be willing to learn codes for data entry, occasional users hate to waste time learning that *tail shape* is character *F* or that *conoid tail* is state *3*.

Data entry is easier with natural language than with codes, but even this can stymie occasional users. While experts may understand the difference between tail shape *more curved dorsally* and tail shape *dorsally convex-conoid*, general practitioners may be baffled by the unfamiliar terminology and thus be unable to make a choice. For them the easiest form of data entry will be accomplished by picking up the right shape out of a table of illustrations.

Nematode Domain

Biological Material. A sample typically includes the roots of a single plant, with the attached soil (about one liter) called the rhizosphere. Aboveground plant parts (either vegetative parts, leaves, stem, or reproductive parts, seeds) are collected separately where the presence of aboveground nematodes is suspected. Aboveground parasites are exceptionally found in the ground; ecto-parasitic species are found only in the rhizosphere, never in the root tissues; endo-parasitic nematodes typically found in the roots may also be present in the rhizosphere in large numbers at certain points of the life cycle, but be almost absent at other times. Often, a species is represented in a field sample by only a few specimens, occasionally by a single specimen.

The nematodes are extracted from the sample by an appropriate method.

Then, they are examined under the dissection microscope for the identification of broad categories (typically at the family to genus levels). Finally, examination with the compound microscope is necessary for identification at genus and species level. This last step can use freshly heat-killed specimens in temporary mounts, or specimens killed, fixed, and mounted in glycerine in permanent mounts. Mounted specimens can be distorted by the fixation process or old age. Identification is made typically on one to half a dozen specimens, and rarely on more than ten specimens. These few specimens are often in poor condition and will not allow observation of all the diagnostic characters.

There are few expert identifiers, because it is particularly difficult to identify nematodes. Identification characters are inconspicuous, ambiguous, variable, and often very difficult to record.

Conspicuity. Plant-parasitic nematodes are very small animals, typically 0.5 to 1.5 mm long and 20 to 30 μm in diameter. Their body envelopes are transparent, and the internal organs can be seen without dissection. Yet, most organs are seen only under the highest powers of optical microscopes, with an oil immersion objective, 1000X magnification, and phase microscopy. Even in the best conditions of observation, most organs are almost at the limit of the separating power of the optical microscope. Some external (cuticular) organs are seen only with a scanning electron microscope (SEM), under magnifications of thousands or tens of thousands.

The inconspicuousness of many morphological structures makes it very difficult for nonexperts to discover their characteristics. For example, in the genus *Pratylenchus* the first character used in all published keys is the number of annuli in the lip region, two, three, or four. Two annuli are easy to see, but three or four annuli crowded together to fit on the low lip region are inconspicuous. Experts know that when you can see the annuli, there are two; when you cannot see them, there are three or four (Loof in Fortuner 1989). Experts have to make good guesses for the character they are trying to capture, and there is a high risk of error attached to this guessing game, even higher for casual users.

Ambiguity. An ambiguous character is understood here in both senses of the expression as (1) doubtful or uncertain, especially from obscurity or indistinctness, and (2) capable of being understood in two or more possible senses.

Characters attached to an organ with low conspicuity will have high ambiguity. High ambiguity also occurs with highly conspicuous organs. For example, spicules are highly conspicuous male copulatory organs, often rose-thorn shaped. Spicule length can be measured along the dorsal edge, along the ventral edge, or along the longitudinal axis of this organ, each with very different results. In another example, found in the genus *Trilineellus*, the very conspicuous lateral fields are composed of two longitudinal ridges running along the lateral sides of the body. In lateral view, the edges of each ridge are seen as two-well marked lines. In most forms with two ridges the lateral fields are usually seen with four lines, but in *Trilineellus* the ridges touch over most (but not all) of their length. It is not clear, in spite of the name of this genus, whether it should be described with three or with four lines.

Variability. Besides the usual normal distribution of measurements found in most biological characteristics, there is a high variability caused by the environment. For example, the length of the body of some *Ditylenchus* species can vary by a factor of two with different host plants (many observations reviewed by Fortuner 1982). Also, when the progeny of a single parthenogenetic female of *Helicotylenchus dihystera* was raised under ten different host plants, all body measurements varied, and the variation was statistically highly significant (Fortuner and Quénéhervé 1980). Qualitative characters also vary, and the variation itself is variable. For example, the shape of the fusion of the lateral field lines is constant in *H. dihystera* (all specimens have a Y-shaped fusion), but it is highly variable in *H. pseudorobustus*, a closely related species (Fortuner et al. 1984). The type population of this species had 20 percent of specimens with a Y-shaped fusion and 80 percent with a U-shaped fusion. These percentages varied in other populations from 0 percent Y-shape and 100% U-shape to 100% Y-shape and 0% U-shape (Fortuner 1986). The character shape of fusion is diagnostic for *H. dihystera*, but it is useless for *H. pseudorobustus*, although these two species are so close that they can be differentiated best by multivariate statistics (Fortuner et al. 1984).

The extent of the variability is known only for a small fraction of described species. A sampling of the database NOMEN revealed that almost two-thirds (61 percent) of the species of plant-parasitic nematodes are described from their type population alone and that another 25 percent of the species have been redescribed from only one other population. Only 5.6 percent of the species in the database sample are known from more than four populations (including their type population). In the genus *Helicotylenchus*, type populations of half the species published since 1972 have been described from ten or fewer paratypes (Fortuner 1984). Many species are known from only a few specimens found in one or two populations. Because of the small sample sizes, characters are often described as constant, whereas well-described species often show extensive variation. Variability exists, even when its existence cannot be revealed by limited studies.

Easy Characters. Easy characters are characters that are easy to record. They are attached to conspicuous organs. They are not ambiguous, and they can be recorded with little risk of error; they are characters that do not vary or that have limited variation in the sense that a definite gap exists between the values in related species. There are very few easy characters in nematode morphology, particularly for identification to the species level. This means that difficult characters, that is, characters that do not meet all or some of these requirements, must be used for species identification and that a high risk of error exists in recording the character values, particularly by nonexpert identifiers.

Because of this, anyone attempting to identify an unfamiliar form will feel some anxiety when reaching an answer. Is this the correct answer, or was a wrong decision made that affected subsequent steps in the process that led to the wrong species? A good identification system must reduce the risks of reaching the wrong answer by a high built-in robustness. Then, one wrong answer will not jeopardize the entire identification session. An even better system will calm the fears of the user by offering several independent means to verify the correctness of the answer.

The list of easy characters varies from taxon to taxon, and a character that is easy in one species or genus may be difficult in a related taxon. A general system for identification of all the species in a category (e.g., all plant-parasitic nematodes or all species in the order Tylenchida) must include all the easy characters for all the forms in the group. In Tylenchida, the order containing 90 percent of the plant-parasitic nematodes, we found that more than 400 characters have been used so far in the literature. It is not yet clear how many of these characters happen to be easy for at least one species in the order. It is expected that their number will reach at least several dozen, if not hundreds of characters. This is at variance with the traditional identification aids for a particular genus that use at most one to two dozen characters.

As a result of inconspicuous organs, ambiguous characters, and large intraspecific variability, there is a high risk of missing data, when the identifier cannot see an organ or cannot decide what is the true state of a character, and of errors, when the identifier disregards these difficulties and tries to provide an answer at any cost. Of course, no system can give an accurate answer when fed totally erroneous data, but a good identification system must have what is called *graceful degradation.* There are several meanings attached to this expression, and here it will be taken to mean that the system should lose its ability to provide an accurate identification only gradually as the number of missing or erroneous characters increases.

Requirements

To summarize the requirements resulting from these considerations, an identification system should be:

1. a universal system, valid for all nematodes, or at least all species in a particular group. For example, a system could include all plant-parasitic nematode species, or a particular order;

2. a flexible system, that can be used by experts and generalists alike and that is consistent with the different ways each type of user works;
3. an easy to use system, and thus accessible to persons who are not experts in identification or not experts in computer use;
4. a fast system, requiring minimum data entry and delivering a speedy answer; and
5. a reliable system, taking into account the reliability of the data depending on the group observed.

In other words, the system must be quick, easy, and reliable. These requirements are more fundamental than the usual bells and whistles of user friendliness. An identification system that does not meet them all will be unable to offer reliable identification, and it will not be used, particularly by occasional users. As we will now see, traditional identification systems fail to meet at least one of these requirements.

Existing Aids to Identification

Dichotomous Keys

Dichotomous keys are deterministic in the sense that their authors assume that occurrences in nature (here, the identification of a particular species) are completely determined by antecedent causes (here, the presence of a particular state or value of a character). At each line of the key, the presence of a particular character state in the unknown specimen determines that the answer will be found in a particular section of the rest of the key. On the other hand, the probabilistic approach assumes that the observation of a particular character state or value only makes the identification of a particular species more or less probable. This statement may surprise some readers because morphological descriptions are typically given in a nonprobabilistic manner. The true nature of biological data needs to be examined in more detail.

Most quantitative characters, measurements (real numbers) and most counts (integers), are normally distributed, and they are probabilistic in nature. For example, if the length of an organ had mean M and standard deviation s in taxon T, 95 percent of the specimens in T have organ length between $M - 2s$ and $M + 2s$. In other words, each value within the range $M \pm 2s$ only has a certain probability to occur in a specimen of the species. Even if the entire range of values is counted as one state, an individual value has only 95 percent chance of belonging to this state.

Some quantitative characters (integers) are constant in a given taxon. Others can take more than one value in a given taxon, but these values are not normally distributed. Such quantitative characters behave like qualitative characters. In published descriptions, most qualitative characters are given as *character C with state S*. This type of statement seems to support the belief that taxonomic data are deterministic. Some will argue that deterministic statements are only an extreme of probabilistic statement with probability $P =$ 100 percent. State *S* always present might be read as state *S* has 100 percent chance of being present. However, most biologists will probably look at such arguments as hair splitting.

Still, even those biologists would agree that apparent determinism is often caused by insufficient data. For example, note the variability of the shape of lateral field line fusion as described above for *H. pseudorobustus*. This fusion is described as Y-shaped in *H. inifatis*, but this species was described from four specimens found in a single population (Fernandez et al. 1980). There is no assurance that other populations of this species (or even other specimens from its type population) will never show the U-shaped fusion.

Still unconvinced, those biologists would then argue that there exist true deterministic statements that are valid in every specimen of the taxon. All *Helicotylenchus* do have four lines in their lateral field. I prefer to say that deterministic statements are valid until specimens are found that contradict them. For example, the genus *Pratylenchus* has always been described, among other characters, with lip area low, flattened anteriorly and with oesophageal glands overlapping the intestine for a medium distance. Most *Pratylenchus* have tail short, cylindroid, with broadly rounded end. The species *P. morettoi* was placed into this genus because it has most of its other diagnostic characters, but *P. morettoi* has lip area dome shaped, oesophageal overlap elongate, and tail conoidal with a somewhat pointed end and a terminal projection (Luc et al. 1986). These characters were deterministic in the genus until the discovery of *P. morettoi* proved that they were probabilistic. Now, it can only be said that 99.999 percent of the specimens in *Pratylenchus* have low lip, medium-sized overlap, and short cylindroid tail. In this optic, no statement will ever be given as completely deterministic (i.e., with a 100 percent probability). We should say that all *Helicotylenchus* have 99.99999 percent chance to have four lines in their lateral field.

When intraspecific variability is given for qualitative characters, it is most often reported in a very imprecise manner. Each state observed in the taxon is said to be mostly present, sometimes present, rarely present, etc. The data in such cases are definitively probabilistic, and it is possible to attach probabilities to them. For example, the nematode taxonomists participating in the NEMISYS International Project were asked what percentages they would attach to such qualifiers. It was found that their answers are normally distributed. There were twenty-two respondents out of the group of fifty-eight expert identifiers in the project. From their answers, a mean percentage of, for example, 76.9 percent can be attached to the term *mostly* with standard deviation of 8.5, and range of individual answers from 60 to 90 percent.

Biological data are probabilistic, whether biologists accept that fact or not.

It is true that many descriptions do not describe the variability, but that does not mean it does not exist. I have shown elsewhere (Fortuner 1986, 1987) that builders of dichotomous keys have to force intraspecifically variable data into the straightjacket of these deterministic identification devices by pretending that intraspecific variability does not occur.

Experts have been using dichotomous keys for 200 years because their knowledge of variability allows them to take this phenomenon into account as they search for the right answer. Also, they often use keys as a quick reminder of diagnostic characters for the species in a well-known genus. When I suspect that I have the species *Pratylenchus penetrans*, for example, I can quickly go through the well-thumbed key of Loof (1978) and verify that the specimen does have all the right characters for this species.

At the opposite, nonexperts and experts out of their field of expertise do not know the extent of intraspecific variability in the group covered by the key. They are unable to supplement the intrinsic limitations of dichotomous keys, and in their attempts to use these devices for identification they either get stuck (no answer) or they run a great risk of reaching the wrong answer. Dichotomous keys offer absolutely no graceful degradation as all the characters in the key must be provided by the user (no missing data) and as a single wrong answer will send the user to the wrong end of the key.

Tabular Keys

Tabular keys offer a better approach to identification because they are polytomous (all characters are considered together instead of a single character at a time in dichotomous keys). This will allow for some errors, as the rest of the data will still point to the right answer. The tabular arrangement of the data makes it possible to note intraspecific variability. Finally, an identification can proceed in spite of missing data, as the identifier can rely on the characters that are not missing. However, graceful degradation in a tabular key depends entirely on the expertise of the user, who must decide how many mismatches and how many missing characters can be overlooked in the final identification. Intrinsically, tabular keys do not have graceful degradation. Printed tabular keys are useful for species identification in a small genus (with no more than two or three dozen species). They are too cumbersome for larger groups of species. The only way to use tabular keys for identification of the 3700 species of plant-parasitic nematodes would be to offer a set of tables, each with a few taxa, at the order, superfamily, family, genus, and species levels. An occasional user would quickly get lost in such a maze, and when an answer is reached, there would be no way to assure the user that he or she did not make a wrong choice somewhere.

Another disadvantage of tabular keys is that they force the user to gather data for far too many characters. A tabular key at the genus level includes at least a dozen characters. Experts can identify most species in a well-known genus by looking at less than half a dozen characters, typically three or four. They will resent having to do twice to four times the amount of work when using a tabular key.

Coefficient of Similarity

Computation of similarity is another polytomous approach. It offers the added advantage of considering all the taxa together, instead of one at a time in the sequential devices described above.

Typically, a coefficient of similarity scores each couplet of characters found in the unknown and in one species as similar (score 1) or dissimilar (score 0). The average score, with or without weighting, gives a measure of the morphological similarity between the unknown and the species. The same computation is made for all the species in the group considered, and a list of the most similar species is presented to the user.

There are many different ways to compute the coefficient of similarity (Sneath and Sokal 1973). One that fits all kinds of characters is the coefficient of general similarity of Gower (1971b). I have modified Gower's formula to account for intraspecific variability (Fortuner and Wong 1984; Fortuner 1986). The modified algorithm is used in a custom-made computer identification application, NEMAID. NEMAID uses different algorithms for qualitative and for quantitative characters.

With quantitative characters, the algorithm compares the mean value C_u of the character in the unknown to the mean value C_s of this character in a species. Also used in the algorithm are the values R and p. The value R is the difference between the highest and the lowest specific values for this character in the genus; p is a percentage of correction that depends on the intraspecific variation observed in the genus. The percentage p is proposed by an expert, but it can be modified by the user. It is used to compute a correction value P for the couplet unknown/species equal to:

$$P = \frac{C_u + C_s}{2} \cdot p. \tag{1}$$

The similarity S for quantitative characters is equal to:

$$S = 1 - \frac{|C_u - C_s| - P}{R - P}.$$

The difference $C_u - C_s$ is taken into account only when it is larger than the threshold value P. The similarity S is taken as equal to 1 when the difference $|C_u - C_s|$ is smaller than P; it starts decreasing as the difference becomes larger than P. It would be equal to 0 if the smallest and the largest species in the genus were compared ($|C_u - C_s| = R$).

Qualitative characters are represented by two or several character states. The percentages of specimens in a given species that exhibit each state of a qualitative character are recorded in the database. For example, the fusion of lateral field lines for *H. dihystera* is recorded as: state Y-shape: 100% percent; state U-shape: 0 percent. For the type population of *H. pseudorobustus* it is recorded as: Y-shape: 20 percent, U-shape: 80 percent. All described populations of a species are entered individually in this manner in the database. As explained above, actual percentages are rarely given, and the imprecise terms (*sometimes, often, rarely*, etc.) that are found in descriptions have to be translated into percentages.

When several populations have been described for one species, two values, M (the average or mid-range) and A (half-range), can be calculated for each character state. This computation uses K_1 and K_2, the smallest and the highest percentages observed for this state in the described populations of the species.

$$M = \frac{K_2 + K_1}{2} \qquad \text{and} \qquad A = \frac{K_2 - K_1}{2}.$$

The values M and A are computed and stored for all the species in the database. The percentage U of specimens that have the same character state is recorded for the unknown population, and the coefficient of similarity s for this stage is computed by:

$$s - 1 - (|U - M| - A).$$

The same process is followed for all the states of this character. If the character has n states, the final similarity is computed as:

$$S = \frac{\sum_{k=1}^{k=n} s}{n}.$$

Here again, the intraspecific variability of the character is taken into account as far as it has been recorded for each species. Species described from a single population have $M = A$. The computation of the similarity does not include the true variability, but only the published variability. This is unavoidable, and this situation will be improved only if authors describe additional populations of existing species.

The similarities S for both types of characters are given a weight w by the experts or by the user. The successive Sw values for all the characters are averaged for the coefficient of general similarity. Missing characters are neutralized.

Similarity provides true graceful degradation. When erroneous data are entered, the accuracy of the coefficients of similarity diminishes gradually, not suddenly. For example, during an identification using ten characters, if an error is made on one character, the coefficients of similarity still are 90 percent accurate. The species to which belongs the specimen will be listed as 90 percent similar to the specimen. Also, missing data are neutralized, and the coefficients are computed only from the characters that are known in both the specimen to be identified and the species to which it is compared.

This approach is safe in the sense that no taxon is ever eliminated. All the species are listed in order of decreasing similarity, and it is up to the user to examine them and to pick one for the identification. Obviously, this requires some expertise, and the nonexpert user may make a wrong choice, particularly if too few characters were used. To work well, the NEMAID algorithm requires entering many characters, typically a dozen or more. Hurried users may be tempted to enter only a few characters, and they may end with the wrong species.

Expert Systems

Rule-based expert systems have limitations from the computer science aspect, but from the user point of view their main drawback is the excessive amount of work they require. A traditional rule-based expert system looks much like a fanciful dichotomous key. The system asks for character after character, and it takes far too much time to gather and enter all these data. Experts like to take shortcuts, and this is difficult when the system is in charge.

Probabilistic Identification Systems

Adapted to biological identification, Bayes' rule gives the probability of having a taxon T if a particular character state C is present. This probability $P(T|C)$ is:

$$P(T|C) = P(C|T) \cdot P(T) \ / \ P(C) \ ,$$

or the probability of observing this character state given the taxon, $P(C|T)$, times $P(T)$, the prior probability of the taxon, divided by $P(C)$, the probability of observing the character state. When we have several taxa, $T_1, T_2. \ . \ . \ , T_k$, that are exclusive and exhaustive, the probability of C in the denominator can be written as a weighted sum of the conditional probabilities $P(C|T_i)$ where the weights are $P(T_i)$, and the formula becomes:

$$P(T_i|C) = P(C|T_i) \cdot P(T_i) \ / \ P(C|T_1) \cdot P(T_1) + \ . \ . \ . + P(C|T_k) \cdot P(T_k) \ .$$

The probability of observing a character state C given taxon T, $P(C|T_i)$ for taxon T_i, and $P(C|T_k)$ for taxon T_k, can be found in a data matrix, provided character C has been described for taxon T. As for NEMAID, it is necessary that, besides the straight data, the database include the frequency of observation of each character state.

The prior probability of each taxon, $P(T_i)$, $P(T_k)$, is the probability of observing the taxon T_i or T_k before any data are known to us. Prior probabilities ultimately rely on observations made on field populations of nematodes. These observations may have been published or not published. If no publication is available, experts familiar with local conditions may be interviewed for their estimate of probabilities of observing the various species in these conditions. Published observations may or may not include the following indices:

1. the absolute density (= abundance) of a species in one sample, or the number of specimens of this species per unit (in weight or volume) of soil or roots in this sample;
2. the absolute frequency (= constancy) of a species, or the percentage of samples where this species was observed (for example during a survey of the fields with a particular crop in a particular region); or
3. the prominence of a species, or its absolute density multiplied by the square root of its absolute frequency.

The prominence index is an absolute value, and it can be made relative by dividing it by the sum of the prominence indexes of the species present.

Bayes' rule is fundamental to identification strategies. All identification methods either are explicitly based on Bayes' rule, or they can be formulated as variations of this rule with different assumptions. For example, in dichotomous keys all species are presumed to have an equal chance of being present in the sample (all prior probabilities are equal), and characters are presumed to be either present or absent (probability of the evidence is either 1 or 0). For another example, in NEMAID, based on a coefficient of general similarity, all species also are presumed to have an equal chance of being present, but the NEMAID algorithm uses the actual probability of the characters, as found in species descriptions. The question remains whether these different assumptions are necessary or, using Bayes' rule in its entirety, including actual estimates of prior probabilities, is more practical and realistic.

A prior probability expresses the analyst's opinion of a population parameter before any new data are available (Iversen 1984). In the Bayes' formula above, the probability of observing a species, $P(T)$, is supposed to be valid for the whole universe. Actually, prior probabilities are always conditioned by some background information, and they are really just conditional probabilities (Horvitz, personal communication). It remains to be decided where to begin the process. The following three cases will be examined below: (1) proposing probabilities valid only in narrowly defined circumstances; (2) proposing a single probability for each species, valid for nematode identifications in all circumstances; and (3) taking all probabilities equal. I will discuss the practical feasibility of proposing probabilities and how realistic they are in an actual identification environment, in all three cases being considered.

Studies of nematode populations are always made under restricted circumstances. Samples are taken in a particular geographical region, on a particular host, and from a particular part of the plant (rhizosphere, roots, aboveground vegetative parts, and aboveground reproductive parts). For example, I know from experience that, in root samples taken from flooded rice fields of northern Senegal, there is a 98.8 percent chance of observing the species *Hirschmanniella oryzae*, a 1.2 percent chance of observing *H. spinicaudata*, and almost no chance of observing any of the 3700 other species of plant-parasitic nematodes (Fortuner and Merny 1974). Such a precise and intimate knowledge of nematode populations is possible only because circumstances have been very narrowly defined.

In southern Senegal, these probabilities become 20.5 percent for *H. oryzae* and 79.5 percent for *H. spinicaudata*. In the same area, samples of upland rice instead of flooded rice would not yield these two species of *Hirschmanniella*. Instead, there would be a 64.8 percent chance of observing *Pratylenchus brachyurus* and a 32.2 percent chance of observing *P. sefaensis* (Fortuner 1975). Thus, depending on circumstances (region, type of rice), the probability for observing *H. oryzae* in Senegal is 98.8 percent, 20.5 percent, or 0 percent. When I was based in Senegal and was studying rice nematodes, I was aware of these different probabilities, and my expectations of observing *H. oryzae* varied greatly depending on the type of sample. In any given set of circumstances there is only one realistic probability.

Proposing probabilities valid only in narrowly defined circumstances is possible and relatively easy when species populations under these circumstances have been studied. Hundred of articles have been published with this type of information. Probabilities of observing species may be extracted from these studies and stored in a database in which rows would be sets of circumstance and columns would be the 3700 species. However, there are four different plant parts that can be sampled, there are dozens of major crops and dozens more of wild plant associations, and there are hundreds of significantly different geographical regions. Consequently, there are tens of thousands of possible sets of circumstances. Only a small fraction of these possible sets of circumstances have been studied and probabilities are unknown in most cases.

Proposing a single prior probability for each species, valid for nematode identifications in all circumstances, would require estimating the probability of observing this species in each possible set of circumstances and the probability of observing each set. Because only a few hundreds of sets of circumstances have been studied, all the relevant knowledge that has been accumulated since research on plant-parasitic nematodes began would still be insufficient.

How realistic the resultant probability would be is also in question. Making very broad estimates and generalizations, I could propose a worldwide probability for a few species. For example, *H. oryzae* is widely distributed in paddy rice but is almost unknown on other plants. Let's suppose that the probability of observing *H. oryzae* in root samples from paddy rice is 0.9 worldwide, and let's pretend that this species has no other host at all. There are about 140,000 hectares of paddy rice in the world, out of 1360 million ha of cultivated land, out of 14,900 million ha of total land surface on our planet, and out of a planet surface of 50,000 million ha. Paddy rice represents only 1 percent of cultivated land, 0.0009 percent of total land mass, 0.00028 percent of the total planetary surface. The prior probability for *H. oryzae* would be 0.9 times 0.01, or 0.009, for sampling on cultivated areas only; it would be 0.9 times 0.0009, or 0.0008, for sampling anywhere on land; and it would be 0.9 times 0.00028, or 0.00025, for the whole Earth, including oceans. All these numbers are far smaller than the probability of 0.9 that would be more appropriate if we knew that sampling was made from paddy rice roots, and than the probability of 0.98 if the sampled field happened to be in northern Senegal. These small numbers would look unrealistic to nematologists.

Taking all probabilities equal would reflect the fact that we have very limited prior knowledge for all species of nematodes on all possible sets of circumstances. For the 3700 species of plant-parasitic nematodes it is easy to calculate that all prior probabilities would be equal to 1/3700 = 0.00027. This number is almost identical to the prior probability for *H. oryzae* for the whole Earth (0.00025), it is not too far from the prior probability estimated above for sampling anywhere on land (0.0008), but it is far too low for the more realistic assumption of sampling limited to agricultural lands (0.009). Also, presuming that all species have an equal chance of being observed eliminates from the general Bayes' rule the information contained in the relative distribution of each species. The only probability that remains is the probability of observing each character in the various taxa, but this corresponds to the assumption for similarity systems such as NEMAID. Taking all probabilities equal simplifies the general Bayes' rule to a point where it is undistinguishable from some other identification strategies.

We are left in a quandary. Estimating prior probabilities valid for the whole planet or taking all prior probabilities equal are perfectly valid operations, but they yield numbers that are far removed from the everyday experience of nematode identification, which is usually performed with at least some knowledge of the origin of the specimens. If this knowledge is taken into consideration, the probability of observing a species may change from less than one-thousandth to almost one.

Realistic prior probabilities can be proposed that are valid for specified circumstances of observation only. Yet nematode populations have been studied only in a relatively small fraction of all possible sets of circumstances. In particular, nematode populations are unknown in most uncultivated areas. This strongly limits the value of calculating Bayesian prior probabilities in an identification system for the study of biological diversity because uncultivated areas usually harbor a richer fauna than cultivated lands (Fortuner and Couturier 1983). Conditional probabilities also cannot be established and used for a nematology laboratory newly established in an area where no sampling has ever been done, for example, in a developing country (Doucet 1989). In such cases, the identifier has very little prior knowledge, and only the simplified strategies may be used.

Bayesian probabilities offer graceful degradation. When erroneous data are entered, the accuracy of the probabilities diminishes only gradually.

NEMISYS

In 1987, the NEMISYS International Project (NIP) was launched, in collaboration with two computer scientists from the University of California, Davis, Jim Diederich and Jack Milton. They were looking for a suitable domain for the practical application of their ideas on expert workstations, and recommended implementation of NEMISYS using object-oriented methodology (see chap. 7 for a discussion of expert workstations in this context and chap. 10 for a presentation of the computer science aspects of NEMISYS). Over the next couple of years more than seventy participants became formally associated with NIP. These participants include nematode taxonomists, computer scientists, and others interested in nematode identification.

In the rest of this chapter I show the NEMISYS approach to satisfying the requirements presented above for a complete nematode identification system. I discuss here only features that are special to NEMISYS. Obviously this system also includes the features found in many modern systems such as help functions, windows, and images.

The prototype system completed to date, NEMISYS 1.0, incorporates the basic functionality of the system. A few important tools have been implemented, the Basic ID tool for typical identifications, the Show Me tool to gather information about the different taxa, and the Promorph Tool, a graphical tool for focusing identifications. The attached database is very limited, with information for a few plant-parasitic nematodes at the genus level only, primarily because: (1) rapid prototyping of tools can be done on a limited set of data, and (2) the building of the NEMISYS database, or Nembase, will be a major undertaking. We are developing computer-based tools to assist with building Nembase.

General Identification

Typically, identification aids are prepared by one author for a small group of species, often the species in one genus. This allows complete control of the data. The author is often able to redescribe all the species considered, making sure there are no missing characters and that the characters are described consistently from one description to the next.

A general system is needed for all the species of plant-parasitic nematodes. We have seen that about 3700 species have been described in this category alone, and it would be impractical, or even impossible, to redescribe them all. Besides the enormous amount of time and money this would require, many species have been found from localities that are no longer accessible. For example, the type locality of *Hirschmanniella belli*, a species common on rice in California, is now the parking lot of a shopping mall! As discussed above, the descriptions of several populations for each species are needed to give a realistic account of the intraspecific variability. This would require an even greater effort.

It is imperative that a general system obtain its data from the literature, where about 8000 descriptions and redescriptions of plant-parasitic nematode species can be found. Extracting data from published descriptions encounters other difficulties, such as descriptions written in foreign languages, the sheer number of characters used by the authors (more than 400 in Tylenchida alone), the size of the database needed for storing these characters, the presence of inconsistencies in the data, etc. A tool under construction, the Terminator, will help reduce this task to a manageable one. The process first requires scanning published descriptions with optical character recognition (OCR) software. The tool is in the testing stage, and data entry from published sources can begin in earnest when funds become available.

The data from the literature must be processed in the sense that redundant characters must be eliminated, fuzzy and qualitative characters must be linked (if possible) to exact measurements, etc. Work has started on rules for doing this translation semiautomatically. This will come handy in later years when the time comes to incorporate image analysis and automatic data capture into NEMISYS. It will become the first step for translating traditional morphological characters into computable characters (i.e., characters that can be automatically extracted by current computer techniques).

Flexibility

The requirement for flexibility of the system for accommodating different users working under different circumstances is met by the concept of a set of tools in an expert workstation (Diederich and Milton, this volume chap. 7). The different tools will be used depending on the user's expertise, on what the user wants to do, from a quick routine verification to the exact identification of a quarantined species, and on the point reached in the identification process, as different tools may be used at different times during a session.

NEMISYS tools help experts do what they do traditionally, but faster and more comprehensively. They help nonexperts approximate the work of an expert.

So far, we have listed about fifty basic functions that will be available through a dozen tools, and more will be defined as the NEMISYS project develops. This may seem bewildering to a prospective user, but in the final version of NEMISYS only a few tools will initially be active, only a few functions will be accessible, and only default settings will be available. This will still be enough for reaching a fast and reliable answer, but not enough for the user to get lost in the complexity of the system. Then, as (and if) the user gains confidence, he or she can start more tools, access more functions, and perhaps change some settings, until the frequent user will enjoy the full functionality of the system.

Even with the limited facilities available at first, the user will control the process, including the ability to request or reject the help offered by the system. A fortiori, the fluent user having access to the complete set of tools and functions will enjoy complete control. A few examples will give a better idea of NEMISYS flexibility.

Data Entry. Data entry is one aspect of NEMISYS where flexibility is at its highest point. The Basic ID window includes a pane where the user can enter a description of the characters seen in the specimens. The user has full and unrestricted access to the English terms found in nematode descriptions. No codes are necessary, and the user enters, for example, *tail rounded* without having to remember tail shape is, for example, character number 246 and that rounded shape is character state number 5, or any such code. The tool reads the entries, compares them to a dictionary of morphological terms used in nematology, and tries to identify the characters and states meant by the user. The user can accept or reject the interpretation made by the system.

The only restrictions when using this mode of data entry come from the limitations of the method, when the language of the user veers too far from normal usage. There are no other restrictions such as fixed field length or order of character entry. The user can enter any character at will or enter the characters suggested by the system. Data entry can be made one or a few characters at a time, allowing the user to assess the situation reflected by each additional entry before looking for more characters if necessary. This limits data entry to the minimum number of characters needed for the identification.

The Basic ID tool allows another data entry method, using a hierarchical arrangement of systems, organs, organ parts, characters, and character states. The user goes down this hierarchy to the desired character state and highlights and accepts it. It is expected that some users will prefer this direct path to the right entry and that others will prefer the freedom of describing the data in their way and having the system attempt to reconcile their entry with the character names in the database schema. Choosing one method versus the other is largely a matter of personal preference, and NEMISYS leaves this choice to each user.

Graphical data entry is being investigated as an alternative to textual entry. Limited graphical entry is already available in version 1.0, where some difficult characters such as tail shape are shown on screen. The user can then compare the shape in the specimens to be identified to the basic shapes illustrated and pick the closest shape. Later this will be offered for all characters, at the discretion of the user.

Another option that may become available in later versions of the system is similar to police composite drawings of crime suspects. The witness composes an image of the suspect's face by choosing each character separately. Similarly, future NEMISYS users could build a composite image of the nematode to be identified by picking the right shapes for basic organs out of a graphical bank of possible shapes.

We are also planning to connect NEMISYS with a commercial data capture software. When available, this option will allow a user to see on-screen the image of the specimen, captured by a video camera installed on the microscope. Measurements and observations will be made on-screen and immediately and automatically made available to the system. The final version of NEMISYS will be provided with an automatic data entry facility through full use of image analysis.

Errors. Our approach to the problem of data reliability offers a second example of system flexibility. A difference between experts and nonexperts in identification is that experts can assess the reliability of their data while nonexperts cannot. This limits the use of many systems that accept at face value all data entered.

Users of NEMISYS can ask the system to assess the reliability of their data. We have characterized the factors that influence reliability, including the nature of the character observed, the circumstances of the observation, and the observer expertise. The factors associated with the characters include conspicuity, ambiguity, and variability of the character (see above). The factors associated with the observation include the biological material (how many specimens were observed; how well preserved they were) and the optics (type of microscope used; magnification). The intuition of the observer, what we call personal intuitive feeling or PIF, is also considered with two components: (1) how clearly the observer saw the character state, and (2) how consistent this observation was in all specimens observed. The expertise of the observer comes into play, as the PIF of an expert is given more weight than the PIF of a beginner. All these factors are used in an algorithm that computes an endorsement percentage, from 0 to 100 percent, giving a measure of the reliability of each piece of data.

This endorsement algorithm can be used in several ways. For example, with a self-professed expert the PIF will be given the greatest weight. The PIF allows the expert observer to tell the system that some data entered may not be correct, while he may be very confident about other data. This imitates the way an expert usually operates, by putting more confidence in some observations than in others. Obviously, the expertise of the observer may be more wishful than real, but the basic philosophy of NEMISYS is to respect the freedom of its users.

For a nonexpert, the algorithm will rely mostly on the other factors, and it will provide an automatic evaluation of each piece of data. This automatic evaluation emulates evaluation by a human expert. Of course, experts and nonexperts alike can turn off the endorsement option and ask the system to accept each observation at face value.

An example of the use of the endorsement algorithm can be found in the computation of the NEMAID similarity coefficient. The NEMAID algorithm includes weights that are preset by experts or modified by the user. In NEMISYS, the weights of the NEMAID algorithm will be the endorsement scores of each piece of data.

The consequences of errors can be reduced by other options in NEMISYS tools, as explained below.

Identification Methods

The flexibility of NEMISYS is most obvious in the variety of identification tools it offers depending on who the users are and what they want to do. A routine check of a well-known species by an expert is different from the verification of the identity of a regulated pest whose verified presence will cause the destruction of 25,000 pothos or 125,000 strawberries, or from the slow plodding of a beginner unable to recognize even the most basic forms. Each tool is using one or several of the identification methods described above, as most appropriate for the tool functionality. During an identification session, the user will be able to select one or more tools, and one or more options within each tools, and use them at will, sequentially or simultaneously, following a hand-crafted identification strategy. Below, the various identification methods used in NEMISYS are presented in the context of some of the tools that use them.

Recognition. It has been said (Pankhurst, this volume, chap. 8) that the best way to identify a specimen is to know what it is already. Obviously, there is no need for an identification aid when the observer can fully identify a species. Still, even the best experts may want to verify their initial identification, or they may have recognized a level higher than the species level and still need to go to species level.

This is where most existing methods fail, because there is no mechanism for the user to provide this type of information. In NEMISYS, the Ask Me tool allows the user to enter the name of a species or a group of species, and the system responds with a list of characters to be checked for full identification. This is a goal-directed procedure, somewhat similar to the backward-chaining procedure of an expert system that also allows verification of a tentative identification.

NEMISYS goes beyond this process in that it allows the user to use shortcuts in the data-directed procedure. Often the user can narrow down the possible identification to a limited area at the family or genus level. In other systems the user still has to plod slowly through many steps to enter what was obvious at first glance. For example, if you see a fish you do not want to go through a long list of questions, Does it have vertebrae? Does it have scales? Does it have gills?, etc., before the system reaches the obvious conclusion that it is a fish.

In NEMISYS, we call *promorph* (*pro* before; *morph* morphology) a form that can be recognized at low magnification powers, before observation of detailed morphology. For example, the promorph *fish* would include all true fishes but also dolphins, whales, and other cetaceans.

When a promorph is recognized, the user enters its name into the system, without having to provide its description. Nonexperts unable to recognize promorphs can ignore the shortcut provided by this option, or they can look at drawings of common promorphs in the hope of recognizing one of them.

Promorph is a concept; it is not a classification. Promorphs are not hierarchically arranged, they are neither exclusive (the same species may belong to two promorphs) nor exhaustive (some species may not belong to any defined promorph), and their effective use depends on the level of expertise of the observer. That is, an expert can recognize many more promorphs than a beginner.

Promorphs are identification concepts. They do not attempt to account for the phylogeny. Promorphs depend on immediate recognition, not on careful consideration of phylogenetic characters. At first glance a whale IS a fish, as any Nantucket whaler would have told you (see chap. 31 of *Moby Dick*).

In NEMISYS, promorphs are used for focusing the identification process toward the area where it is most likely that the right answer will be found. If I see a fish, I know it can be a true fish or a cetacean, but it cannot be a sea urchin.

Deterministic Method. It has been argued above that most morphological data are probabilistic. Yet, the deterministic approach is the fastest way to eliminate taxa that obviously do not fit the unknown specimens.

NEMISYS eliminates unsuitable taxa by relying only on what I call primary identification criteria. A primary identification character is a morphological characteristic that is both useful for the differentiation of a species or a group of species and easy to observe in this group of species. It has high conspicuity, low ambiguity, and low variability. A primary character is one that even a beginner can identify without great risk of error.

In NEMISYS, the elimination process relies on the ad hoc concept of *nest of species* that are groups of species sharing the same set of primary identification characters. As for promorphs, nests are not units of a classification. They are heuristic concepts that rely entirely on clearly visible phenetic characters, and they do not follow established phylogenetic classifications.

A primary character C for the nest N_1 may not be primary for another nest, N_2. When an unknown specimen is compared to N_1 and to N_2 using this character C, the elimination process is different for N_1 and N_2. During the comparison with N_1 (C is primary for N_1), two things can happen:

1. the specimen belongs to N_1, and it will have the correct state; N_1 is selected as it should be;
2. the specimen does not belong to N_1; C may still have the correct state for N_1, in which case N_1 is kept by error on the list of possible candidates; or the specimen may differ from N_1 in the value of character C, and N_1 is rejected as it should be.

During the comparison with N_2 no action is taken, and N_2 remains in the list of possible nests whether the specimen has the correct state or not for character C. If the specimen does not belong to N_2 after all, the nest is kept on the list of possible candidates, and it will be eliminated later, using other characters that are primary for N_2.

Using nests there is a risk of keeping the wrong answer, at least temporarily. On the other hand, the risk of rejecting the right answer is very slight because elimination of taxa relies on the best, most reliable characters for each nest considered.

A prudent user may reduce the risk of wrongful elimination even further by allowing one or two mismatches on primary characters before eliminating a nest. This option provides some measure of graceful degradation. It is found in the Basic ID tool, which is already available with version 1.0. This tool is described in detail by Diederich and Milton (this volume, chap. 10).

Rules. There is no rule-based expert system in NEMISYS. However, little rule-based modules will be used in several ways and in various tools in the system. For example, after an answer is reached, rules will allow a quick verification of the accuracy of the answer by pointing to the most obvious mistakes. For each species, rules will make sure that all the diagnostic characters are confirmed and in particular that the characters that differ between that species and related species have the correct value or states. During the building of NEMISYS, the rules will be extracted primarily from the species diagnoses (the list of differentiating characters proposed by the authors of each new species). The experts participating in the NEMISYS International Project will be asked to check these rules and, if necessary, to provide additional rules for selected species.

When a dead end is reached in the identification process because all taxa have been eliminated and none fit the data entered, other rules will instruct the system on the best way to get back to the last reliable part of the process, assess the situation, and suggest a different path. When every attempt has failed, more rules will be triggered for deciding that a form represents a new taxon, never described before.

Similarity. At times, the user may want to rank the remaining candidates. This can be done by calculating the coefficient of similarity. NEMISYS uses the NEMAID algorithms described above. A variation of the algorithm has been implemented in version 1.0 for comparing nests. The extended version of the algorithm requires access to data at the level below the level being investigated. Species data are needed for estimating NEMAID similarity at the genus level; population data are needed for estimating similarity at the species level. The extended algorithm will be available when the species and population database Nembase is completed.

As explained above, the similarity approach has built-in graceful degradation. In addition, the NEMAID algorithm used in NEMISYS takes into account the intraspecific variability, thereby reducing the risk of error.

Probabilistic Approach. It is possible to propose realistic Bayesian probabilities only in some well-defined circumstances of region, host, and plant part being sampled. Still, most identifications are made in circumstances where the nematode populations are known from published studies. This is due to the fact that nematode populations are known under the circumstances that are the most commonly studied (important crops, areas located near a nematology laboratory, etc.). Many future identifications will also be made under the same circumstances. Consequently, an identification strategy explicitly using prior probabilities is quite justified.

We are considering incorporating Bayesian probabilities in a future version of NEMISYS. First, a specialized database will have to be created for storing probabilities from published descriptions of nematode populations found in given circumstances. Experts can provide additional information and offer rules that would extend our knowledge of populations in known circumstances to circumstances that have never been studied but are somewhat similar. For example, if the probability for observing *H. oryzae* in northern Senegal is 0.988 and if it is 0.205 in southern Senegal, a probability of 0.59 could be proposed for central Senegal. When circumstances are too far removed from all published experience to allow for extrapolations, all nematode species might be considered to be of equal likelihood. The user of the Bayesian probability function would specify the circumstances of the current identification session. The corresponding probabilities would be extracted from the specialized database, and they would be used according to Bayes' rule.

Some biologists will say that, if they cannot reach a firm answer in an identification, for example, because some diagnostic characters cannot be observed in the specimen, they do not want any answer at all. This has not been my experience. I recently had to identify a *Pratylenchus* sp. from California. Using the dichotomous key of Loof (1978) as a reminder of diagnostic characters for the species in this genus, I found that the form was somewhat intermediate between *P. thornei*, a cosmopolitan species known from California, and *P. delattrei*, a species described from Madagascar thirty years ago and never reported since. My final identification was "*Pratylenchus* sp., probably *P. thornei*," because I implicitly took into account the probability of observing this species in California versus the probability of observing *P. delattrei*.

Using prior probabilities does not allow the swift elimination of the obviously wrong choices, which is a characteristic of deterministic methods. On the other hand, once deterministic elimination using safe primary characters has reached its limits, Bayesian probabilities offer a new strategy for pursuing the identification to its conclusion at the species level. Other strategies, such as deterministic with unsafe characters or estimation of overall similarity, can also be used to reach the species level. If all approaches point to the same species, the user confidence in the answer would be greatly increased, provided the various methods are really independent, of course.

NEMISYS users will have the choice of tools, each tool being based on one or the other identification methods, or a combination of these methods. They probably will use several tools in a single identification session. The identification process may be quickly focused toward a small group of species using a tool that allows proposing a promorph (recognition method). Then the Basic ID tool can eliminate some nests (deterministic method). This elimination process can be made less stringent by allowing one or two mismatches before eliminating a nest. A browsing tool can show the user why a nest was eliminated. Then the remaining nests can be assessed by calculating a NEMAID coefficient of similarity with the unknown specimens. Once an answer is reached, rules can help with checking for possible errors. What tools are used, and in what order, depend on circumstances. An expert can use a goal-directed strategy, through the Ask Me tool, while a beginner may prefer step-by-step guidance by the system that will be in full control in a Help Me tool (to be added to future versions). Each tool relies on the identification method or methods that were deemed to be the most appropriate for achieving the tool functionality. The users have entire control on tool selection, and by selecting tools they also have access to several identification methods (whether they know which methods are used in each tool or not). This is a unique characteristic of NEMISYS.

Most of the other identification systems are based on a single identification method, using probabilities, similarity, deterministic elimination, or some other method. Each system has or does not have graceful degradation depending on the method selected. In the contrary, NEMISYS tools rely on several identification methods. Each method can be said to have graceful degradation or not, but the system as a whole goes beyond this concept. The rich environment provided by NEMISYS allows many options for recovering from errors. A worse-case scenario might be using the Basic ID tool set at zero tolerance (no mismatch allowed). Then the nest to which belongs the unknown might be eliminated if the user makes one mistake on a primary character for that nest. However, this is unlikely to happen because primary characters are easy to record. If it does happen, the right nest is eliminated, and the user may eventually be left with an empty set of nests. He can still switch to the coefficient of similarity option in the Basic ID tool. This would provide the intrinsic graceful degradation of similarity coefficients. Or the system may arrive at the wrong species, but then rules would point to the closely related species, which would probably include the right answer. The user could also go to a browser and look at descriptions and illustrations of possible taxa. These examples show that the concept of graceful degradation is actually irrelevant for expert workstations such as NEMISYS.

Conclusions

A review of the requirements for a comprehensive identification system listed above shows that NEMISYS possesses:

1. universality: the system is valid for all plant-parasitic nematode species because of the reliance on published data; it will later be extended to other categories of nematodes and then to other biological groups;
2. flexibility: the system can be used by experts and generalists alike, and it fits with the different ways each type of user works; flexibility is most visible in the concept of set of tools, in the different options available for data entry, and in the different identification strategies;
3. ease of use: the system is accessible to persons who are not experts in identification or not expert in computer use; the concept of expert workstations and the building of NEMISYS as a fully integrated interactive system promotes ease of use;
4. speed: this is realized mostly through shortcuts and the use of promorphs and host and geographical origin;
5. reliability: the system estimates the reliability of each piece of data entered and uses it accordingly.

NEMISYS helps experts do what they have been doing before, but better, faster, and more comprehensively, and it also allows nonexperts to do what experts do. This philosophy will guide the future development of the system in which we will add more tools and more functions to attain the full functionality described in this chapter.

Discussion

Calabrese: I share your uneasiness about prior probabilities, but I wonder if you could not let the system accumulate them over time, from the results of the successive identifications, or find them from your nematode museum by looking at, say, slides from California.

Fortuner: In either case, I would still be finding priors that are valid only under certain conditions. Slides from California obviously would give priors for California only, and as for recording my identifications, I am not identifying species from all possible plants and locations. It is easy to find prior probabilities for species in certain conditions, as they are regularly published in the nematological literature. What I cannot do is give a prior valid in any circumstances, because the knowledge just does not exist. A universal system will have to deal with this fact.

Marcus: Your priors are conditional probabilities depending on region and other exogenous characters. I think you are talking about the dependency on the priors. I would be open minded and let the people get the priors out of you by interview techniques.

Fortuner: Any prior probability depends on exogenous circumstances. At the very least, you must assume that you are on Earth! You can define the universe to be the whole Earth, or to be paddy rice roots in northern Senegal—it does not matter. I agree that in theory each species has a prior valid for the whole Earth, but our knowledge is so limited for most species that it would just be impossible to offer even the wildest estimate. No interview technique can extract a knowledge that does not exist.

Jeffrey: I strongly agree, be empirical, see if it works, but it would be a methodological mistake to assume that these priors must be somehow buried in your subconscious and then go digging for them.

10

NEMISYS: A Computer Perspective

JIM DIEDERICH AND JACK MILTON

This chapter gives a more detailed discussion of NEMISYS, as a good example of one kind of expert workstation (EWS), discussed in Chapter 7 (this volume). We wish to look at some domain considerations, some resulting system requirements, and some successes and problems during the development of the current NEMISYS prototype. Finally, we will illustrate how some of the specific tools work to solve the domain problems.

The Identification Process and NEMISYS

A Standard Approach

The identification process can be defined as the sequential process of data entry, followed by selection of a taxon matching these data, followed by confirmation. That is, in nematode identification we begin by entering data about specimens under the microscope, we see if we can find appropriate taxa that match these observations, and then we make certain of our conclusions through some sort of verification. Obviously this process is not purely sequential, as we might engage in subloops as we proceed to the final identification. For example, verification would almost certainly involve entering more data to serve as a check on what has previously been entered. At a high conceptual level, however, these three activities seem to characterize the process. Thus it certainly makes sense to build a system that is based on this paradigm alone. In NEMISYS, however, we want to add much additional functionality, focusing and browsing, for example, for a variety of reasons.

Advances in Computer Methods for Systematic Biology: Artificial Intelligence, Databases, Computer Vision, ed. Renaud Fortuner (Baltimore: Johns Hopkins University Press).

A Nonlinear Approach

Figure 10.1 shows a few of the higher level activities a user might perform at different stages of an identification. The basic identification process is indicated by the vertical arrows down the center of the figure. Again, in an actual identification session those three activities would typically be mixed, beginning perhaps with data entry, then making some selections, perhaps some more data entry, etc., prior to any verification. In NEMISYS examples of additional functionality are found in the FOCUSING and BROWSING functions indicated in Figure 10.1. All five represented activities might then be done in many different orders, depending on such parameters as the reason for the identification, the knowledge of the user, etc. For example, a user may have decided that "I strongly suspect what this species is, so I may only need quick confirmation," perhaps during a routine survey of a field experiment station with known nematode populations. Such a user might take the route at the left of the figure to go directly to the verification facility. This user will not use any of the other illustrated activities. Another user might begin with the attitude, "I do not want to look at all possible taxa; rather, I want to focus on the most likely candidates." This user might enter some data, use focusing tools, and then move back to the data entry, selection, verification triad. After a significant amount of focusing, the data entry task might be reduced considerably. Somebody else might say, "I really do not have a lot of sense of what I am doing, so I would like to look at possible taxa or characters." This user would go directly to a browsing tool, as illustrated with the arrow to the far right. From there, the user might do some focusing, some data entry, a

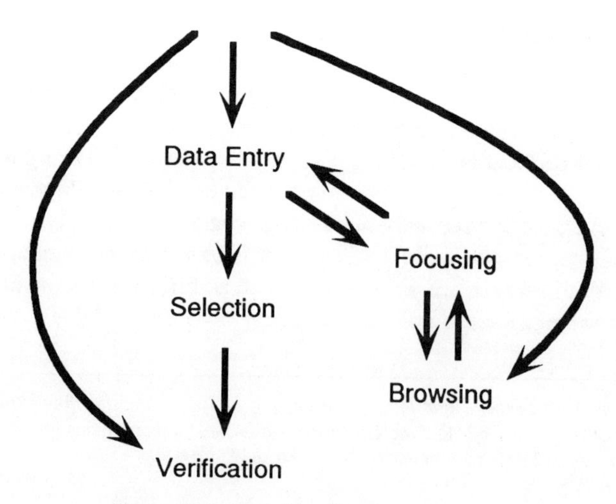

Figure 10.1. Basic identification activities.

combination of the two, etc. Other situation parameters might dictate other orders, and we have not included all possible arrows. We leave the choice to the user, which fits well the concept of EWS. We would not presume to know how different users might go about the task of identification. We would even expect experts with similar backgrounds to identify differently depending on factors that cannot be accounted for in any system. Unless the activity can be very well defined and limited, it may be extremely difficult to accommodate a range of users with an expert system. The strength of an EWS is that it allows this kind of flexibility.

NEMISYS Tools

The Basic ID Tool

One of the primary tools of NEMISYS illustrates the type of tool we use in an EWS. Figure 10.2 shows the Basic ID tool. This is the same tool as Figure 7.1 in Chapter 7, but it shows a more advanced stage of an identification session, and we wish to discuss it in more detail. Each of the lists in the top five panes is an instance of a Selection-In-List. Selection of a system in pane 1 prompts a listing in pane 2 of all organs in that system. Selection of one of the organs then prompts a listing of all features in that organ in pane 3, and so on, until a list of possible values or states is displayed in pane 5. Pane 6 is the main action pane of the tool. This is where the user can enter data from the keyboard based on observations that have been made at the microscope. This entry is free-form and there are no language restrictions, as long as the terminology is not too unlike that used in nematode descriptions. When the user selects the button marked 7 (by moving the cursor over that button and clicking the mouse) the system interprets the data entered by comparing it to the content of the character database. If possible correspondences are found, the system lists those characters for confirmation in pane 8. Each such character is chosen according to heuristics and a scoring mechanism, but the user can request a list of alternatives in pane 13 and replace the system-chosen character with one of them (or simply fail to accept the character in question). Alternatively, the user could select a character in the panes across the top, panes 1 through 5, showing system, organ, feature, character, and state, and enter the character in that way.

Eventually we plan to offer different views (e.g., by body region) of the nematode in these panes, but since a given user will not necessarily be familiar with any of the decompositions, we anticipate that the free-form data entry in pane 6 will be the primary means. When the system has trouble finding a given character, it might be possible for a user with some experience with the system to enter it via the five top panes. Also, there will be other tools that will assist the user to find such a character.

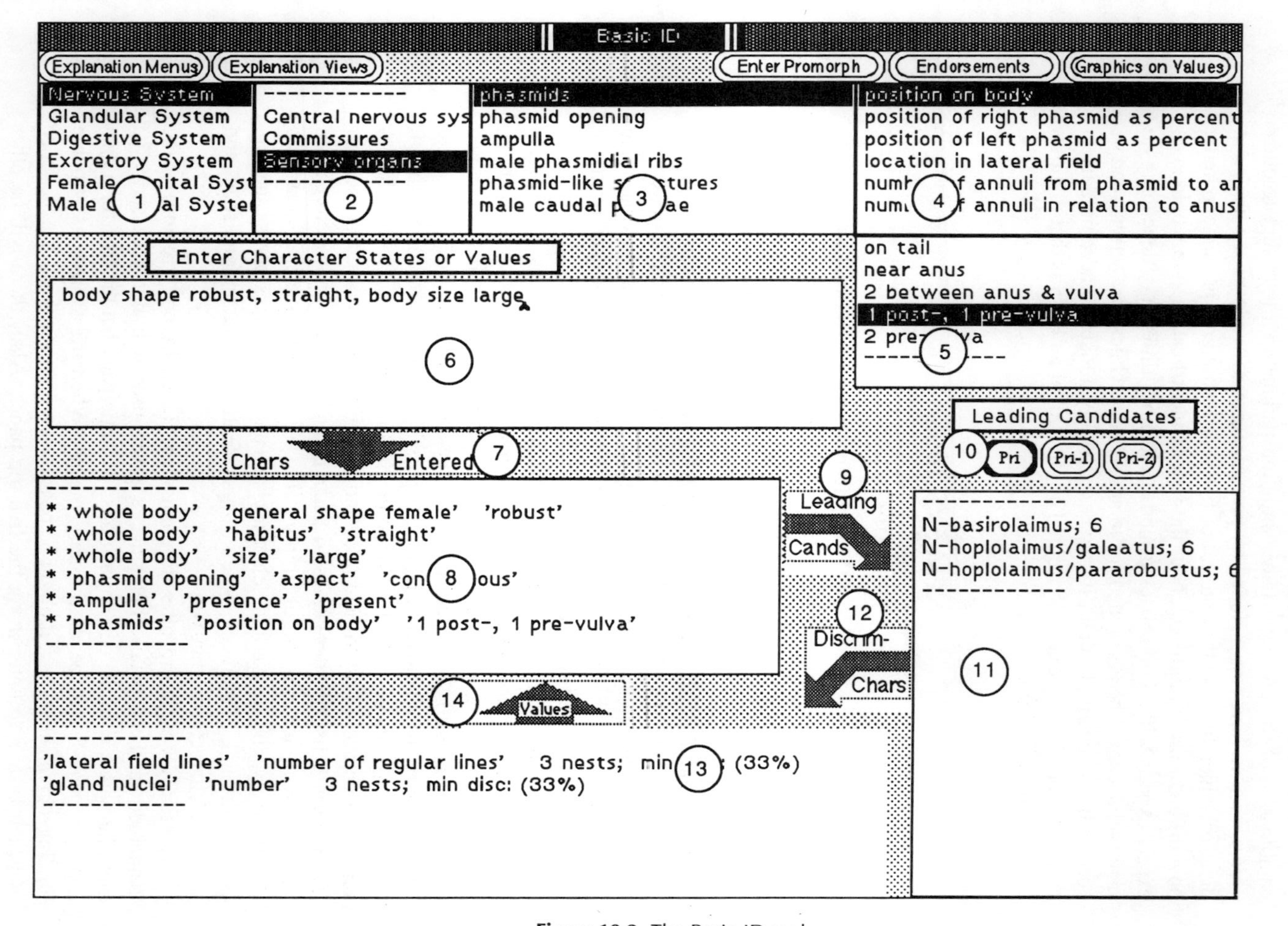

Figure 10.2. The Basic ID tool.

After entering a few characters, and having them accepted by the system, the user can select the *Leading Candidates* button, 9. This causes a list of the most likely candidates to appear in pane 11. If the user cannot tell what characters best discriminate between the displayed candidates, the system can provide some assistance. Pushing button 12 causes a list of the most discriminating characters to appear in pane 13, and the user can then attempt to enter states for any of them via free-form or through the top 5 panes. For example, button 14 can be pushed, which will cause a pop-up menu to be displayed, from which the user can choose the correct state or value. Alternatively, a menu item on pane 13, *get path*, will cause the path of the selected character to appear across the top, and the appropriate state or value can be entered from there.

As system builders add functionality to a tool and fine tune it for different uses, it can easily get fairly complex. It seems that a point of diminishing returns can be quite easily reached in design and implementation of a specific tool. Added features increase functionality, but too much functionality makes it more and more difficult to use a tool and remember how to use it, as well as to fashion a reasonable interface for it. This is one important reason why we insist in our definitions of basic elements of EWS that a tool need not do the entire job, but something meaningful along the way to job completion.

We feel that this Basic ID tool is near that point of diminishing returns in terms of complexity. For example, pane 13 can be used for three different things. It can be used to display either alternatives to the system-selected characters, discriminating characters, or the results of a *find*. To use *find*, you select or enter a word or phrase and ask the system to find all character paths in its database that contain that word or phrase. Thus at any time, several possible things might appear in this pane. You should *spawn* a window if you are getting ready to wipe out the contents of a pane and you wish to refer to those contents later. Spawning a window is a system-supplied facility that allows you to bring up a new plain window, in which the contents of the current window automatically appear. While facilities such as spawning windows ease the user's burden in dealing with complexity, the Basic ID tool has now reached the point that it has become difficult for the system builders to manage the complexity when even small changes are made in the tool design as discussed below.

After going through the Basic ID loop a few times a user might need to examine some collateral information to proceed with the identification. This might be a good time to bring up a browsing tool, for example, and it can be brought up off of different panes, focusing on the contents of the chosen pane. So, you could save lists you might want to refer to later, you could bring up new tools such as browsing tools, and all of these windows could exist on the screen at the same time. There is no conceptual limit to the number of windows open at once, but there is a limit in terms of the power of the machine, and there are practical limits in terms of the amount of clutter you can deal with on the screen at a given time.

Finally, while there is much functionality of the Basic ID tool we have not discussed, we note that one menu item allows you to duplicate the window in any given state. This would be helpful if you were torn between two choices, and you wished to explore both in parallel. The current window could continue, and the new window could involve the other choice, which gives a very rudimentary form of what might be called *versions* of the same identification. This capability also allows you to drive the identification forward and backward by undoing some steps and moving forward with other input.

A Graphical Tool

Let us examine another tool that is quite different. Figure 10.3 shows what we call the Promorph Tool. Promorphs have been defined (Fortuner, this volume, chap. 9) as forms that are fairly easy to recognize. The left column is what we call level-1 promorphs. When one is selected, for example, dorylaims and tylenchs (Fig. 10.3), a sequence of images appear to the right, with all of the level-2 promorphs associated with the selected level-1 promorph. You can select any number of the level-1 promorphs you like, or any number of level-2 promorphs for a selected level-1 promorph. Selected promorphs can be deselected, and the final collection is used by the system to focus the identification. That is, the subsequent identification will concentrate on the nests contained in the selected promorphs. This could eventually be overridden for the purpose of verification, for example, or a user could later decide to ignore this focusing entirely, but the tool provides a relatively quick means of narrowing the nest candidates.

Subtools

Rather than illustrate more tools, we mention a few subtools. A subtool is an important element of a tool, but is something that does only a part of a task. Several subtools may be pieced together to form a tool. For example, we have implemented graphical menus in addition to the typical textual menus. One use of a graphical menu would be to present a collection of choices, such as all possible tail shapes, and allow the user to make the selection by choosing the shape closest to what is observed, rather than insisting on attaching a name to the particular state. We also have implemented what we call pixel menus, that is, menus that are activated at different parts of a pane containing an image, depending on which part of the image is closest to the cursor at the time the mouse button is pushed. We are in the process of building a combination graphical/textual browser, in which the user can switch back and forth

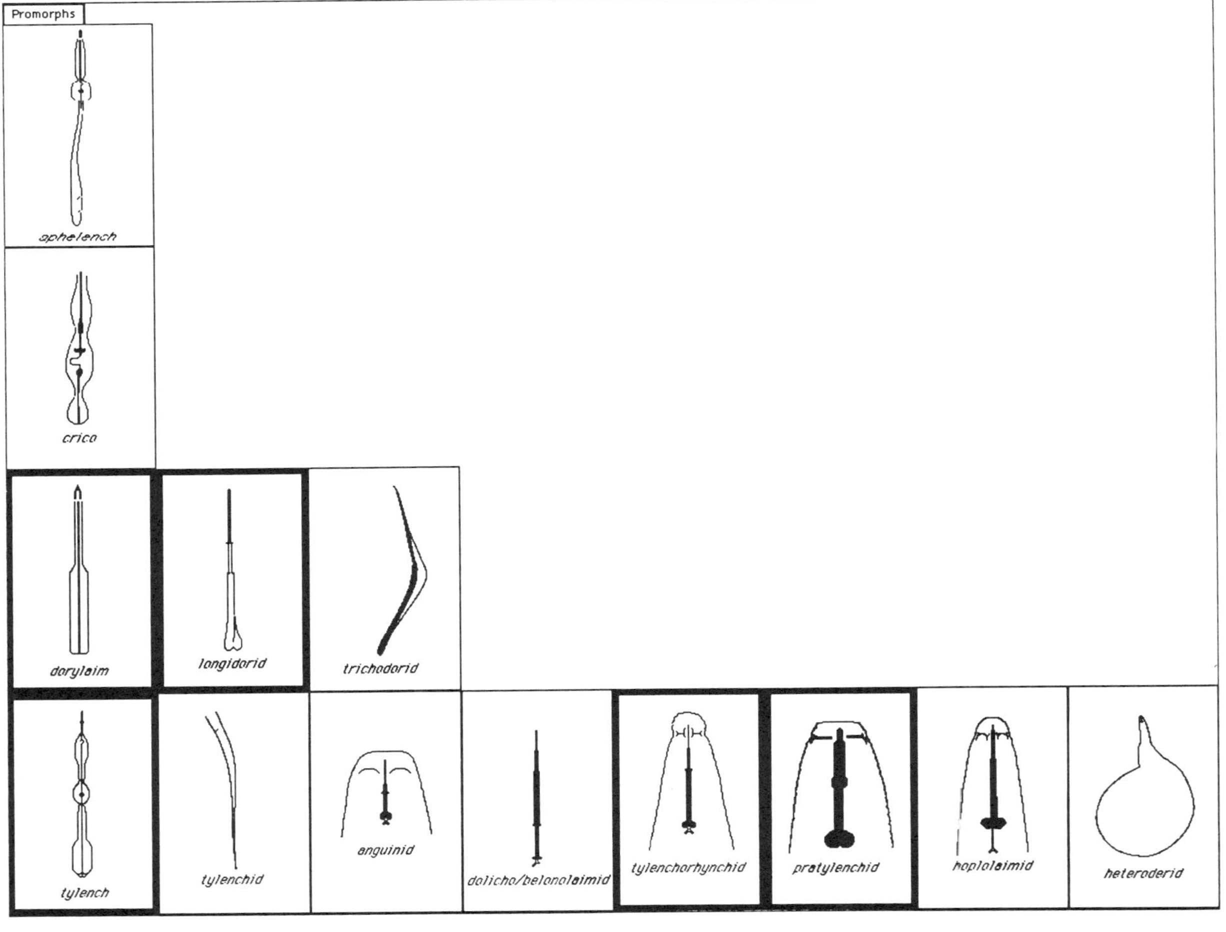

Figure 10.3. The Promorph tool.

between graphical and textual descriptions, depending on which are most familiar at any given point.

User Interaction in NEMISYS

In Chapter 7 we indicated that we have all of the five main types of user interaction in NEMISYS, which are used as appropriate in a given context. The most difficult of the types of interaction to implement is, of course, that of natural language (Schneidermann 1987). Our use of natural language is given by the interaction in pane 6 in the Basic ID tool. As indicated above, while the user will not use completely free-form English, the language used in nematode descriptions is a restricted form of natural language. The nematologist can enter whatever she or he wants at the keyboard and expect a reasonable interpretation by the system, as long as it does not stray too far from this language of nematode descriptions. If the system cannot interpret an entry, no character will be proposed, and the user can revise the entry and retry, perhaps after doing some exploration in the database.

Icons are used for direct manipulation. Pop-up menus are ubiquitous in Smalltalk, and it is rare to find a pane on which some sort of menu does not appear. Form fill-in is not used to a very large extent, except for such things as entering a profile of the specimen(s) or of the user. That information is then used by the system automatically. Finally, while beginners would undoubtedly not wish to make much use of command language, we have workspaces that contain typical actions that might be taken, and any user can bring up such a workspace and execute the commands.

One major advantage of being able to use all five types of interaction is that the interaction can be suited to the task. In addition, being able to accomplish certain things in several ways both increases the power of the user and makes it easier for the casual user to remember how to use the system in productive ways. We have found that our development language, Smalltalk-80, is a rich developmental environment that has greatly facilitated the construction of the interactive aspects of the system. Fortuner (this volume, chap. 9) discussed criteria he felt were important for NEMISYS. These criteria were then translated into a collection of developmental principles, and we wish to mention a couple of them.

First, from the early days of the work we felt that it was very important to keep interface considerations at the forefront and have these considerations drive the underlying architecture. This is not the typical way of developing systems, but after our experience to date we feel that this approach has some distinct advantages for the kind of system we envision, and it is quite promising for general system development if a set of tools can be identified for a given domain. Early expert systems basically took over the process and directed it. Many users find this very annoying, as they do not see how some of the questions asked by the system are connected. In such a system it makes much more sense to build the architecture first and then develop the interface. In an EWS, however, the interface plays a much more prominent role, and we feel it quite appropriate, if not mandatory, that interface considerations drive the architecture. We might view an older expert system as saying "Do as I say" to the user, through its style of interaction. To the contrary we see an EWS saying, "Do what you want, when you want!" From comments in other chapters and from the discussions it is clear that the biologist wants to be in control in these systems. With an EWS, you do what you want; we are not going to control you.

Second, while rapid prototyping is a well-known aid to conquering the complexity of large software systems, we believe it has played an important role in conquering interdisciplinary difficulties in collaboration. We could implement a module, perhaps along with a visual interface, for a new concept and discuss this concept at the computer in the context of how it would be used in the system. Although this always led to a better understanding of the concept, it sometimes exposed basic misconceptions about the system, the underlying database, or a given paradigm that were due to misunderstandings of an unfamiliar discipline and were unknown up to that point.

Utilizing Object Orientation

The Terminator

To better examine the nature of object-oriented systems and tool building in that paradigm, we consider a tool we are building that was conceived as completely separate from NEMISYS and identification. Because it was built in an object-oriented system, we were easily able to incorporate it as a key subtool in NEMISYS when the need arose. An extensive bibliography of object-oriented concepts and systems can be found in the March 1991 issue of *SIGMOD Record* (Vossen 1991), and the *Communications of the ACM* of September 1990 is devoted to object orientation (CACM 1990).

This tool is the *Terminator*, a name that derives from the tool function of finding target terms or expressions in published nematode descriptions, extracting data attached to these terms, and storing these data in a database. The process involves scanning descriptions of plant-parasitic nematodes in the literature, transferring these textual descriptions (and associated graphics, if desired) to the Terminator, and using it to help place different items of information from the descriptions in a database.

It became clear early in our project that we would need some automated assistance in building our character database, as there are estimated to be on the order of 10,000 published descriptions of plant-parasitic nematodes, and in our system we use over 400 morphological characters to represent the data contained in these descriptions. Given this large character set, reading the printed description and placing characters and appropriate values in the database would require a good deal of time. Without assistance for finding where in the database you are going to place a particular piece of data, data collection could be an overwhelming task. In Figure 10.4 we see the interface for the Terminator. In the large pane at the bottom is a scanned description of a nematode species, *Tylenchorhynchus gladiolatus*. The Terminator reads a portion of the text, usually a sentence or less, and makes suggestions as to what character this sentence refers and where in the character set to place it. In Figure 10.4, the Terminator is considering the text, "The body is ventrally curved in the shape of a widely open C," and the operator has highlighted *widely open C*. In the pane where the items are listed under SUGGESTED, the Terminator has given possibilities for placement of this information in the database. The operator scans these quickly and sees that *whole body, habitus* is the correct character and that the value *widely open C* is related to some of the suggested values such as *weak C*. At this point, the operator has selected with a click of the mouse *widely open C*, the value found in the text, as a synonym of an existing character state, as shown in the pane marked ACCEPTED above the description.

There are a number of bells and whistles in the Terminator that allow the operator to enter remarks, to search for alternate characters, to determine if something has been used as a synonym before, to indicate that the data are possibly invalid (i.e., the author of the description may be mistaken), etc., but our intention here is not a detailed description of the Terminator. This tool is currently in prototype stage, so it has not been extensively tested. In the tests to date, however, the correct character is usually in the top few suggested by the Terminator, which confirms that typically the characters are fairly easy and quick to find. Thus most of the operations carried out by the operator involve only picking, clicking, and highlighting, which is similar to spell-checking.

On the architectural side we note that the Terminator is driven by the character set of the domain, using its words as clues to the correct terms. Consequently, the user does not have to create a large lexicon of terms, the system does not do any parsing, and so a lot of work that would be required for natural language processing is avoided. Instead, we have used keyword selection in combination with some heuristics that we developed.

Turning a Tool into a Subtool: Leverage of Object Orientation

Our original intent in NEMISYS was for the user merely to pick and click to enter observations made during an identification session. But a practical problem arises regarding how the user is going to pick and click among 400 characters, especially when the important characters vary from taxon to taxon

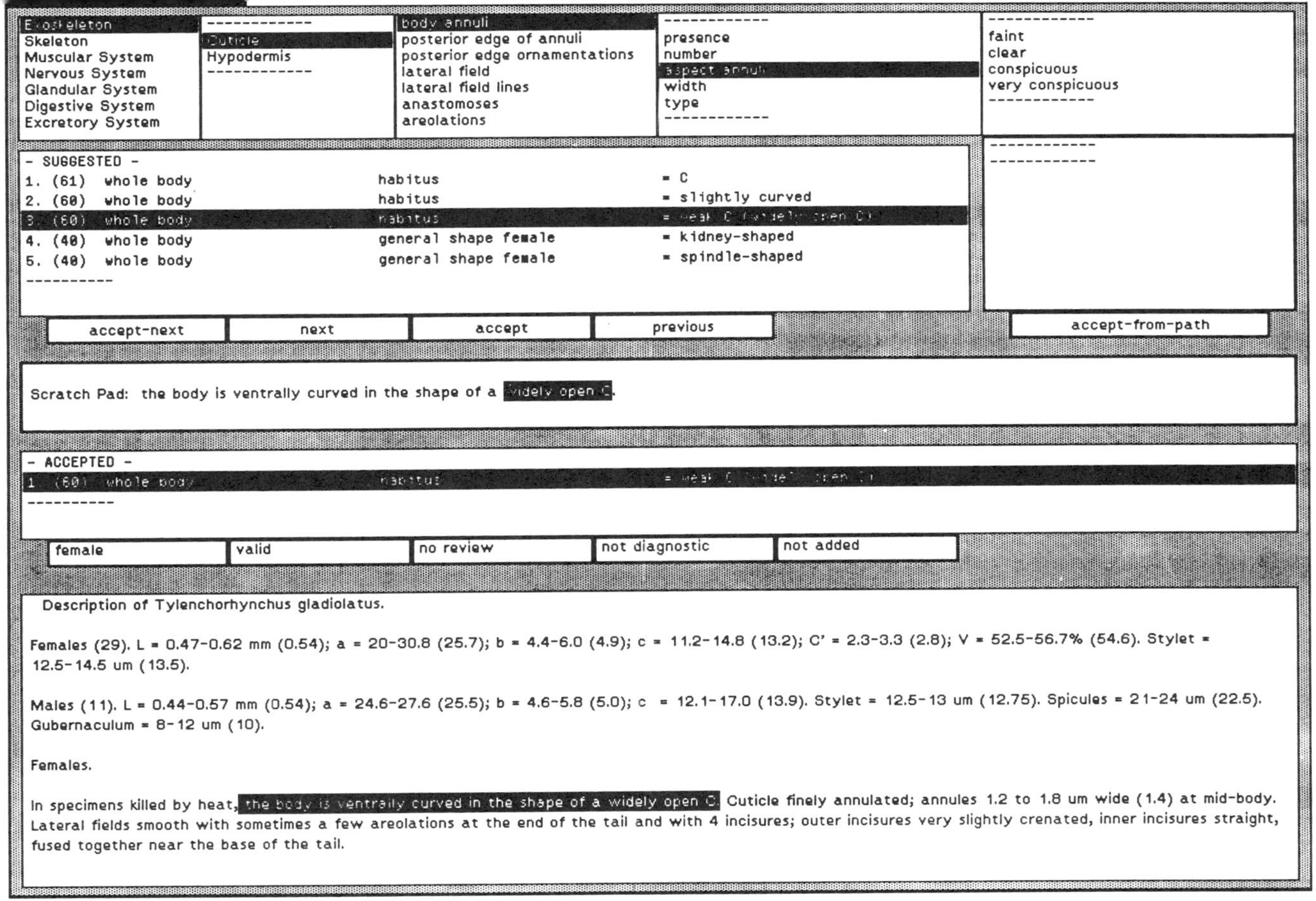

Exoskeleton
Skeleton
Muscular System
Nervous System
Glandular System
Digestive System
Excretory System

Cuticle
Hypodermis
----

body annuli
posterior edge of annuli
posterior edge ornamentations
lateral field
lateral field lines
anastomoses
areolations

----
presence
number
aspect annuli
width
type

----
faint
clear
conspicuous
very conspicuous

- SUGGESTED -

1. (61) whole body habitus = C
2. (60) whole body habitus = slightly curved
3. (60) whole body habitus = weak C (widely open C)
4. (40) whole body general shape female = kidney-shaped
5. (40) whole body general shape female = spindle-shaped

----

accept-next | next | accept | previous | accept-from-path

Scratch Pad: the body is ventrally curved in the shape of a widely open C.

- ACCEPTED -

1 (60) whole body habitus = weak C (widely open C)

----

female | valid | no review | not diagnostic | not added

Description of Tylenchorhynchus gladiolatus.

Females (29). L = 0.47-0.62 mm (0.54); a = 20-30.8 (25.7); b = 4.4-6.0 (4.9); c = 11.2-14.8 (13.2); C' = 2.3-3.3 (2.8); V = 52.5-56.7% (54.6). Stylet = 12.5-14.5 um (13.5).

Males (11). L = 0.44-0.57 mm (0.54); a = 24.6-27.6 (25.5); b = 4.6-5.8 (5.0); c = 12.1-17.0 (13.9). Stylet = 12.5-13 um (12.75). Spicules = 21-24 um (22.5). Gubernaculum = 8-12 um (10).

Females.

In specimens killed by heat, the body is ventrally curved in the shape of a widely open C. Cuticle finely annulated; annules 1.2 to 1.8 um wide (1.4) at mid-body. Lateral fields smooth with sometimes a few areolations at the end of the tail and with 4 incisures; outer incisures very slightly crenated, inner incisures straight, fused together near the base of the tail.

Figure 10.4. The Terminator interface.

and our decomposition of the nematode will be relatively unknown to new users of the system. At the same time the domain expert expressed the desire to have a variation of a natural language interface at which the user could simply enter data at the keyboard using familiar terminology, as discussed previously in conjunction with Figure 10.2. The answer was to take the Terminator and embed it in the tools in NEMISYS as a subtool.

Because the Terminator was constructed in an object-oriented system, it was possible to carry it over, making a small number of modifications and additions. This was done by creating a subclass in Smalltalk, which we called the EmbeddedTerminator, that had most of the behavior of the Terminator and inherited that behavior from the Terminator. This subclass was easily modified to handle things in a special way when this subtool was embedded in the Basic ID tool in NEMISYS. The amount of code that had to be modified and added to create the EmbeddedTerminator was minimal due to the inheritance and to the use of polymorphism in object-oriented systems. That is, different kinds of objects, such as Terminators and EmbeddedTerminators, can receive exactly the same commands, usually called messages in the object-oriented world, and respond to them differently.

As an aside to illustrate polymorphism, in a standard language without polymorphism one might write a function called *area_of_circle(C)* and have another called *area_of_rectangle(R)*, which calculate the area of circles, *C*, and rectangles, *R*, respectively. In code that deals with circles, if we wish to deal with rectangles instead, we would have to replace all function calls to *area_of_circle(C)* with function calls to *area_of_rectangle(R)*, perhaps throughout the code. Note that solving this problem is not just a matter of having a good library; it is a matter of having to change the code wherever that code is used. But in object-oriented systems we could just use the generic message *area* and send it to any object. Each object will then respond based on its class (type) as appropriate, so anywhere we use the message *area* we do not have to change the code if we use rectangles rather than circles.

Standard computer languages are generally used by decomposing the functions of an application rather than focusing on the objects in an application. The functional decomposition in one application is usually different from that of even closely related applications, so it can be a problem to take the decomposed functions to another application, though libraries can be helpful. On the other hand, object-oriented systems decompose according to the objects found in an application or area, and these objects and object definitions can be used and reused, usually with far less difficulty than with libraries of functions.

An anecdote further illustrates the benefits of object orientation in this context. One of us kept working on and revising the Terminator, while the other worked on the Basic ID tool. Concerns that subsequent work on the Terminator might clobber the work on the EmbeddedTerminator, since the latter had its own special behavior, were proved unfounded. The Embedded-Terminator was not affected by the changes in the Terminator where they shared common behavior. It was quite easy to keep these separated. In the few instances where there was some impact, the changes were minor. The main point here is that object orientation provides a solid foundation for tool development and subtool integration within an EWS.

Data Requirements in NEMISYS

Turning to the database side, in chapter 7 we mentioned that biological data types have not been adequately supported in commercial DBMSs. We pointed out that there was no support for the kinds of metadata we encountered, and we mentioned that general and specific states were not supported. We want to mention a few other observations regarding character data that requires support in a DBMS.

One example is a summary character. A summary character is one that summarizes a number of other characters and states. A summary character is a high-level abstract characterization and is shorthand for other character values. For example, the statement that "the type of the stylet is hoplolaimid" implies that a number of other characters have certain states: *stylet size* is medium to long, *stylet aspect* is robust, *cone size* is the same as shaft size, *cone shape* is conoid, *knobs* are true knobs, and *knob size* is medium to large.

Another kind of inter-character-state-based relationship is what we call a dependent character, that is, a character that is applicable only if another character has a particular state. A simple example of dependent characters is the set of all the characters attached to a morphological structure that are applicable only if this structure is present in a specimen. Other cases are more complicated. For example, the character *shape of body behind neck* is applicable only if the whole body type is nonvermiform. Summary and dependent characters affect many aspects of data use. Clearly, queries will be affected by the presence of such relationships in the schema. Also, the quality of an evaluation during an identification depends on whether character states are independent. There are other relationships that affect queries and evaluation as well, for example, what we call *fuzzy characters* (i.e., characters that express a subjective appreciation of a feature). We have identified four kinds of fuzzy characters. One we call *standard fuzzy*, where a qualitative character corresponds to some measurement, as in, "The stylet is very short, short, average, long, or very long." Another we call *subject fuzzy*, where there is no underlying measurement to which the fuzzy values can be mapped. Naturally, these fuzzy values are subject to ambiguities arising from the point of reference, such as *short* for the genus or *short* relative to all known species. A third kind we call *comparative*, in which one character is compared with another as being smaller, about the same, or larger. And finally, as we have seen, some values are a little fuzzy and only *approximate*, as in "about a dozen."

Conclusion

In sum, we have built the NEMISYS prototype, and we are testing the concepts to see how far we can push them. We have found that object orientation has helped a great deal with prototyping and that prototyping is critical in development of an EWS. Smalltalk has facilitated rapid prototyping very well. Trying to design a system like this and build it from scratch would be extremely difficult, and we had the added disadvantage of needing to learn the biology at the same time. Our methodology allows us to exercise the system, learn some of the biology in parallel, and be able to reformulate the system without having to completely rewrite it as we learn. It is safe to say that we simply would not have attempted to build the system in Pascal, C, or other standard languages. Undoubtedly we would have pursued other avenues of research, possibly on entirely different projects. With Smalltalk, however, we took on what has turned out to be a massive project, and we are delighted with the results to date.

This is not to say that the object-oriented approach solves all the problems. For example, the logic of the system and interface together can become quite complex and hard to manage. Not only must the interaction be understood, but it has to be coordinated. For example, pane 13 in the Basic ID tool, with its three different uses, has been getting increasingly difficult to manage. This is part of the reason we think that this tool is beginning to reach the point of diminishing returns in complexity. Perhaps this point can be extended a bit if we can develop some powerful metatools to use in our system development. So, while Smalltalk and object orientation have been vital to our system development, there are miles to go in interactive system development, and we believe that building tools to build other tools is one of the more important areas of research in EWS and interactive systems.

At this point in the project two possible directions to explore are most evident. One is to examine in detail the prototype and see what we have learned about developing an EWS. From there we would probably try to generalize the EWS for use in other domains. The other would be to examine the database needs for identification, perhaps for other activities, and move in the direction of building a Bio-DBMS, a system layered on top of an existing DBMS, that supports many of the data types and relationships found in biological domains.

Acknowledgments

This research was partially supported by NSF grant IRI 88-07475.

Discussion

Dunn: How long did it take you to get up to speed using Smalltalk?

Milton: Smalltalk is a programming environment, as opposed to just a language, and both the language and environment are complex. It thus takes some time to learn, but the payoff was well worth it to us, the system builders, and may be worth it to others. On the other hand, in an EWS (here built in Smalltalk) we try to minimize what has to be learned by a new user before the system can be used productively.

Diederich: There are more complex software management systems for managing large commercial projects, but as a pure programming environment Smalltalk may be the best. We use the Smalltalk model for our notion of EWS. The programmer can do some programming, start debugging, start modifying code, go back to debugging, modifying, etc., and do all of this from "under" the program (that is, suspending execution, making changes, and continuing). Compared to standard computer environment where you do these things sequentially, and you have to get in and out of editor and compilers and so forth, it is a wonderful paradigm. But it is extremely rich, extremely complex. In experimenting with EWS, we have tried to create a rich set of tools without overwhelming the user. We have come up with some concepts like the Paradigm Tool Set, which is a limited set of tools for all users, but probably the maximum set for beginning or occasional users. The person who is using the system more regularly might want additional tools used in more sophisticated ways. To conquer the complexity of the EWS we also developed the Visual Plan (e.g., the set of arrows in Figure 10.2). According to the characterization of expertise into five levels (Dreyfus and Dreyfus 1986), the first three levels use plans. At the first level, the beginner is given a plan, with explicit instructions. At the second level more experienced users can make slight modifications, and at the third level the user has some responsibility and can formulate a plan. We thought of making this plan visual for the users to see. Their level of expertise then will dictate to what degree they vary from that plan.

11

Object-centered Representation and Fish Identification in Antarctica

NICOLE GAUTIER, ALAIN PAVÉ, AND
FRANÇOIS RECHENMANN

Introduction

There is increasing concern about the deleterious effects of perturbations in the marine environment upon sustained marine harvests in Antarctica. To monitor the situation, fish populations in the Antarctic Ocean are assessed by regular research cruises (Kock et al. 1985) with multispecies sampling in well-delimited areas such as the Kerguelen Islands shelf. Identification of the many species of fish collected offers various levels of difficulty. Some species are recognizable at first glance whereas others are extremely difficult to differentiate. Our research attempts to provide a computational aid to nonexpert identifiers of species. The identification approach presented in this chapter utilizes artificial intelligence (AI) and expert systems.

In specimen identification, information almost always consists of qualitative (i.e., symbolic) morphological characters, even though it may include some quantitative observations. As a consequence, algorithmic computation is not suitable. Our approach to identification is based on traditional classification systems and on characteristics that are easy to observe and collect in the field.

For identification, an unknown *object* is compared to known ones in a reference network. The unknown object is said to be closest to the known object with which it has the fewest differences. If there are no differences at all, the two objects are said to be identical. Structural models based on an object-knowledge representation and hierarchical relations between these objects are widely available. For our application, we chose an object-centered

Advances in Computer Methods for Systematic Biology: Artificial Intelligence, Databases, Computer Vision, ed. Renaud Fortuner (Baltimore: Johns Hopkins University Press, 1993).

representation rather than object-oriented systems. While object orientation mixes knowledge and control and uses a procedural programming language, object-centered representation stores knowledge separately from its exploitation and it uses a declarative language (Nebel 1985).

Our identification system is based on SHIRKA, an object-centered knowledge base management system that was developed for the selection of mathematical models in biology (Pavé and Rechenmann 1986). It includes the functionality of an expert system, in particular an inference engine and a mechanism for providing explanations.

The relevance of the object-centered representation approach and of SHIRKA characteristics to identification problems was verified earlier for trees of tropical forests of the Ghat region in India (Gautier and Pavé 1990). The second example presented here uses identification of fish of the Kerguelen Islands.

SHIRKA

SHIRKA was developed by Rechenmann (Rechenmann 1985; Rechenmann et al. 1989). The knowledge model is based on schemes, a concept similar to frames (Fikes and Kehler 1985). In contrast to frames where the object is linked to the notion of prototype, there is a difference between class schemes and instance schemes in SHIRKA. SHIRKA includes the functionalities of an object-centered knowledge representation: slots inheritance, default values, and procedural attachment. In addition, it includes the following functions:

1. The knowledge model is totally uniform. Any scheme is an instance of an upper-level class scheme: a metascheme. In the same manner, any slot, facet, or value is an instance or a reference to an instance.
2. The slot type is mandatory, either basic (integer, real, character string, symbol, Boolean) or complex (pointing to another class by the name of its definition scheme).
3. The procedural attachment, which associates one or more procedures to a slot, uses external descriptions of procedures as schemes. The call procedure is thus completely controlled by SHIRKA.
4. The inference of unknown slots values uses pattern matching rather than rules. Pattern matching, as a crucial inference mechanism, is designed to be efficient; an original technique for compiling a knowledge base has been successfully implemented.
5. The network of classes and subclasses supports a identification mechanism to determine potential classes of an instance.
6. SHIRKA makes it possible to manage instances consistency through procedures attached to slots, which are activated when values are modified,

deleted, or added to the slot, a class scheme being an instance of a meta-scheme. Management of consistency is extended to the classes themselves. This makes knowledge base restructuration and the definition of knowledge acquisition mechanisms easier.

SHIRKA is written in Le_Lisp from I.N.R.I.A. (Institut National de Recherche en Informatique et Automatique) (Chailloux et al. 1986). Le_Lisp is a widely implemented Lisp dialect.

The Knowledge Model

Classes and Instances. A scheme describes a class of objects as well as a specific element of a class, an instance. Here, a class is an taxon (e.g., a species), and an instance is an individual fish. A class and the related instances are defined by slots, which are similar to characters. In turn, a slot is defined by a list of facets, which represent knowledge about these characters. A facet is defined by a list of values; a value may be a scheme or a reference to a scheme (Fig. 11.1), or it can be the range of possible states for the character. Species are described as class scheme and unknown specimens as instance scheme. In Figure 11.1, any two-dorsal-fin fish is specified by the dorsal fin and the first dorsal/second fin types to which it belongs. The slot **is-a** defines an instance of the *Two-dorsal-fish* class.

Classes are organized as an acyclic network of classes and subclasses. A subclass describes objects that are more specific than those described by its parent classes (Fig. 11.2). A subclass inherits the knowledge of the slots defined for its parent classes, and it contains specific knowledge stored in additional slots. Specific knowledge must be consistent with inherited knowledge.

To use SHIRKA, classes and subclasses are first defined (inclusion linkage is made by the slot **a-kind-of**). Then instances are created using the slot **is-a** to link the instance to its class). For example, a taxon is created as a subclass, and an unknown specimen is created as an instance. SHIRKA attempts to answer requests for the unknown slot values, using the knowledge attached to the class.

Creation of new classes and of new instances are two distinct tasks, with distinct complexity. The former is usually made by the designer of the knowledge base, the latter by the user.

Slots. The semantics of the knowledge model are specified to a great extent by the available facets. A slot type is defined by the facets *$one* or *$list-of*, according to the admissible number of potential values. The type is either basic (integer, real for numbers, character string, symbol, Boolean) or com-

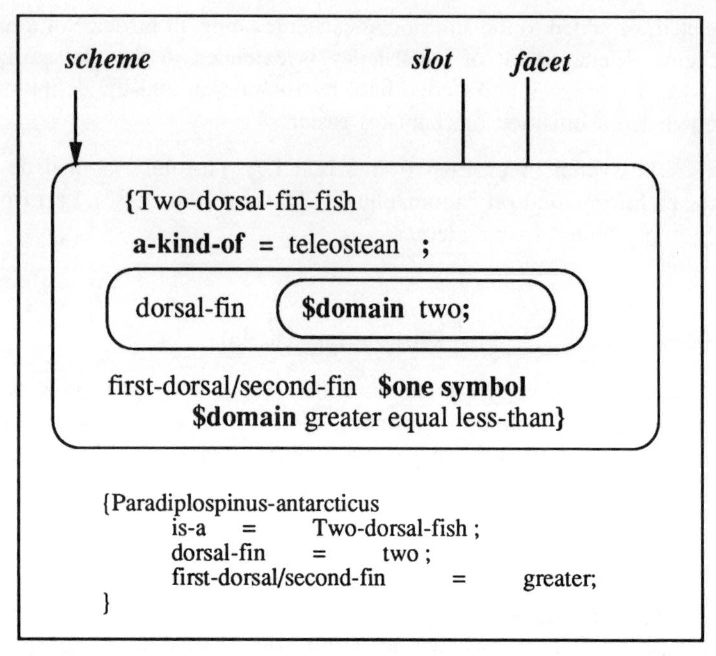

Figure 11.1. Structure of the scheme defining the class **Two-dorsal-fin-fish.**

plex (referencing another class by the name of its definition scheme); in this case, particular rules are applied for coherence maintenance.

Some facets are used to restrict the admissible values for a slot. *$domain* defines a complete list of admissible values, and *$range* a list of ranges of admissible values for a slot with type = ordered. *$card-min* and *$card-max* specify the minimum and maximum number of values for a multivalued slot. Complex predicates may be associated to a slot with the facet *$check*.

Other facets define how to infer the unknown value of an instance slot. *$value* introduces a class value, that is, a value assigned to a slot for every instance. *$default* is similar, but the relative value may be redefined in an instance.

Procedural Attachment

The procedural attachment in SHIRKA differs from that in classical frame systems. Each procedure is externally defined as a class, where the name of the Lisp function (its internal definition) follows the facet *$value* in the slot **fct-name**. The procedure call consists in instantiating the subclass scheme

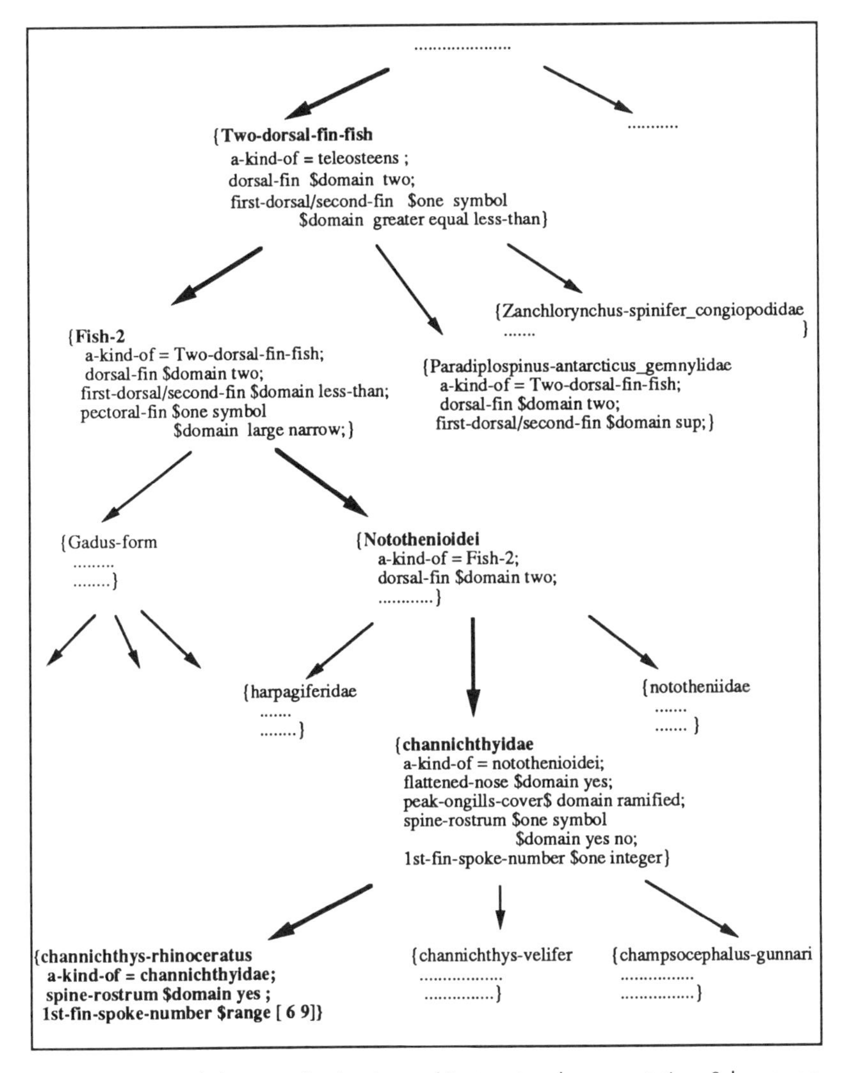

Figure 11.2. Knowledge organization in an object-centered representation: Schemes are linked following the hierarchy via the predefined slot **a-kind-of.**

and passing the resulting instance to the function. The function accesses the input parameters and carries out the necessary computation. When procedural attachment is used in a facet *$ifn-exec* (if needed, execute), the function provides the instance with the values of the slot attached to the output parameter. The value is passed to the slot that uses this procedural attachment by means of the variable attached to the output slot by the facet *$var->*. When used in the facet *$ifn-exec*, this controlled procedural attachment infers unknown slot values. It is fired by the inference engine and also by the facet *$check*.

Pattern Matching

Pattern matching is an inference mechanism specific of SHIRKA. First, the instances that match a description, (i.e., a pattern) are selected; then, among these instances, values for the unknown slots are selected via the variables. The facet *$ifn-match* (if needed, match) implements this mechanism; it may include several patterns, which are sequentially tried.

Pattern matching consists of instantiating the pattern and comparing the instances of specialized scheme in the base with the pattern instance. When an instance matches the set of conditions, the value of one of its slots becomes an admissible value for the slot to which the pattern-matching facet is attached.

SHIRKA allows the definition of constant and variable conditions in a pattern. Constant conditions are introduced by the facets *$value, $range, $excet*, and *$domain*. Variable conditions are described with the facets *$check, $var←* and *$var-list←, $var→* (attached to the predefined slot **self**), *$ifn-match*, and *$ifn-exec*. For identification, the classification mechanism uses the facets *$range, $domain, $except* to introduce constraints attached to slots. The first two facets are used depending on the type of each slot, integer, real, character string, symbol, or Boolean. The third facet, *$except*, is available in all cases.

Identification Mechanism

Since object classes are organized in a hierarchy (Fig. 11.2), in which the more general classes dominate classes that are more specific, a very natural reasoning mechanism consists of looking for the possible locations of a given object in this hierarchy. The only purpose of the root of the hierarchy, the *class object*, is to transmit to every class some bookkeeping slots that are required for proper operation of the system. It has no meaning regarding the domain which is modeled in the knowledge base. The knowledge base is made up of several hierarchies that are rooted in the subclasses immediately below the class object.

The object to be identified is initially known to belong to its instantiation class in one of these hierarchies, that is, in the class that was used to create the instance. This class may be different from the root class of this hierarchy. It is supposed that the object satisfies all the constraints attached to this class, even when the object is incomplete (i.e., when some of its slot values have not been assigned during its instantiation). The identification algorithm is recursive and makes use of a *breadth-first* scheme. At the beginning, it attempts to match the object with every immediate subclass of the instantiation class (Fig. 11.3).

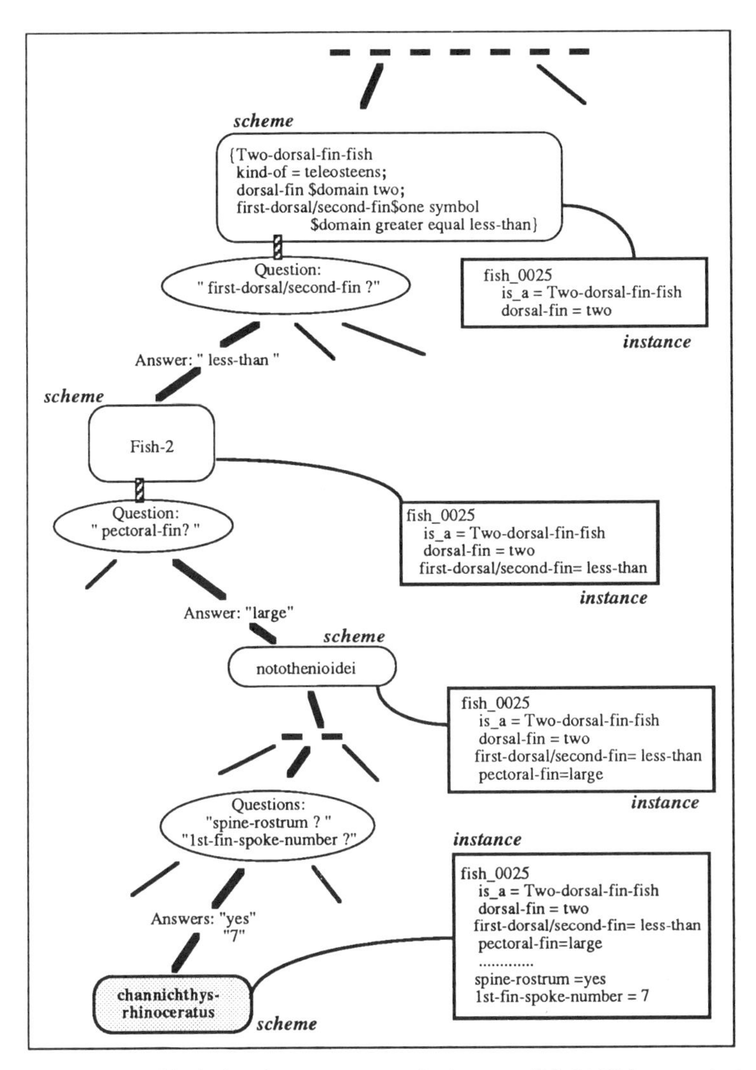

Figure 11.3. A simplified identification process. An instance (fish_0025) is created; then slot values are set by asking the user or by inference.

The constraints expressed in these subclasses and attached to their slots are checked against the slot values of the object instance. If all the constraints can be verified for a class, this class is said to be "certain" for the instance. As soon as one constraint in a class is not satisfied, this class is said to be "impossible" for the instance. When a slot that has not been inherited from the parent class appears in a subclass, it is necessary to get its value for the instance being matched with the subclass. The value cannot be obtained using knowledge from the parent class since it is a new slot. Moreover, procedures or default values attached to the slot in the current subclass should not be used, since the instance is being matched against this class in order to know whether it might belong to it. In our application, the user is asked to provide this value. If he is unable to do so, the slot value remains unknown, and the current subclass is said to be "possible" for the instance.

When all the immediate subclasses have been tested, the matching process is recursively applied to every certain or possible subclass, since these subclasses are not supposed to be exclusive. The process terminates when terminal classes are reached or when only impossible classes remain. The output of this identification process is a list of certain classes, a list of possible classes, and a list of impossible classes in the hierarchy.

To explain why a particular class is possible or impossible, SHIRKA provides the name of the first slot for which the value was unknown or the name of the first slot whose constraints were not satisfied. No explanation needs to be provided for a class given as certain since obviously all its constraints have been satisfied.

Our knowledge of an object increases through this identification process since the classes to which it could belong are more specialized than its instantiation class. This reasoning mode can be used in any diagnostic-oriented knowledge-based system (Puvilland et al. 1989).

A possible extension of the algorithm would consist in computing some measure of the distance between a potential class and the object under consideration. Checking a constraint would no longer be made in a binary (*yes/no*) manner, but by computing the distance. This mechanism is used in CLASSIC (Anonymous 1988b), a knowledge-based system development shell. Another problem to consider is the practical inability for an expert, or a group of experts, to define a unique hierarchy of classes and subclasses. Conflicting classifications systems can be used to define several hierarchies. These hierarchies certainly have several classes in common, but they differ in the respective locations of these classes and in the slots which are taken into account (Marino 1990).

Metaschemes

The consistency of a knowledge base must be maintained whether it uses production rules or object-centered representation. A Truth Maintenance System (TMS) solves this problem in rule-based systems for factual knowledge, either inferred or provided by the system (Doyle 1979). SHIRKA provides value consistency control mechanisms through the use of event-driven facets together with an adapted TMS when actions (e.g., addition, deletion or modification of values of a slot) are required (Euzenat and Rechenmann 1987).

The concept of metascheme (Fig. 11.4) makes it possible to consider a class scheme as an instance of the metaclass scheme or of one of its specializations. The class scheme has the following slots:

ako (a-kind-of), to link a class with its superclasses;
spec, to link a class with its specializations;
inst, to link a class with its instances;
slot-list, the list of the slots of a class.

This metadescription level provides a straightforward solution to the problem of class consistency. Specializations of scheme may be created in order to define specific event-driven methods such as adding, deleting, or modifying values of a slot, which is an item of the list **slot-list**. Such event-driven methods may also be defined in the class scheme itself in order to provide default event-driven behaviors, or in other slots (e.g., **ako** or **spec**) in order to deal with the modifications in the inheritance lattice.

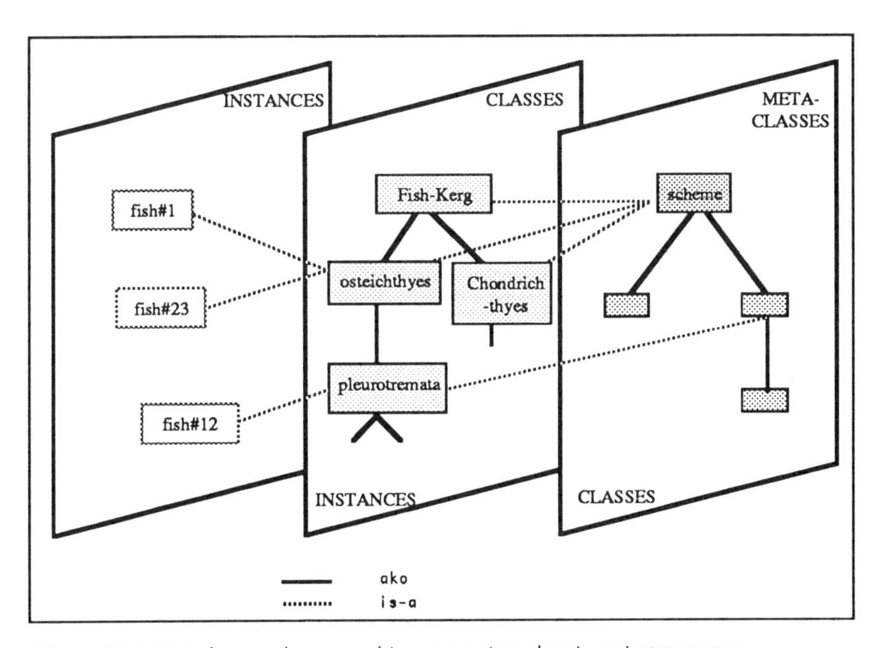

Figure 11.4. Metaclasses, classes, and instances. Any class is an instance or a specialization of the metaclass scheme.

Development of a Knowledge Base for Antarctic Fish

Capture of identification data in fishing boats must be simple, and it must rely mostly on conspicuous morphological characters. Identification should mimic the expert-reasoning process with a global approach using all the features simultaneously but also allowing instant recognition of specific forms.

To create a knowledge base for Antarctic fish we must find the most suitable class hierarchy and, for each scheme, we must find the slots that allow fine discrimination of objects. In addition, the hierarchy and the slots must allow updating, adding, deleting, or modifying the knowledge base without reprogramming the whole application. The traditional classification system (with orders, genera, species) seems to be a good model to build the knowledge base.

A first knowledge base has been built with the systematics characters used in the Food and Agriculture Organization (FAO) guide for the identification of Austral Ocean fish species (Fisher and Hureau 1985). These characters were used in a dichotomous key for thirty-four species (Sieffer 1988). Then, a second knowledge base was built with multiple criteria (i.e., multiple slots) at each node (i.e., each scheme).

Description of Slots

The slots describe morphological characteristics: shape of head, snout, and fins (dorsal, caudal, pectoral, pelvic, and anal), number of dorsal fins, relative lengths of fins, number of thorns, position of eyes, head scales, etc. To describe slots, those characters are selected from the FAO guide that can be used to differentiate taxa or groups of related species. This means that these characters are grouping species into families or genera in the various orders. For example, the presence or absence of prickles on fins would seem at first to be a valid identification criterion because it splits the thirty-four species into two groups of equal size. However, this character was discarded because species belonging to different orders may have the same value for this character.

Another difficulty is that the characters that are the easiest to observe (numbers, lengths, relative lengths, fin base) are often similar in different groups or, at the opposite, are variable between species in the same group or, even between individuals of the same species. Also, quantitative characters (size) were discarded because they depend of ecological or age conditions (although some species are consistently larger than others).

It appears that users have difficulties with the terminology for morphological characters. The term definitions are very precise, and there is a great risk of giving erroneous answers (not counting spelling mistakes!) if the user is not quite familiar with the specialized vocabulary. This is particularly true if the correct character state must be picked up from a list of several closely resembling expressions. To avoid some of these ambiguities, the set of possible answers has been connected to an image base. The user can choose the slot value in a list of possible answers by selecting the corresponding picture (Piraud 1988).

Building a Hierarchy

The first version of our expert system was based on the traditional dichotomous principle, but this concept is outdated and virtually unmanageable.

The passage from a parent node to its children depended only on presence or absence of one character (e.g., presence or absence of chin-barbel). This does not take into account character variability (multivaluate slots), and it is much too crude. It results in the choice of common criteria that differentiate individuals without due consideration of their taxonomic relationships. For example, *Channichthys rhinoceratus* was identified on a spin-upon-premaxilla character. This does not take into account the fact that this species belongs to the "osteichthyes teleostean gadiform notothenioidei channischthyidae" taxonomic series, and it does not follow the taxonomic model chosen to develop the knowlcdge base. Also, the lack of taxonomic efficiency made it necessary to use too many intermediate nodes: one node for each value of a slot domain. Closely related taxa present very few obvious differences. Two fish species of same group might be distinguishable by only one character. Then, a second character differentiates a third species, and so on. Additional characters are used to distinguish additional species in the group. This complexity makes it difficult to choose the best hierarchy, and programming this *yes/no* process is almost impossible.

The SHIRKA language can solve these difficulties. The passage from one hierarchical level to the next is based on restrictions on the values of parent-class slots and on additional slots. Child classes inherit several slots, and they can have some values in common for some slots. In scheme writing, the number of slots and the number of slot-typing values are not limited. At the opposite, a *$except* facet can be used to restrict the admissible values for a slot. Combining these possibilities, a class scheme allows describing an object in fine details. The choice of a node on a lower level in the hierarchy scheme is made by analyzing the intersect between domains of the different slots of this node scheme. This mechanism is also available to solve ambiguity of value answers for an particular slot. We can explain this by the example in Table 11.1. Table 11.1 represents an object 0 and its five children, objects I to V. The child-object I inherits specifications a11 and a12 to slot 1, a21 to slot 2, etc. It is easy to see that objects II and IV are not distinguishable if the specimen to be identified has values a11 on slot 1, a22 on slot 2, and a31 on

Table 11.1. Example of Slot Values Domain for an Object and Its Five Children

		Slot Values Domain				
Object 0	Slot 1	a11	a12	a13	a14	a15
	Slot 2	a21	a22	a23		
	Slot 3	a31	a32	a33	a34	
Object I		a11	a12			
		a21				
		a31		a33	a34	
Object II		a11				
			a22			
		a31				
Object III				a13	a14	
				a23		
			a32			
Object IV		a11				a15
			a22			
		a31			a34	
Object V		a11	a12	a13	a14	
		a21	a22	a23		
		a31	a32	a33		

slot 3. In such case, it is necessary to consider an additional slot (slot 4), either at level 0 or at level 1 as shown in Figure 11.5.

It is better to attach slot 4 to the parent level (Fig. 11.5, strategy A) because this leaves the children at the same level. This fits quite well with the domain when object 0 is a genus and the child-objects are species belonging to this genus. It is obvious that the dichotomous approach that uses one character at a time is not suitable here, because the child-objects are differentiated by all the characters taken simultaneously (polytomous approach). The example shown in Table 11.2 represents the lower part of hierarchy schemes for Nototheniidae family in the Kerguelen fish knowledge base.

The SHIRKA Knowledge Base

During the inference process, the system goes down the hierarchy as far as possible using the values given to the slots. At each node, the choice depends on several answers linked to several slots. This allows one to choose among a great number of possible directions at each node. This mechanism mimics the

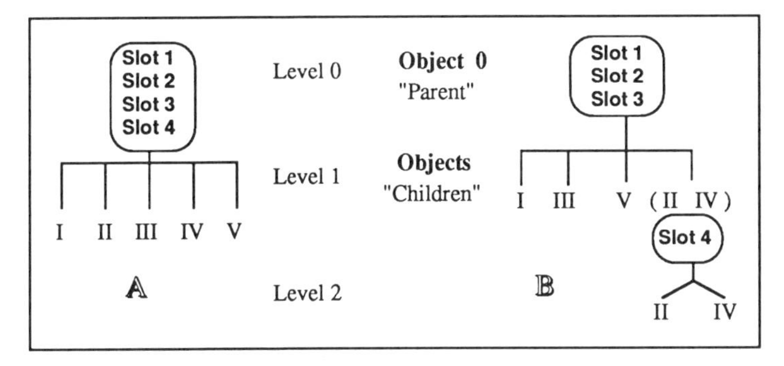

Figure 11.5. Comparison of two strategies for the placement of slot 4.

reasoning of a human expert who uses all relevant features simultaneously. The dichotomous approach can be seen as the most simplistic form of this process where a scheme has only one slot presenting a single alternative.

For the identification of individuals present in a well-delimited geographical area where the number of species is rather limited and where they belong to several taxonomic families, the best strategy for building the knowledge base is to select the criteria that best discriminate species into existing taxonomic units. The knowledge base is organized as follows:

1. The top of the hierarchy uses characters that allows the differentiation of large groups of species, for a clear separation into taxonomic groups. Each scheme needs only one or two slots.
2. At the intermediate level, these large groups of species are divided into smaller groups, each subgroup including closely related species.
3. Schemes at the terminal nodes of the hierarchy allow the identification of every species. For discrimination of related species, the selected characters (i.e., slots) must allow for the description of the smallest morphological differences. Scheme complexity increases, and it may be necessary to include multivaluate slots.

This organization facilitates updating the knowledge base and ensuring its coherence. New taxa can be added without major restructuring of the knowledge base. New species are introduced in the lower part of hierarchy with additional slots or modification of the values domains. A new family (or a new group of families) can be incorporated in the upper part of hierarchy, often by creating a new scheme.

Table 11.2. Example of Lower Part of Hierarchy Schemes for Nototheniidae Family in the Kerguelen Fish Knowledge Base

Parent-scheme

```
{nototheniidae
   a-kind-of = notothenioidei;
   self $com"family : nototheniidae, order perciform, suborder : notothenioidei";
   flat-noose $domain false;
   thorns-cover $domain without;
   scales-on-head$a symbol
      $domain all-the-head except-preorbital-space except-top-head without-scales;
   Pecto-fin-width>pelv-fin-width $a Boolean;
   ray.-pecto.-fin $a symbol
      $domain>23 22>>19 <19;
   %-interorbit/head-width $a symbol
      $domain %<15 15<%<25 25<% }
```

Children-schemes

```
{notothenia-squamifrons
   a-kind-of = nototheniidae;
   scales-on-head $domain all-the-head}
{nototheniops-mizops
   a-kind-of = nototheniidae;
   scales on-head $domain except-preorbital-space;
   ray.-pecto.-fin $domain 22>>19;
   %-interorbit/head-width $domain %<15;
   pecto-fin-width>pelv-fin-width $domain false}
{notothenia-acuta
   a-kind-of = nototheniidae;
   scales-on-head $domain except-preorbital-space;
   ray.-pecto.-fin $domain 22>>19;
   %-interorbit/head-width $domain %<15;
   pecto-fin-width>pelv-fin-width $domain true}
{notothenia-coriiceps
   a-kind-of = nototheniidae;
   scales-on-head $domain except-top-head;
   ray.-pecto.-fin $domain <19;
   %-interorbit/head-width $domain 15<%<25;
   pecto-fin-width>pelv-fin-width $domain true}
{notothenia-rossii
   a-kind-of = nototheniidae;
   scales-on-head $domain except-top-head;
   ray.-pecto.-fin $domain <19;
      (i.e., $domain >23 22>>19)
   %-interorbit/head-width $domain 25<%;
   pecto-fin-width>pelv-fin-width $domain true}
```

(*continued*)

Table 11.2. *(Continued)*

```
{paranotothenia-magellanica
  a-kind-of = nototheniidae;
  scales-on-head $domain without-scales;
  %-interorbit/head-width $domain 25<%;
  ray.-pecto.-fin $domain <19;
  pecto-fin-width>pelv-fin-width $domain true}
{notothenia-cyanobrancha
  a-kind-of = nototheniidae;
  scales-on-head $domain without-scales;
  %-interorbit/head-width $domain 15<%<25;
  ray.-pecto.-fin $domain 22>>19;
  pecto-fin-width>pelv-fin-width $domain true}
```

Conclusion

SHIRKA, although still at the prototype stage, is a complete development tool for knowledge-based systems. It includes an inference engine, a scheme editor, a spreadsheet-like user interface allowing any operation on instances, an explanation module, and a truth maintenance module. However, the user interface is very primitive, using a *line per line* mode. Icon selection would be a useful aid, and other facilities must be developed. The succession of questions asked to the user must be optimized so that the questions follow each other logically. If the user suspects he made an error in character entry, backtracking is not easy with the current system. A single erroneous answer makes the inference process fail, or it downgrades the correct answer from "certain" to "possible." In case of missing or imprecise characters, the system does not provide the option to explore alternative paths as an expert would do. The system has some limitations in difficult cases where experts would use rules of thumb, backtracking, and alternative answers when their first intuition has proved to be false.

In its present version, SHIRKA offers the essential characteristics of an object-centered knowledge base management system for identification of species that are known in a well-defined ecological environment. Implementation of a knowledge base a few kilobytes large is not difficult. The Kerguelen fish knowledge base with 34 species and the knowledge base on tropical trees from the Ghat forest with about 300 species have been created on a Macintosh II.

12

Information Processing with Neural Networks

ROBERT ZERWEKH

Neural networks are information-processing structures that depart significantly from the manner in which information is processed in traditional, programmed computing environments. Under the traditional paradigm, a problem is solved by coding an algorithm in software that specifies precisely the tasks that have to be done and the order in which they must be performed. Neurocomputing, by contrast, does not require that algorithms or rules be developed for solving a problem. Instead, information-processing structures, called neural networks, "autonomously develop operational capabilities in adaptive response to an information environment" (Hecht-Niclsen 1990).

A neural network can be described as an information-processing system that is composed of a number of interconnected processing elements. These processing elements are also called nodes, cells, units, or neurons. Each processing element computes its activity locally based on (1) the activities of the nodes to which it is connected and (2) the strengths of these connections, or its weights. A specific transfer function then determines the output of the processing element given its input (Rumelhart and McClelland 1986; Grossberg 1988).

Comparison between Conventional Programming and Neurocomputing

Conventional Programmed Computing Model

The conventional model of information processing, programmed computing, has been used since the advent of digital computers. Programmed computing solves a problem by breaking it down into its component parts and analyzing

Advances in Computer Methods for Systematic Biology: Artificial Intelligence, Databases, Computer Vision, ed. Renaud Fortuner (Baltimore: Johns Hopkins University Press).

these parts. An algorithm or set of rules is developed that solves the problem, and this algorithm is then coded in software. Despite the success of this paradigm, a number of difficulties are associated with it.

Software development is a time-consuming process and can often be costly, particularly when extensive revision and testing of the software is required. Furthermore, this paradigm assumes that an algorithm to solve the problem is already known or can be invented. Inventing a new algorithm to solve a problem will take additional time even if the effort is blessed with the creative power of geniuses. This effort still may not be successful since there often is no clear procedure or algorithm to solve a problem, especially in knowledge domains characterized by complex mental skills, fuzzy logic, or uncertainty.

Moreover, when a procedure has been developed to solve a problem in a complex domain, it typically requires huge databases of information and large libraries of production rules like those commonly found in traditional expert systems. As the size of these databases and rule libraries increases, the computations required to solve a problem become so great that either we must give up hope of reaching a solution within a reasonable time or it becomes apparent that the solution is computationally intractable.

Neurocomputing Model

Neurocomputing can significantly reduce the impact of these bottlenecks for some problem areas (e.g., pattern recognition) since it does not require algorithms or rule-based procedures for information processing. Robert Hecht-Nielsen (1990) has defined a neural network as follows:

> A neural network is a parallel, distributed information processing structure consisting of processing elements . . . interconnected via unidirectional signal channels called connections. Each processing element has a single output connection that branches ("fans-out") into as many collateral connections as desired; each carries the same signal—the processing element output signal. The processing element output signal can be of any mathematical type desired.

Figure 12.1 shows a schematic of a simple two layer network where each input unit (J) is connected to each output unit (I). Each connection from an input unit to an output unit has a numeric value associated with it, which is called a weight, and this value is used by the output unit in calculating its net input. Weights are initially set to small random values before processing begins and are changed during processing according to one of a number of specific learning laws. Which learning law is used is in part determined by the architecture of the neural network (i.e., the number of processing units, the number of layers of units, the connections between layers) and in part by the kind of processing task one needs to perform. Each output unit produces an output signal that may fan-out to other units in other layers, or, in some neural

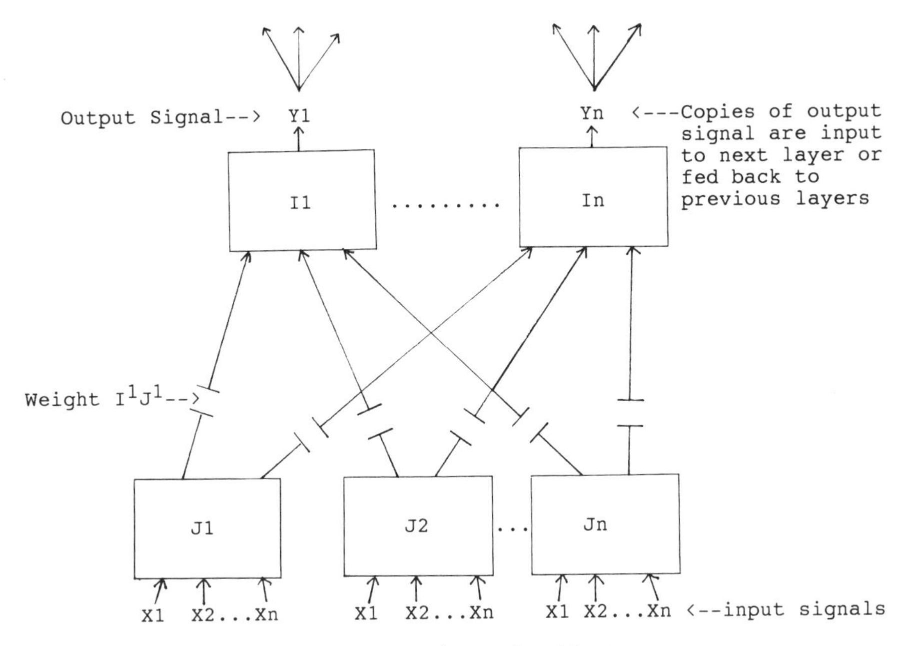

Figure 12.1. Neural network architecture.

network architectures, may be used as a feedback signal to previous layers. The signal produced by each output unit is a function of the net input it received.

Figure 12.2 illustrates an individual processing unit. The unit receives some input data which are usually an ordered list of numbers called an input vector. The input data may originate from outside the neural network or may come from a previous layer of processing units. Upon receiving its input data, a processing unit multiplies each input value by its corresponding weight and sums these products. This value constitutes the net input to the unit.

The net input is then sent through a transfer function that determines the output signal given its current net input level. Several variations of transfer functions have been proposed and used in different neural network architectures. The most commonly used transfer functions are: (1) linear: the output signal is equal to the input signal; (2) linear threshold: the output signal is set to one if net input exceeds zero (otherwise it is set to zero); (3) stochastic: the output signal is set to one with a probability given by a logistic function; and (4) continuous sigmoid: the output signal is compressed to lie within a continuous range between zero and one.

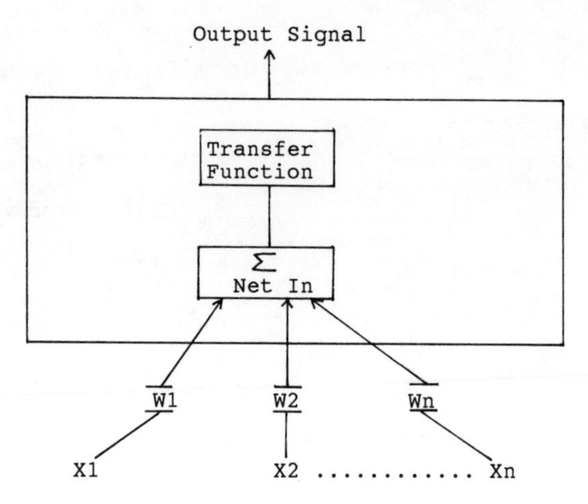

Figure 12.2. An individual neuron.

Implications of Parallel and Distributed Processing

As Hecht-Nielsen's definition notes, computation is parallel in a neural network because each processing unit is carrying out its internal computations simultaneously with every other processing unit. To achieve true parallelism, a neural network must be implemented on a special computer that contains multiple processors. Much of the current research in neural networks, however, simulates parallel processing on sequentially programmed computers. Computation is also distributed in a neural network because what the network *knows* does not exist at a single location in memory as in conventional computers. Rather, knowledge exists in the strengths of the connections that exist between processing units. It is helpful to think of knowledge in this sense as consisting of a pattern of activity among various processing units (Rumelhart and McClelland 1986). The parallel and distributed computation that characterizes neural networks bestows a number of highly attractive features on this model of computing.

Reasoning with Multiple Constraints

For one thing, such a system can deal with very large numbers of constraints and variables and still process information efficiently. Whereas people reason faster when they are able to exploit additional constraints, algorithmic computation gets slower as additional constraints, propositions, and rules are added. Since neural networks learn by being shown examples of the types of information processing they are to perform, no explicit coding of rules and constraints is required. Whatever constraints are implicitly embedded in a particular example of conceptual reasoning or information processing are detected and learned by the network through its internal self-adaptation of connection strength weights. The learning process continues until the neural network is eventually capable of duplicating that kind of information processing.

Graceful Degradation

Since knowledge is not stored in one memory location, and since processing is distributed throughout the system, damage to the system or to processing units does not result in complete loss of knowledge or the inability to continue to function. Instead, damage to the system simply results in some partial or gradual reduction of processing power. There is a graceful degradation of the system, not a complete shutdown. It is therefore possible to retrieve and process information even when part of the network has been compromised.

Content-addressable Memory

With conventional computers, retrieving a desired piece of information requires that the exact memory address of the item is known. Since most users of computer systems rarely, if ever, know the exact memory location of any item, methods have been developed to let users query databases on the basis of their contents. Instead of requesting the data record at hex location FC210, a user can ask for the data record that answers the query, *Who is Clark Kent's manager*?

Implementing this kind of content-addressable, or associative, memory on conventional computers, where specific items can be retrieved based on partial descriptions of their contents, has high costs associated with it. Either 20 to 30 percent excess memory is needed to support an efficient hash-coding scheme that enables information to be retrieved quite rapidly (and then only if the content descriptions are completely error free), or extra overhead is required to support the construction and maintenance of multiple indexes. In general, content-addressable memory schemes on conventional computers require some kind of massive and sophisticated search strategies to locate the item or items that best fit the description.

A content-addressable memory can be implemented quite easily in a neural network. Since knowledge resides in the connection strengths of the processing units in a neural network, and not at any specific memory address, activating or stimulating one or more of these units, by providing a partial description of a desired memory, leads to the activation of the other units involved in the memory in accord with the learned connection strengths. Consequently, we can produce the desired representation or memory merely by providing some feature or combination of features. Since memories reside in a neural network as patterns of activity over a number of interconnected processing units, a complicated search strategy does not need to be devised to locate a memory.

Incomplete Data

A noticeable advantage of neural networks over rule-driven systems is that they are capable of dealing with input vectors of data that are partially incomplete or partially incorrect. The discussion of how a neural network learns anything at all is treated in more detail later. For the moment, suffice it to say that a neural network learns to associate a particular output response with a particular input vector of data. It is trained to perform this association by repeatedly being exposed to the input information and, in one learning paradigm, being instructed that a certain output response is the correct one when confronted with a particular input vector. Once this association is learned, a neural network can produce the correct output response even when presented with an input vector that is incomplete (some of the numbers in the ordered list are missing) or that contains incorrect data (some of the numbers in the ordered list are wrong). A neural network is capable of doing this because what it learns about one input pattern generalizes to other similar patterns. This is an important feature that distinguishes neural networks from production rule systems that typically will falter if not presented with complete and accurate input data.

Learning Models

One of the most remarkable things about neural networks is the manner in which they learn how to process information. Two fundamental learning paradigms are currently used in teaching neural networks how to process information. Each aims at getting the network to produce some consistent or desired output when exposed to an input vector of data that originates from outside the network. This is accomplished by having the network adjust its weights, the connection strengths that exist between units, until presentation of the input vector produces the desired, or consistent, output response.

Supervised Learning

In supervised learning, a neural network is exposed to a vector of input data and allowed to calculate its output response. This response is then compared against a desired, or target, output response, and a calculation of the difference, or error, is made between the network output and the target output. This calculated error quantity is then fed back through the processing units so that the network can adjust its weights to minimize the error. In supervised learning, therefore, the network is shown what the correct output is supposed to be for every input pattern of data to which it is exposed. The network adjusts its weights according to the learning law that governs weight changes until the error between its response and the target response reaches zero or some specified minimum. At this point the network is said to be trained and can consistently produce the correct output for each of the input vectors that it has reviewed, or for input vectors that are similar to the ones it has reviewed.

Unsupervised Learning

The other learning paradigm, unsupervised learning, does not employ any desired or target responses in training a network. This paradigm is particularly suited for situations where it is not possible to predict the number and kinds of possible input patterns the network will be exposed to. The network is exposed only to input vectors of data, and it changes its weights according to its learning law so that it produces consistent output representations of the input patterns. In other words, once the network has learned the training set patterns, applying the same input vector, or applying an input vector that is sufficiently similar to it, will cause the network to produce the same output response. In this sense, the network is allowed to construct its own representations of the most significant features of the input patterns (Kohonen 1988). Since the network is not shown what the correct response is to be, unsupervised learning simply extracts the statistical properties of the training set input patterns and groups similar patterns into classes or categories.

Application to Pattern Recognition

Example of Application: Primate Foot Anatomy

One of the problem areas in which neural networks excel is that of pattern recognition. I used the pattern recognition capabilities of a neural network to identify primate foot bones and bone fragments and to assign these bones and bone fragments to various primate families in a manner similar to the way in which a physical anthropologist would perform the task. The learning law employed in the network (or the law that specifies how weights are to change during training) was the back-propagation algorithm described by Rumelhart et al. (1986).

Back-propagation is a supervised learning procedure that attempts to teach a neural network correct output responses by minimizing the error between the actual output response of the network and the target response. The target responses taught to the neural network in this case were correct identifications of various primate taxa when exposed to data that were descriptive of foot anatomy. The network consisted of three layers of processing units. Thirty-nine nodes at the input layer were all fully connected to twenty-seven middle layer nodes (called hidden units). The twenty-seven hidden units were in turn connected to fourteen output units.

The input to the network consisted of a list (vector) of thirty-nine physical characteristics descriptive of primate foot anatomy. Examples of some of these characteristics are:

Talo-fibular facet:
- sloping
- steep-sided

Calcaneo-cuboid joint pivot:
- deep
- moderately deep
- shallow
- flat

Original Data

The network was to identify one of fourteen families and subfamilies of primates from a given input vector of foot-bone characteristics. The families represented were those of tooth-comb prosimians, tarsiers, New World monkeys (e.g., squirrel and spider monkeys), Old World monkeys (baboons and rhesus monkeys), and the apes (gibbons, orangutans, chimps, and gorillas).

The raw information used to train the network to identify these taxa was provided by Dr. Daniel Gebo, a primatologist at Northern Illinois University. Dr. Gebo was given a chart with thirty-nine rows for the foot characteristics and fourteen columns representing each of the fourteen taxonomic groups of primates. He was asked to rate on a scale of 0 to 8 how indicative each characteristic is for each of the taxonomic categories. For example, a sloping talo-fibular facet is highly indicative of the prosimians (a score of 8) and not indicative at all (a score of 0) for anthropoids (monkeys, apes, and humans). A partial representation of this chart and ratings is shown in Table 12.1.

Training the Neural Network

The network was trained with this information in two ways. First, it was presented with input data vectors describing the foot-bone characteristics of each of the fourteen taxa. For example, an input vector consisting of characteristics 1, 3, 5, 7, 9, etc., is descriptive of the cheirogaleids, or dwarf lemurs. An input vector consisting of characteristics 2, 4, 6, 7, 10, etc., is

Table 12.1. Indicativeness of Foot Characteristics for Fourteen Taxonomic Groups of Primates

Foot characteristics	Taxonomic groups of primates[a]													
	A	B	C	D	E	F	G	H	I	J	K	L	M	N
Talo-fibular facet														
1. sloping	8[b]	8	8	8	8	8	0	0	0	0	0	0	0	0
2. steep-sided	0	0	0	0	0	0	8	8	8	8	8	8	8	8
Talo-tibial medial facet														
3. Broad and extending plantar surface	8	8	8	8	8	8	8	0	0	0	0	0	0	0
4. Narrow and elevated above plantar surface	0	0	0	0	0	0	0	8	8	8	8	8	8	8
Flexor groove position (posterior talus)														
5. Lateral position	8	8	8	8	8	8	0	0	0	0	0	0	0	0
6. Midline position	0	0	0	0	0	0	8	8	8	8	8	8	8	8
Subtalar joint														
7. Continuous distal facets with long proximal facet	7	7	7	4	7	7	7	7	7	7	0	0	7	7
8. Separate distal facets with a short proximal facet	0	0	0	0	0	0	0	0	0	0	8	8	0	0
Calcaneo-cuboid joint														
9. Moderately deep pivot	7	0	0	7	7	7	0	7	7	7	0	0	0	0
10. Deep pivot	0	8	8	0	0	0	0	0	0	0	0	0	8	8
11. Shallow pivot	0	0	0	0	0	0	0	0	0	0	8	8	0	0
12. Flat surface	0	0	0	0	0	0	8	0	0	0	0	0	0	0
. . . etc. . . .														
Talonavicular joint														
35. Laterally rotated and oblong talar head shape	0	0	0	8	8	0	4	7	7	7	0	0	8	8
36. Medially rotated and oblong talar head shape	8	0	8	0	0	0	0	0	0	0	0	0	0	0
37. Laterally rotated and very round talar head shape	0	0	0	0	0	8	0	0	0	0	0	0	0	0

(*continued*)

Table 12.1. *(Continued)*

Foot characteristics	Taxonomic groups of primates[a]													
	A	B	C	D	E	F	G	H	I	J	K	L	M	N
38. Laterally rotated and asymmetrical talar head shape	0	0	0	0	0	0	0	0	0	0	8	8	0	0
39. Medially rotated and flattened talar head shape	0	8	0	0	0	0	0	0	0	0	0	0	0	0

[a] A-Cheirogaleidae B-Lorisidae C-Galagidae D-Daubentoniidae
E-Lemuridae F-Indriidae G-Tarsiidae H-Callitrichidae
I-Cebidae J-Atelidae K-Cercopithecinae
L-Colobinae M-Hylobatidae N-Pongidae.

[b]Numbers represent indicativeness of taxonomic group on a scale of 0 (not indicative) to 8 (highly indicative).

descriptive of the pongids, or great apes. For each input vector presented, the network response was compared against the desired response, and weight changes were made according to the back-propagation learning law in order to minimize the network error until it could correctly associate each of the fourteen input vectors with the correct primate taxon.

Next, the neural network was further trained on each of the thirty-nine foot-bone characteristics individually and taught to report which of the fourteen categories of primates exhibited that characteristic. For example, if presented with the information that a bone fragment had a steep-sided talo-fibular facet, the network would respond that such an animal belonged to the suborder Anthropoidea. If told that the calcaneo-cuboid joint of a bone fragment had a flat surface, the network would report that such a characteristic is indicative only of the tarsier family.

Testing the Neural Network

Once the network had completely learned all of the training set input/output pairs, it was time to test its approximation accuracy and generalizing ability. The real value of using a neural network to perform recognition and identification does not lie in its ability to associate fourteen foot-bone characteristics with a particular primate family. Conventional programming techniques can do this just as easily. Rather, its value lies in its ability to generalize beyond the training set examples and to produce correct answers when presented with incomplete data or with data that have been saturated with noise.

To test performance of the network, Dr. Gebo was asked to describe the characteristics exhibited by foot-bone fragments selected at random from his laboratory. The responses of the neural network to these test set descriptions were compared against Gebo's expert judgment. One bone fragment, for example, had only the following five characteristics: numbers 2, 4, 6, 7, and 35. Faced with such a paucity of input data, the network nevertheless correctly identified the bone fragment as indicative of the cebids, or squirrel monkey group. In actuality, the network responded that the bone fragment might be indicative of four families, but these judgments were weighted ones with the cebid family receiving the highest weighting (cebids: 0.999; atelids: 0.976; hylobatids: 0.921; pongids: 0.977). The other three choices, incidentally, were atelids (a New World monkey similar to the cebids), and the two categories of apes (also very similar to atelids). Note that all of these are anthropoids and that the network did not indicate that this set of characteristics was indicative at all of the tooth-comb prosimians or the tarsier family.

To test the ability of the network to tolerate noisy data, or data that contain contradictions, the following characteristics were input to the network: 1, 2, 4, 8, and 11. Note that characteristics 1 and 2 are mutually exclusive; 1 is characteristic of prosimians while 2 is characteristic of anthropoids. Feature 4 is true of all anthropoids, while features 8 and 11 are true only for the Old World monkeys. The network assessed such a bone fragment as belonging to the two subfamilies of the Old World monkeys (Cercopithecinae: 0.999; Colobinae: 1.0). Despite the conflicting information over features 1 and 2, and the presence of a general anthropoid characteristic in feature 4, the network correctly identified the bone fragment due to the uniqueness of features 8 and 11.

Conclusions

The results of these and other experiments indicate that certain neural network configurations, employing specific learning laws, are capable of recognizing biological specimens and associating these specimens with their correct taxonomic category. The network was able to learn successfully the descriptive characteristics of fourteen families of primates and was able to associate descriptions of bones and bone fragments with their correct taxa based on incomplete descriptions and descriptions deliberately tainted with noise. It performed these tasks without having been told how to do it using rules or any other kind of programmed algorithm. This in itself represents a considerable savings in development time since the alternative, coding a rule-based system capable of performing the same tasks, would have taken far longer to implement and even then may not have performed as quickly or correctly as the neural network.

Future research plans call for constructing an Intelligent Tutoring System (ITS) that will assist in teaching the principles and techniques of comparative anatomy. Although functioning ITSs are a relatively recent phenomena, enough success has been achieved in their design and deployment that it is possible to identify a system configuration, or architecture, that is common among them all. One of the main components of an ITS is the expert module. This is the part of the system that contains the expert knowledge about the teaching domain and simulates the expert's reasoning and judgment. The ITS will be the vehicle used to construct an expert module using neural networks.

One of the problems that must be addressed in this context can loosely be called the input problem. That is to say, how can we effectively communicate to the expert module those characteristics of bone fragments (shape, texture, size, density, facet angle, etc.) that are necessary to associate the animal's anatomical characteristics with a particular family or subfamily? To date our expert has simply keyed in the relevant characteristics. But an autonomous ITS must be able to make these discriminations on its own so that its judgment can be compared with those of students. One possibility is to use three-dimensional imaging or scanning techniques to capture the anatomical images. The neural network component of the expert module would then be trained to recognize the visual images and associate these images with different taxa.

Discussion

Milton: What data structures did you use to build your neural network? Did you use any special languages? Can I build a neural network tool or a set of tools that might work in conjunction with a standard environment?

Zerwekh: Although I have spoken of neural networks as parallel and distributed processing structures, my simulations were run on a 386 computer and were written in Turbo Pascal. You can easily simulate connections between processing units in any high-level language by using two-dimensional arrays. To fully exploit the power of parallel and distributed systems, however, you should implement a neural network on a parallel processor. Can you use a neural network as a tool in conjunction with a standard environment? Absolutely. I do not think that a neural network should be used as the sole information-processing structure, but rather as a preprocessor of information or, alternatively, as a postprocessor of information that a standard environment might issue as output. In fact, I think that the next phase of serious neural network applications will find these structures working in concert with expert systems.

Pankhurst: How do you decide how many processing units you need, whether they should all have the same or different transfer functions, and how many connections you need?

Zerwekh: The number of input and output processing elements required is often determined for you by the nature of the information-processing task you want to perform. In the case discussed here, thirty-nine input nodes were required because each input vector of data could contain data in any of thirty-nine ordered positions. And there were thirty-nine possible positions for input values because that is the number of foot-bone characteristics that my expert deemed appropriate for making discriminations. Similarly, fourteen output units were used because there were fourteen primate taxa, and the network was trained to indicate which of the fourteen taxa provided the best fit for the input information. The choice of twenty-seven middle-layer units was a product of experimentation with the number of units at this level. There really are no clear design principles at this point that one can turn to for guidance in selecting the correct number of middle-layer units, although a rule of thumb is that you start with a number halfway between the number of input and output units. Once that is done, you just have to experiment to find a number that gives you the best performance.

In terms of setting up the connections between processing units, many of the established neural network paradigms specify how units at different levels are to be connected. In a back-propagation network, for example, all input units are connected to all hidden units, and all hidden units are connected to all output units. The same thing holds true for deciding on a transfer function. Most back-propagation applications use a nonlinear sigmoidal transfer function for all of its units. In general, multilayer networks will have no more computational power than a single-layer network unless one does use a nonlinear transfer function. Researchers are continuing their investigations into transfer functions and connections between units. Some have experimented with networks where connections can be created and destroyed and transfer functions can be replaced with new ones as the network processes information.

Walker: Whenever you have a feature space, you have the factors that you have measured, and in one way or another, whether you use a neural net or discriminant analysis or rules, you are partitioning that feature space. Then you say that if the unknown falls in this region of the feature space you will call it X, and if it falls in that region of the feature space you will call it Y, etc. You would like to be able to examine those partitions, particularly when you have a nonlinear transfer function, because then you do not have a straight line dividing surface but some complex dividing decision surface. Are there commercial software packages that show you the decision surface that you are getting?

Zerwekh: To my knowledge, there are no products that show you the decision surface per se. A number of products allow you to examine the values of the weights that connect the units. From these, you can get an idea of what features in the input pattern the network is responding to and how strongly it is responding.

Meacham: Do you have to input all of your known characters at once, or can you enter a few characters, see the result, then enter more characters if necessary?

Zerwekh: The manner in which my present system is set up requires that you enter all of the known characters at once. It cleans itself out after having made its judgment. The system could easily be modified, however, to handle the scenario you envision. Alternatively, you could enter one character, note the network's judgment, then enter that same character plus an additional one and see if the result changes or stays the same. You could continue in this matter, simply increasing by one or two the number of known characters.

Fortuner: You are using very simple data, simple numbers. Are neural nets able to handle any kind of data including uncertain data?

Zerwekh: The data used in this system were actually real numbers. Neural networks in general can handle data of any mathematical type desired. If by uncertain data you mean data that represent guesses, or data that have a probability associated with them, the answer is *yes*. As I mentioned previously, neural networks tend to generalize what they learn about one pattern to other similar patterns. If you enter data that you believe are descriptive of some pattern, but you are not certain about all of the data, the network will locate an output response that best fits the input description.

——: What happens if I enter erroneous observations?

Zerwekh: If you enter one erroneous observation, say you thought a bone fragment had a sloping talo-fibular facet instead of a steep-sided one, then the network will report that such an animal is a prosimian, when in fact it is an anthropoid. But as far as the network knows, a sloping talo-fibular facet is indicative of prosimians. However, if you are entering a number of observations, and a number of them are correct and some of them are incorrect, the network will probably be closer to identifying the correct category than not. As my last example showed, conflicting information does not paralyze a neural network. It will still go ahead and make its best judgment based on the information that it does have. This is unlike many rule-driven systems where exact and complete input is required in order for the rule-base to function properly.

It takes expertise to be able to identify the bone characteristics. As I mentioned earlier, our research aims at moving away from relying on the expert's judgment and toward teaching the network to make these discriminations about bone features on its own.

13

Judgment-Simulation Vector Spaces

H. Joel Jeffrey

Introduction

A judgment-simulation vector space (Judgment space or J-space) is a technique for producing software that has the equivalent of human judgment in a specific area. The technique was originally proposed by Ossorio (1964) and has been used successfully in several applications, including medical diagnosis (Johannes 1977) and document relevance judgment (Jeffrey 1991). A J-space is produced by factor analyzing a data matrix in which the elements represent expert human judgments in the area needed. The procedure for creating this matrix, and for the encoding of judgments as numbers, is described below.

Systems based on J-spaces exhibit several desirable characteristics that are not usually found in traditional rule-based expert systems, such as recognition of ambiguous cases, recognition of cases the system is not prepared to handle, and graceful degradation of performance (the ability to give a partial or less certain answer when insufficient data are present, rather than simply failing to give an answer).

Unlike traditional expert systems, J-spaces do not require breaking down the problem into rules followed by human experts. Human judgment data are gathered and represented, but they are different in kind from condition-action or inference rules. For example, in the foot-bone identification space described in this chapter, the judgments gathered represent simply the degree to which each of a set of characters is indicative, in the expert's opinion, of each of a set of families or subfamilies. Further, these data are much more straightforward to collect, not requiring the time-consuming in-depth interviews of

Advances in Computer Methods for Systematic Biology: Artificial Intelligence, Databases, Computer Vision, ed. Renaud Fortuner (Baltimore: Johns Hopkins University Press).

traditional knowledge engineering. Collecting the judgment data is also discussed in the following section.

Any matrix, including a judgment matrix, with *M* rows and *N* columns may be considered to be a set of *MN*-tuples. Further, by analogy to the two- and three-dimensional case, each row (which is an *N*-tuple) is a point in an *N*-dimensional space. Here, *N dimensions* simply means that we need *N* coordinates to specify a point. Equivalently, we can consider each row to be a vector pointing from the origin, which is the point with coordinates (0, 0, . . . , 0), to the location given by the *N*-tuple. Such a vector is called an *N*-vector. Accordingly, one frequently finds an *M*-row, *N*-column matrix treated as a set of *M* points (or vectors) in *N*-space. However, this is not quite correct. It takes only three coordinates to determine any point in ordinary three-dimensional space; a set of 5-tuples would be overspecifying the points. If we had such a set of 5-tuples, it would be desirable to find out how many dimensions were actually needed to describe our data. Further, it is desirable (for reasons to be discussed below) to be able to describe our data in coordinates that are independent of each other, as the three coordinates of three-dimensional space are. Such a set of coordinates is called an orthogonal basis. The *x* and *y* axes of an ordinary graph are an orthogonal basis of 2-space; for 3-space, the *x*, *y*, and *z* axes are a basis. Thus, the basis can be thought of as the set of coordinates needed to describe points in the space. In constructing a J-space, the factor analysis is used to find an orthogonal basis for the data, so that it is represented in a true vector space in the mathematical sense, as contrasted to simply a set of tuples, the usual case (this is discussed in detail below.)

In this space, the mathematical distance between two elements corresponds to similarity of the two elements, as judged by persons with the relevant skill at the judgment. Further, a J-space exhibits generalization, the ability to simulate the human judgment of items, items not previously analyzed or judged by human experts, which is the sine qua non for a skill as contrasted with a device for retrieving known cases.

Construction of a Judgment Space

General Principles

Describing the construction of a J-space can be done either by giving the general form, in abstract terms, or by giving a particular case. In this chapter we have chosen to follow the second of these options. We believe that the case of primate foot bones is sufficiently representative that it will not be difficult for the reader to generalize to other data and other types of judgment. The work we are describing was done as part of an effort to produce a software system to identify primate foot bones. The space we used was an Identification space, a J-space in which the judgment to be reproduced is the identification of an item as a member of one class or another, based on characteristics used by the human experts. In more detail, it is an n-dimensional Euclidean vector space in which: (1) there is an orthogonal set of basis vectors (i.e., set of coordinates); (2) each basis vector represents a distinct type, or group, of taxa; and (3) each component of an item's n-component location in the space represents the expert judgment of the degree to which the item is identified as one of that group of taxa. As we shall see below, an item may be either a character or a specimen.

Each item has associated three coordinates, much as any point in ordinary three-dimensional space has three coordinates, typically written as a triple, for example, (6,2,4). Each coordinate represents the degree to which the item appears to belong in each group of taxa. Here 6, the first coordinate, represents how much the item appears to be in group 1; 2 represents how much the item appears to be in group 2; and 4 represents how much the item appears to be in group 3. The notion of distinct groups of taxa will be clarified below.

Step 1: Developing a List of Taxa

The Identification space is intended to identify a specimen as a member of one or more taxa. Sometimes, a specimen cannot be identified by human experts as belonging to one and only one taxon. For example, it may be that the expert reports, "This specimen very much appears to come from an organism of taxon A, but could also be from taxon B." The Identification space is intended to recognize such cases as the human expert does. The first step is to develop a list of taxa of interest. This is done by asking an expert, or a representative group of experts, for the taxa, or distinctions, of interest. These can be traditional taxa (genera, families, species, etc.), or any other classes of interest to this expert. In this particular case, we interviewed an expert physical anthropologist (Dr. Daniel Gebo of Northern Illinois University) and asked him to list the kinds of primates he wanted the software to able to identify. The resulting list of fourteen primate families is given in Table 13.1.

Step 2: Selecting Characters

The second step is to develop a list of the characters used by human experts in classifying specimens. In this case the task was to reproduce identification by foot-bone characters, so we had the expert look at a set of sample bones that he commonly used in teaching identification and simply note any characters that he considered at all useful or significant in identifying the bones, whether minor, barely relevant characters or major, highly significant ones.

This form of interview is quite different from an ordinary knowledge-engineering interview. In many ways it is a good deal simpler. Traditional

Table 13.1. Families and Subfamilies of Primates

Cheirogalidae
Lorisinae (Lorisidae)
Galaginae (Galagidae)
Daubentoniidae
Lemuridae
Indriidae
Tarsiidae
Callitrichidae
Cebidae
Atelidae
Cercopithecinae (Cercopithecidae)
Colobinae (Cercopithecidae)
Hylobatidae
Pongidae

knowledge-engineering interviews are quite time consuming and very concerned with answering *how* and *why* questions. Such interviews are based on two fundamental assumptions: (1) the expert is very good at articulating his or her reasoning; and (2) the expert's reasoning is a mental process that fits the model of if-then rules (either backward- or forward-chained). Producing a J-space requires neither of these assumptions. This is a strength, because these assumptions are rarely verified, and in fact it can be extremely difficult to find an expert who is also an articulate (and accurate) describer of his or her reasoning.

The only assumptions made in the data gathering for a J-space are that: (1) we have one or more experts; and (2) these experts can, when presented with a sample case, respond meaningfully to the questions, What is important here?, What should someone look at here?, or If you were teaching me to do this, what would you say about this sample? This is very much simpler than asking someone to explain his or her reasoning, and as a result is much less time consuming than usual knowledge-engineering interviews. For the primate foot-bone system, this procedure resulted in thirty-nine characters, some of which are shown in Table 13.2.

Step 3: Gathering Expert Judgments

If we consider the set of taxa in Table 13.1 to be column labels and the characters in Table 13.2 to be row labels, we have an empty matrix. This matrix is filled by asking human experts to rate the degree to which each of the characters indicates each taxon. This degree of match is expressed numerically, on a scale of 0 to 8, as follows:

Table 13.2. Selected Characters

Character number	Name
1	Sloping talo-fibular facet
2	Steep-sided talo-fibular facet
3	Broad, extending talo-tibial medial facet
4	Narrow, elevated above plantar surface talo-tibial facet
5	Lateral flexor groove position
6	Midline flexor groove position
7	Continuous distal facets, with long distal facets, on the subtalar joint
8	Separate distal facets, with a short proximal facet, on the subtalar joint
9	Moderately deep pivot on the calcaneo-cuboid joint
13	Long, narrow calcaneal shape
14	Short, wide calcaneal shape
15	Plantar heel tubercle present
19	Moderate (30 to 45%) distal calcaneal length
35	Talonavicular joint laterally rotated, oblong talar head shape
38	Talonavicular joint laterally rotated, asymmetrical talar shape

0: Not indicative at all, or counter-indicative;
1–2: Not likely, but marginally possible;
3–4: Could be, but not really indicative; not what you would think of at first;
5–6: Straightforwardly indicative;
7–8: Highly significant and indicative.

Missing values generally do not arise, because the expert is in each case being asked for a judgment he or she is competent to make, as an expert in the field. Should there be such, other experts can supply ratings. If a character is irrelevant to a particular taxon, it is not indicative, and so is rated 0. Selected rows of the rating matrix are given in Table 13.3. In making the ratings, judges are instructed to take no account of the relations between taxa or to analyze implications, but simply give their overall expert opinion as to this case, their first reaction.

The amount of labor required for rating every character with respect to every taxon is significant, but not overwhelming. Three to four hours is sufficient time to rate approximately 200 characters with respect to twenty taxa. The fact that the expert's first reaction is sought, rather than the result of detailed analysis, is important; it takes a few seconds, on the average, for each judgment. Further, the judgments can be done in batches of a half-hour or one hour at a time.

Table 13.3. Selected Judgments

Characters[a]	Taxa[b]													
	Ch	Lo	Ga	Da	Le	In	Ta	Ca	Ce	At	Cer	Co	Hy	Po
1	8	8	8	8	8	8	0	0	0	0	0	0	0	0
2	0	0	0	0	0	0	8	8	8	8	8	8	8	8
—														
9	7	0	0	7	7	0	7	7	7	7	7	0	0	0
—														
13	6	0	8	4	4	4	8	4	4	0	4	4	0	0

[a]See Table 13.2.
[b]See Table 13.1.

For the foot-bone system, many of these judgments were dichotomous, being 0 or 8. This is by no means universal; it is, if anything, more common to have many expert ratings in the range of 3 (*might be*) to 5 (*reasonably*). This was the case, for example, with the judgments involved in the thyroid diagnosis system of Johannes (1977).

Step 4: Factor Analyzing the Judgments

Step 3 above results in a great quantity of expert judgment about the relation of characters to taxa, in the form of a matrix. One would like to use this information for identification of new bones, for the point is to develop a system that can identify new items, not simply retrieve information about known samples. Doing that with the data in their current form would be very difficult, because the taxa (the columns of the matrix) are not independent; they will exhibit varying degrees of overlap and correlation. In other words, the columns representing the taxa may have similar values. When this is the case, it means that the characters used are not sufficient to provide a good discrimination between the two taxa. For example, the Lemuridae (Le) and Indriidae (In) columns in Table 13.3 have many identical or close values, indicating that these characters are not adequate for distinguishing between these two Lemuridae and Indriidae. Mathematically speaking, the columns do not form an orthogonal basis for the data.

In a bit more detail, the problem is as follows. With N taxa, we have an N-vector (i.e., an N-tuple) for each character (for three taxa, we would have a point, or vector, in three-dimensional space). Mathematically, it is of course a simple matter to calculate the distance between any two such N-vectors, or to mathematically combine a set of N-vectors representing a set of characters, thus producing a single N-vector for the set as a whole. However, it is difficult to tell whether the N-vector would closely match an expert's hand-assigned vector, without knowing how similar the taxa are. The original N taxa were deliberately chosen simply by asking the expert for taxa of interest. No analysis of similarity or overlap of taxa was done, and when the expert judgments are gathered, judges are specifically instructed to give their first reaction, which is essentially their initial expert judgment of the connection between character and taxon, with no analysis of statistical or logical connections between taxa. The data matrix thus represents, as far as possible, pure expert judgment (or expert recognition) data, with no consideration of overlap of characters or taxa.

To solve this problem, rather than use the expert judgment matrix directly, we find an orthogonal basis for it, in which each basis vector represents a set of correlated taxa, as measured by the correlation coefficients of the judgment matrix. The basis is found by factor analyzing the judgment matrix. Measurable common factors, plus any necessary unique factors, are the basis vectors. The factor loadings give the relationship between the original taxa and the factors, in two ways: (1) the loading is the cosine of the angle between the factor and the taxa, and (2) the square of the loading is the portion of the variance of the taxa accounted for by the factor. The basis may be thought of as the *basic set* vectors that can be used to label any point in the space. With a flat piece of graph paper, we have a basis set of two vectors (usually labeled x and y, or horizontal and vertical), and these vectors can be used to label any point on the paper, in the usual fashion: *This point is 3 over and 4 up*. For three-dimensional space, we have a similar situation, but we have three basis vectors that we might label x, y, and z, or *length*, *width*, and *depth*, and a point is labeled by noting its length, width, and depth.

Due to the meaning of the original data, the basis vectors represent distinct groups of taxa. Correlated taxa have been grouped into common factors, and the troublesome redundancy in the original knowledge basis is removed. The resulting vector space is the J-space. Since it represents judgments about characters as they related to identification of taxon membership, we term it an *Identification space*.

Step 5: Finding Orthogonal Axes

The factors extracted in Step 4 describe the data in terms of separate groups of the original taxa; each factor corresponds to a group. These factors are almost the basis set (i.e., the coordinates) for the space. To avoid computational anomalies the factors are manipulated as follows, to produce the actual basis vectors.

1. Factors with no loading over 0.7 are considered unmeasurable and are discarded. Intuitively, a loading of less than 0.7 means the column (the original taxon) points more than 45° away from the factor.

2. Any taxa (columns) with a loading below −0.7 are separated from the factor and treated as a separate vector, with loadings equal to the absolute values of the significant negative loadings. Intuitively, a loading below −0.7 means that the taxon points in the opposite direction from the factor; a loading of −1.0 would indicate a direction in the factor space perfectly opposite the factor. This means that taxa with negative loadings represent a distinct group. Recall that the factors, due to the meaning of the data, represent distinct groups of taxa, as indicated by the characters. Additionally, it is informative to consider the case in which an item has two characters, one of which is highly rated on a taxon with a high loading on the factor and the other highly rated on the taxon with the large negative loading. Since the value on this coordinate of the item location is an average of the ratings on the taxa composing the factor, the resulting value on this component would be approximately 0. This is a net loss of significant information, namely, that the item strongly resembles two separate types of taxon (those with positive and those with negative loadings on the factor). Separating the taxa into distinct basis vectors preserves this ambiguity, so that the system can report it to the user.

3. Each taxon has a loading that is the cosine of the angle between it and the basis vector. If the sum of the squares of the loadings (the commonality) of the taxa composing a factor is significantly less than 1.0, this means that taxon is not well represented in the space. Such a taxon (a unique factor, in the language of factor analysis) is included as a separate basis vector.

Step 6: Locating Characters in the Space

The space is initially populated by calculating the location of each of the characters via a formula that combines the original *N*-vectors, weighting each judgment by the loadings of the taxa that make up the basis vector (i.e., the coordinate):

$$C\,L_k = \frac{\sum_{i=1}^{n} r_i \cdot l_{ik}^3}{\sum_{i=1}^{n} l_{ik}^3} \cdot L_k$$

where

$C\,L_k$ is the *k*th component of the character location;

r_i is the rating of the character with respect to taxon *i*;

l_{ik} is the loading of taxon i on factor k;

L_k is the taxon with the maximum loading on factor k;

n is the number of taxa composing factor k;

The loadings are cubed to emphasize the contribution of the taxa with the highest loadings; the effect of the term L_j is to deemphasize the more diffuse factors. The Identification space is now complete and populated with the original characters.

Use of the Identification Space

To identify a new bone, a list of characters is produced. Each character has a location in the Identification space. That set of vectors is combined into a single vector for the bone. A wide range of algorithms could be used for this calculation; prior J-space-based systems have used the following formula, and we used it in the primate-bone identification system:

$$B\,L_k = \log_b \frac{\sum_{j=1}^{m} b^{c_j^k}}{m}$$

where

$B\,L_k$ is the kth component of the bone location;

c_j^k is the value of component k of character j location in the space;

b is base for the log average, typically $2 \leq b \leq 10$;

m is the number of characters identified;

Here we see the crucial contribution of orthogonality of the axes: we are guaranteed that the value of component k of the bone vector (location) can be calculated entirely from the kth components of the characters. A log average is used for ease of experimentation. By varying b in the above formula, the contribution due to lower values of the component is increased or decreased.

Having located the item in the Identification space, actual identification may be done in via two methods:

1. Projection: Find the component of the item location with the maximum value.
2. Nearest neighbor: Find the other item in the space closest to the new one and (within some threshold distance) identify the new item as being of the

same taxon as the previously identified close one. This algorithm of course assumes an existing base of identified items. In this experiment we used only the projection method.

Producing the list of characters may be done in many ways. For the work reported here, we used simple expert selection of features, much as a medical diagnosis systems often begin from a list of selected signs and symptoms. Physicians refer to this as *recognized* or *identified* signs and symptoms. Clearly, this is an area in which work is needed, exploring the possibilities of automated techniques for producing the list of features.

One such possible technique that we find very intriguing is functional composition of J-spaces. It is quite possible that, as the bone identification system grows, or as the approach is applied to other areas, some characters used by experts will not be easily, or perhaps at all, algorithmically recognizable. Table 13.2 shows several characteristics that appear to be of this sort. In such a case, another Character J-space could provide a means of bootstrapping the system. A Character space is produced by factor analyzing a matrix of judgments in which the columns represent characters to be recognized and the rows represent aspects of the sample that tend to indicate one or more character. The aspects, of course, are characteristics of the sample, or an image of it, that are algorithmically recognizable. An item to be identified is then processed by scanning it for algorithmically recognizable characteristics, identifying characters with the Character space, and using these characters as input to the Identification space.

Explanation of J-Space Characteristics

It is now easy to show why J-spaces exhibit the desirable characteristics mentioned at the outset of this chapter, namely, graceful degradation of performance, recognition of ambiguity, and recognition of cases the system is not prepared to handle.

Graceful Degradation

A location in the space is calculated for the item to be identified, based on the recognized characters. If an insufficient number of characters are supplied, the system still calculates the location, which means that in effect it still judges what the item most appears to be, based on supplied information. As the list of supplied characters is reduced, the identification, of course, becomes more questionable, just as it does with a human expert identifier, but the system does the best it can with the data at hand.

Ambiguity

It can happen that a set of characters is supplied to the system that results in the item location having a high value on more than one axis. This in fact happened with the test cases described in the following section. Such a vector indicates that the item, based on the supplied characters, strongly looks like two or more distinct types of taxon. The system can report the degree of ambiguity. For example, if component 1 is 7.5, and component 2 is 6.3, the identification is that the item strongly appears to be type 1, but also definitely appears to be type 2. Refinement of the system, by adding characters that resolve this ambiguity, may then be done.

Cases beyond the System Knowledge

It can also happen that all the characters supplied for a set have rather low values on all axes. In such a case, the resulting vector for the item will have low values on all components, thus representing in easily identifiable numerical form an identification of the item as one which is beyond the scope of the system, as determined by the characters and taxa given. In effect, the Identification space is reporting that the item only marginally appears to be any of the types of taxa known to it.

Results

We began with fourteen families and subfamilies of primates as listed in Table 13.1. Thirty-nine characters were then compiled, as listed in Table 13.2. Factor analysis, using the principal components method and varimax rotation, yielded the factors shown in Table 13.4. All of the four common factors are measurable, and none have significant (i.e., over 0.7 in absolute value) positive and negative values. Two taxa (Tarsiidae and Atelidae) are unique factors, having no loading with an absolute value over 0.7. Thus, the resulting vector space had six dimensions. The basis is shown in Table 13.5. The Identification space was then tested on three fossil bones and its identifications compared with those of the human expert. Fossil 1, for example, had the following characters:

1. Steep-sided talo-fibular facet (No. 2)
2. Talo-tibial medial facet narrow and elevated above the plantar surface (No. 4)
3. Midline flexor groove on the posterior talus (No. 6)
4. Continuous distal facets on the subtalar joint (No. 7)
5. Lateral rotated and oblong talar head (No. 35)

Table 13.4. Factors from the Character-Taxon Judgment Matrix

	Factors			
Taxa[a]	I	II	III	IV
Cheirogalidae	0.85	−0.13	0	0
Galaginae	0.70	−0.22	−0.16	0.30
Lorisinae	0.74	−0.15	0.24	0.38
Daubentoniidae	0.82	−0.08	−0.07	−0.30
Lemuridae	0.93	−0.05	0.01	−0.19
Indriidae	0.90	0.06	−0.04	0.07
Tarsiidae	0.49	−0.15	−0.06	0.41
Callitrichidae	0.03	0.28	0.17	−0.87
Cebidae	0.03	0.41	0.31	−0.8
Atelidae	−0.18	0.18	0.59	−0.66
Cercopithecinae	−0.15	0.95	0.08	−0.22
Colobinae	0.15	0.95	0.08	−0.22
Hylobatidae	0.04	0.03	0.91	−0.24
Pongidae	−0.12	0.00	0.95	−0.09

[a]See Table 13.1.

Table 13.5. Basis Vectors (Coordinate System) of Identification Space

	Factors					
Taxa[a]	I	II	III	IV	V	VI
Cheirogalidae	0.85	0	0	0	0	0
Galaginae	0.70	0	0	0	0	0
Lorisinae	0.74	0	0	0	0	0
Daubentoniidae	0.82	0	0	0	0	0
Lemuridae	0.93	0	0	0	0	0
Indriidae	0.90	0	0	0	0	0
Tarsiidae	0	0	0	0	1.0	0
Callitrichidae	0	0	0	0.87	0	0
Cebidae	0	0	0	0.80	0	0
Atelidae	0	0	0	0	0	1.0
Cercopithecinae	0	0.95	0	0	0	0
Colobinae	0	0.95	0	0	0	0
Hylobatidae	0	0	0.91	0	0	0
Pongidae	0	0	0.95	0	0	0

[a]See Table 13.1.

The location of Fossil 1 in the Identification space is: (4.9, 7.49, 7.86, 7.71, 7.26, 7.7). The most highly indicated group was Vector III (Hylobatidae and Pongidae), with Vector IV (Callitrichidae and Cebidae) and Vector VI (Atelidae) next. The expert anthropologist was asked to evaluate this judgment as he would that of a human identifier (student or colleague). He agreed that it was an appropriate identification of the item, including the most likely group membership and the ambiguity of other possible, though less likely, identifications. Similar results were obtained for two other fossils, with the following lists of characters:

Fossil 2: Characters 2, 4, 6, 8, and 38.
Fossil 3: Character 9, 14, 15, and 19.

Relation to Other Work

Historic

The original work on J-spaces was done by Ossorio (1964), who validated the factor-analytic techniques and produced spaces for judging subject matter relevance, attributes, and category membership of text documents. Johannes (1977) produced a thyroid diagnostic system using two J-spaces. The initial diagnosis from patient symptoms was produced by a space based on judgments of *the degree to which this symptom is indicative of, or looks like, this disorder*; a separate Test Result space refined the initial diagnosis based on the outcome of diagnostic tests ordered. Thus, the fundamentals are well understood and have been successfully applied in several different contexts. However, J-spaces have not previously been applied to pattern recognition. A working document retrieval system based on this technique was produced by Jeffrey (1991).

Expert Systems

In the sense that J-spaces are intended to reproduce expert judgment, they could be considered a type of expert system. This general definition is not, however, the one that is generally understood by the term *expert system*. That term is most often reserved for a system based on condition-action rules, and a separate inference engine doing either forward or backward chaining (Luger and Stubblefield 1989). Comparison of judgment-based systems with rule-based systems shows that the type of data is different, the methods for gathering it are different, no knowledge base of IF-THEN rules is produced, and the software to produce the simulated judgments is completely unlike an inference engine.

Factor Analysis

Factor analysis and the related Karhunen-Loeve analysis are standard techniques in statistical applications, where they are used for a purpose that is, to some extent, similar to ours: redescribing the data, so as to isolate certain more fundamental aspects of them that can be used in various ways. This process is often described as dimensionality reduction, in which a reduced number of dimensions adequate for describing the data is sought. In image processing, for example, Karhunen-Loeve analysis is used for data compression and image rotation (Gonzalez and Wintz 1987).

The use of factor analysis in J-spaces is unusual. In almost all cases the factor analysis is in effect a cluster analysis, and the factors are used as clusters; items are identified by searching for the nearest cluster, etc. In J-spaces, the factors (clusters), as we have seen, are used as a basis (using the term both formally and informally here) for describing the item to be identified.

Neural Networks

Neural networks, in effect, are another approach to clustering, not subject to the linearity constraints of factor analysis. Saund (1989) has shown that neural networks are also performing dimensionality reduction, that is, finding a new coordinate system (a basis) with fewer coordinates. With the exception of Zerwekh and Jeffrey (in preparation) no work has been done comparing the use of J-spaces with neural networks. Zerwekh (this volume, chap. 12) discusses primate foot-bone identification using neural networks.

Underlying Process Models

J-spaces directly address the problem of reproducing human judgment, rather than assuming that judgment is the outcome of an assumed unconscious process, that is to be modeled. No assumptions are made as to the nature of the process by which judgments are formed. In fact, the approach does not assume even that there is or is not any such process, cognitive, neural, or otherwise. The use of factor analysis should not be taken as implying that we believe that a similar process happens when a human being makes a judgment.

IV

Database Systems

14

Taxonomic Databases: The PANDORA System

RICHARD J. PANKHURST

This chapter is intended as a sequel to the companion chapter on identification methods (Pankhurst, this volume, chap. 8).

The normal practice of taxonomists when preparing an account of the plants of a region (a flora) or animals of a region (a fauna) or a monograph of a particular genus or family has been to collect the relevant data over a long period, and at the end to publish only some of the most vital facts together with various conclusions or assertions. The main collection of data on which the research was based is then just shelved, and often lost or forgotten, or even deliberately destroyed. The time period between revisions of a flora or monograph may be as long as a hundred years, so that the next worker to take up the subject often has to reassemble the data from scratch. This is of course tremendously wasteful, and may be poor scientific practice as well. Other workers should be able to reconsider the conclusions reached by a scientist during his research without having to gather original data afresh to see whether the same or different results may be reached.

The kind of taxonomic research just described may be intended only for internal use by other taxonomists. There is an increasing awareness, however, of external communities of users of taxonomic data, such as conservationists, legislators, horticulturalists, and agronomists who have genuine needs, but who, with good reason, have little interest in the academic aspects of taxonomy. The outward-looking design of database projects such as ILDIS (Bisby 1984) and ALICE (Allkin and Winfield 1989) illustrates this point of view. In this chapter, I shall concentrate more on databases for taxonomists than for users.

Advances in Computer Methods for Systematic Biology: Artificial Intelligence, Databases, Computer Vision, ed. Renaud Fortuner (Baltimore: Johns Hopkins University Press, 1993).

The Nature of Taxonomic Data

Taxonomic data may be classified into five main groups, which are all interconnected:

Specimen or Label Data—Data which relate to specimens in museum collections and which may be thought of as essentially a copy of the information on a specimen label. This will typically include the name of the collector, when and where the specimen was collected, and possibly notes on the habitat and the appearance of the specimen at the time it was collected. There may be none to several taxon names written on the label by various persons who have identified the specimen at different times. Type specimens, which form part of the published definition of a species, are of particular importance. It is important also to know in which institutions type specimens are kept, so that they can be borrowed or visited for study.

Nomenclatural Data—The published scientific names of taxa, the place and date of the publication of names and their author(s), and information on the status of names (accepted, provisional, synonymous), and, for synonyms, an indication of what name(s) are equivalent. The nomenclature usually implies a classification, which states which taxa are to be grouped together. There may be more than one classification in use.

Distribution Data—Data showing where taxa are or have been found, at various levels of geographical precision (via names of places and/or geographical coordinates), ranges of altitude or depth, and possibly status data (e.g., native, introduced, cultivated). Distribution information is derived from specimen labels, field notes, and literature.

Bibliography and Literature Data—Books and journal names, dates of publications, authors and editors, and page and volume numbers, for published sources of all kinds of information, especially for nomenclature, but also for uses, ecology, breeding, etc.

Morphological Data—The descriptions of the actual appearance of the organisms (i.e., their characters) at various taxonomic levels from the individual to high-level categories such as families.

It is probably true to say that the ultimate source of most taxonomic data is the preserved specimen, with the addition of field observations. Much of this information will in fact be derived indirectly from publications, however.

Data Structures

As a rather broad generalization, it may be said that the first four of the above categories of data can all be accommodated without difficulty in relational data structures as rectangular files (tables) with records containing fields, in the familiar way. These have been frequently referred to as *flat files*. Many examples of databases of this kind already exist, especially in botany; see the symposium volume on databases in systematics (Allkin and Bisby 1984). This volume contains descriptions of a number of databases that cover one of the above areas. The difficulties associated with such projects seem to have been principally managerial rather than technical (Pankhurst 1984).

There are also database projects that have attempted to combine these different types of data (e.g., Pankhurst [1988a]). The nomenclatural data files, and to some degree the geographical files also, involve the expression of hierarchical information in relational form. Let us consider a nomenclatural file as an example. It is frequent practice to have one record for each species and to have fields within each record for other categories such as family, genus, or subspecific taxa. This is not actually wrong, but it does have an important disadvantage, which is that it becomes awkward to introduce other levels of taxonomic category as an afterthought. A better way to express this kind of data is instead to have one record for every name, regardless of its taxonomic level. Each record will contain at least the following three fields (see the chapter appendix for an example):

1. The actual name string;
2. The category level (e.g., species, family). This may be expressed in various ways (e.g., as an integer), but what is important is to be able to decide whether one name is at the same or a lower or higher level than another. The actual set of categories that are in use can in fact be stored in a separate table of their own, together with their relative levels;
3. A pointer to the next highest taxon to which this name belongs. For a species, this might point to the genus, and for the genus, it could point to a family. Eventually, this chain of pointers will reach the root of the hierarchy, and the last pointer will be null (i.e., it will point to nothing).

In order to record, for example, the members of a genus, there will be a record to describe the genus and then a series of records, one for each species. A small number of extra records are needed for this representation, but flexibility is gained. In order to add a new category, all that may be needed is the addition of one record to the file of categories.

Nomenclatural data do present one special difficulty, which is that there may be relationships between the records of the same file, rather than between different files. If, for example, one record holds a name that is a synonym of a name in another record, then some means is needed of expressing this relationship. This can, of course, be done in a proper relational fashion by having separate link files whose records express the accepted name to synonym relation, or by putting accepted names in one file and synonyms in another, but it would be more elegant to be able to keep all the information about names in one file.

Morphological Data

In chapter 8 an explanation has already been given of the DELTA format and how it is used to express morphological data. DELTA, however is just a data format, and not a database. Data for descriptions of taxa in conventional relational database management systems present special difficulties, which will now be discussed.

Mixed Data Types. It is extremely common for a combination of qualitative, integer, and real quantitative characters to be considered together. It is, of course, possible to express quantitative characters as qualitative characters by dividing up numerical characters into ranges, but this is often an excessively crude approximation, and it is better to able to treat all the variables in the same way. Hence a suitable database system will be flexible about data types and ideally should allow the database designer to define his own.

Variability. It is very usual for characters to vary, for example, *flower color = white* or *pink*. Such variability occurs at all levels from the individual to high-level taxa. At the level of an individual the variation is within one organism, such as different shapes of leaves on one tree. At higher levels such as species the variation can also be between individuals that belong to the category, as well as within them, but this distinction does not normally have very much practical consequence.

If a character is represented as a field in a file, there is an immediate problem because there are very few database systems that allow the contents of a field to vary. It is possible to use more than one record, or to define special character states for the combinations, but neither of these approaches is very practical. Alternatively, variable character states may be represented as a string (a list of states) of somewhat arbitrary length. This can be done if the database system allows fields that have no type or allows special data types to be defined.

Character Dependency. In most DELTA data sets there are rules (not in the expert system sense) that connect characters. For example, if a plant has no stem, then characters of the stem are not possible. If there is a stem, and if stem leaves are present, then it is possible to say whether these leaves are simple or compound. If the leaves are compound, the leaflets may be described, but not otherwise. In other words, there is a hierarchy of rules connecting characters. If characters are stored as fields in records, then some means is needed to express relations between records of the same kind. The same remarks as already made in the context of nomenclature apply here also.

Image Data. Very little image data have yet appeared in taxonomic databases, but it is clear that these will become very important in the near future (Gómez-Pompa and Plummer 1990). It is not very much to ask for a database system to be able to display images that have been previously stored. All that is needed is to store the name of the picture file and to have a routine that will display it.

Another desirable feature of the database system, in view of the considerations just given, and of the presence of much character string data of arbitrary length, is variable field length. A great many database systems operate only with fields of fixed length (e.g., the remarkably popular dBASE systems). There does not seem to be any very good reason for this restriction, but it does cause problems with character data, such as titles of articles, for example. Either an excessively long maximum field length has to be set, to allow for the longest possible string, thus wasting storage space, or the contents of fields have to be abbreviated or truncated in order to fit.

A taxonomic database that included images would be likely to display pictures of type and other specimens, color slides of organisms in the wild and their habitats, monochrome and color drawings, and distribution maps. Rather than attempt to convey the excitement and potential of this kind of development in words, an eighteen-minute video was shown describing the Q-TAXA project (Gómez-Pompa 1989).

Choice of Database Management System

The Hamburger Model

From the point of view of the database designer, and in the view of the author, the ideal database system would follow the hamburger model, that is, a system with three layers of code.

The uppermost layer (the top slice of bread) would be the developer's own code, written in some language suitable for software development, such as C.

This code would call into a package of subroutines that constitute the proprietary database system, which would handle all the regular features of a database, data capture, browsing, indexing, reporting, and so on. This is the middle layer, the meat so to speak.

This package would also be provided with "hooks" so that the developer's own code could be called to provide data checking of special data types, checking of consistency between records, special-purpose report programs, and custom features in general. This would again be written in the developer's choice of language and would constitute the bottom level (the lower slice of bread).

Existing Database Management Systems

One database system that actually fits this model, and that allows variable field length, is EMPRESS (RHODNIUS 1986). The author has not yet used this system, but has found Advanced Revelation (COSMOS 1987) to be one widely used system that fits many of the requirements. Revelation is, in fact, closely related to the PICK operating system, while not actually being directly derived from it.

Revelation allows both variable field lengths and variable numbers of entries within fields. It also permits the display of images. Revelation does not actually fit the hamburger model, but instead might be described as an *open sandwich.* The developer can write his own code below the level of the database system in order to customize his database application, but Revelation is a free-standing system running under MS-DOS that cannot be called as a subroutine package from an upper level of code. Revelation has another disadvantage in that it has its own special programming language known as R/BASIC (equivalent to the ACCESS language of PICK), which is a dialect of BASIC. R/BASIC is, however, well designed for string handling.

Taxonomic Database Applications

PANDORA Database for Taxonomic Revisions

The author is using Revelation in order to develop PANDORA, which is a general-purpose database application for taxonomy. It handles and integrates all five types of taxonomic data and, in particular, includes full input and output and editing of DELTA data files for morphology.

Morphological descriptions occur as *items* (in the sense of DELTA), that is, descriptions extracted from literature, descriptions of specimens, and descriptions of taxa. Descriptions of taxa are, of course, derived from descriptions of specimens and from literature, so a routine is provided so that a selection of descriptions of items can be merged to form the description of a taxon. This is presumably what a taxonomist does in his head when writing a taxon description based on a selection of specimens. There is probably also another step in the process in which the taxonomist eliminates untypical or aberrant characters or specimens, so there is also an *edit* type of item that records character changes made when polishing the description of a taxon. This has the important property that it can easily be repeated whenever the concept of a taxon changes, as happens upon the arrival of new and better material, or by the renaming of a specimen that had previously been included.

Another important feature of PANDORA is that, although it reflects only one classification of the group being studied, the PANDORA user can continuously change his working conception of what the currently correct classification is supposed to be, and the output obtainable from PANDORA changes continuously in order to reflect this. If, for example, a taxon is moved into synonymy when it was previously thought to have been an accepted name, the reports on nomenclature and the checklists and specimen lists will change to fit the new situation.

PANDORA also assists with collecting data for the descriptions of items. Once a list of characters and states has been set up, a new item can be described via either a menu of characters from which the user chooses the character(s) he wishes to score or in an automatic mode. In the latter, PANDORA cycles through the available unscored characters, presenting a menu of states for each, inviting the user to enter descriptions. In this way, descriptive data can be captured in an orderly fashion. If data are not available for a character, the character can be skipped. If a character is dependent on the state(s) of another character, the controlling character has to be satisfied first before the dependent character is requested. The cycle can be repeated until the user has no more data that can be entered. At this stage, a report can be printed listing missing characters for the item, if any. This procedure can also be followed when it is desired to alter the description of an item that has already been scored.

PANDORA is currently being used to prepare an account of the family Rosaceae for the Flora Meso-Americana. A full description of the features of PANDORA is given in the chapter appendix.

Other Taxonomic Databases

The ALICE system (White et al., this volume, chap. 19) is also suitable for handling taxonomic data, but is oriented more toward users of taxonomic data than toward taxonomists themselves. With this in mind, ALICE puts less emphasis on data about specimens. The HyperTaxonomy system for Macintosh microcomputers (Skov 1989) has a lot in common with PANDORA. It uses the HyperCard system, which handles files of records as though they were cards in a filing cabinet. HyperCard has a good graphics interface and is immediately suitable for handling image data, which Skov has put to good use for the production of distribution maps. On the other hand, HyperCard is not well designed for indexing. Like Revelation, it has its own special programming language, called HyperTalk, which is object oriented in nature.

Database Standards

A database standards group exists for botanical databases, the International Union of Biological Societies (IUBS) Commission on Taxonomic Databases, also known by its previous name of the Taxonomic Databases Working Group (TDWG). This has been preparing standards for the interchange of botanical garden data, for the description of taxon status, for botanical nomenclature, for abbreviations of the names of books and journals, and for the abbreviations of the names of authors of taxa. TDWG standards will be published by the Hunt Institute for Botanical Documentation, Carnegie Mellon University, Pittsburgh. The DELTA format is already a TDWG standard, and a standard general-purpose set of descriptors (characters) is being considered. In the zoological world, the *Zoological Record* is the standard database for zoological taxonomic literature and is made available online via the BIOSIS organization.

Appendix: Notes on the PANDORA Database

PANDORA is a database system for the preparation of monographic or floristic accounts. It handles data about nomenclature, specimens, literature, geographical distribution, and morphological descriptions—in fact, all types of data needed for specialist taxonomic revisions. It is envisaged as a tool for an individual researcher, but data can be imported from and exported to other databases. PANDORA is intended to produce text for inclusion in a flora or monograph.

The storage and input/output of descriptive data are completely compatible with the DELTA format for the production of keys and descriptions. PANDORA is implemented as an application of Advanced Revelation (AREV) relational database system under the MS-DOS operating system for IBM PC–compatible microcomputers. PANDORA is derived from the database for the Flora Meso-Americana described by Pankhurst (1988a) and is influenced by other databases, particularly Hypertaxonomy (Skov 1989). An AREV database for data of the Flora Europaea shares some details with PANDORA. The ALICE software for the ILDIS database has some similar features, but is oriented toward taxon data rather than specimens.

PANDORA requires the Advanced Revelation database management system, which must be purchased separately (runtime version available), and an IBM PC 286 or 386.

Scope of the Database. Certain principles are followed in the design of PANDORA. Every screen is provided with online help, so that the database is largely self-documenting. As a matter of principle, it is possible to support each item of data by a source, which is often a reference to a specimen or a publication. This can be very important when it is necessary to know where data are derived from. There exists an international commission of IUBS for standards in taxonomic databases (TDWG), and again as a matter of principle, its standards are incorporated and followed. Every effort in programming has been made to check the consistency and form of data at the time of entry and to use a relational database structure. This means that, wherever possible, any data item that can be used more than once is stored only in one place (e.g., a personal name). Also, all data fields that contain fixed character strings are presented to the user as menus from which choices are made. For example, with personal names again, at each instance when a personal name is referred to, a menu pops up with all known personal names presented in alphabetical order. This makes it unlikely that such names will be duplicated by carelessness or misspelling. Many of the records in the various files are identified by arbitrary record numbers, but a means is always provided so that the user may quickly search for a particular record or group of records. In other words, there are numerous indices of different kinds, and these are automatically maintained by the system. In practice, the user may completely ignore the numbering scheme. There is also a *browse* feature that allows selected viewing of data in a file.

The data screens are greatly nested within each other; in other words, if, at any time, data are being entered and it becomes apparent that it is necessary to refer to other data that are not already present, then it always possible to call up another data screen to add something, and then return to the original task. For example, in the middle of entering a new name, an authority string may be wanted. If this is not already stored, then a screen for authorities can be called up in order to put in the new authority. If this in turn requires the name of a person that is not already available, yet another screen can be called to do this. The user may then escape back to the authority screen, and then back to the name screen, where all the partly entered data are frozen and still awaiting completion. Alternatively, of course, the user may try to anticipate such situations by entering the low-level data first, but there is no obligation either way. A selection of reports for output is provided, but there is also a means for the user to design his/her own reports. It is likely that a user will be working on one revision at a time and that the database will hold all the relevant data for one project. It is nevertheless possible to dump all data from one project and reload with another.

At the time of writing, the plan of the database includes features that are implemented and others that are not.

People. Personal names arise in many contexts, such as collectors and determiners of specimens and as authors and editors of books and journals and as authors of species. Since often the names of the same people occur in several of these ways, each name is stored in full just once, and then used repeatedly. The life dates of persons are also included, as from the Kew draft index, so as to make it possible to cross-check the dates connected with specimens and literature.

Nomenclature. The basic nomenclatural record in PANDORA is a taxon name with its taxon level and its relation to higher taxa. An example should make this clear. Consider a name such as *Taraxacum officinale*. There will be a record for the name *officinale* at the level of species, with a pointer to another record that gives the next highest taxon to which this belongs. This will be another record for the name *Taraxacum* at the level of genus. This record, in turn, may point to another record for the family Compositae (Asteraceae). Many levels are allowed for, from class down to cultivar, including intermediates such as tribes, subgenera, etc.

Each name has a status, according to the researcher's current opinion, and is set to be accepted, synonymous, provisional, or unknown. An accepted name can have no to many synonyms, and a synonymous name is associated with one or more accepted names. A name with provisional status will probably be converted to accepted or synonym status at some later point. Various kinds of synonyms are allowed; in particular there are *pro parte* synonyms that express the one-to-many relation between a *sunk* aggregate species and its accepted splits (e.g., the historical *Orchis latifolia* versus various modern species of *Dactylorhiza*). The database keeps track of the relation between accepted and synonymous names at all times, as when the status of names is changed by editing. The database keeps track of the parents of hybrids and of homonyms. The authorities of names are abbreviated and stored according to the Kew draft list of author names (Meikle 1980). The place and year of publication of the taxon (the citation) are visible in the taxon record.

Specimens. A record for a specimen is, roughly speaking, a copy of the standard fields of information from a herbarium label. Data from other kinds of museum specimens (e.g., zoological specimens) will be very similar. The unit record is a sheet, rather than a collection, so that a specimen is identified by collector, collector's number (if any), and the owning herbarium. This is to allow for differences between duplicates, such as (1) different determinations and other annotations; (2) differing type status; and (3) different morphology (e.g., some sheets with or without flowers or fruit). Determinations or identification of specimens is stored in a different file, so as to permit specimens to have more than one determination by different experts. Data on types, with reference to relevant specimens, are stored in the taxon file, and there can be more than one type specimen (isotypes, syntypes) for a name. Notes from the label can be stored and whether it is flowering or fruiting and reported uses.

Geography. Geographical data are closely connected with specimen label data. There is a four level-hierarchy for geographical areas such as country, state or department, county or town, and finally locality details. It is possible to define a study area so that distribution of records from inside the area and outside it are distinguished. Distribution data can come from either specimens or literature, and include geographical coordinates, and there can be used to display and print maps. There is a gazetteer of place names and coordinates that can be cross-checked with distribution data. The status (native, introduced, etc.) of each taxon in a recording area can be stored and retrieved.

Literature. Each literature reference is split into information about which book or journal it is, and then the individual references by volume number, page, etc. Books and journals are identified by an abbreviated and full title, according to BPH (*Botanico Periodicum Huntianum*) or TL2 (*Taxonomic Literature*, 2d ed.). It may be necessary to allow for more than one standard abbreviation. Literature references for particular taxa may be indexed.

Descriptions. It is intended that the processing of descriptive data (i.e., the construction of keys, generation of descriptions, clustering, etc.) will be carried out outside PANDORA using various programs for the purpose, but based on the DELTA format. Hence PANDORA can load, edit, and export DELTA data in much the same way as the DEDIT editor. DEDIT is an editing program for DELTA format. The description of taxa by their characters can be carried out completely within PANDORA with minimal knowledge of the details of the DELTA format itself.

Descriptions of taxa depend ultimately on descriptions of specimens and descriptions from publications, and these can be combined as required. The currently valid description of a taxon is based on a combination of characters from validly identified specimens merged with literature descriptions. If specimens are reidentified or if taxa are lumped or split, new correct descriptions can be regenerated. It is also possible to distinguish observed characters (interpreted from looking at specimens or illustrations) from recorded characters (from specimen labels or published descriptions).

Reports. The variety of reports that might be generated in the context of a flora or monograph is virtually endless, especially since the system has its own built-in report generator available to users. Some report programs have been provided, which have various formatting options included. These are as follows;

- report on status of type specimens (not located, located in which herbaria, loaned or not, seen or not) as needed for the revision;

- list of accepted names with synonyms;
- checklists of taxa for a region, or of regions for a taxon;
- specimen lists for a taxon or for a region;
- generation of distribution maps;
- retrieval of taxa satisfying given criteria;
- literature list for given taxa;
- flower and fruiting times and altitude ranges for a taxon.

15

Hierarchic Taxonomic Databases

JAMES H. BEACH, SAKTI PRAMANIK,
AND JOHN H. BEAMAN

Database development for biological diversity information is now being actively pursued by many individuals and institutions throughout the world. Biological databases cover several domains spanning from nucleotide sequences and micromolecular characters to morphological data, specimen occurrence, and ultimately classification and taxon data from the hierarchical level of kingdom to that of species and below. Abbott et al. (1985) recognize five major types of taxonomic databases, namely, curatorial, biogeographic, bibliographic, nomenclatural, and descriptive.

An area of biological database management that has not yet been effectively optimized is the processing of specimen data and taxon-level information within the context of the taxonomic hierarchy, allowing for multiple alternative classifications and nomenclatural data processing at all hierarchical levels. New classifications are usually built by sharing, changing, and tuning taxonomic concepts of existing classifications. Developing an information model for taxon information should be based on the important criterion of sharing and adjusting taxon concepts of existing classifications when new classifications are proposed. Any information-processing model for classifications must capture both the taxonomic concepts and the relationships among them as data entities. We describe an information model here that will capture taxon concepts, their relationship to other concepts in the taxonomic hierarchy, and alternative taxon names (synonyms) resulting from interacting classifications, as well as the historical modification or evolution of taxonomic concepts.

This chapter does not deal with the theory and process of producing a taxonomic treatment or a classification. Instead, it describes the types of

---

Advances in Computer Methods for Systematic Biology: Artificial Intelligence, Databases, Computer Vision, ed. Renaud Fortuner (Baltimore: Johns Hopkins University Press).

information present in classifications as an initial step toward the implementation of a taxonomic database management system (DBMS).

Biological Classification

Classification Data Elements

Published taxonomic information about organisms traditionally has been organized in four major forms: (1) comprehensive classification systems that arrange families, orders, or other higher-level taxa according to putative evolutionary relationships; (2) monographs and revisions that present in a systematic manner all of the taxonomic information known about a particular group throughout its geographic range or for a specified part of the world; (3) floras and faunas that summarize the taxonomic information about the plant or animal species of a particular area; and (4) descriptions of one or a small number of recently discovered taxa, usually new species, subspecies, or varieties. Specimens collected from the wild are the original source of information on the taxa to which they belong.

Relatively few comprehensive classifications for the higher hierarchical levels (above the level of family for plants) are currently in use, and new ones appear only at infrequent intervals owing to the enormous effort involved in their development. The taxonomic hierarchy, however, is not fixed at these higher levels; there can be several competing views of the relationships within and between the taxa. Systematists classify and rearrange concepts at the higher levels based on various types of research into the evolutionary relationships of the groups.

Most research on the classification of plants and animals is performed at the level of the species. Taxonomic revisions and monographs are common vehicles for the communication of classification data. They typically analyze ten to forty species belonging to one genus, or in some cases an equivalent number of species within several small genera. Floras and faunas are the principal mechanism in which plants and animals of a given geographical area are identified. They may include formal descriptions of new taxa. Most taxa "new to science," however, particularly at the level of species and below (subspecies, varieties, forms, etc.), are first recognized in short articles in botanical and zoological journals.

The principal information elements of monographs, revisions, floras, and faunas are (using an analysis of species within a genus as an example): (1) the genus name; (2) a summary of the higher-level taxa (subfamily, family, order) to which the genus belongs; (3) the original author (describer) of the genus; (4) the bibliographic reference to the original description; (5) a list of the synonymy, that is, other names and taxonomic concepts which have been used to describe the genus in full or in part; (6) a description of the biological characteristics of the genus; (7) a diagnostic key to included species; and (8) a formal systematic treatment of each species and any other subordinate taxa such as subspecies and varieties.

Formal treatment of each species would then usually include: (1) the species name (comprising the generic name, specific epithet, and author of the species, e.g., *Andromeda calyculata* Linnaeus); (2) a reference to the publication which first described the species (if it is an existing taxon and not being described for the first time in the current work); (3) a reference to specimens which have been designated by the author of the species to typify the species (*type specimens* of various kinds); (4) a citation to the herbaria or museums in which the type specimens are located; (5) a list of names and concepts considered as synonyms to the accepted species with their bibliographic and specimen citations; (6) a description of the biological characteristics of the species, such as morphological, molecular, anatomical, and ecological features; (7) specimens examined in the study; and (8) discussion.

Classification is the hierarchical arrangement of groups by rank and position according to artificial criteria, phenetic similarities, or phylogenetic relationships. Generally, classifications provide a system of named and circumscribed taxa for information storage, retrieval, and use (Radford 1986). The conceptual bases supporting classifications of living organisms pertain to each rank in the hierarchy, but far more attention has been devoted to species concepts than to the other ranks. We refer the reader to Mishler and Donoghue (1982) and Stuessy (1990) for theoretical considerations of plant species concepts.

A new classification of a biological group is an attempt to replace previous classifications on the basis of new data and the results of more sophisticated analyses. New classifications are never completely accepted by everyone, and conflicting classifications, for example, of the species delimitations within a genus, can coexist for many years. Multiple overlapping classifications are widespread in biological systematics. The formal taxon definitions of earlier works provide points of reference and must be analyzed in order for the names to be correctly applied to taxa in new classifications. For these reasons, each of the distinct layers of interpretation set down for a group of organisms by successive authors must be recorded, managed, and consulted during the classification process.

Nomenclature and Classification Methodology

After a new classification has been created, the international codes of nomenclature (botanical, zoological, bacteriological) specify protocols for designating the correct name for any new or modified taxonomic concepts. The rules largely deal with the availability of names for a taxon and, for the available names, the precedence for those that have been applied to the same group in the past. The rules do not constrain a classification, but they do specify which names can be applied to taxa, once a classification is established. Successive taxonomic revisions of a group result in the creation of various classes of synonyms, or alternative names for taxa. The rules of nomenclature consider the status of accepted names and their synonyms in considerable detail. Tracking and handling accepted names and their synonyms are nontrivial tasks for classification authors.

The names of taxa at the rank of species and below are fixed through the designation of particular specimens that serve as *types*. At the rank of genus, one component species is designated as the *type species* for the genus; thus the type in this case is a concept rather than a specimen. Likewise, a component genus serves as the type for a family.

The challenge for a DBMS is to store and manage all the name and type specimen information associated with taxonomic concepts in a flexible manner accurately representing diverse overlapping classifications. Table 15.1

Table 15.1. Taxonomic Hierarchy for Flowering Plants and Approximate Number of Taxa

Level[a]	Use[b]	Approximate number of taxa
Kingdom	1	1
Subkingdom	2	1
Division	1	1
Class	1	2
Subclass	2	30
Superorder (used by some in place of subclass)	2	
Order	1	50
Family	1	390
Subfamily	2	?
Tribe	2	?
Genus	1	36,000
Section	2	?
Species	1	250,000
Subspecies	2	?
Variety	2	?
Form	2	?

[a]Additional ranks may be added to this hierarchy.
[b]1 = mandatory, 2 = optional (whether or not the entity must be present in the classification of a plant species).

indicates for flowering plants the approximate number of currently recognized taxa at each hierarchical rank that must be accommodated. With the inclusion of synonyms and homonyms, the number of names to be managed at each of the ranks shown is considerably greater than indicated in the table.

Example of a Classification

Table 15.2 gives an example of the current classification of a single plant species. The table shows the taxonomic hierarchy for *Chamaedaphne calyculata* (L.) Moench, a plant common to Northern Hemisphere bogs. It can be seen from the table that the concepts of several different authors contribute to the classification of this plant as we currently conceive it. The history of the classification and nomenclature of *Chamaedaphne calyculata* is complex. The genus name *Chamaedaphne* was first validly published by Mitchell in 1769 for a species of partridge berry in the coffee family, Rubiaceae. Linnaeus, however, had used the name *Mitchella* for the same genus of plants in 1753.

Moench, in 1794, used the name *Chamaedaphne calyculata* for leatherleaf, a species in the heath family, Ericaceae, that Linnaeus had originally placed in the genus *Andromeda* as *A. calyculata*. Moench was breaking up Linnaeus's heterogeneous genus *Andromeda*, but in the process created a homonym by using the name *Chamaedaphne*, which had already been used by Mitchell for a species of partridge berry. According to the rules of botanical nomenclature the name *Chamaedaphne* was unavailable for use because it had already been applied to another group of plants, even though the name was reduced to a synonym of *Mitchella*.

Don in 1834 also concluded that certain elements should be segregated from the genus *Andromeda* as construed by Linnaeus in 1753. Don provided

Table 15.2. Example of a Classification of One Species

Kingdom: Regnum vegetabile (or Plantae)
  Subkingdom: Embryobionta Cronquist, Takhtajan and Zimmerman, 1966
    Division: Magnoliophyta Cronquist, Takhtajan and Zimmerman, 1966
      Class: Magnoliopsida Cronquist, Takhtajan and Zimmerman, 1966
        Subclass: Dilleniidae Takhtajan, 1966
          Order: Ericales Lindley, 1833
            Family: Ericaceae A. L. De Jussieu, 1789
              Subfamily: Vaccinioideae Endlicher, 1839
                Tribe: Andromedeae de Candolle, 1839
                  Genus: *Chamaedaphne* Moench, 1794
                    Species: *Chamaedaphne calyculata* (L.) Moench, 1794

the new generic name *Cassandra* for this segregate taxon and, like Moench, derived it from Linnaeus's species *Andromeda calyculata*, to create a taxon with the name *Cassandra calyculata* (i.e., the "type species" of the genus *Cassandra* Don).

Apparently without knowledge of Don's work, Spach, in 1840, independently used the name *Cassandra* for a genus based on a different species in the heath family that had been originally described by Linnaeus as *Andromeda racemosa*. Because the type species of Spach's genus *Cassandra* was different from that of Don's *Cassandra*, Spach, by applying the same name to a second distinct group of plants, had created an illegitimate homonym of the name *Cassandra*. The validity of Don's generic name *Cassandra* was not affected by Spach's later homonym.

The rules of nomenclature permit appeals for the use of names for taxa that would otherwise not be permitted for reasons of priority. Because the name *Chamaedaphne* had first been used for species of Rubiaceae, it could not be validly applied to species in the heath family, Ericaceae, but because the name had been relegated to a synonym of *Mitchella* describing the same genus of Rubiaceae, Ingram, in 1967, proposed that *Chamaedaphne* Moench be "conserved" over *Chamaedaphne* Mitchell, in order to make it available for Ericaceae. After several years of controversy, the proposal was eventually approved at the 1975 International Botanical Congress, and *Chamaedaphne* Moench was recognized as a conserved name in the 1978 edition of the *International Code of Botanical Nomenclature*.

Taxonomic research on taxa below the species rank has also contributed to the information-processing complexity of the taxon *Chamaedaphne calyculata*. The variety *Andromeda calyculata* var. *angustifolia* was described by Aiton in 1789. It was subsequently raised to a species-level taxon as *Andromeda angustifolia* by Pursh in 1814 and transferred to *Cassandra* as *Cassandra angustifolia* by Don in 1834. The same taxon was recognized as *Chamaedaphne calyculata* var. *angustifolia* by Rehder in 1900. Another entity that has been recognized in the species *Chamaedaphne calyculata* was originally described by Aiton in 1789 as *Andromeda calyculata* var. *latifolia*. This was recognized as *Chamaedaphne calyculata* var. *latifolia* by Fernald in 1945 and accepted by Wood (1961). Rehder published another variety, *Chamaedaphne calyculata* var. *nana*, in 1914.

The name conflicts in the preceding paragraphs involve either *nomenclatural synonyms* or homonyms which result from the mistaken application of names to taxa, due to faulty taxonomic judgment, ignorance of other published works, or various types of historical accidents. To complicate matters further, but in order to represent the complete history of the nomenclature and taxonomy of this one species, *Chamaedaphne calyculata*, another class of names would need to be considered. *Taxonomic synonyms* are duplicate names for taxa which result from classification decisions that combine two or more originally distinct taxonomic entities. "Lumping" and "splitting" taxa on the basis of new data or new analyses are commonplace operations on classifications. Examples of taxonomic synonyms are given below.

Also, we have not addressed the details of the classification of *Chamaedaphne calyculata* above the level of genus, but many more relevant names and concepts could be identified in the classifications at higher taxonomic levels shown in Table 15.2. If all of the classifications that have ever treated this species and the taxa above it in the hierarchy were considered, the complexity for data management would be much greater than what we have discussed so far.

Management of Taxonomic Concepts

Constraints of Taxonomic Databases

Managing and tracking formally published names yield only part of the history and relationships of taxonomic concepts, because the rule systems of biological nomenclature were not designed to uniquely identify all taxonomic concepts. The rules of nomenclature specify rigorous and conservative protocols for adjusting the names of taxa after changes are made in the classification of a group, but they do not respond to many modifications to taxon circumscriptions (i.e., changes to taxon definitions or logical boundaries). DBMSs that work with names and their associated nomenclatural type concepts as indices to taxonomic concepts are constrained in two major ways: (1) they are unable to record any changes to taxonomic concepts that do not involve the reassignment of a name; and (2) they are unable to perform cross-classification comparisons to any extent.

With regard to the first constraint, consider the two common operations on a classification shown in Figure 15.1. Figure 15.1a shows a hypothetical example of four congeneric species that were taxonomically combined or lumped into a single taxon as part of a new classification. The rules of botanical nomenclature specify that the names *Species two*, *Species three*, and *Species four* are taxonomic synonyms of the concept named *Species one* in the revised classification. Most taxonomic database systems would record these names and establish the logical link between *Species one* and its synonyms. It is obvious from the figure, however, that this nomenclature does not uniquely identify the new species concept as indicated by the reuse of the oldest acceptable name (*Species one* Harris) for the new taxon. Taxonomic database systems that use names as indices to taxa are unable to distinguish between two concepts with the same name.

A similar situation occurs when new taxonomic concepts are created as a result of splitting up a broader, previously recognized taxon. Figure 15.1b diagrammatically illustrates this commonly occurring operation with a tax-

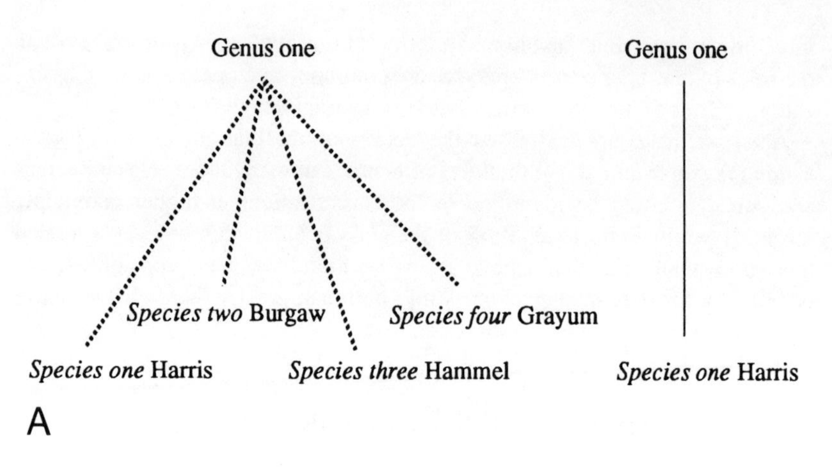

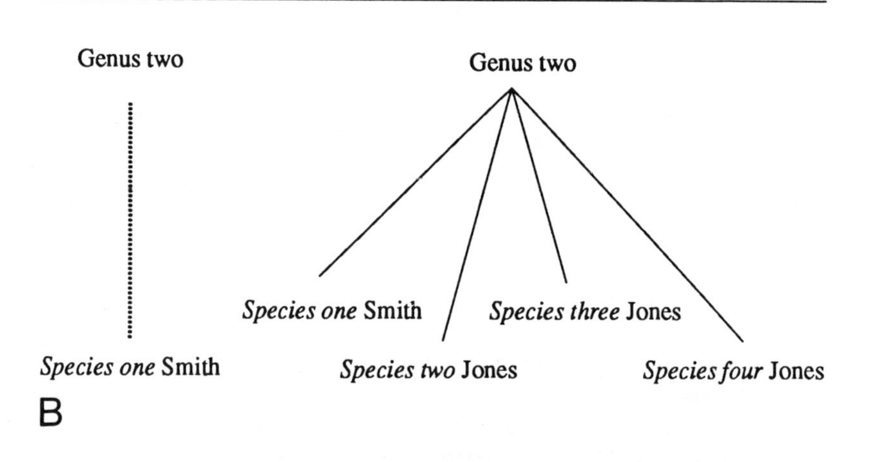

Figure 15.1. (a) The nomenclatural result of combining four species into one, where the oldest acceptable name for the new species concept is that which was previously applied to a narrower taxonomic concept. (b) The nomenclatural effect of splitting a species into multiple taxa. Note particularly that there is no change in the name *Species one* Smith even though the taxonomic concept has been substantially modified.

onomist named Jones splitting a species previously proposed by Smith into four separate species. The distinction between a preexisting concept and a newly narrowed concept with the same name is lost in the scientific nomenclature.

In the second constraint, cross-classification comparisons of taxonomic concepts are essentially impossible when taxon names and not concepts are the basis of a taxonomic database system. This is true for comparisons between classifications on those concepts that lose their unique identity (i.e., when their name is used for a revised concept as in the examples above), but it is also true for comparisons of concepts that are uniquely indexed by the nomenclatural system. The difficulty for automated comparisons of classifications (see queries below) is largely due to an incomplete data model resulting in inadequate data structures.

Processing Ad Hoc Queries

Processing and analyzing data in a DBMS can be effected in several modes. It is theoretically possible to analyze classification information in any type of file or database system, given enough time and programming input. A distinction must be realized between developing one-time programs for specific queries and a more general approach to managing classification information that would permit ad hoc queries over any number of taxonomic and nomenclatural data elements. The distinction is important because taxonomic DBMSs that require repeated, customized programming to access data are not practical for taxonomic research.

The conceptual data structures outlined below could facilitate processing taxonomic information in several ways, but none is more illustrative than the manner and extent to which information in classifications could be queried. We will describe three example queries that could be handled by our approach. High-level algorithms for query processing are discussed in the section on "Processing UNIC Structures to Answer Ad Hoc Queries."

> *Query 1.* Which species of the genus *Andromeda* described by Linnaeus in *Species Plantarum* in 1753 were subsequently transferred to another genus? (Who did the transfers? In which publication? For each of the species involved, was the entire concept moved to a new genus or just some of the biological elements it contained?)

To answer this question, a database system would first need to locate the species of *Andromeda* described by Linnaeus. It would then need to identify all of the species-level concepts in the database that were transferred from the Linnaean species of *Andromeda* as the result of later taxonomic revisions (classifications).* The query compares a single classification against an indefi-

*Here, we apply the term *classification* to indicate any publication that discusses a taxonomic concept (e.g., taxonomic revisions, monographs, floras, faunas, and new species descriptions). Traditionally, systematists would not consider a list of species or a single new species description a "classification," but we use that term to signify any formal publication that contributes new information or taxonomic opinion to our knowledge of a group.

nite number of unknown classifications. It is a meaningful, practical query because taxa are frequently remodeled by later investigators. An understanding of the taxonomic status and delimitation of revised concepts is a necessary part of taxonomic research.

Query 2. What is the history of the nomenclature and taxonomy of the species *Andromeda calyculata* Linnaeus?

This query represents a typical question posed by systematists when they undertake a monograph or revision. To find the answer, a database system must find the species concept with this name and author as well as a classification under which it is currently recognized. It would locate other names applied to the same taxonomic concept by other authorities in other classifications and indicate the relationship of those names to the present one as nomenclatural or taxonomic synonyms, as well as their respective authorities[‡] and publication information. The system would locate any uses of the name in any unrelated taxa by any other author or authority and report those names as homonyms.

The system would also navigate through the species-level concepts and determine which earlier species concepts circumscribed all or part of the same taxon. The system would then report all of the relevant author, publication, and date information on those concepts.

Query 3. How does the concept of *Hieracium abscissum* Lessing of Robinson and Greenman (1904) compare with the concept of *H. abscissum* Lessing of Beaman (1990)?

This query seeks whatever information is available in the database that would distinguish two interpretations of the same taxon by two different authorities. The query would locate the two taxonomic concepts in the database within the two classifications indicated. It would then compile a list of differences between the two, starting with an enumeration of taxon concepts that are subsumed within each classification (e.g., Beaman considered fourteen other names to be taxonomic synonyms of *H. abscissum*). The query processing would look for differences in the delimitation of infraspecific taxa, their number in each case, and the differences between them. The type of information that could be found by this query would include: (1) differences in the number of infraspecific taxa; (2) inclusion or exclusion of other named taxa within these two concepts; and (3) inclusion or exclusion of unnamed (implied) concepts that may have been included or excluded in the circumscrip-

[‡]Here, an *author* is the person or persons who published the original name for a taxon. An *authority*, in contrast, is the person or persons who published a classification (in a broad sense) that may or may not affect the name used for the taxon or the taxonomic placement of that group. For example, Niedenzu (1890), Wood (1961), Stevens (1970), and Judd (1979) could be regarded as authorities in the overall classification of *Chamaedaphne calyculata*.

tion of the taxon by either authority. If morphological attribute information were stored in the database, aggregate morphometric statistics might be generated.

The essential point is that a DBMS architecture allows concepts and their attribute information to be tracked, thus permitting comparisons of taxonomic concepts across classifications. Systems that track only concepts that have unique names (that have not been reapplied to a revised broadened or narrowed concept) could not process this query, because the name *Hieracium abscissum* would be used to indicate (index) both Beaman's and Robinson and Greenman's concepts of the species.

Approaches to Processing Hierarchic Classification Data

Existing Database Systems

Several taxonomic DBMSs have been designed and implemented for processing the taxonomic data elements that are based on classifications (cf. Allkin 1988; Allkin and Winfield 1990; Beaman and Regalado 1989; Crosby and Magill 1988; Gibbs Russell and Arnold 1989; Gómez-Pompa and Nevling 1988; Humphries et al. 1990; Skov 1989). All of these systems employ a simple approach to managing the taxonomic hierarchy and associated taxon data entities, principally by tracking nomenclatural changes and the resulting synonymic links as indices to the underlying classifications.

Some commercial database systems allow direct representation of hierarchical data. IMS of IBM Corp. and System 2000 of Intel Corp. are two such hierarchical database systems available on large computers. Although these systems are able to represent parent/child relationships through hierarchical schemata, they lack the flexibility to represent a variable number of levels in the hierarchy. They are also unable to support multiple, overlapping hierarchies for the same nodes in the tree structure. A requirement for taxonomic database systems is the ability to modify parent/child relationships, dynamically including changes that would link new levels. Such requirements are not easily implementable with traditional hierarchical database systems.

In relational systems, tree-structured binary relations for recursive retrievals are difficult to accomplish (Celko 1989; Date 1990, p. 665). There has been a significant amount of research for database systems where data elements represent parent-child relationships. Much of the work involves computing transitive closure of binary relations (Valduriez 1987; Larson and Deshpande 1989; Naughton et al. 1989; Agrawal et al. 1989; Pramanik and Vineyard 1990). Valduriez and Boral (1986), for example, proposed the implementation of recursive joins by using join indices, which are binary relations representing the hierarchical structure of the database.

New System Research and Design

Our objective is to represent a classification by logical fragments of a unified hierarchic structure representing all interacting classifications. Pramanik and Vineyard (1990) have investigated fragmentation of hierarchic databases for optimizing distributed database queries and for minimizing transmission costs for fragmented recursive relations in distributed database systems. We are concerned here in designs that will logically partition a unified hierarchic database for efficient accessing of individual classifications and allow navigation through the interacting classifications.

Representation of Taxonomic Hierarchies. We propose using directed graphs to represent the structure for classifications and interactions between classifications. A directed graph is a finite nonempty set of vertices (nodes) and a set of edges, such that the edges are ordered pairs of vertices. A set of directed graphs will represent the taxonomic hierarchy as well as the relationships between hierarchies of related classifications. The main idea is to combine various related classification hierarchies into a unified structure that we call UNIC (UNIon of Classifications), which is a directed graph with each node representing a taxon concept based on a distinguishable set of character data. A node includes the taxon name, author or authority, date, and a reference to the publication.

Information in a parent node can be derived from its children in the UNIC graph structure. Use of this UNIC structure has important advantages over systems that do not use directed graph structures. For example, attribute data at higher ranks can be derived automatically from the attribute data of the corresponding specific and infraspecific categories. Alternative classifications can be related through the shared UNIC nodes, facilitating navigation through various classifications.

The UNIC structure can represent the union of taxonomic hierarchies of all classifications. To locate the subgraphs representing individual classifications in a UNIC structure, a CLassification INdex (CLIN, as we call it) to the UNIC structure is needed. This index points to individual hierarchies within the UNIC structure, representing various classifications. For most classifications this index can be represented by a set of root nodes and the number of levels in the hierarchy.

UNIC-Node Lineage Information and Evolution of Taxonomic Concepts. UNIC and CLIN structures store various classification information and the taxonomic hierarchies within classifications. The proposed UNIC structure also allows sharing of nodes between classifications. However, the UNIC and CLIN structures do not provide any information about the progression for an evolving taxonomic concept within a particular rank. Although systematists recognize the identity of distinct interpretations of a given taxonomic concept by different classifications, these evolving "concept lineages" must be represented by additional information in the UNIC and CLIN structures. One approach for representing evolving taxonomic concepts is to use another directed graph structure within the same UNIC structure. Thus the same type of UNIC structures is used to represent both hierarchical and lineage information. To differentiate between hierarchical and lineage structure, two types of links are used between the nodes.

Figure 15.2 illustrates a UNIC structure for the example classification shown partially in Table 15.2. In Figure 15.2 the hierarchies are represented by lines without arrows. The lineage information is implemented by arrows indicating the direction of the transfer in a particular publication. The nodes <*racemosa*, Linnaeus, 1753, Pub 1> and <*calyculata*, Linnaeus, 1753, Pub 1> represent *Andromeda racemosa* and *Andromeda calyculata* of Linnaeus's *Species Plantarum* (Publication No. 1). The node <*angustifolia*, Pursh, 1814, Pub 2> represents the species *Andromeda angustifolia* (Ait.) Pursh (Pub 2). The figure shows that *Andromeda calyculata* L. was transferred by Moench in 1794 (Pub 4), resulting in the name *Chamaedaphne calyculata* (L.) Moench. The transfer can also be between different ranks. For example,

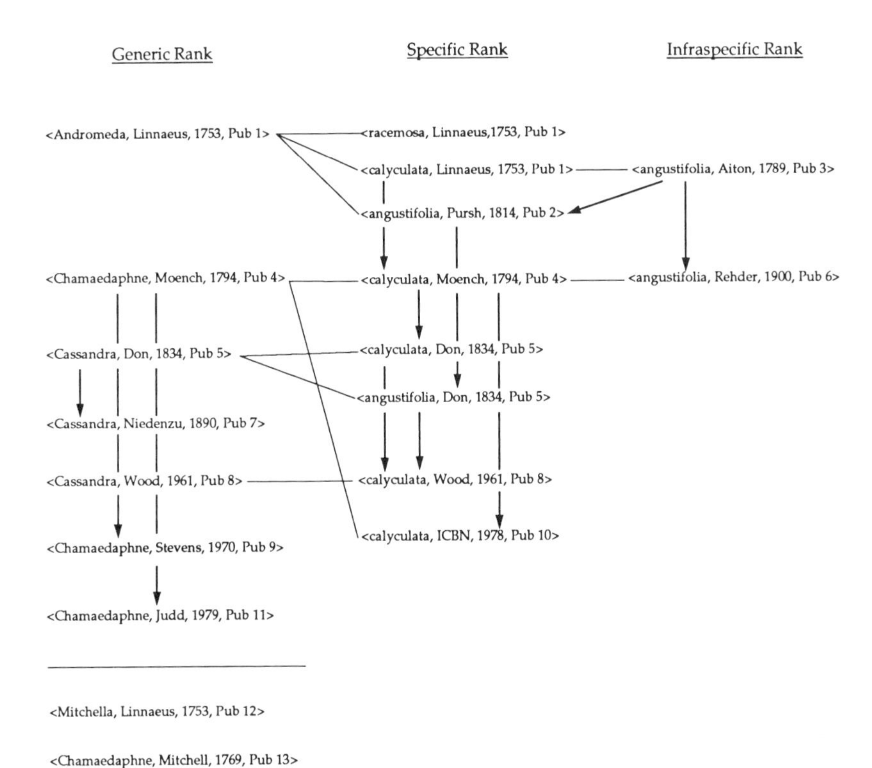

Figure 15.2. A UNIC structure for part of the example classification of Table 15.2.

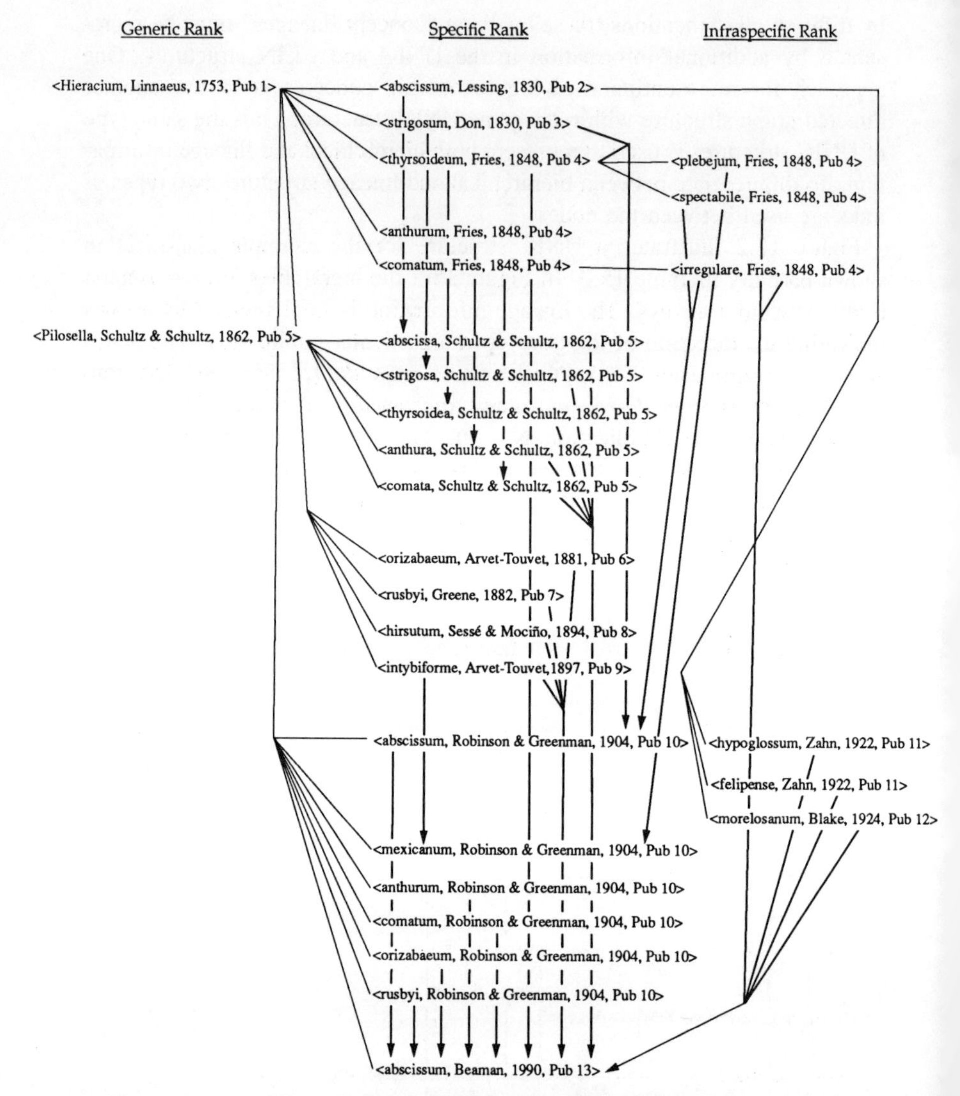

Figure 15.3. A UNIC structure for classifications of *Hieracium abscissum* Lessing.

the variety *Andromeda calyculata* var. *angustifolia* Aiton was elevated to the species *Andromeda angustifolia* by Pursh.

Processing UNIC Structures to Answer Ad Hoc Queries. In this section a conceptual framework is presented for implementing Queries 1, 2, and 3 (cf. "Processing Ad Hoc Queries" above) by processing the UNIC structures of Figures 15.2 and 15.3.

Query 1 would be answered by locating the node <*Andromeda*, Linnaeus, 1753, Pub 1> in the UNIC structure and then finding the child nodes with author Linnaeus. This will find the species *Andromeda racemosa*, *Andromeda calyculata* (and others not shown in Figure 15.2). Only *Andromeda calyculata* will be chosen for the answer because it is the only species (in this data set) that is transferred, as indicated by a transfer arrow, to another genus.

To answer Query 2, first the species node <*calyculata*, Linnaeus, 1753, Pub 1> would be located in the UNIC structure (Fig. 15.2). By following the transfer arrows, all the other species nodes belonging to other generic concepts would be derived. The generic information for these species is given by their parent nodes. The following history information will be created as a result of this query.

> *Andromeda calyculata* Linnaeus, 1753; *Cassandra calyculata* (L.) Don, 1834; *Cassandra angustifolia* (Ait.) Don, 1834; *Cassandra calyculata*, Niedenzu, 1890; *Cassandra calyculata*, Wood, 1961; *Chamaedaphne calyculata*, Stevens, 1970; *Chamaedaphne calyculata*, ICBN, 1978; *Chamaedaphne calyculata*, Judd, 1979.

The classification information for Query 3 is shown in Figure 15.3. To answer Query 3, the database system would first locate the two nodes corresponding to *Hieracium abscissum* Lessing of Robinson and Greenman and *Hieracium abscissum* Lessing of Beaman. This could be achieved by locating the two nodes at the specific rank, <*abscissum*, Robinson & Greenman, 1904, Pub 9> and <*abscissum*, Beaman, 1990, Pub 13> in the UNIC structure. By following the arrows we would then derive the two sets of taxa, at the specific and infraspecific ranks, that have been merged into these two species concepts. These two sets are then compared to derive the desired results of the query.

Functional Requirements and Future Research Directions

Research is needed on the various conceptual models for the UNIC structure. Models would need to support the algorithmic development of nomenclatural and taxonomic synonyms, store evolving taxonomic concepts (concept lineages), and be implementable in a distributed database environment. Some of the important data parameters to be considered are the number and size of classifications, the number of taxonomic concepts, and characteristics based on various requirements such as the ability to create partial classifications and to share taxonomic databases.

An efficient implementation of UNIC and CLIN structures is vital for the development of the taxonomic DBMS because they are the major structures controlling the taxonomic hierarchy and the history of successive taxonomic revisions. The implementation must be flexible, allowing the UNIC and CLIN structures to be easily modified and extended. The same basic UNIC and CLIN structures should be capable of representing the biological hierarchies across taxonomic disciplines.

Performance of competing models for implementing UNIC structures must be examined. In particular, performance should be measured in the context of modifying UNIC structures to add classifications and new ranks, accessing various parts of the UNIC structure, and navigating through the taxonomic database collecting information on the nodes within the UNIC structure. Various auxiliary access structures will probably be needed to facilitate the navigation process. A set of fundamental operators will be needed to manipulate the UNIC and CLIN structures.

An easy-to-use interface to the database is necessary. The type of queries that should be handled include searches on databases using completely or partially specified keys, the capability to navigate through hierarchies of interacting classifications, and the ability to insert new and revised classifications into the database.

It should be possible to copy any classification that has been entered into the UNIC and CLIN structures and transport it into other independent UNIC and CLIN structures. Whole classifications would probably be most often shared, but copy and transport of partially entered classifications should be allowed. For example, an investigator may enter UNIC data that include only a segment of a hierarchy, representing only a part of a more comprehensive classification. Some users may require the sharing of segments of UNIC structures that may be incomplete at the higher-level nodes of a classification.

In addition to mechanisms to export and import UNIC data between systems, there is also a requirement for a mechanism to maintain update consistencies between databases. That is, if the data in a particular classification in a UNIC structure are updated at one site, a mechanism would be needed for a second site to integrate that information into a second UNIC structure without any loss of information already in the UNIC structure at the second site. There should be a mechanism allowing the user to choose nodes or a fragment of the incoming hierarchy or to equate incoming nodes with existing nodes, etc. Research is needed on primitive UNIC structure operations that will handle the update decisions and actions that arise with shared data.

Acknowledgments

We would like to acknowledge the financial support of NSF grant BSR-8822696 and NSF grant DIR-9021656.

16

New Database Technology for Nontraditional Applications

MARIANNE WINSLETT

Introduction

A database has traditionally been thought of as a large computer-based collection of data, typically a million or more bytes; having a regular format, usually tabular; usually shared between multiple users who often cause conflict by their desire to update the same piece of data at the same time; usually having high performance requirements (up to a thousand interactions per second between the database and users); and requiring the use of a special language for access to data, both to make it easy to pose complicated questions about the data and also to insulate the users of the system from the details of how the data are stored and accessed.

Current database management systems (DBMSs) can be adapted to meet some of the needs of systematic biology, and emerging DBMSs hold the promise of meeting many more of these needs. In this chapter we survey the facilities available from current commercial DBMSs, discuss new features that are relevant to systematic biology, and close with a few words of advice for systematists who are planning on working with computer scientists.

The Advantages of Using a DBMS

Traditionally, research and development work in the database community has focused on the problem of efficiently managing large amounts of persistent, reliable, shared data.

The key words in this definition are: *persistent*, meaning that the data survive across invocations of the programs that access them; *large*, meaning megabytes or gigabytes of data; *reliable*, meaning that the data should survive

---

Advances in Computer Methods for Systematic Biology: Artificial Intelligence, Databases, Computer Vision, ed. Renaud Fortuner (Baltimore: Johns Hopkins University Press).

across software and hardware crashes; *shared*, meaning that multiple users should be able to access the data simultaneously, without fear of introducing anomalies through a strange interleaving of data reads and writes; and *efficient*, meaning that such large quantities of data will require special access techniques in order for programs using the data to be able to find the data records they need in a reasonable amount of time (e.g., without searching through the entire database in order to find one needed record).

Traditionally, database technology has been aimed at the needs of business data processing: programs for payroll processing, billing, banking transactions, and online reservation systems, for example. If one uses ordinary files rather than a DBMS as a repository for data, then data will survive invocations of the programs that access it, as would happen if a DBMS were used; but with ordinary files there will be no built-in support for sharing, reliability, or efficiency of access, all of which are provided by a DBMS.

Another advantage of using a DBMS rather than ordinary files is *data independence*. With ordinary files, when the data format of the files changes, one must rewrite every program that accesses those files. It is much better to insulate the programs that access the data (called *applications*) from changes in data format, to the maximum extent possible; otherwise, in the long run this maintenance overhead becomes enormous. Special DBMS techniques ensure that applications generally do not have to be rewritten when simple changes are made to the format of the data.

DBMSs assure the automatic enforcement of *integrity constraints* for the data. An integrity constraint is a rule specifying that certain combinations of data values are illegal. For example, an integrity constraint might decree that for all descriptions of chickens in the database the number of legs and the number of wings be exactly two; another constraint might require that all managers make at least as much salary as do their employees; another might insist that all reported egg weights be between 1 gram and 1 kilogram and that the egg be no heavier than the bird that laid it. Then the DBMS can automatically detect and reject descriptions of three-legged chickens that lay 10-pound eggs.

If one uses ordinary files rather than a DBMS, then each program that accesses the data must guarantee that all integrity constraints are satisfied. This makes it difficult to construct applications in a modular fashion: the writer of a program who wants to change one part of the data must understand all the integrity constraints that might possibly relate to that data. Further, this architecture requires replication of code for checking constraints inside many different application programs, resulting in unnecessary coding effort. It also becomes difficult to change or add constraints, as every program that enforces those constraints must be tracked down and changed. For these reasons, it is preferable to have a centralized facility for managing constraints, rather than spreading that functionality across all applications programs.

Using ordinary files also makes it difficult to combine data generated by different persons, using possibly different hardware and software. Often one would like to be able to access the data as though they resided in a single common unified repository, even though in fact the data may be scattered across many different machines at different sites, using different data formats and having incompatible semantics. Ordinary files do not provide any special help in giving a unified front to these disparate data; some DBMSs will help in giving an integrated appearance to the data. In particular, most standard commercial relational DBMS companies sell products that allow one to access data residing at other sites, providing that all the data are stored using the company's own product. Current products vary in how easy they make this access (e.g., to access remote data, does one have to specify the sites at which the data live, or will the system find them for you automatically?); commercial products are rapidly improving in this area. One can already use these commercial products to, for example, create a combined database for a dozen nematology museums, each with tens of thousands of microscope slides with nematode specimens. Each museum can maintain its own collection records locally, and a user can query all museum records simultaneously, as long as all museums agree to use the same data structure, same nomenclature, and same DBMS. For combining data with widely different formats or with different semantics, however, current commercial relational systems do not provide enough support to meet the anticipated needs of biological applications.

Commercial DBMSs provide a number of other facilities that are important in business applications, but probably not so important in biological applications. These include facilities for maintaining the integrity and consistency of the data when multiple programs access the data simultaneously, persistence of the data across software and hardware catastrophes so that not even one millisecond of completed work can be lost, and security measures to keep users from seeing and/or changing information to which they should not have access.

As it is extremely difficult to write a program that solves all the problems listed above, it is best to write that code just once and share it among all application programs; such a program is called a DBMS. Ordinary heavy-duty commercial systems deal fairly well with all the problems discussed above, at least in the form that those problems appear in managing business data. But these DBMSs are not built to deal with the problems that arise in biological applications, so, as we discuss below, there is a mismatch between what commercial DBMSs are good at and what biological applications need.

One can also buy commercial DBMSs that are not heavy-duty: DBMSs to run on personal computers. These DBMSs tend to be slow when there is a lot of data (megabytes), and they often do not provide many facilities for sharing of data, since personal computers are intended for single users. Some of these DBMSs lack a powerful language for use in accessing data; for example, it may be hard to combine data from two tables and put it out in tabular form.

It is easy to confuse a DBMS with a tool for presenting data, such as HyperCard and other hypertext systems. HyperCard is not a DBMS; it solves none of the problems listed above. For example, Hypercard does not provide a data access language in the sense we described above (*Give me all experiments that showed evidence of . . .*); it does not provide any data independence; and it is not practical for use with large amounts of data. On the other hand, HyperCard is an excellent *user interface* for a separate DBMS, that is, it would provide easy-to-use screens for the user interaction with this DBMS. As most biologists would not want to learn the details of a DBMS data access language, it will be very important in biological applications for the user to have a high-quality user interface attached to the DBMS. A number of DBMS companies already sell a combination of a DBMS with a hypertext-based or otherwise user friendly user interface. It is very tempting, when one sees a demonstration of a high-quality user interface, to think that the main functionality of the system resides in that piece of software, but in fact the invisible pieces of the system, such as a DBMS back end, are necessary and very important.

Relational Databases and Systematic Biology

Description of Relational Databases

Almost all commercial DBMSs sold today are relational, which means that all data must fit into simple tables (often called *relations*), such as the two shown in Figure 16.1. Nonrelational systems include a few hangers-on from earlier

BIRD_SIGHTINGS

EXPERIMENT_ID	OBSERVER	DATE	BIRD	NEST_ID
34	Mendez	1-1-91	Warblecock	64
34	Smith	1-1-91	Crested Bleater	7
59	Smith	10-5-90	Warblecock	64
64	Liu	3-6-90	Tufted Harlpet	21

NEST_INFO

NEST_ID	MATERIAL
64	Grass
64	Straw
64	Paper clips
21	Twigs
21	Grass

Figure 16.1. Two examples of simple tables appropriate for relational database management systems.

days (*hierarchical* and *network* DBMSs), but these are not relevant for the vast majority of applications, including scientific ones. The remaining non-relational systems are *object oriented*.

The tables in relational DBMSs must have a completely regular format, with exactly one entry in every intersection of a row and a column. If the correct entry is unknown or is unavailable, many systems allow the use of a special value to represent this situation, which can be printed out as a blank or as a special symbol. In the first relation, BIRD_SIGHTINGS, we have a unique ID for each experiment, then the name of the person who made the observation, the date the observation was made, and the kind of bird sighted. Apparently every sighting is supposed to include the viewing of a nest, because the final column of BIRD_SIGHTINGS asks for a unique ID for the nest observed. Information about those nests is kept separately, because the same nest can be observed in multiple sightings, and one would not want to repeat the same nest description for each sighting. In NEST_INFO we have the unique ID for the nest, and then the materials that the nest is made of. We could not have a single column MATERIALS for all the materials in a single nest, because in the relational model one can put only one value in each intersection of a row and a column. Therefore we flattened out the information about nest composition and used several rows to describe nests that were made of several different materials.

The relational model also includes a data-access language, in which one can ask for such information as, *Please give me the names of all the observers who saw warblecocks in 1989*, or *List all the materials seen in nests in experiment 35*. The data used to answer a query are a set of tables, and the answer to a query is yet another table; tables are the basic unit of operation. This allows one to cascade queries upon queries.

The advantages of the relational approach are its simplicity; its clean, solid, mathematical theoretical foundations; its data independence, where all the details of storage are hidden; and the relative ease with which one can write new programs that access the data, using a *nonprocedural* language (one describes the data one wants to find, not how to find them on disk). The relational model offers data independence at three levels: first, the way data are actually stored on disk is much more complex than just a simple table, but the user is unaware of this; second, one can change the schema for the data (i.e., the choice of tables, table columns, and constraints on the values that may appear in them; one can think of the schema as a description of the data format or structure) without having to rewrite the programs that access the data, for a wide range of changes; third, one can reorganize the way the data are actually stored on disk, without having to rewrite any of the programs that access that data.

Problems arise if one tries to apply relational technology to many biological problems. While it is possible to force some types of biological data into a tabular format, it may be awkward and difficult to do so. For example, protocols, algorithms, pictures, and hierarchies do not have a natural tabular structure. In contrast, business data fit relatively easily into a tabular format. Once data are in tables, data access in biological applications will often be slow. This is because business software typically accesses only a few records per interaction with the DBMS, with a very high volume of interactions. Current DBMSs are built to run quickly for business programs, not for biological programs that want to analyze and mull over the data. And finally, one will still need a lot of software on top of the DBMS, because current DBMSs offer relatively little support for incomplete and imprecise data and for checking constraints, and this capability is needed for biological applications. We will discuss each of these problems in more detail below.

From the point of view of a traditional relational DBMS, there are many troublesome characteristics of biological data, described in the subsections below. Most of these troublesome characteristics can be handled, with varying degrees of difficulty, in the object-oriented model. One can also address many of these problems by adding new features to the relational model, an approach often overlooked by proponents of object-oriented DBMSs. We will discuss some useful DBMS features that are relevant to both relational and object-oriented approaches to this problem. We will also present another approach to the problem: use a programming environment intended for artificial intelligence (AI) applications, where many of these problems have been addressed, and have a DBMS as a back-end just to store data (see Karp, this volume, chap. 17, for a discussion of the use of AI frame representation languages in systematic biology).

Data with Complex Structure

In biology, one often has data that do not fit easily into simple tables: sequences (i.e., ordered lists of data items, such as readouts from monitoring equipment), hierarchically ordered data, partially ordered data, maps, protocols (plans of experiments), etc. Although one can store these types of data as ordinary files or as long, unstructured, single entries in a database, doing so eliminates the possibility of having a database understand the internal structure of the data. For example, if one needs to look up all recent experiments that have used a particular piece of equipment, the DBMS cannot handle that query efficiently unless protocol information is stored in a structured manner, rather than lumped together in a single, uninterpreted field of a record.

Imprecise Data

Biological experiments often generate imprecise data: a value may be expressed as a range, or with a probability attached, or as a fuzzy value, a qualitative judgment (*not very large*), etc. Different observations of the same item may use different ways of handling imprecision, making the observations quite difficult to compare for a computer (is *between 0.7 and 0.9* greater than, less than, or equal to *not very large*?). In order to piece together this kind of information to reach conclusions, one needs approximate operations, which are not supported very well by traditional DBMSs.

Inconsistent Data

Traditional approaches to performing inference on data assume that the data are consistent, which biological data usually are not: the data usually contain errors and also descriptions of aberrant specimens. Inconsistency creeps in because measurements are inaccurate, because the same data can be interpreted in different ways, because of experimental error, and through anomalous biological events (e.g., actual three-legged chickens). One can address these problems by adding support for imprecision: intervals, probabilities, partial orders, the AI technique of *truth maintenance systems*, and so on.

Incomplete Data

Traditional relational DBMSs do not offer much support for handling missing values (e.g., observations omitted from experiments). DBMSs do often allow the use of a special null value whose meaning is *value exists but is unknown*, but then that null value requires special support during query processing: a null does not mean the same thing as an ordinary value. For example, one should not assume that two null values in a table are equal to one another, yet two occurrences of the same ordinary value do have that property. Special support for missing data is needed in biological applications.

Although one can code in special support for missing data in the application, this is a tedious and error-prone process. It is much better to be able to tell the DBMS that certain values are to be treated as missing data and then let the DBMS incorporate that information into its routines for computing statistics. Even when the application layer provides special support for declaring missing data (as in the popular statistical packages used in the social sciences), it is all too easy to forget to include the needed declarations of missing data. As a result, debugging a program written in one of these statistical languages usually becomes an exercise in tracking down undeclared missing values. It would be much better to make a permanent record of exactly which values were missing, and to keep that information with the data, so that the user only had to include code for handling missing data if she or he wanted to override the default treatment provided by the DBMS.

Rapid Changes of Type of Data

A biology lab is not like a bank. Different experiments keep different data, and ideas on what is worth recording change rapidly. Traditionally, this would mean revamping the schema for the data whenever there are changes in the information being kept, or declaring new relations whenever the information recorded is changed. Neither approach is very workable when the schema changes rapidly: it is hard and tedious to force old data into a new format, and it is difficult to issue a uniform query over a lot of different versions of relations.

Multiple Versions of Data

Replicated experiments give multiple versions of data, to increase reliability. Biological variability requires different versions of information (e.g., male versus female, variation between individuals of a population, variation between different species). Different versions of derived data (e.g., gene maps) represent competing hypotheses. Management of multiple versions of engineering design objects (e.g., car doors, wheels) is fairly well understood; biological versions have not been studied. The main difficulty is that the data access language needs to understand the versions, so that the user can refer to a generic object that exists in many versions and have the DBMS translate that reference into something sensible for the current context.

Sharing Data from Very Different Databases

Traditional database technology does not offer means of issuing a single query to access data from multiple databases that use inconsistent terminology, operate under incompatible software, have incompatible data formats, and so on.

Database Access Languages and Query Processing

A new data access language is needed to solve many of the problems listed above, so that the user does not have to worry about versions, inconsistent terminology, multiple databases, evolution of schemas, etc. Unlike traditional data access languages, the new language must be able to answer questions that require traversal of structures of arbitrary size, such as hierarchies, iterative protocols, and partial orders.

Object-oriented Databases and Systematic Biology

Object-oriented databases have been proposed as the solution to the needs of many new applications. Though rarely mentioned by proponents of the object-oriented approach, it is worth noting that most of the beneficial features of an object-oriented DBMS can also be added to the relational approach, and with an appropriate user interface for the DBMS, the user need never know whether the underlying DBMS is thinking in terms of objects or of extended relations.

For those readers well versed in computer science terminology, an object-oriented DBMS (OODBMS) can be thought of as an object-oriented programming language (object identity, a class hierarchy, methods, encapsulation) with persistence for objects. For those readers not already familiar with object-oriented concepts, we will present the central ideas through a series of examples, rather than trying to be technically precise.

The Object-oriented Approach to Databases

A relational database is a set of tables; an object-oriented database is a set of objects. The objects are organized into a hierarchy of classes, which are sets of objects that have similar properties. For example, suppose we want to talk about a set of living animals, as in a database about a zoo. Animals can be classified according to their kind as shown in Figure 16.2a. To represent the relationships between the different kinds of animals, in the relational world one would convert the hierarchy into a table as shown in Figure 16.2b. The problem with this schema is that the structure of the data no longer matches how one would intuitively think about the structure of the data, which makes further manipulation of the data more difficult.

In the object-oriented world, one can represent the hierarchy directly as a hierarchy of classes. Animal, bird, fish, robin, penguin, peacock, etc., can each be a class, arranged in the hierarchy shown in Figure 16.2. Further, most varieties of the object-oriented model assume that all pairs of classes are mutually disjoint, except those pairs with an ancestor-child relationship. For example, no fish can be a robin, according to the hierarchy above; but every robin must be a bird and must be an animal. An object-oriented database contains not just a class hierarchy, but also information about the objects that are members of those classes.

At the time we choose a class hierarchy, we usually also decide what kinds of information we want to keep about the members of that class and about the class itself. For example, suppose that for all animals we want to record their diet and means of locomotion:

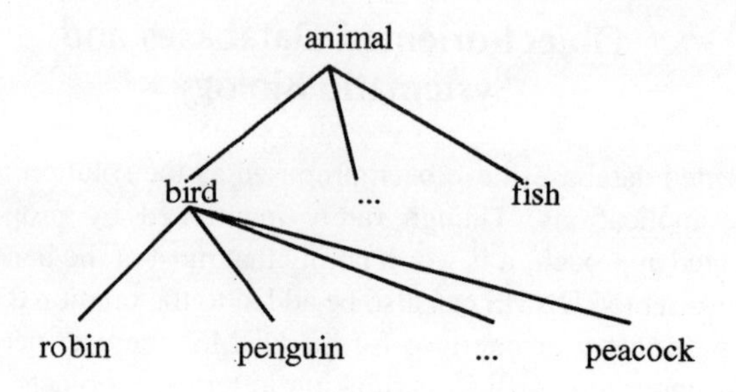

KINDS	
PARENT	CHILD
animal	bird
animal	fish
bird	robin
bird	penguin
bird	peacock

Figure 16.2. Representation of relationships between different kinds of animals. (a) Hierarchical; (b) tabular.

```
class animal
   has attributes
      diet
      meansOfLocomotion
```

In practice, we would also declare the type of data for *diet* and *meansOfLocomotion*: Are the acceptable values strings, integers, or objects in their own right? In our example we will omit the details at that level.

Once we have declared that we are recording the diet of each animal, we are automatically keeping that information for every bird, fish, robin, peacock, etc., because those classes are descendants of the animal class. In other words, the children of a class inherit the properties of their parent class(es), and so on down the class hierarchy. Whenever we define a property, an operation, a default, etc., for a class, that information automatically propagates to all the children of that class, unless we explicitly override the inheritance in those children. The advantage of inheritance is that, in a large application, it becomes very easy to specify what information is to be kept, using stepwise refinement down the class hierarchy. Also, one can represent information having a complex structure quite easily.

One can also decide to keep information lower down in the hierarchy that is not relevant higher up.

```
class bird
   inherits from class animal
   has attributes
      featherType
      hatchDateThisYear
```

One can also supply default values for information, which can be overridden at lower levels:

```
class animal
   has attributes
      diet
      meansOfLocomotion (default: {'walks'})
class bird
   inherits from class animal
   has attributes
      featherType (default: 'spinaceous')
      hatchDateThisYear (default: 'June 1')
      eggWeight
   overrides default for
      meansOfLocomotion (default: {'flies'})
class peacock
   inherits from class bird
   has attributes
   overrides default for
      featherType (default: 'extrinaceous')
class penguin
   inherits from class bird
   has attributes
   overrides default for
      hatchDateThisYear (default: 'January 1')
      meansOfLocomotion (default: {'swims', 'walks'})
```

One can also define defaults and attributes that are not simply stored values. For example, one can give a function for computing the *hatchDateThisYear* for each class of bird. If robins are a special focus of research, we might have a special, very accurate formula for estimating the hatch date for particular robins, involving the geographical location of that robin, when it arrived there from its migration, what the weather has been like in that spot, and perhaps the actual date of hatching if that event has already occurred for this year. Note that, in the relational model, one cannot store functions or procedures, only data; there is no notion of what functions it makes sense to apply to data. One can store functions and procedures, with instructions on how to use them, in a program that sits on top of a relational database; but it is easy for that information to get separated from the data and lost, and there is no guarantee that users will actually make use of the supplied procedures, nor any way to prevent them from concocting their own, nor any way to keep track of all the programs written for the data. In the object-oriented model, one can code the semantics of the data right into the data, through the procedures and functions associated with the data. Those procedures and functions are called *methods* in object-oriented terminology.

Another feature of the object-oriented model, called *encapsulation,* gives a measure of data independence to the object-oriented approach. The idea behind encapsulation is that objects have two parts: what is publicly visible and what is private. Only what is publicly visible is accessible to programs that want to make use of that object; the private part of the object is the data structures and functions used internally, inside the object, in order to provide the functionality that is visible to the outside world. An analogy to object-oriented encapsulation in the relational world is that the internal data structures and routines used to implement relations and to provide concurrent access and other features are not visible to application programs; application programs cannot directly invoke, access, or make use of any of that functionality. Instead, they have access only to the top levels of software of the DBMS. This enables one to change code inside the DBMS, change the storage techniques used for particular relations, etc., without affecting applications programs. This same division into public and private functionality permeates the object-oriented world and gives the same advantage: one can change the underlying details of implementation, as long as the publicly visible functionality remains the same.

Encapsulation gives some subtle advantages. For example, suppose a user wishes to compare two data values: Is *bird1.eggWeight* less than *bird2.eggWeight*? In a traditional DBMS, *bird1.eggWeight* would refer to a specific value stored in the database and available to any applications program that wishes to access it. In an object-oriented approach, one can have *bird1.eggWeight* be the raw value stored in the database; but one can also have *bird1.eggWeight* be a publicly accessible method that calculates a corrected value for egg weight, based on the private information *bird1.rawEggWeight* and perhaps other information, such as the altitude at which the egg was laid, the health of the laying bird, what experiment the measurement was part of, the identity of the measurer, the ambient temperature, whether the stored value is clearly erroneous, etc. The user does not know what functionality might underlie the simple phrase *bird1.eggWeight* (unless that information is specifically made public) and that functionality can be changed without effects visible to the user, as, for example, one derives better functions for correcting egg weight biases. This gives a measure of data independence.

We have seen that data values can have hidden semantics attached to them. One can also attach special semantics to, for example, the comparison operation "*bird1.eggWeight* < *bird2.eggWeight*," the user can invoke a great deal of complex machinery without effort. Of course, in order to gain this ability to attach additional semantics one must write the code that corrects raw measurements of egg weight, and the code that gives special semantics to "<."

One could also put this complex functionality into a layer of software running on top of the DBMS, rather than putting it into the DBMS itself; we will discuss this approach in a later section. The advantages of placing the functionality in the DBMS are that the resulting system is seamless and better integrated than a hybrid system and that the integrated system will probably run faster than the hybrid approach, because in the hybrid approach one must constantly pass data back and forth between separate programs, which is expensive.

Disadvantages of the Object-oriented Approach to Databases

The first disadvantage of the object-oriented approach to date is that, in contrast to the relational languages, it is necessary to have a fairly good understanding of how the data are organized in order to be able to write a query in the currently available object-oriented query languages. Thus major changes in the publicly visible portions of the data organization will require rewrites of existing programs that access the data. Probably this shortcoming of OODBMSs will be remedied in the next few years.

In addition, in the OODBMSs implemented to date, the organization of data on the disk is such that programs that were not anticipated at the time that the schema was designed may run very slowly. This is not such a problem in the relational model.

In contrast to the relational model, the mathematical theory to underlie the OODBMS approach is only in its infancy.

Finally, and perhaps most important, we do not yet know very much about how to implement an OODBMS efficiently. On the other hand, a great many researchers and developers in industry and academia are working very hard on this problem, so great improvements in performance are likely in the near term.

New Model-independent DBMS Features

A number of newly popular DBMS features are independent of the data model chosen (e.g., relational or object oriented).

The first is the ability to store and manage extremely large objects whose internal structure is not of interest—for example, bird songs, images, videotapes, large compiled programs. The second is support for schema evolution, as the type of data kept evolves. The third is a feature called *triggers*. A trigger is a definition stored in the DBMS, specifying that a particular action is to be taken whenever a certain type of event occurs or when the database reaches a certain state. For example, if someone enters experimental data that mention a chicken with three legs, a trigger could invoke a program that, say, begins a dialogue with the user about whether the chicken really had three legs. Triggers are useful for enforcing complicated integrity constraints and for monitoring the status of the database. The last new feature involves integrating data from different groups, where each group insists on its own autonomy and its own conventions regarding nomenclature, schemas, data semantics, etc. The goal is to have a universal data access language and a means of translating automatically from the nomenclature of one group to that used by another group. This is a popular topic of research in the database community right now, under the rubric of *federated databases*. Currently no commercial systems offer support for federated databases, but in the next decade one can expect to see some commercial offerings in this area.

Adding Functionality through an Additional Layer of Software

One could also choose to purchase a programming environment intended for use in artificial intelligence applications, with a DBMS running underneath to store the data. Under this approach, the DBMS is just a repository for the data, fetching and storing data that the upper layer of software has requested, and most of the "smarts" of the system lie in the upper layer of software. Such systems can be purchased today from the companies that market AI programming environments (e.g., Symbolics, ART, KEE) and will help with many of the characteristics of biological data that cause trouble for relational DBMSs. On the other hand, such systems tend to run rather slowly and have trouble with schema evolution.

Biologists Working with Computer Scientists

A key to successful interdisciplinary research is the construction of a bridge between researchers from two different disciplines, so that one researcher is able to understand how the other researcher thinks. This is more than a matter of learning a new vocabulary; one must also learn how the other person conducts research. In practical collaborations, we usually do not find two people building a bridge that meets in the middle; instead usually one person builds a bridge to the other's side. Bridge-building is hard to do and takes a lot of time and commitment.

I suspect that many people not employed in the computer industry have the illusion that if one uses computers then one knows computer science. In fact, use of a good program will not give any insights into computer science, because a good program will have a user interface tailored to the way that humans think, and not at all exhibiting the quirks of computer thought.

In fact, knowing how to program, in the sense of being able to write small programs, does not give one much insight into computer science either. In particular, it does not equip one with the skills necessary to build large or complex systems: the construction of a large system is not a scaled-up version of the construction of a small program. The special skill possessed by computer professionals is their ability to think extremely systematically and algorithmically about a large, irregular, and complex artifact, to reduce a seething mass of exceptions to something with some semblance of regularity, so that a program can manipulate a representation of the artifact efficiently and somewhat intelligently.

As a closing comment, if common sense is required to make sense out of one's data, then it will be very difficult to build a program to reason about that data, for example, if it is obvious when one looks at the data that these two particular data values cannot be compared with one another, because this value was measured on a nematode with one tail and this value came from a nematode with two tails. Computers are notoriously lacking in common sense, and it is nearly impossible to program it into them.

Discussion

Allkin: To what extent can a hierarchy be changed once it has been established?

Winslett: If you have chosen to represent the hierarchy as ordinary data, then it can be changed as often as desired. If the hierarchy is built into the schema (as in the earlier examples about birds), then changing the hierarchy means changing the schema, and this is schema evolution rather than just updating data. Each system differs on how much schema evolution it supports. When you change a class, the DBMS has to make matching changes in all the objects that belong to that class, and this can be difficult and time consuming, whether the DBMS makes those changes at once or waits until the user issues a query against the affected objects. Because of this performance penalty, most current OODBMS products do not offer very much in the way of schema evolution. Incidentally, if establishing the hierarchy is the focus of your research, keeping that hierarchy as data will allow you to keep multiple versions of the hierarchy (no products currently allow multiple versions of a schema, only of data.) The price you pay for treating the hierarchy as data is that you lose the automatic support for inheritance that comes with the class hierarchy. For example, we could encode the taxonomy as data with the following schema:

```
class taxonomicClass
   inherits from class allObjects
   has attributes
```

name
subTaxonomicClasses

This definition of a taxonomicClass says that each taxonomicClass object has a name and a set of subTaxonomicClasses (indicating its place in the taxonomic hierarchy).

Humphries: The concept of inheritance corresponds to how systematists make phylogenetic trees, so that property is highly appealing. Is it possible to include multiple taxonomies within a system and still have inheritance?

Winslett: Let us suppose that the taxonomy is kept as data, and not built into the schema. You can program support for inheritance down hierarchies other than the class hierarchy. In fact, programming these extra types of inheritance even if the taxonomy is put into the schema can be very useful (e.g., to inherit properties down a *part-of* hierarchy). For example, bugs are made up of parts: wings, a body, legs; those parts are in turn made up of other parts. It is nice to be able to say that the bug is brown, and know automatically through inheritance that the wings and the body are also brown. This differs from inheritance down the class hierarchy, because *wing* is not a subclass of *bug*; the subclasses of bug will be particular kinds of bugs.

Not all object-oriented DBMSs will support default values, incidentally. Those that are intended for use with AI applications probably will, however, becausc defaults are useful in AI.

Hendy: When the inheritance is built into the hierarchy, what happens if, for example, you define animals as having four legs but some animals have only two legs? If you query the higher level, does it think there are only four-legged animals, or does it know that there are two-legged animals?

Winslett: Object-oriented queries are usually issued over all members of a particular set. For example, you could ask the query, *How many legs do animals have*? and the DBMS would run through all the objects that belong to the class *animal* and return a list of the different values it found for the *numberOfLegs* attribute. In other words, the question would be answered not by looking at the definition of the class *animal*, but at all the objects that happened to be animals. There may be many such objects, but the DBMS will be built to find those objects quickly. If the query about numbers of legs is answered too slowly, and if it will be asked often in the future, one can ask the DBMS to create some additional structures to speed up answering of queries involving *numberOfLegs*. You can also ask a query about the class definitions themselves (e.g., *What is the default value for* numberOfLegs *for class* animal? or *What are all the default values for* numberOfLegs *for class* animal *and all subclasses of* animal? These queries would be answered by looking only at the class definitions, not at the objects that are members of those classes.

Allkin: Given the financial resources available to most systematists, is it possible for us to use modern DBMS products?

Winslett: DBMSs for personal computers are quite inexpensive, and most companies have special pricing for educational institutions (e.g., 90 percent discount for the relational DBMS Ingres). Most of the major relational DBMS companies offer a system that runs on a PC and accepts standard query language (SQL) queries. The commercial extended relational DBMS Orion (Xidak Corp., Palo Alto, California) is free to educational institutions and offers many of the features associated with object-oriented DBMSs. One can also buy a DBMS intended just for personal computers and not available on larger computer systems. However, this latter approach is probably not a good idea if you want to network the world's systematists together: probably some of the capabilities offered only by the more powerful systems will be required as you scale up. In addition, you may find that your institution already has a site license for a relational DBMS, so that you can install a copy of the DBMS on your machine for a nominal fee; a site license is a good purchase for a large organization that will need many copies of a DBMS in order to satisfy its business data-processing needs.

17

Frame Representation and Relational Databases: Alternative Information-Management Technologies for Systematic Biology

PETER D. KARP

This chapter introduces a type of information management system that has been developed in the artificial intelligence (AI) community, called frame knowledge representation systems (FRSs) (Fikes and Kehler 1985; Finin 1986). The chapter will also compare and contrast FRSs with relational database management systems (RDBMSs) (Date 1983; Schmidt and Brodie 1983; Ullman 1982). For other comparisons of these two types of information management systems, see Freundlich (1990); Brodie (1988); Brodie and Manola (1988); Brachman (1988); and Brodie and Mylopoulos (1986).

The chapter will review the advantages and disadvantages of using these two different classes of information management systems. More specifically, I will compare traditional FRSs with traditional RDBMSs. There exist non-traditional FRSs and RDBMSs that will not exactly fit all of the generalizations herein. In fact, some systems will not possess every capability ascribed to the general class of systems to which they belong. In addition, current computer science research that attempts to merge these types of systems is blurring these distinctions even further. The approach taken in this chapter is to compare the different types of capabilities that these two classes of systems provide, both to the users and to the designers of information management applications. I assert that the best technology for building such applications for systematists will be a hybrid of FRS and RDBMS technology because at

Advances in Computer Methods for Systematic Biology: Artificial Intelligence, Databases, Computer Vision, ed. Renaud Fortuner (Baltimore: Johns Hopkins University Press).

the moment neither type of system provides the full range of capabilities that systematists require.

These capabilities will be grouped into three different levels, as defined by Brachman and Levesque (1986). The first level is the *symbol level*. It encompasses the low-level implementation details related to the efficiency and the low-level data integrity features of the information management system (I use the term *information-management system* to subsume both FRSs and RDBMSs). The second level is the *system engineering level*. It includes the capabilities provided to the designers and maintainers of information management applications that affect the understandability and the maintainability of applications in the long term. The third level is the *knowledge level*. It includes the functions that users can execute with respect to a frame knowledge base or a database, as well as the types of knowledge that can actually be recorded by that knowledge base or database. Note that the term *knowledge base* subsumes a number of different AI techniques, such as production systems, logic programming, and frame knowledge representation. Similarly, the word *database* subsumes technologies such as network databases, relational databases, and object-oriented databases. Because this chapter is concerned only with FRSs and RDBMSs, the reader can assume that henceforth the terms *knowledge base* and *database* refer specifically to FRSs and RDBMSs.

The Symbol Level

The capabilities provided at the symbol level are not generally visible to the users of an information system because they form a substrate on top of which higher-level functionality is constructed. The information management capabilities included in the symbol level are the following.

The capability of *persistence* refers to whether or not an information base can survive events such as the termination of a particular program and operating system crashes. Although most information management systems provide persistence, an important issue is whether the system automatically stores updates to persistent storage at the end of every transaction or if the user must explicitly request updates to be stored. An additional concern is whether only updates must be transmitted to permanent storage or if the entire information base must be stored to update the persistent storage. A related capability is *reliability*, which means that, if the information base becomes corrupted so that it is unreadable (due to hardware or software failures), the system is able to detect that corruption and perhaps recover from it automatically. Recovery might be implemented through backup or transaction-logging techniques. Another way of increasing reliability is by replicating a given information base at multiple computers, so that, if one of the sites goes down or its information becomes corrupted, the dataset is still available from other sites.

Replication requires special techniques to ensure that updates are received by all sites. Another approach is to distribute different parts of the overall information base to different sites so that if one site goes down, a subset of the data is still available.

Another capability is the ability to *share information* among multiple users at the same time. It is often desirable to allow many people to read from and perhaps write to a large, central dataset from a number of networked computers. Sharability requires techniques such as locking to prevent inconsistent updates to a shared information base.

An important attribute of an information system is the *maximum amount of data* that system can reasonably manage. As a dataset becomes bigger and bigger, can users access the information in a reasonable amount of time? To provide timely access to large amounts of information, a system must employ special data structures and access algorithms. Replicating data across multiple computers can also improve performance by distributing the processing load across multiple computers and by decreasing the distance that query responses must travel to reach users.

Generally speaking, the symbol level is where database researchers have spent most of their efforts over the past twenty years, and where AI researchers have spent relatively little effort. Thus, RDBMSs are extremely good at providing fast, reliable access by multiple users to large amounts of persistent information. RDBMSs automatically save the results of individual transactions in persistent disk storage and are sometimes able to detect database corruption and to recover from it. FRSs have few of these capabilities. Generally, a frame knowledge base (KB) resides in the virtual memory of the host computer. Transactions in a FRS modify information in virtual memory only, and users must explicitly cause a KB to be saved to persistent storage. Therefore there is persistence, but at a coarse time scale. FRSs have little or no ability to detect or recover from corruption. There is generally no ability to share information among multiple users as most FRSs are single-user systems. In addition, FRSs cannot handle huge amounts of information, for example, in excess of 10,000 frames (defined below).

The System Engineering Level

Fundamental Data Primitive

At the system engineering level we will first examine the fundamental data primitive that is used in RDBMSs and FRSs. For RDBMSs, the fundamental data primitive is the relation. A relation is analogous to a table that has rows (also called tuples) and columns (also called attributes).

Table 17.1 shows a sample relation from a hypothetical taxonomic database describing the character states observed in several fish taxa.

Table 17.1. The Taxa Relation, in which Each Row of the Table Is One Tuple and Each Column of the Table Is One Attribute

Taxon	Foramen.exoccipitale	Suprabronchial.air.chamber/ buccopharyngeal.cavity
Belontidae	Cartilaginous	Separated
Mugiloidei	Uncovered	
Splendens	Bony	Separated

In the tuple describing the Belontidae (Table 17.1), the value of the *taxon* attribute is the string *Belontidae*, and the value of the *foramen.exoccipitale* attribute is the string *cartilaginous*. The *taxon* attribute is a *key* for this relation because this attribute has a unique value in every tuple of the relation.

One important aspect of the relational data model is that the information stored within each column must be a primitive item such as a number or a string of characters. A column cannot store data types such as lists or sets of items, or complex structured records of items. This is a serious limitation, and one that is not shared by the FRS.

The FRS primitive is called a frame. A frame describes either an individual thing or a class of things. A frame consists of a set of properties that can take on values; these properties are called *slots*. Frames are similar to, and were developed at the same time as, the objects of object-oriented databases. A user can attach an arbitrary number of different subproperties to each slot, which are called *facets*. Every slot has a value facet that stores what one thinks of as the primary piece of data for that slot.

Table 17.2 shows a frame corresponding to one tuple from Table 17.1. The value facet of the slot called *foramen.exoccipitale* holds the state of that character in the *Belontidae* taxon. A facet might also be used to store a comment for a given slot value, a data type for the slot, or constraints on the possible values of a slot.

In Table 17.2, the *foramen.exoccipitale* slot has the value *cartilaginous*. The slot is restricted to have only one value, and inheritance for this slot is computed according to a procedure called *override.values*, meaning that if values for the slot are present in a child frame, these values override values defined in ancestor frames. This example was created using KEE, an FRS from IntelliCorp.

Whereas different relations exist independently of one another, frames are arranged in a taxonomic hierarchy. Here the word *taxonomy* is used in a more general sense than a biological taxonomy. When encoding knowledge from any problem domain in a FRS, frames high in the hierarchy represent more general concepts than do their descendants in the hierarchy. An operation

Table 17.2. A Sample Frame Called Belontidae, a Child of Class Frame Called Anabantoidei

Frame: BELONTIDAE in knowledge base ARTITAXA
Subclass of: ANABANTOIDEI

Slot: FORAMEN.EXOCCIPITALE from ANABANTOIDEI
inheritance: OVERRIDE.VALUES
valueclass
cardinality.min: 1
cardinality.max: 1
comment: "The character state of the foramen exoccipitale."
values: CARTILAGINOUS

Slot:
SUPRABRONCHIAL.AIR.CHAMBER/BUCCOPHARYNGEAL
CAVITY from ANABANTOIDEI
inheritance: OVERRIDE.VALUES
valueclass:
cardinality.min: 1
cardinality.max: 1
comment: ""
values: SEPARATED

called inheritance propagates information such as slot definitions and values that are present in frames high in the hierarchy, down to frames lower in the hierarchy.

Figure 17.1 shows a hypothetical frame hierarchy of fish taxa. This display was generated by the KEE FRS. Lines to the right of the name of a frame connect that frame to its child frames. In this hierarchy, inheritance would propagate the value *cartilaginous* of the slot *foramen.exoccipitale* as defined in the *Belontidae* frame to the children of *Belontidae* (*Betta* and *Colisa*). Thus the inheritance operation in a FRS mirrors the evolutionary inheritance of

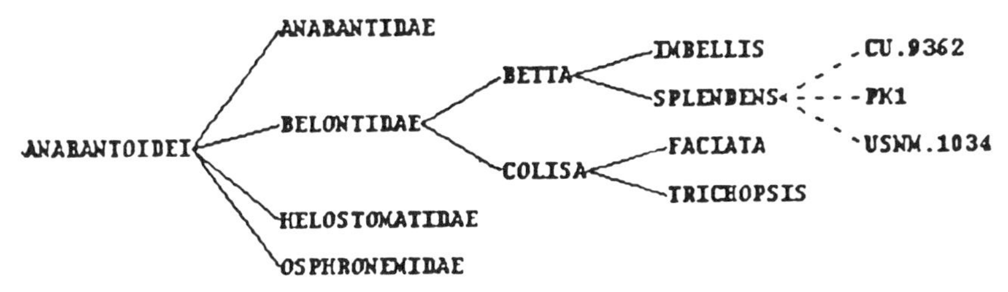

Figure 17.1. A hypothetical hierarchy of fish taxa. Lines connect a frame to its child frames. Dashed lines connect instance frames (representing specimens) to their parents.

characters by the descendants of a taxon. FRS inheritance can be used both to store information about known character states and to generate phylogenetic hypotheses as to the characters that are likely to be observed in child taxa. Figure 17.2 shows a different taxonomic hierarchy. This hierarchy was created by the authors of the PIR protein-sequence database. I loaded it into the KEE FRS for display.

Inheritance is valuable for a number of other reasons. In very complex domains, an information system will manage many different, but similar, types of entities. Users of an RDBMS must define a new relation for each new type of thing from scratch. Each relation is independent of every other relation. But inheritance allows users to define new frame classes by modifying existing frames. Thus inheritance provides compact representations. Inheritance also facilitates KB modifications because it systematically propagates modifications to a frame to all descendants of that frame in a KB.

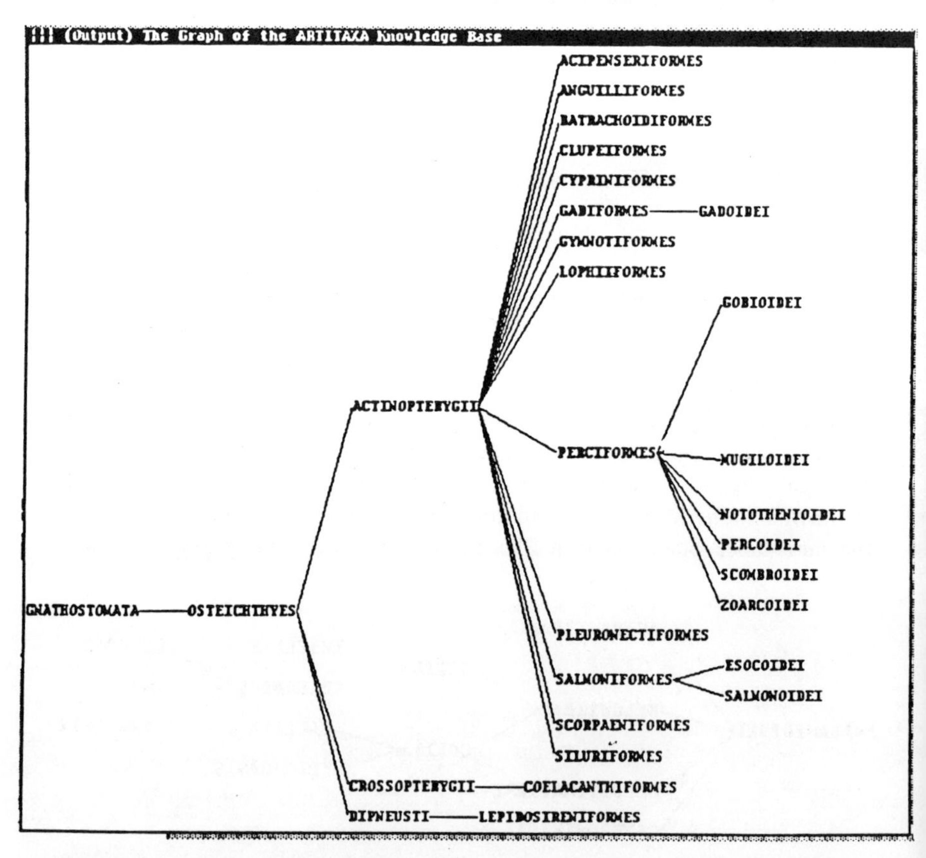

Figure 17.2. A small part of the taxonomic data that accompany the PIR protein-sequence database, displayed using the KEE FRS.

Flexibility

We now return to a discussion of other capabilities at the system engineering level. The capability of run-time flexibility allows a user to redesign a KB dynamically, that is, to modify the set of slots within a frame (adding, deleting, or changing slot definitions) or to rearrange the inheritance hierarchy. Generally, FRSs allow a user to perform all of these operations at run time (i.e., at any time during the operation of the system). In contrast, although RDBMSs support some of these operations in a transparent manner, other operations require the user to dump out the entire contents of the database, modify the database schema, and then load back all of the data. (The schema defines the set of relations in the database and their attributes.) These operations can be quite time consuming and may be required frequently in a problem domain that is as complex as systematics because the information base designer's conceptualization of the information base is bound to change frequently over time. It is important to note that run-time flexibility can cause a significant decrease in performance. An RDBMS can achieve faster performance by assuming that certain aspects of a database are static.

Classification

Many FRSs provide an operation called classification, namely, the ability to automatically position a newly defined frame in the taxonomic hierarchy. Classification positions frame *F* under the most specific frames in the KB that are more general than *F*. Classification ensures that the ordering of the taxonomic hierarchy by degree of generalization is preserved as a knowledge base evolves. RDBMSs do not provide classification.

Note that the FRS operation of classification bears some similarity to the notion of classification used in systematics, but there are significant differences. FRS classification is strictly based on the definitions of two concepts (frames) as stored within a knowledge base, and thus has no uncertain or probabilistic component. FRS classification also applies to any concept, not just to biological organisms. For example, an FRS could classify the concept of a black-and-white television as being a specialization of the concept of a television based on the definitions of the frames *television* and *black-and-white-television*. The definition of the latter would restrict the values of the slot *picture-colors* to the values *black* and *white*—a subset of the values allowed for color television.

Security

A capability provided by an RDBMS but not by an FRS is security. RDBMSs allow a database manager to specify which users are allowed to perform what types of operations (e.g., retrieval versus modification) on what tables of the database.

Version Management

Another capability provided by some RDBMSs but not by FRSs is version management: the ability to track old states of a database. Users can examine old data and revert the database to older states by undoing a series of transactions.

Application Programming Language

Another way to compare FRSs with RDBMSs is to ask what programming languages are used to implement them and what languages may be employed by users who write application programs that interact with the information system. RDBMSs are typically implemented in the C language, which users can also utilize. FRSs have traditionally been implemented in the Lisp language, although some recent FRSs have been implemented in C.

The Knowledge Level

We now turn to the knowledge level to ask what types of information can be represented in these information management systems and what functional operations these systems provide to their users. For example, are there types of knowledge that can be encoded in one type of system that cannot be encoded in the other?

Expressiveness

An FRS user can store information about individual things (particulars) and about classes of things (generalizations). Class frames encode generalizations. For example, if we encode a value of *true* in the *flight* slot of the class frame *birds* to represent the assertion that all birds can fly, we have made a general assertion about the ability of birds to fly that will be propagated by default to all particular birds in the knowledge base (unless explicitly overridden). In a RDBMS, in contrast, users can make statements only about particular birds. It is not possible to encode the statement that all birds fly. Therefore there is a difference in the expressive power of these two types of systems: RDBMSs can represent only particulars, whereas FRSs can represent both particulars and generalizations.

Inference

A significant operation that FRSs provide that RDBMSs do not is inference: an FRS can infer new conclusions from existing facts in the information base.

One form of inference is the propagation of default information via inheritance: generalizations such as "all birds can fly" are propagated to specific birds, allowing the FRS to deduce that specific birds fly.

A second form of inference is carried out by a production system, which is a rule-based reasoning system that allows users to construct an expert system using a set of IF-THEN rules. These rules specify situation-conclusion patterns that can be executed to produce expert problem-solving behavior in tasks such as diagnosis and identification. One of the advantages of production rules is that programs can manipulate them in multiple ways. For example, once the FRS has derived a conclusion, the system can explain its reasoning by stepping back through the set of rules that was used to reach that conclusion.

A third form of inference involves the classification operation discussed earlier. Classification is itself a form of problem solving since the act of positioning a frame in a taxonomic hierarchy often gives us new information. Every FRS supports at least one of the preceding types of inference, whereas RDBMSs do not support any inference mechanisms.

Conclusions

In general, RDBMSs have superior symbol-level capabilities, whereas FRSs have superior knowledge-level capabilities. More specifically, RDBMSs excel at providing simultaneous users with fast, reliable access to large amounts of persistent information. In contrast, FRSs are able to express both particulars and generalizations and provide several types of inference mechanisms to support expert problem solving. The two types of systems provide different sorts of system-engineering-level capabilities: FRSs provide run-time flexibility that allows dynamic reconfiguration of a knowledge base schema, automatic classification of new frames into the taxonomic hierarchy, and inheritance of properties within that hierarchy. RDBMSs provide version management and security.

Although the capabilities of existing FRSs and RDBMSs are sufficient for many applications in systematics, some future information systems for systematics will require all of the preceding capabilities. Multiple users will require fast access to large amounts of data such as museum collection datasets, taxonomic and feature data, geographic data, and images. The concepts that systematists wish to describe will often be very complex, requiring the ability to express generalizations and inheritance and the ability to alter the information base schema dynamically as the designer's conceptualization of the information base evolves. Protection and version management will be important in these systems, and inference and classification will play an important role in application programs for systematics, such as taxonomic classification and identification.

Computer science researchers are now working to improve the capabilities of their information systems. For example, database researchers are adding inferential capabilities to RDBMSs, and AI researchers are adding many symbol-level capabilities to FRSs. The information management needs of systematics will provide important challenges for this new generation of information management systems.

Acknowledgments

The examples in this chapter were derived with help from Paula Mabee and Julian Humphries.

Discussion

McGranaghan: Is it a fair characterization to say that you build semantic networks and belief networks from frames, and that frames could be conceived of as pieces in semantic networks?

Karp: Semantic nets and frames are essentially the same thing, although some subtle historical distinctions could be drawn between them. Belief networks are similar to frames and semantic nets in that all three organize information in a network of connected nodes and provide inferential capabilities for reasoning about that information. However, belief networks are specifically designed for probabilistic inference, and FRSs are not. Thus, the information represented at belief-network nodes are probabilities only. Belief-network nodes do not have the rich structure of frames, and belief networks do not have many frame capabilities, such as inheritance. Similarly, although frames can easily store probabilities, FRSs do not contain the probabilistic-inference procedures found in belief networks. Belief networks are not built on top of frames for efficiency reasons.

Buckup: It looks like there is a polarization between relational databases and object-oriented systems such as FRSs. I wonder how much that is necessary. I understand that you have to make these distinctions to explain the differences between the two approaches, but can't you have hybrid systems?

Karp: Researchers have built hybrid systems, for example, a commercial system called KEEconnection couples an FRS to an RDBMS. However, there are many possible ways of coupling the two, and the question of which approach works best for different types of problems is an unsolved research problem. From a practical standpoint, biologists will probably find hybrid systems to be less stable now than are nonhybrid systems. But sometime in the future hybrid systems should be superior.

18

DELTA and INTKEY

MICHAEL J. DALLWITZ

DELTA (Dallwitz 1980a; Dallwitz and Paine 1986; Partridge et al. 1988) is a multipurpose format for generating identification keys. It is not geared, as are many formats, to the requirements of one particular type of program (e.g., Dallwitz 1974; Rohlf et al. 1981; Swofford 1984). It was designed to be easy for people to use. On the other hand, a degree of complexity was necessary to avoid loss of significant information, and the complexity has increased over the years in response to requests from users. I wrote a program called CONFOR to translate the format into natural language and into formats used by various other programs. This makes the data accessible to programs that carry out key generation, phenetic and cladistic analysis, and interactive identification and information retrieval. CONFOR also helps with maintenance of the data, such as keeping the data tidy and changing the order of the characters.

The DELTA Coding System

The DELTA format is based on ordinary text files (sequential files of ASCII characters, with records of up to 120 characters). These files may be created and modified with any text editor or word processor (we will soon be writing a new system based on random-access files, with an integrated editor). The data are in free format; that is, they do not have to be positioned in fixed fields in the records. The examples below are taken from a small subset of one of Leslie Watson's data sets (Watson and Dallwitz 1981; Watson et al. 1988). We distribute this subset with the programs.

Figure 18.1 shows a DELTA character list. Five types of character are available: text characters (e.g., *1*); multistate characters, which can be either ordered (e.g., *7*) or unordered; and numeric characters, which can take either

---

Advances in Computer Methods for Systematic Biology: Artificial Intelligence, Databases, Computer Vision, ed. Renaud Fortuner (Baltimore: Johns Hopkins University Press, 1993).

#1. <Synonyms: i.e. 'genera' included in the current description>/
#2. <Longevity of plants>/
    1. annual <or biennial, without remains of old sheaths or culms>/
    2. perennial <with remains of old sheaths and/or culms> <Figs 1, 2, 18>/
#3. <Mature> culms <maximum height: data unreliable for large genera>/
    cm high/
#4. Culms <whether woody or herbaceous>/
    1. woody and persistent/
    2. herbaceous <not woody, not persistent>/
#5. Culms <whether branched above>/
    1. branching <vegetatively> above <Fig. 2>/
    2. unbranched <vegetatively> above <Figs 1, 7>/
#6. <Culm> nodes <whether hairy or glabrous>/
    1. hairy <Figs 4, 33>/
    2. glabrous <Fig. 4>/
#7. Leaf blades <shape: data very incomplete>/
    1. linear/
    2. linear-lanceolate/
    3. lanceolate/
    4. ovate-lanceolate/

Figure 18.1. Part of a DELTA format character list.

real (continuously variable) values (e.g., *3*) or integer values. Comments enclosed in angle brackets can be placed anywhere; they are omitted from most kinds of output. There are no restrictions on the numbers of characters or states or on the amount of text.

Figure 18.2 shows a coded taxon description. The name of the taxon is at the top. A typical *attribute* consists of a character number, a comma, then a state number (e.g., attribute 6: *6,2*). Text attributes are slightly different, consisting of a character number and text within angle brackets (e.g., attribute 1: *1*<Czerya . . . >).

In more complex cases we can have several states separated by "/" (meaning *or*), and we can have comments associated with any of those character

Phragmites <Adans >/
1<*Czernya* Presl, *Miphragtes* Nieuwland, *Oxyanthe* Steud., *Trichoon* Roth, *Xenochloa* Roem. & Schult.> 2,2 3,80–400(–1000) 4,1–2<often somewhat persistent> 5,1<especially when main culm damaged>/2 6,2 7,2–3 8,1 9,2 10,1 11,3 12,2 13,5 14,1<20–60 cm long, plumose, the fertile lemmas surrounded by long white silky hairs> 15,2 16,2 18,– 19,2 25,9–16 26,1 27,1<at least above the L1> 28,2 29,1 30,1 31,2 32,1 33,1 34,2 35,2<rounded on the back> 36,2 37,3 38,1 39,1 40,2 41,(2–)3–10 42,1 43,1<acute to acuminate or aristulate> 44,1/3<narrow-attenuate, muticous to aristulate> 45<(if lemmas aristulate)>,1 46,3 47,1 48,1 50,1–3 51,1 52,2 53,1 54,1 55,1/2 56,3<or two in the lower floret> 57,1 58,2 59,3 60,1 61,2 62,1 63,2 64,2 66,1 67,1 68,2 69,1 70,2 71,2 72,2 73,2 74,3 77,4 82,3 83,1&2&3&5&6

Figure 18.2. A description coded in DELTA format.

states (e.g., attribute 44). We can have ranges, denoted by "–" (e.g., *7*), and we can have states separated by "&" (meaning *and*) (e.g., *83*). The three separators /, –, and & can be combined within the same attribute. Ranges of values of numeric characters can include parentheses to indicate values outside the normal range (e.g., *3*).

Output Produced from DELTA Data

Natural-Language Descriptions

We can use CONFOR to translate a coded description into natural language, as shown in Figure 18.3. This was produced and typeset automatically from the data, without manual intervention. It corresponds to the data in Figure 18.2. Notice that parts of the description are in italics. These parts constitute a diagnostic description. The diagnostic characters were selected by the program INTKEY and then fed through to CONFOR, which was instructed to italicize the parts of the description corresponding to these characters.

Phragmites Adans.

Czernya Presl, *Miphragtes* Nieuwland, *Oxyanthe* Steud., *Trichoon* Roth, *Xenochloa* Roem. & Schult.

Habit, vegetative morphology. Perennial. *Culms* 80–400(–1000) cm high; *woody and persistent to herbaceous (often somewhat persistent)*; branching above (especially when main culm damaged), or unbranched above. Nodes glabrous. Leaf blades linear-lanceolate to lanceolate; broad. *Adaxial ligule a fringe of hairs.*

Reproductive organization, inflorescence. Plants bisexual, with bisexual spikelets. *Inflorescence paniculate*; open (20–60 cm long, plumose, the fertile lemmas surrounded by long white silky hairs); not comprising 'partial inflorescences' and foliar organs. Spikelet-bearing axes persistent. *Spikelets not in distinct long-and-short combinations.*

Female-fertile spikelets. *Spikelets* 9–16 mm long; compressed laterally; *disarticulating above the glumes (at least above the L1)*; disarticulating between the florets; with the rachilla prolonged apically. *Glumes* two; *very unequal*; decidedly shorter than the adjacent lemmas; awnless; not carinate (rounded on the back). Spikelets with incomplete florets. The incomplete florets both distal and proximal to the female-fertile florets. Proximal incomplete florets 1; male; awnless. *Female-fertile florets (2–)3–10.* Lemmas entire; pointed (acute to acuminate or aristulate); awnless, or awned (narrow-attenuate, muticous to aristulate). Awns (if lemmas aristulate) 1; apical; non-geniculate; much shorter than the body of the lemma. Lemmas 1–3 nerved. Palea present; conspicuous but relatively short. Lodicules present; fleshy; ciliate, or glabrous. Stamens 3 (or two in the lower floret). *Ovary glabrous.* Stigmas 2; brown.

Fruit. Fruit small; smooth. *Hilum short.* Pericarp fused.

Photosynthetic pathway, leaf blade anatomy. C_3. XyMS+. Mesophyll with arm cells; without fusoids. Midrib conspicuous; with a conventional arc of bundles; without colourless tissue adaxially. All the vascular bundles accompanied by sclerenchyma.

Taxonomy. Arundinoideae; Arundineae.

Distribution. 3 species. Holarctic Kingdom, Paleotropical Kingdom, Neotropical Kingdom, Australian Kingdom, and Antarctic Kingdom.

Figure 18.3. The description in Figure 18.2 translated into natural language.

Identification Keys

In Figure 18.4 we have part of an identification key produced by first translating the data into an intermediate format, then passing them through our key generation program, KEY (Dallwitz 1974; Dallwitz and Paine 1986). Again, everything is completely automatic, including the typesetting. However, the user has a lot of control over the structure of the key, by changing parameter values.

1(0). Spikelets disarticulating above the glumes. 2
Spikelets falling with the glumes. 11
Spikelets not disarticulating. 13
2(1). Female-fertile florets 1. 3
Female-fertile florets 2 or more. 8
3(2). Inflorescence of spike-like main branches; lodicules fleshy; C_4. 4
Inflorescence paniculate; lodicules membranous; C_3. 5
4(3). Glumes very unequal; lemmas awned; stigmas white; biochemical type PCK. . . . . **Chloris**
Glumes more or less equal; lemmas awnless; stigmas red pigmented; biochemical type NAD–ME. . . . . **Cynodon**
5(3). Ovary glabrous. 6
Ovary hairy. 7
6(5). Spikelets with female-fertile florets only; stamens 3; hilum short; mesophyll without arm cells; midrib with one bundle only. . . . . **Agrostis**
Spikelets with incomplete florets; stamens 5 to 6; hilum long-linear; mesophyll with arm cells; midrib with complex vascularization. . . . . **Oryza**

Figure 18.4. Part of a computer-generated key.

Foreign Languages

The character list can be translated into other natural languages. This is done manually, but then all the products (descriptions, keys, and interactive identification) are available automatically in that other natural language (e.g., French [Watson et al. 1986], Greek [Watson et al. 1988], Spanish and Portuguese [Webster et al. 1989], and Chinese [Xu Zhu and Dallwitz, in preparation]). Figure 18.5 shows part of a key in Greek, and Figure 18.6 shows a description in Chinese.

The programs themselves (directives, error messages, and manuals) have recently been translated into Spanish (Valdecasas et al. 1989) and Chinese (Xu Zhu and Dallwitz, in preparation). Future versions of the programs will be much more convenient to maintain in different languages, because all text will be in files separate from the program files.

1(0). Θηλυκά-γόνιμα σταχύδια αποκοπτόμενα πάνω από τα λέπυρα. 2
Θηλυκά-γόνιμα σταχύδια αποκοπτόμενα μαζί με τα λέπυρα. 96
Θηλυκά-γόνιμα σταχύδια μη αποκοπτόμενα. 141
2(1). Θηλυκά-γόνιμα σταχύδια με ατελή ανθίδια στη βάση. 3
Θηλυκά-γόνιμα σταχύδια χωρίς ατελή ανθίδια στη βάση. 8
3(2). Ταξιανθία με φύλλα ή μέσα σε σπάθη· ελάσματα φύλλων με εύκολα ορατές κάθετες νευρώσεις· μεσόφυλλο με ατρακτοειδή κύτταρα. . . . . **Arundinaria**
Ταξιανθία όχι με φύλλα ή μέσα σε σπάθη· ελάσματα φύλλων χωρίς εύκολα ορατές κάθετες νευρώσεις· μεσόφυλλο χωρίς ατρακτοειδή κύτταρα. 4
4(3). Καρπός με μικρή ουλή. 5
Καρπός με μία μακριά-γραμμική ουλή. 7
5(4). Γλωσσίδα μεβρανώδης χωρίς κροσσό από τρίχες· θηλυκά-γόνιμα σταχύδια χωρίς επιμηκυσμένη στην κορυφή ραχίλλα· εμβρυο μικρό· φύλλα χωρίς ωτία· ελάσματα φύλλων μη αποκοπτόμενα από τους κολεούς. 6
Γλωσσίδα κροσσωτή· θηλυκά-γόνιμα σταχύδια με ραχίλλα επιμηκυσμένη στην κορυφή· εμβρυο μεγάλο· φύλλα με ωτία· ελάσματα φύλλων τελικά αποκοπτόμενα από τους κολεούς. . . . . **Phragmites**

Figure 18.5. Part of a computer-generated key, produced in Greek by translation of the character list.

Typesetting

By default, CONFOR and KEY produce plain ASCII files, suitable for viewing on a computer screen or printing on an ordinary printer. However, they can be instructed to put typesetting marks in their output, which then may be processed by our typesetting program, TYPSET (Dallwitz and Zurcher 1988). The input data may also include typesetting marks (e.g., superscripts and subscripts, font changes). The programs normally pass these through to TYPSET, but they can be made to remove them, for example, to produce

9. 肥披碱草 Elymus excelsus Turcz.

茎140厘米高，基部具白色粉层．叶鞘基部光滑，或被微柔毛．叶扁平，20-30厘米长，10-16毫米宽，叶两面粗糙．穗状花序稠密，直立，15-22厘米长，绿色．穗轴被短硬毛，节间膨大，光滑．小穗4-5，12-25毫米长．颖狭披针形；10-13毫米长，等长，脉5-7，粗糙，背部光滑，芒7毫米长．外稃披针形，背部光滑，基部光滑，脉5，8-12毫米长，芒15-40毫米长，反折，粗糙．内稃短于外稃；截形，脊全部具毛，脊间无毛．

染色体基数，X= 42．地理分布 东北，华北，四川；山披，草地，路边．

饲用价值：优良牧草，适口性好，产草量高，具有较高的营养价值．

Figure 18.6. A computer-generated natural-language description in Chinese.

plain text for display on a screen (Dallwitz 1984). CONFOR and KEY were designed so that they would be easy to adapt to other typesetting or word-processing systems. A cruder way to convert to other typesetting systems would be to edit the typesetting marks in the intermediate files.

The Interactive Identification Program INTKEY

Introduction

Our interactive identification program, INTKEY, was developed from version 3 of Richard Pankhurst's ONLINE program (Pankhurst and Aitchison 1975), which we got in 1982, modified, and eventually completely rewrote as INTKEY. We are currently completely rewriting it again, to add new features suggested by experience with the earlier versions. The new version was released in October 1991. Pankhurst has also continued to enhance ONLINE, which is now in version 6.

INTKEY provides tools for identification and information retrieval. It does not provide a fixed sequence of actions: it lets you choose what the actions are to be, and you are free to follow a quite complex path through it. It is a complex system, but users can easily be instructed in the use of simple sequences of operations.

INTKEY has complete online help, and "?" can be entered at any prompt to get some information about the required response. The new version will be completely menu driven, although a command-line interface will still be there and will be preferred by experienced users.

Database systems are now readily available off the shelf, and there is a growing tendency to think that it should be easy to use these for interactive identification. After all, is there anything more to it than finding which taxa have a certain value in a certain data field? We have spent many years enhancing Pankhurst's ONLINE (which was already quite a powerful program), adding features that we thought to be essential in the light of experience with nontrivial data sets (several hundred characters or taxa). Figure 18.7 is a current list of the program commands, taken from the menus. The asterisks mark features that were present in ONLINE version 3, though all of these have been enhanced (with the possible exception of the FINISH command). Figure 18.8 is the Help for one of the commands, as a detailed illustration of the type of facility that is needed for a practical system.

Examples

Watson et al. (1989) and Dallwitz (1989a,b) give extensive examples of the use of the program. However, I may be able to give you some idea of the flexibility of the program by describing some of the possible courses of action

*Best: display the best characters to separate the remaining taxa
*CHaracters: display names and numbers of characters
COmment: ignore text
DAta: read main data files
DEFine: define a keyword to represent a set of characters or taxa
*DELete: delete a previously used character
*DEScribe: display the description of a taxon
DIAgnose: generate diagnostic descriptions of taxa
*DIFferences: display the differences between taxa
DISplay: set screen display and prompts
EXAct: specify characters not subject to error
EXClude: exclude characters or taxa
FILes: menu for file input/output, display, and prompts
*FINish: exit from the program
FIX: retain the current character values when restarting
Help: display information about commands
INClude: include characters or taxa
INPut: read commands from a file
Keywords: display keywords
Log: send input and output to a file
MAtch: set criteria for matching of taxon descriptions
MEnu: return to main menu
OMit: omit inapplicable or unknown characters from descriptions
*OUtput: send output to a file
Parameters: menu for setting or displaying parameters
RELiabilities: set character reliabilities
REMark: copy text to the output file
*REStart: restart an identification
SAve: generate files for input to other programs
*SEParate: display the best characters to separate a taxon from the rest
*SET: set autobest, *autotaxa, rbase, stopbest, *tolerance, varywt
SHow: display text on the screen
SImilarities: display the similarities between taxa
STatus: display parameter values
SUBset: generate files containing subsets of the data
SUMmary: display a summary of the data
*TAxa: display names and numbers of taxa
*Use: use a character to describe the specimen

Figure 18.7. The INTKEY commands, with short descriptions. The asterisks mark features that were present in ONLINE version 3.

once a tentative identification is made, that is, once the program has indicated that only one taxon matches the specimen description that you have entered. Actually, any of the commands below might be useful at *any* stage of the identification, and we feel strongly that programs should allow this kind of flexibility, rather than leading the user along predetermined pathways. This certainly means that some effort is required to learn to make the best use of the program, but this should be acceptable to professional users wanting to achieve professional results. (By "professional," I mean not just taxonomists,

MATCH options

where options is one or more of the letters
I - inapplicables
U - unknowns
S - subset
O - overlap
E - exact

This command specifies which character values are to be regarded as equal – i.e. 'match' – in the commands USE, DIFFERENCES, SIMILARITIES, or TAXA. MATCH I and MATCH U specify respectively that 'inapplicable' and 'unknown' match any value. MATCH S specifies that two sets of values match if one set (usually the values of the specimen) is a subset of the other. (E.g. 1/2 matches 1/2/4 but not 2/3; 2–5 matches 1–6 but not 4–10). MATCH O specifies that two sets of values match if they overlap, i.e. if they have any values in common (e.g. 1/2 matches 2/3; 2–5 matches 4–10). (S and O cannot be used together.) MATCH E or MATCH without parameters specifies that two sets of values match only if they are identical.

The default setting is MATCH O U I, which is usually the most appropriate for identification. For information retrieval, the most appropriate setting is usually MATCH O.

Figure 18.8. The Help text for the INTKEY command MATCH.

but everyone who needs identification or information retrieval as part of their job.)

DESCRIBE SPECIMEN

Recapitulate the specimen description that you have entered, so that you can check it.

DESCRIBE REMAINING

Display the full description of the "remaining" taxon. REMAINING is an example of an automatically defined "taxon keyword" representing a set of taxa. At this stage of the identification, it represents a single taxon, but at earlier stages it would represent several.

DESCRIBE REMAINING HABIT DISTRIBUTION ECOLOGY

Display the description of the remaining taxon in terms of its habit, distribution, and ecology. These are examples of user-defined "character keywords" representing sets of characters. They would generally have been defined by the person who prepared the data.

DIAGNOSE REMAINING

Generate and display a diagnostic description of the remaining taxon, in terms of characters not used in the identification. This description will distinguish the remaining taxon in at least one respect from all the other taxa, and so provides an independent check.

DIFFERENCES (SP 6)

Display the differences between the specimen description and taxon 6. (Maybe you thought your specimen was taxon 6. What is the evidence that it is not?)

MATCH EXACT
DIFFERENCES (SP REM)
Set exact matching (see Fig. 18.8) and display the differences between the specimen description and the remaining taxon. If the MATCH setting were left as it was during the identification (normally Overlap, Unknown, Inapplicable), no differences would be shown, because the remaining taxon is, by definition, the one that matches the specimen. Setting MATCH EXACT allows the DIFFERENCE command to pinpoint characters where the specimen and the remaining taxon differ because of variability or because the character is unknown or inapplicable for the remaining taxon.

SET TOLERANCE 1
Set the "tolerance" parameter to 1. This brings back as "remaining" taxa all those that differ in not more than one respect from the specimen description. You can then continue with the identification as before. This is particularly useful if you suspect or know that there has been an error; for example, if the number of taxa remaining is 0, or if the description of the remaining taxon does not fit the specimen.

ILLUSTRATE TAXA REM
Display illustrations of the remaining taxon. This is not available in the current version (1990), but is implemented in the new one (1991).

Conclusion

We are aiming to produce practical tools, not just to develop methods: we want to put the methods in the hands of a wide variety of users. We support the programs, and they evolve through feedback from the users. We aim to avoid manual manipulation of data wherever possible, so we provide pathways from one program to another. We want the programs and the data to have depth and flexibility, without ad hoc restrictions built in, so that people can use them in ways we did not anticipate. We want the programs to be able to benefit both the compiler of the data and end user. Perhaps these aims are rather ambitious, but I think we are succeeding to some extent.

Acknowledgments

I trained in physics and mathematics and joined the CSIRO Division of Entomology to do ecological modeling, but my work evolved into general computing, and I developed a particular interest in taxonomy.

In about 1971 I started writing a program for generating identification keys (Dallwitz 1974), and a few years later I met Leslie Watson of the Australian National University, who was using Richard Pankhurst's KEYGEN program for the same purpose (Pankhurst 1970). In collaboration with Peter Milne, of the CSIRO Division of Computing Research, Watson had devised a more flexible format for preparing data for this program (Watson and Milne 1972). Leslie and I talked about it and decided that the idea could be improved, and this was the origin of the DELTA format. I have worked closely with Leslie ever since, and the development of the programs would not have been possible without his collaboration.

19

Systematic Databases: The BAOBAB Design and the ALICE System

RICHARD J. WHITE, ROBERT ALLKIN, AND PETER J. WINFIELD

We are biologists by training, and we have been exposed to the frustrations of biologists trying to use existing software for biological purposes. Our experience has shown us that the solutions proposed by computer scientists *can* be used to meet biological requirements, but only after extensive consultation and design work involving both biologists as users and people with backgrounds in both biology and computing as designers.

Recognizing the needs and present difficulties of biological users of computerized tools, we set ourselves the goal of designing and providing database software tools for biologists.

We have in mind two categories of users, neither of which consists of computer specialists. In the first category are the taxonomists who handle systematic data for various uses, perhaps to publish taxonomic revisions, monographs, floras, and faunas. In the second category are ordinary biologists, and users outside biology, who are trying to store, manipulate, or retrieve data of various kinds, but with a basic systematic organization to it. For example, conservation studies frequently require checklists of species, which often are organized according to a taxonomic hierarchy, refer to bibliographies, and contain geographic or descriptive data and many other data elements similar or identical to those handled by professional taxonomists.

The design of software tools obviously involves a number of processes. Here we do not wish to dwell on the requirements, specifications, program structure, processing algorithms, user interface, and so forth. Instead we will

Advances in Computer Methods for Systematic Biology: Artificial Intelligence, Databases, Computer Vision, ed. Renaud Fortuner (Baltimore: Johns Hopkins University Press, 1993).

concentrate on the structure of the tables needed in a database to hold comparative systematic data and its taxonomic organization.

The Complexity of Systematic Data

Such databases may contain descriptive, taxonomic, nomenclatural, bibliographic, biogeographic, or curatorial data, in almost any combination. There are many complexities in these kinds of data. An example of a simple artificial data set composed principally of descriptive data with a taxonomic organization is represented in Table 19.1, as a biologist user might view it. The table illustrates five types of complexity of the kinds of data we wish to handle. These five features might be taken as characteristic of systematic data in biology.

Characters of Different Types

For example, flower color is a multistate character, corolla length is a quantitative measurement, and leaflet presence is a binary character. In many applications, the users will determine the set of descriptors to be included and will want to be able to define or redefine them and their properties.

Variability

The flowers of the species *Vicia cracca* can have two different colors; therefore, flower color is a variable multistate character in this species. There is a range of values for corolla length in all four taxa, which makes corolla length a variable quantitative character.

Table 19.1. An Example of a Conventional Taxon by Character Data Table

Taxon	Flower color	Corolla length (mm)	Leaflet	Leaflet hair
Vicia cracca	Blue or violet	8–12	Present	?[a]
Vicia sativa				
Subsp. *sativa*	Purple	18–30	Present	Present
Subsp. *nigra*	Purple	10–19	Present	Present
Lathyrus aphaca	Yellow	6–18	Absent	X[b]

[a]Observation unknown.
[b]Inapplicable.

The Taxonomic Hierarchy

In the Taxon column of Table 19.1 there are three species listed, but the middle species, *Vicia sativa*, is represented by two subspecies, *sativa* and *nigra*. This gives a hierarchical structure to the objects.

Missing Values

The question mark in the last column of Table 19.1 represents the fact that for some reason the person compiling these data did not record whether leaflet hairs were present in the species *Vicia cracca*.

Character Dependence

The concept of character dependence has been explained by Pankhurst (this volume, chap. 14). In the species *Lathyrus aphaca*, the letter X implies that the character *leaflet hair presence* cannot be recorded at all. Since leaflets are absent in this species, the question of whether there are hairs present on the leaflets is meaningless.

We will concentrate on this last kind of complexity, character dependence, to demonstrate what effect it has on the design of a database. Clearly, appropriate data structures must be designed in order to represent the information and the relationships between data items explicitly and allow efficient algorithms to be developed. Table 19.1 showed a simple character dependency, where one character, leaflet hair presence, is dependent on another character, presence of leaflets, but it can be more complicated than that. Character dependencies may extend to a second level, or beyond, and form a whole chain of dependencies or even a directed graph.

For example, in Figure 19.1, the character length of the hairs on the petioles (leaf stalks) can be measured only if the petioles are actually hairy.

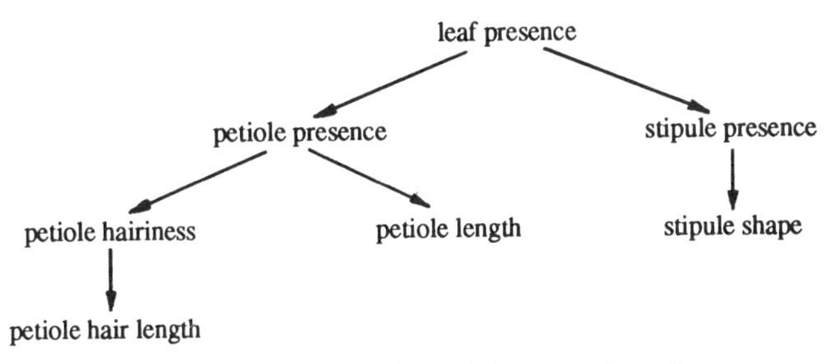

Figure 19.1. An example of serial character dependence.

We can determine whether the petioles are hairy only if the leaves actually possess petioles. If leaves are absent then we cannot decide about petioles.

The BAOBAB Database Design

In 1988, we began designing a complete database structure that could be used by a set of data management tools for systematically organized databases (Allkin and White 1988). We adopted the relational database design approach (Codd 1970).

We call this database structure design BAOBAB. In principle, since BAOBAB is a relational design, which specifies certain tables and their records and fields, a database using this design can be created and queried using any adequate general-purpose relational database management system (DBMS).

However, the complexity of the data and therefore of the database design is such that direct data entry and retrieval would be a highly complex process involving a deep understanding of the design of the database. This is not realistic for a biologist wanting a useful software tool. The user's view of the data is too different from the internal structure of the database. What is needed is a special-purpose database manager for biologists, which would act as an interface to the underlying database structure and DBMS, translate a user's requests into commands appropriate to this structure and DBMS, and convert the results of retrievals back into the user's view of the data.

No program to manipulate a database using the whole BAOBAB design structure has yet been completed. Our aims and activities remain to produce software tools suitable for a taxonomist's or other biologist's work-bench data management system, which Diederich and Milton call an expert workstation (this volume, chap. 7).

In this chapter, we distinguish where necessary between three related uses for the name BAOBAB: the database design research *project*; the existing, though evolving, *design*; and the concept for an as-yet unimplemented biologist's database management *program*.

Character Dependence in BAOBAB

We will discuss the way character dependence is handled in BAOBAB as one example of this database design. A table contains a record for each dependent character, such as the presence of leaflet hairs, which specifies the *master character* that it depends on, in this case the presence of leaflets, and the controlling data value. An example of the contents of this table is shown in Table 19.2. Row one means that petioles have to be present before you can record the presence of petiole hairs. The second line takes us to a second

Table 19.2. The BAOBAB Character Dependence Structure

Master character	Data value	Dependent character
Petiole presence	Present	Petiole hairiness
Petiole hairiness	Hairy	Petiole hair length

level. It means that we can measure the length of those hairs only if the petiole hairs themselves are present. Each line represents one of the dependent links shown in Figure 19.1.

Special program code has to be written into the application software to check that these restrictions, as described in this table, are not violated. The mere presence of a relation like this does not, in a conventional relational database design such as this, automatically enforce that particular constraint.

Other BAOBAB Database Design Features

Character dependence as exemplified above is one of a number of complex features of systematic data. There are quite a number of other complicated aspects of the design. These are issues of the organization of the data and the design and implementation of the database structures, rather than aspects of the objective data as in Table 19.1.

1. We need structures to accommodate at least one taxonomic hierarchy, preferably several. Many applications in professional systematics require that alternative opinions be expressed or that the taxonomic hierarchy be adjusted after the other data have been entered.
2. The sheer complexity of handling the nomenclatural data is very significant. This can be a major portion of the design of the whole database. A full understanding of the nomenclature requires information on the taxonomic hierarchy it is describing, and vice versa, with links from each nomenclatural item to the appropriate part of the taxonomic hierarchy it is labeling (Beach et al., this volume, chap. 15). These authors explain the importance of having different views of the nomenclature, together with cross-linking between them, and preferably a record of how these views have changed through time.
3. Characters may be grouped into hierarchical or overlapping sets for convenience in data entry or retrieval. A user preparing a report might, for example, want to issue a simple command to retrieve all the biochemical data for a given species of plant. Another use would be to indicate economically that the whole set of floral characters would be inapplicable if flowers were absent or not recorded for a particular taxon.
4. The properties of each type of character, such as the allowable range of

data values, need to be defined using a differently structured set of tables within the BAOBAB database design, because different types of character require different sets of metadata to describe them. This information is needed by a biologist's DBMS, such as the BAOBAB program, to interpret its interaction with the user correctly, for example, to validate input or format tables of results.

5. The set of characters included in a database may change as the use of the database develops. For example, the user may wish to add new kinds of data to the database as they become available or to refine the way some existing data are treated. Because the user's view of the data is so different from the internal database structure, users are not in a position to alter these internal data structures directly. A user's requests for adding or refining characters must be interpreted by the BAOBAB program, and the interpretation or definition of the characters must be stored in a flexible system of data structures for defining the user's view of the characters.

6. As well as defining their own characters of *existing types*, individual users or developers addressing a specialized set of users may possibly wish to add *new types* of data such as images or DNA sequences. New types of data need new types of database structure to accommodate them, and in designing a set of general tools we have to allow for the addition of data and algorithms using such new kinds of structure. This is a much more challenging task than satisfying the previous design requirement.

7. Characters might have different values in different parts of an organism. This increases the number of dimensions in the user's view of the data set. For example, with data on plant chemical constituents, you may have a number of chemicals and you may wish to record their presence separately in a number of different organs of the plant, leading to a two-dimensional array of combinations of chemicals with organs for each species.

8. In some applications it may be important to handle the correlated variation of two or more characters, such that certain combinations of character states are less likely or absent (Allkin 1984). This applies particularly to data that have been aggregated, say from a set of individual specimens to a description of a species (Maxted et al., submitted). Such data may be more manageable if correlations are explicitly stored.

9. There may be associated or curatorial information to manage things like notes on the source and accuracy of the data, who entered the data, references to publications, dates on which the data have been updated, and so on.

Implementing the BAOBAB Design

During the BAOBAB project described above, we created, discussed, modified, and discarded numerous alternatives for entity relationship diagrams and relational database structures. We achieved a much clearer understanding of the nature of taxonomic data and of taxonomic data management requirements. We experimented, for example, with alternative data structures adequate for tracking and presenting different nomenclatural views (Beach et al., this volume, chap. 15) and for incorporating alternative taxonomic hierarchies within the system. Following full normalization procedures, the whole system contained in excess of 160 relations.

Using the nomenclatural component of the BAOBAB design, we built a prototype program based on the commercial relational DBMS MISTRESS (now EMPRESS) running under the UNIX operating system. However, we had insufficient resources to write the software necessary to manage such complex data structures for all the components of the BAOBAB design and to provide an easily used working tool with the necessary functionality.

Taxonomists, meanwhile, continue their work. Systematics is information management. Systematists are involved in data management when managing their collections, in undertaking their systematic research, and in disseminating their results to the scientific public. They work largely without access to or awareness of software tools suitable for their use (Allkin 1984; Allkin and White 1988). This bars most taxonomists from the benefits of available technology, although an increasing number, convinced of the exciting advantages that databases offer, spend their time learning to use commercial database packages and designing and developing their own systems (Allkin 1988; Allkin and Winfield, 1992).

These personal systems, however, often use inadequate data structures. Moreover, they frequently duplicate the efforts of colleagues elsewhere. It is often painful for us to witness the results, both databases and software, which, although they sometimes satisfy the originator, could have been made far more widely useful with the same amount of effort.

The ALICE Database Management System

In this chapter we have concentrated on database structure. When implementing software tools based on this database structure design, it is necessary to design the algorithms that will process the data, and, most important, the user interface. Some of these issues are discussed below in the context of the implementation of the ALICE program suite.

Description of ALICE

Although BAOBAB was a fruitful research initiative, it provided taxonomists with no immediate help. Winfield and Allkin were, however, simultaneously developing a biodiversity database management system called ALICE. Our aim was to produce a set of reliable and easy to use tools for biologists. Initially we were prompted by the urgent needs of a multiinstitutional project, the International Legume Database and Information Service (ILDIS). ILDIS is coordinating systematists working on the plant family Leguminosae and aims to provide the scientific community with consistent biodiversity information about species in the family (Zarucchi et al. 1992).

The ALICE system comprises a suite of seven related programs designed to meet the common needs of a range of biological data management projects (Allkin and Winfield in press; Allkin et al. in press). ALICE is now used in various species checklist projects, conservation directories, and ethnobotanical projects and is suitable for zoologists and botanists alike. Version 1 of the main database management program, ALICE itself, was first released in 1985. The current version 2 was released in 1989. Version 3 is in preparation, while minor amendments and improvements continue to be made to version 2.

Together with documentation staff and some support programming, Winfield and Allkin have taken about seven years to develop the ALICE system to date. The remainder of this chapter summarizes the constraints upon development of ALICE and the lessons we have learned.

Constraints upon the Implementation of ALICE

Contributors to the ILDIS project required working tools within a reasonable time, and the work time available to us was very restricted. Since ILDIS aims to provide an information service outside of taxonomy it has, in common with conservation organizations, ethnobotanists, and any other class of applied scientists, an interest in a single consistent taxonomy. Thus its information system need not handle the complexities of tracing nomenclatural histories or providing alternative taxonomic views. In this sense the design of the nomenclatural component of ALICE could be significantly less complex than the corresponding part of the full BAOBAB design.

Nevertheless, any single taxonomic view summarizes and must reflect previously held nomenclatural systems and, far from being static, will continually evolve. The former requires that the system reflect the sometimes tortuous relationships between all published names and between those names and the taxa they represent. The latter requires that the system provide simple and flexible editing facilities.

A final constraint was the limited computing power of the target machines. Consequently, the original version of the system was developed to run under MS-DOS on machines with 256K of random access memory.

Lessons Learned from the Implementation of ALICE

Resources Required. The first lesson was the sobering realization of the amount of resources required for a professional level of software development. The number of hours spent writing, testing, debugging, documenting, and maintaining code as well as producing user documentation cannot be overemphasized, particularly if programs are to be genuinely reliable and easy to use. A taxonomic database management system requires taxonomic database structures, taxonomically intelligent algorithms, and an interface that can be used by taxonomists.

Normalization. The ALICE DBMS uses about seventy different tables to contain and support a biodiversity database. This relational design is not fully normalized, even for the subset of data types permitted by ALICE. We have had to compromise in the production of an engineering solution. We merged some of the original 160 relations to make it easier to implement.

Nomenclatural "Intelligence" Embedded in Code. The main ALICE database management program contains more than 25,000 lines of code. Approximately 60 percent of that code handles the central nomenclatural system components. Such a large body of code has proved necessary to implement all of the necessary logical and validation rules associated with biological nomenclature. Given the database technology available to us these taxonomic rules need to be "hard-wired," that is, written into the ALICE program code. They cannot yet be stored as part of the data, as might be preferred. Note that this large volume of code was necessary to handle the nomenclature even though ALICE implements only a subset of the BAOBAB design.

Importance of the User Interface. Although the majority of the ALICE code handles the nomenclatural modules, our experience coincides with that of other program designers, for example, Diederich and Milton (this volume, chaps. 7 and 10), namely, that the majority of our design and coding efforts have gone into developing an easily used user interface. A well-designed interface requires a great deal more than a few windows. The words that appear on the screen, the order in which they appear, the order in which subtasks are presented to the user, and the relative priorities for different activities all require considered attention, design, and testing of the resulting interface. The interface must be intuitive to users who are not accustomed to using computers.

Different Tasks and User Interfaces. The ALICE system was developed with a software tools philosophy (Allkin and Winfield 1992) similar to that of the *expert workstation* toolkit of Diederich and Milton (this volume, chap. 7). The system offers functionality as a number of software tools implemented as discrete programs that communicate one with another (Fig. 19.2). Each fulfills a well-defined set of tasks and is designed for a particular class of user. We distinguish, for example, the database manager or administrator from the end user of the database, although in some cases a database will be con-

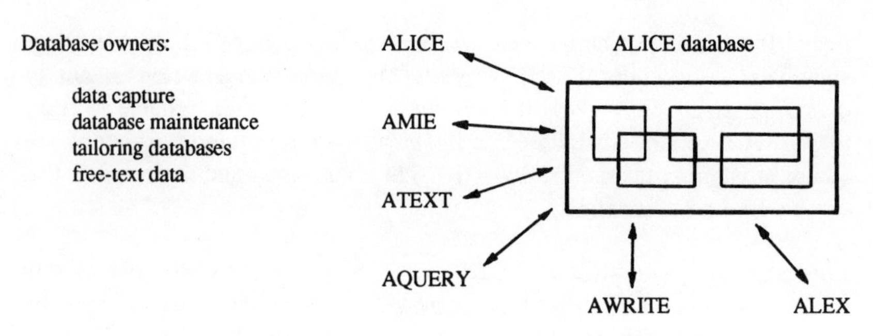

Figure 19.2. The ALICE program suite.

structed and used by a single person. The former needs tools to shape the database to a particular application (e.g., to define a set of morphological characters), to capture data, or for routine database maintenance. The end user wishes to answer particular questions (e.g., using the AQUERY program; see Fig. 19.2) or print simple reports.

These two groups of users need quite different user interfaces. Database managers working regularly with the system find repetitive keystrokes burdensome; some end users, by contrast, use the database only infrequently, have little more than a superficial knowledge of what sort of information the database contains, and need their hands held to a greater extent.

Another striking difference in interface is how may we best access taxa when undertaking different tasks. Sometimes a nomenclatural search is most appropriate, enabling use of any name, be it a synonym, misapplied name, or whatever. For other tasks, however, it is more appropriate that users be offered the current taxonomic view with only the accepted names displayed.

Need for Communication with Other Systems. We now have evidence for the importance of information exchange to the users of ALICE. This has occurred not only among collaborators within the ILDIS project who have agreed to use standard data definitions, but also among other ALICE users with independent databases who wish to import data from their colleagues' systems or export subsets of their data to other databases or to application programs such as expert systems. ALICE now supports a range of formats for the export of files of data to be read by other programs. These export formats range from simple free-field and fixed-field record formats like SDF and dBASE to complex specifications involving the detailed syntax and interpretation of data exchange files, such as DELTA (Pankhurst, this volume, chap. 8; Dallwitz, this volume, chap. 18).

These formats are not sufficient in themselves to allow us to transfer data between ALICE databases without loss of information. We have developed XDF, a data definition language (Allkin and White 1989; White and Allkin 1992a,b) for exchanging more complex biological data sets such as are now being stored in databases. These exchange data sets can range from a simple data table together with its associated definitions, possibly a subset of a larger database, right through to the complete contents of a complex database.

XDF has been recognized as a standard data exchange technique by the Commission on Taxonomic Databases of the International Union for the Biological Sciences (IUBS). XDF is extensible, and a data set expressed in XDF can contain the definition of its own extensions, like a program written in an extensible programming language. It can therefore be employed to define specialized data exchange or *transfer formats* for particular application areas or groups of users. XDF has been used to define a particular transfer format for the biodiversity data used by the ILDIS project. The expectation is that in future the IUBS Commission will define a number of transfer formats for particular application areas using XDF.

The development of XDF was prompted by the need of ILDIS contributors to exchange data among themselves and our own frustrations at writing once-off translation programs every time we needed to use a particular application program or exchange data with a particular colleague. Our goals and the resulting XDF language share some features with the Abstract Syntax Notation One (ASN.1). XDF was developed without any knowledge of ASN.1 despite our attempts to discover what was occurring outside of biology. XDF has greater flexibility than ASN.1 in ways that we have already found necessary in biological data sets. Nevertheless, ASN.1 has the important attraction of being an international standard supported by the International Organization for Standardization (ISO) (ISO standards 8824 and 8825 [ISO 1987a,b]), and we are considering the relationship between XDF and ASN.1.

Importing complex data sets while maintaining database integrity is not straightforward. Consequently, while ALICE exports data in many different formats, including XDF, it currently only imports directly from XDF. The assumption is that a set of records in some other format whose fields are delimited by fixed widths or by delimiter characters can be included within a data set whose syntax and interpretation are expressed in XDF. The difficulty lies not in the syntactic format of the data records but in their correct interpretation. A partial solution is that XDF permits certain forms of automated restructuring and rearrangement of a data set as it is read into an application program such as ALICE. We are actively investigating these problems at present.

Choice of Development Environment. We wrote ALICE in the dBASE programming language. The Ashton-Tate commercial dBASE products are recognized to have some limitations as DBMSs. However, the widespread popularity of dBASE has meant, to our good fortune, that the language offers us a wide range of future development options. ALICE is not interpreted by the Ashton-Tate dBASE software; it is compiled by the Wordtech Systems Quicksilver dBASE III language compiler.

A number of compilers are available across a wide range of platforms. Using only the Quicksilver compiler, we may run multiuser ALICE databases under UNIX or across MS-DOS networks; alternatively we can run ALICE on top of the ORACLE DBMS. Other compilers offer other options. In addition, recently developed sophisticated relational DBMS products, such as RECITAL, provide for importing not only dBASE files but also dBASE program language code. Should we choose to move into C programming, then several products exist for translating our code, and several C program libraries exist for manipulating dBASE databases. ALICE already incorporates some C code for lower-level functions.

Choice of a commercially supported database programming language that is widely used has, by commercial pressure, resulted in a secure present environment and a marvelously rich set of future development options. This would not be the case had we used a computer science research tool, as we had been urged to do, or even one of the many more sophisticated relational DBMSs.

This environment is being improved all the time, again by commercial pressures, but most significant has been our own commitment to creating our own development environment within dBASE. We have built our own tools, for example, for ALICE database maintenance, for routine data verification, and for building, editing, and maintaining menus.

Conclusion: Collaboration with Computer Scientists

Given our own desire to develop reliable software tools for biologists, the advantages of a well-supported commercial development environment are obvious. Though the dBASE database manager is of limited value for complex databases, as a development language dBASE has the advantages of reliability, support, and a secure development path.

We have learned to distinguish between computing research tools and commercially supported products. From the computer scientists we seek immediate advice and the establishment of longer-term research ties. We need help, for example, to navigate our way through the bewildering array of theoretical models in the computing literature and to interpret their relevance to us. Most of all, however, we need practical advice. In particular, we seek answers to questions such as:

1. Whether the apparent fit between object-oriented database design models and taxonomic hierarchically based information systems is in fact as con-

venient as might at first appear, or whether it is a spurious superficial similarity;
2. What techniques exist, such as frame-based logic, to assist in reducing the amount of hard-wired "intelligent" code we need to write into our database applications to encapsulate some of the taxonomic knowledge about the meaning of the data;
3. Whether advances in *federated* database systems (Winslett, this volume, chap. 16) will ultimately reduce the need for users to be concerned with details of data import and export; and
4. What reliable and supported tools are available now for us to build biological software tools benefiting from advances in object-oriented or frame-based representational techniques. Are such tools available in computing environments that will be available to most biologists within a five-year period?

Our concern is to ensure that future generations of systematists make better use of information technology than today's.

Acknowledgments

Our ideas have benefited greatly from discussions with Jim Beach, Frank Bisby, Mike Dallwitz, Nigel Maxted, Richard Pankhurst, and many others, especially within ILDIS. We thank the Science and Engineering Research Council, the University of Southampton, and the Royal Botanic Gardens, Kew, for financial support.

Discussion

——: You said you were looking for suggestions. Can you explain what are the problems you are having or what are the areas for improvement?

Allkin: When you bring data sets together you need to be aware of differences not only in syntax, but also in internal consistency of the data. We are not expecting the system to make decisions. AI people are excited by expert systems, and they want to emulate or model what taxonomists do in their heads. I think they are a million miles from this achievement. It is very difficult to describe these sorts of constraints.

White: We would like help with the problem of managing the complexity of the design: the data themselves, the data structures, the algorithms, and so on. We have many ideas awaiting evaluation, but we do not have the time to look at them properly. The present lack of support is a major problem.

Pankhurst: I suppose that you are not specially worried about defects in the software tools from what you said. What is your ideal tool?

Allkin: I do not believe there is one. If I were starting developing now, I would not choose dBASE. However, we have invested seven years in the dBASE language in developing tools for the end user and for ourselves. We wrote many programs to help us with menus, interfaces, handling taxonomic data structures, and handling nomenclature. We started slowly, but we are now very efficient in generating new products around the system.

Portyrata: You can now purchase source code for maintaining dBASE files, to rearrange your whole structure, or tailor the DBMS to your own use if you wanted to.

Allkin: This is one of the advantages of dBASE. We can port it, we can compile it directly under UNIX, we can recode it to run on top of more sophisticated relational databases, and so we can change the foundations for the implementation. We hesitate to do it now, but it is quite exciting to us to think we do have such flexibility to respond to future opportunities.

Thompson: Is this new XDF format officially accepted by the Taxonomic Databases Working Group?

White: Yes, a first version was accepted as a standard at the fifth meeting of TDWG at Las Palmas in the Canary Islands, in November 1989. It has not been published yet, but it will be soon.

Thompson: Who writes the interfaces to XDF?

White: The great advantage of XDF is that the developers of a particular database, who are familiar with its structure and conventions, write the interfaces. For example, Bob Magill wrote a program to get the data from his TROPICOS database into a form of XDF. One could then, if need be, have the XDF software manipulate that into another form of XDF that more closely resembles the internal structure of a recipient database.

Buckup: When you incorporated the data from two or three databases, how did you handle discrepancies, for instance, differences in definitions?

Allkin: The ILDIS project involves the collaboration of about thirty participating experts. The first stage was the rough and ready merging of data sets for several of the major continents. The importing software wrote reports on the discrepancies in the data, and decisions were taken jointly by the authors of the databases with full knowledge of the differences. The assembled data set was then sliced taxonomically rather than geographically, and we are sending it to the experts in each taxonomic group. Each expert will have the final say on the taxonomic and nomenclatural decisions.

Buckup: Are there any problems in the collaboration of the experts?

Allkin: Things run surprisingly well. It must be said that the legume community is very special. There have been several international conferences, and the classification decisions agreed on by the legume taxonomists are widely accepted. It would be much more difficult to agree on one classification for, say, the grasses, where there are many views.

20

MICRO-IS: A Microbiological Database Management and Analysis System

DAVID A. PORTYRATA

The Microbial Information System (MICRO-IS) is a database management and analysis system designed for use by microbiologists. The system provides them with tools for data collection, data standardization, data quality control, and data analysis. Additional tools are also provided for use of the data in the probabilistic identification of bacteria, yeast, algae, and other microorganisms. It was developed under a contract with the Microbial Systematics Section (MSS) of the National Institute of Dental Research (NIDR). Unlike more general database management systems, MICRO-IS is specifically designed to facilitate the management and analysis of microbiological data. This system has been used for many diverse applications, including taxonomy, ecology, regulatory microbiology, construction of probabilistic identification schema, epidemiology, and culture collection management.

History

An early prototype of the system was developed by the American Type Culture Collection (ATCC) on a PDP-11 using FORTRAN in 1974–75. The prototype provided functions for data entry and probabilistic identification of bacterial strains.

The prototype was ported to an IBM 370 running MVS/TSO in 1976 using PL/I. Several enhancements were made to the system for data management, report generation, and integration to other software tools such as the Statistical Analysis System (SAS) and TELLAGRAF. It has also been installed on

Advances in Computer Methods for Systematic Biology: Artificial Intelligence, Databases, Computer Vision, ed. Renaud Fortuner (Baltimore: Johns Hopkins University Press).

other IBM 370, 3080, and 3090 computer systems located in Japan and France.

Work began on an enhanced version of the system for microcomputers in 1987. At that time, I did an initial survey and found that many existing database management systems (DBMSs) were lacking power, were too expensive, or required royalties. One of the criteria was that no royalties or additional costs were to be incurred by paying off third-party software vendors. Another more important criterion was portability over a wide variety of computers and operating systems (VMS, UNIX, and DOS). Advances in microcomputer technology and software made it feasible to develop a version of the system for users who have access to microcomputers. The microcomputer version is being implemented using C to facilitate future ports to UNIX operating systems.

Software and Hardware Environment

The original mainframe version of the MICRO-IS is housed at the National Institutes of Health (NIH) Division of Computer Research and Technology (DCRT), Bethesda, Maryland. It is running on a cluster of six IBM 3090-300JS computers under the MVS/TSO operating system and WYLBUR.

A microcomputer version of the system has been developed to provide access to MICRO-IS for individuals and agencies who are unable to use the computers at DCRT. The microcomputer version currently has utilities for entering and editing data, querying data, generating reports, doing basic statistical analysis, importing data from other computers, and performing probabilistic identification of bacterial strains. The microcomputer version is designed to run on IBM compatible microcomputers running MS/PC-DOS 2.1 or higher.

Current Usage of System

Over the years, MICRO-IS has been used for a broad spectrum of applications. However, it is primarily used in collection and analysis of bacteria strain data. Some of the more significant uses of the system include both mainframe and microcomputer versions of the system.

ATCC is currently developing several databases in collaboration with the U.S. Environmental Protection Agency (EPA) for studying the use of probabilistic identification matrices in the identification of bacterial strains altered by genetic engineering. The EPA has developed several probabilistic identification matrices for parent strains that may be the focus of genetical engineering. They are interested in determining at what point a strain altered by genetical engineering could not be identified using an identification matrix developed for the parent strain. The EPA is worried that matrices developed for current strains will fail to pick up on a strain that has been altered via genetic engineering. They may have to generate new matrices or expand the existing matrices. These databases contain strain data for several species of bacteria.

The British Antarctic Survey (BAC) is using the system to store data about bacterial strains collected in Antarctica. The system is also being used by the National Oceanic and Atmospheric Administration (NOAA) to store data on estuarine algae.

The U.S. Food and Drug Administration (FDA) is using the system to maintain regulatory and historical databases on *Salmonella* and food container surveys. These databases are used to write annual reports and monitor the incidence of *Salmonella* in various foods, cosmetics, etc..

The NIDR is using the system for data collection and taxonomic analysis of bacteria and yeast. Taxonomic analysis includes the development and maintenance of the probabilistic identification matrices which are provided with the system.

MICRO-IS is also being distributed by the Microbial Strain Data Network (MSDN) as freeware (a nominal charge is made for the disks, postage, and handling) to members of the network. It is offered as a tool for users to convert their data recorded in laboratory notebooks and other printed media into an electronic format which is more conducive to exchange of information among members of the network.

System Description

A general overview of MICRO-IS is given in Figure 20.1. It depicts the major components of the system and their interaction with each other. The box labeled MICRO-IS references the software component of the system. Note that users are considered to be a major component of the system, as they are required to operate it.

MICRO-IS is designed around a six-digit code number assigned to each unique feature or characteristic of an individual microbial strain. These code numbers are obtained from a RKC code scheme developed by Rogosa et al. (1986). The RKC coding system has an open-ended, statement-oriented vocabulary in which each statement is a description of a possible strain feature (character). Each statement is assigned a unique code number. According to RKC code, characters can be quantitative (called *numeric* in MICRO-IS), qualitative with either four states (called *binary* in MICRO-IS), or multistates (called *character* in MICRO-IS). The binary character is actually a four-state (true/false/blank/unknown) character but is referred to as binary in MICRO-IS and in the RKC coding scheme. MICRO-IS is designed to be somewhat flexible in how people use it or describe it. The RKC coding schema are

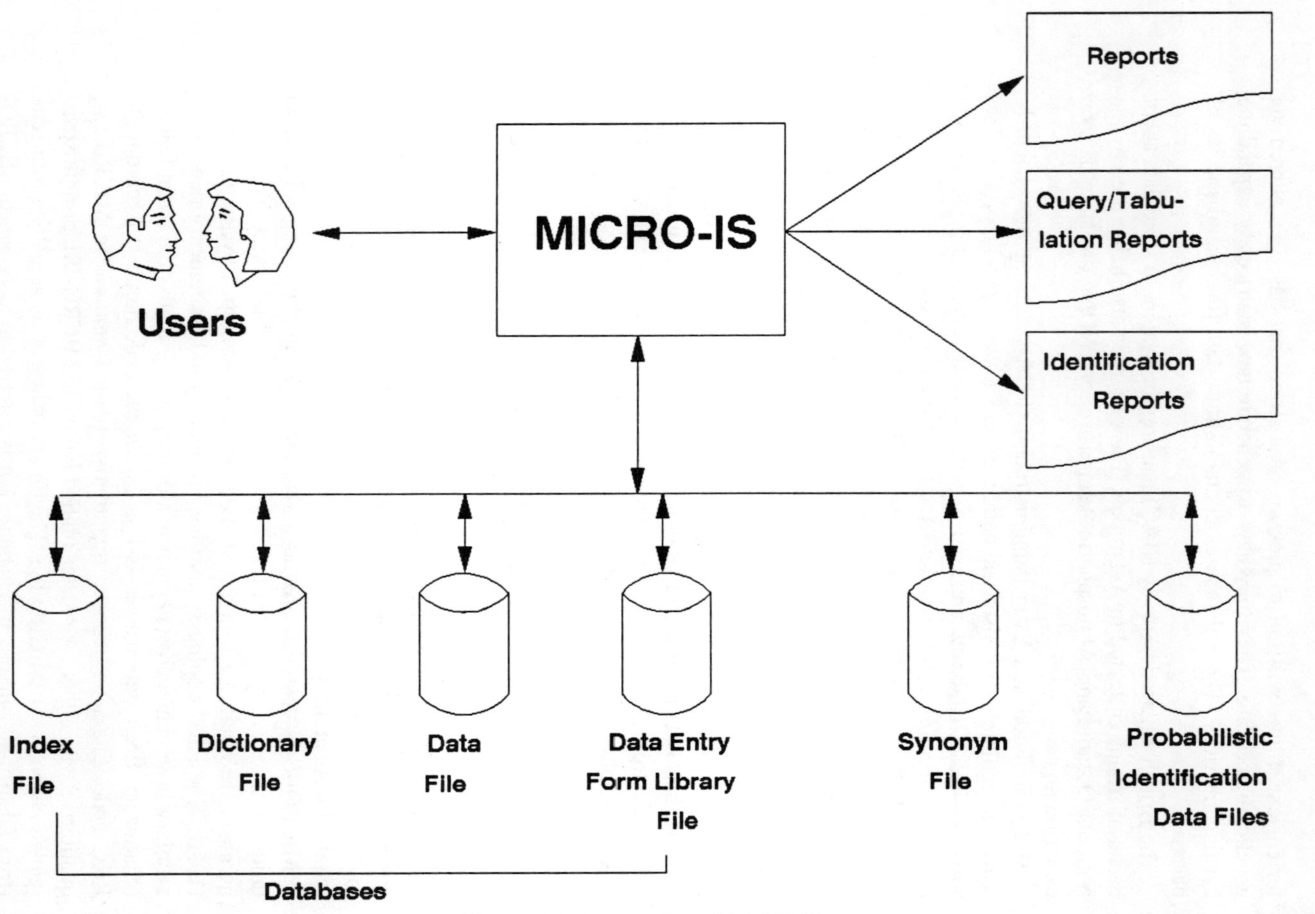

Figure 20.1. An overview of MICRO-IS.

currently maintained by MSS support personnel at NIDR.†

The RKC character code number indicates the data type of the character. Binary characters have RKC code numbers ranging from 1 to 499 999, inclusive; numeric from 500 001 to 999 999, inclusive; and qualitative multistate "characters" from C00001 (1 000 001) to C99999 (1 099 999), inclusive. Binary data can indicate presence, absence, presence/absence unknown, and no answer. Unknown is synonymous with "it is not known if anything was observed for the character." The unknown state was added rather recently to address use of the RKC coding scheme in epidemiological studies. No answer means that it is known that nothing was observed for the character. Numeric data are the actual numeric values for a character. It can be an integer or real number. Unknown and no answer can also be represented in numeric characters. Numeric data can have a maximum precision of seven digits with a magnitude range of 10^{-37} to 10^{38}. Qualitative multi-state characters are items of alphanumeric text and can be a maximum of 1024 characters long.

For example, the binary character code 017017 is *Growth occurs at 45°C*; the numeric character code 535135 is: *Inhibitory zone diameter for bacitracin disk concentration 2 units is _ mm*; the qualitative multistate character C02012 is: *What was the specific source of isolation, e.g., kind of water, soil, etc., species and organ of plant, animal, etc. (CODE SHEET II, columns 29–70)*. As long as the user adheres to the RKC coding scheme, the RKC character codes 017017, 535135, and C02012 will reference these characters.

Database Structure

The database structure of the mainframe version of MICRO-IS is based upon a hierarchical model. However, it is not a full implementation of the model in that it allows for only one level of child records. A true hierarchical database normally allows for a child record to point to one or more other child records (sometimes called subchild records). The database structure is actually based upon RKC data entry forms used to enter bacterial strain data. It is, in fact, a partial implementation of a hierarchical (semihierarchical) database. This inability to support subchild records places limitations on the amount of data (maximum record size is 3156 bytes) and number of characters that can be stored for a strain record.

While designing the microcomputer version of the MICRO-IS, it was decided to use a relational database model. It was believed that the relational model would be much more flexible and would address many of the problems encountered using the semihierarchical database structure of the mainframe

†For more information on the RKC coding scheme, contact: Dr. M. I. Krichevsky, National Institutes of Health, Department of Health and Human Services, Park 5 Building, Room 451, Bethesda, MD 20892.

version. During implementation and early beta testing (when users are allowed to use the system as a test), some of the usual problems associated with the relational database model were brought to light. These included excessive space consumption due to repetitive key fields (each table must be linked by a key field), data integrity assurance, and potential for excessive joining of tables for data retrieval, which can seriously degrade system performance. However, these were offset by the ability to easily update strain data and to add or delete strain characters and associated data.

During several large projects involving large databases containing bacterial and yeast strain data, there was some evidence that this type of data is easier to organize and access when using a hierarchical database model. This is particularly true in the case of bacterial strain data. Microbiologists use a composite field called a strain designator which contains a sample ID, subsample ID, and isolate. A sample ID can have a number of subsamples, and each subsample can have a number of isolates. Although a relational model can handle this situation, the potential for time-consuming searching and joining operations exists when dealing with large tables. It was also found that specifying certain characters at the sample level reduced the incidence of data redundancy.

MSS support personnel found that many microbiologists indicated that they would like to implement databases which would allow them to retain the hierarchical relationships intrinsic in their strain data. FDA and ATCC microbiologists were particularly interested in being able to develop databases with hierarchical relationships. Consequently, the microcomputer version of MICRO-IS will in the future incorporate both hierarchical and relational capabilities. Users will be able to develop databases based upon the relational model or the hierarchical model. One problem is to develop a language syntax that a microbiologist can easily use to specify hierarchial relationships.

Currently, the microcomputer version of MICRO-IS is based upon the relational model. Data are stored as a row and column matrix, referred to as a table. The columns in a table are referred to as characters, which are synonymous with RKC characters. Each table contains a list of RKC character codes of the RKC characters for which it has data. Each row in a table contains data for the RKC characters in the table character list. To facilitate data compression, data are stored in each row in the following sequence: numeric data, binary data, and multistate character data. Tables containing related data are grouped into a database. Figure 20.2 illustrates the structure of a table in a database.

Each database has a dictionary which is used to store the character RKC code number, field name acronym (field name), and, in the case of qualitative multistate "characters," the length (maximum length of character string which can be stored in the field). The RKC character code number is the link between a field in a table and its entry in the database dictionary. This removes

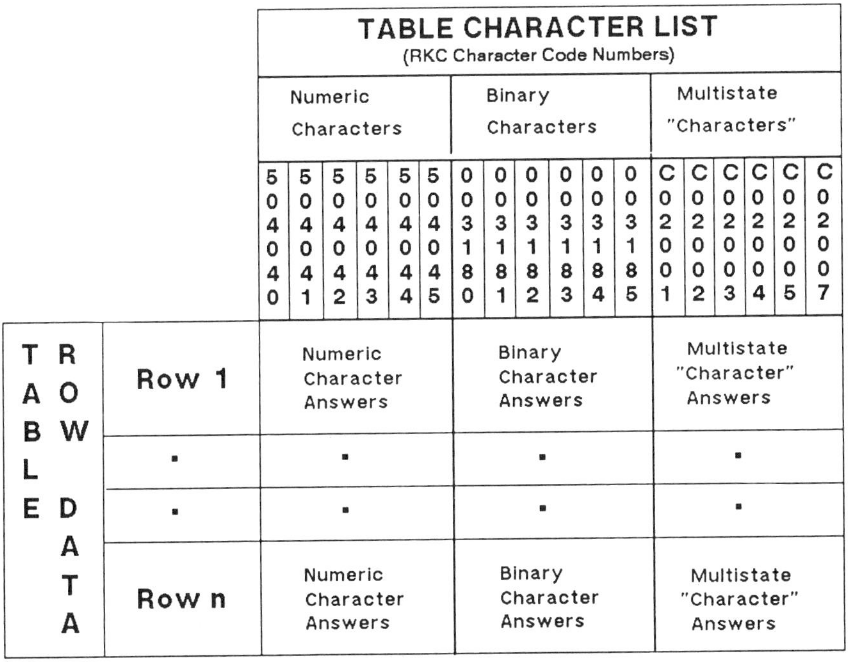

		TABLE CHARACTER LIST (RKC Character Code Numbers)		
		Numeric Characters	Binary Characters	Multistate "Characters"
		504040 504041 504042 504043 504044 504045	003180 003181 003182 003183 003184 003185	C02001 C02002 C02003 C02004 C02005 C02007
TABLE ROW DATA	Row 1	Numeric Character Answers	Binary Character Answers	Multistate "Character" Answers
	.	.	.	.
	.	.	.	.
	Row n	Numeric Character Answers	Binary Character Answers	Multistate "Character" Answers

Figure 20.2. MICRO-IS database table structure.

the need for redundant storage of the field name in the list of fields for a table. Data compression is utilized to maximize storage efficiency, which is critical in the microcomputer environment. Binary data are stored using two bits, and character data are stored with leading and trailing blanks removed from the string.

The microcomputer version of MICRO-IS allows users to assign alphanumeric acronyms or abbreviations to characters for easier reference. These acronyms are limited to a maximum length of twenty-eight characters. The system allows users to reference a character by more than one acronym (synonyms). One of these synonyms can be defined as the preferred synonym or field name. This allows for variation in spellings, abbreviations, different languages, and usage of uppercase and lowercase character sets. The preferred synonym allows users to define at least one entry as the standard acronym for the character. The limit of twenty-eight characters is imposed to conserve auxiliary storage requirements. At this time, the RKC coding scheme does not have a list of standard acronyms for each RKC code. Therefore, users can assign their own field names based upon the RKC code descriptive text. For example, a character with a RKC code of 17017 can be assigned the field name GROW.45C and the synonyms 45C, grow.45, and

GR45. A user can reference the character using the RKC code number, field name, or any one of the synonyms.

Currently, MICRO-IS does not allow users to reference a character by the RKC text description. For example, character 17017 cannot be accessed by its RKC code descriptive text *Growth occurs at 45°C*. The majority of the text descriptions of RKC code characters are quite long (exceed eighty characters) and may not be suitable for keyboard entry by the user.

To allow users to reference a character by its RKC code number, field name, or synonym, a centralized synonym file is used to store the RKC code numbers, field names, and synonyms for all the databases. Acronyms for a specific character are linked by the RKC code number, which is consistent across databases. Entries are added to the synonym file whenever a new field is added to a database or a new acronym for an existing character is defined by the user.

System Functionality

MICRO-IS can be divided into four functional components: database management, report generation, statistical analysis, and probabilistic identification.

The database management component of the system contains utilities for creating databases and tables, data entry, synonym file maintenance, general database file maintenance, probabilistic identification matrix data file maintenance, taxonomic analysis tools, and other miscellaneous functions. The utilities provided in the database management component are as follows:

DEFINE: Create databases and tables and define fields;
EDIT: Create data entry forms, enter and edit table row data;
SYNUTIL: Create and maintain synonym files;
DBUTIL: File compression, data import, and database scheme reports;
TAXUTIL: Probabilistic identification data file maintenance and taxonomic analysis tools;
SELECT: Select an existing database and/or table.

The report generation component of the system is centered around the Query Utility. The Query Utility has a number of commands which are summarized below:

EDIT: Used to edit current command or last command entered or to invoke query editor;
RUN: Used to execute a query command;
GET: Used to read in a query command from a file;
CLIST: Used to list current command or last command entered;
SYSTEM: Used to temporarily exit Query Utility to run other programs or submit DOS commands;
SELECT: Used to Select a database or change the current directory;

SET: Set options such as the output destination, terminal display line width and page size and TABULATE command options;
SHOW: Display the current options settings, current directory current database, tables in database, fields in table, etc.;
HELPME: Obtain information on commands and options;
LIST: Display table row data;
TABULATE: Obtain statistics on table row data;
END: Terminate and exit Query Utility.

The two commands most important for report generation are the LIST and SHOW commands. The TABULATE command can be used for report generation and is also useful as an analysis tool. Information on the database structure, characters stored in a database and tables, and other relevant information are obtained using the SHOW command. This command basically allows a user to obtain a table of contents of a database. The LIST command is used to display character data in one or more tables in a database. Data are displayed in a tabular format similar to a data matrix. The syntax of the LIST command is:

LIST (Field List) (Query)

The Field List can consist of individual RKC character code numbers, field names, synonyms, character ranges, or special keywords. Several examples of character ranges are given below:

504040 TO 504045
003180 TO 003185
MUTANT TO GC

The first range is expanded to include all numeric characters with RKC character codes from 504040 to 504045, inclusive. The second range is expanded to include all binary characters with RKC character codes from 003180 to 003185, inclusive. The third range is expanded to include all qualitative multistate "characters" with RKC character codes from C02002 (RKC character code for Mutant) to C02007 (RKC character code for GC), inclusive. Special keywords are *, ALL, and NOTHING.

A query is used to specify a set of selection criteria for the data to be displayed. Some of the terms which can be include in a query are: RKC character codes, field names, or synonyms; single, list, or ranges of characters; Boolean operators AND, OR, NOT; comparison operators =, >, <, ≥, ≤; special keywords ANY, BLANK, UNKNOWN, and ?. There are three types of queries: binary character queries, numeric queries, and qualitative multistate "character" queries.

Several examples of binary character queries are:

(017012) OR (grow.10c)
(NOT 017012) OR (NOT grow.10c)
(BLANK 017012) (UNKNOWN 017012) OR (? 017013)

(NOT (017012))
(NOT (NOT grow.10c))

The first query asks for all rows in a table in which the answer for characters 017012 or grow.10c is positive. The second query asks for all rows in a table in which the answer for characters 017012 or grow.10c is negative (NOT implies negative of). The third query asks for all rows in a table in which there no answer for character 017012 (BLANK implies no answer). The fourth query asks for all rows in a table in which the answer for characters 017012 or 017013 is unknown (UNKNOWN or ? implies that it is not known why the character was not observed). The fifth query asks all rows in a table in which the answer for character 017012 is negative. The sixth query asks for all rows in a table in which the answer for character 017012 is unknown, positive, or blank.

Several examples of qualitative multistate "character" queries are:

(SPECIES = 'LACTOBACILLUS')
(C02001 = '00001')
(COMMENTS = /BERGEY'S/ COL 14)
(C02005 = /BERGEY'S COL 14/40)
(NOT C02004 = 'LACTOBACILLUS')
(BLANK SPECIES)

The first query asks for all rows in a table in which the answer for character SPECIES contains the string LACTOBACILLUS. The second query asks for all rows in a table in which the answer for character C02001 contains the string 00001. The third query asks for all rows in a table in which the answer for character COMMENTS contains the string BERGEY'S starting at column 14 in the answer (the slashes, /, are string delimiters which mark the boundaries of the string). The fourth query asks for all rows in a table in which the answer for character C02005 contains the string BERGEY'S anywhere in columns 14 to 40 of the answer (this illustrates the substring searching capability). The fourth query asks for all rows in a table in which the answer for character C02004 does not contain the string LACTOBACILLUS. The fifth query asks for all rows in a table in which the answer for character SPECIES is blank.

Several examples of numeric character queries are:

(NO.DAYS < 6)
(596010 = 40E-3)
(598321 > 4.6E12)
(AMP10ZONE > = 29)
(NOT 597005 < 6)
(BLANK 597005)
(UNKNOWN NO.DAYS) OR (? 597005)

The first query asks for all rows in a table in which the answer for character NO.DAYS is less than 6. The second query asks for all rows in a table in which the answer for character 596010 is equal to 40E-3 (0.04). The third query asks for all rows in a table in which the answer for character 598321 is greater than 4.6E12 (4.6×10^{12}). The fourth query asks for all rows in a table in which the answer for character AMP10.ZONE is greater than or equal to 29. The fifth query asks for all rows in a table in which the answer for character 597005 is not less than 6 (rows which have blank answers or unknown answers for the character will also be selected). The sixth query asks for all rows in a table in which the answer for character 597005 is blank. The seventh query asks for all rows in a table in which the answers for characters NO.DAYS or 597005 are unknown.

In addition to the simple queries described, a user can construct complex queries in which all three types of simple queries can be utilized. This is done using the AND or the OR operator as follows:

(597005 < 10 AND 597005 > 1)
(BLANK 025212 OR SPECIES = 'LIST')
(025212 AND 598001 > 10 AND SPECIES = 'LIST')

The first query asks for all rows in a table in which the answer for numeric character 597005 is less than 10 and greater than 1. The second query asks for all rows in a table in which the answer for binary character 025212 is blank or the answer for qualitative multistate "character" SPECIES contains the string LIST. The third query asks for all rows in a table in which the answer for binary character 025212 is positive and the answer for numeric character 598001 is greater than 10 and the answer for qualitative multistate "character" SPECIES contains the string LIST.

Another type of complex query involves usage of the ANY operator as follows:

(ANY (2 019036 NOT 019037 UNKNOWN 019038))

The query asks for all rows in a table in which any two of the following conditions are true: the answer for binary character 019036 is true, the answer for binary character 019037 is false, the answer for binary character 019038 is unknown. A similar query using a number of simple queries and the OR and AND operators would be entered as:

(019036 AND NOT 019037) OR (NOT 019037 AND UNKNOWN 019038) OR
(019036 AND UNKNOWN 019038)

The statistical analysis component of the system is currently centered around the Query Utility's TABULATE command and a third-party software package called TAXAN. TAXAN provides several tools which are used to perform numerical taxonomy on a strain by character matrix of binary data (positive/negative). It calculates and stores interstrain similarities and re-orders strains based on their similarity triangle with similarity values replaced by symbols. It also calculates a dendrogram.†

The Query Utility's TABULATE command is used to obtain frequencies, percentages, cumulative frequencies, cumulative percentages, standard deviations, maximum value, minimum value, and mean and geometric means on selected subsets of table row data. The syntax of the tabulate command is:

TABULATE (Field List) (Query)

The Field List can consist of individual RKC character code numbers, field names, synonyms, character ranges, or special keywords. Special keywords are *, ALL, and NOTHING.

The tabulate command is somewhat cumbersome in that the user currently has to use the SET command to indicate which statistics are to be calculated by the tabulate command. Future releases of the query utility will address this problem. Some of the statistics are data dependent.

Statistics that can be calculated for binary and qualitative multistate "characters" are frequencies, frequency percentages, cumulative frequencies, and cumulative percentages. Statistics that can be calculated for numeric characters are standard deviation, mean, geometric mean, maximum value, and minimum value. These statistics can be generated for the entire database, for each table, or for each table and the entire database.

The last major component of MICRO-IS is the Probabilistic Identification Utility.† This program is based upon work done by Willcox and Lapage (1975). The program uses probability identification matrix data developed by ATCC for FDA bacterial identification. The following identification matrices are included with MICRO-IS:

Aerobic Gram Negative Bacilli (51 taxa)
Facultative Anaerobic Gram Negative Bacilli (92 taxa)
Fastidious Bacteria (32 taxa)
Pseudomonas Species (33 taxa)
Lactobacillus Species (25 taxa)
Bacillus Species (18 taxa)
Streptococcus Species (27 taxa)
Gram Positive Non-spore-forming Bacilli (31 taxa)
Aerobic Gram Positive Cocci (20 taxa)
Mycobacterium Species (27 taxa)
Anaerobic Cocci (16 taxa)

†Details for obtaining the software package can be obtained from Dr. M. I. Krichevsky at the address given above.

†Details for obtaining the software package can be obtained from Dr. M. I. Krichevsky at the address given above.

Anaerobic Gram Negative Bacilli (30 taxa)
Anaerobic Gram Positive Non-spore-forming Bacilli (28 taxa)
Clostridium Species (39 taxa)
Enterobacteriaceae and Vibrionaceae (112 taxa)
Slow Growing *Mycobacterium* Species-Taxonomic Matrix (14 taxa)
Slow Growing *Mycobacterium* Species-Clinical Matrix (14 taxa)

These matrices are distributed in a library file (along with the text files) as part of the probabilistic identification program. Each matrix contains a list of species, characters used to identify the species (characters can be independent, linked, and multistate) and a probability value for each unique characteristic/species permutation. Normally, each matrix is limited to a specific number of species or subset of species of bacteria as indicated in the list of identification matrices above.

A utility is provided to allow users to create their own probabilistic identification matrices that can be used with the Probabilistic Identification Utility. However, the process for creating these matrices is complex and not fully automated. Users are usually encouraged to work with the support taxonomist at MSS when attempting to create a probability matrix for the first time.

The Probabilistic Identification Utility uses row data stored in a table in a MICRO-IS database and a user-specified probability matrix. The program is capable of handling independent, linked, and multistate characteristics in the identification of unknown strains. If it is unable to make an identification, it will suggest additional tests (characters) to be performed that might help to identify the strain. These additional tests are capable of helping to identify the strain only as one of the species listed for the matrix. It is assumed that the user (usually a microbiologist) has a pretty good idea as to the genus of the strain.

Future Plans

Future plans include providing integration with other tools for taxonomic analysis. Whenever possible, the integration will be such that the tool can be called without any kind of intermediate file conversion. The short-term solution to integration of these tools is to provide an export function that will allow conversions of data to and from a generic data file format. There are a number of existing data file formats that seem to have gained widespread support due in part to their commercial success. This include dBASE III, ASCII text files, and DIF (Data Interchange Format).

As with most computer-based tools, the user interface presents the most problems for MICRO-IS. Currently, the system assumes that the user has explicit knowledge of what is in the system or what the system can do. Future developments of the system will present the user with a selection of what is available and, where applicable, a Query By Example (QBE) type interface.

Further enhancement of the data entry component of the system is required to encourage users to computerize data collected in the field and laboratory and from existing sources (publications). This can be done by incorporating scanners, interfacing to laboratory equipment, etc..

Users frequently want to map the distribution of data both in two- and three-dimensional coordinates. It would be very useful to display the distribution of organisms in three dimensions, particularly when dealing with the marine and freshwater environment. The mainframe version of MICRO-IS is used to report the incidence of *Salmonella* in the United States and other countries of the world. There is a need for a similar capability for the microcomputer version of the system. Developing an interface to a geographical database and graphics system would facilitate such mappings.

Discussion

Fortuner: You said that relational DBMSs take more space due to repetitive key fields. I thought the relational model saves space.

Portyrata: The claim made by relational DBMS proponents that this type of DBMS reduces thc need for repetitive entries is somewhat misleading. You have to link the tables with some type of key field. These key fields can consume large amounts of space, particularly if they contain alphabetic and numeric characters. Also, remember that, if you have to link a large number of tables in order to get your data, relational databases have been known to take a long time to display the data when accessing large databases.

Fortuner: How big are your tables? How critical is data compression?

Portyrata: Many people still have microcomputers with 10-MB hard disk drives, and no matter how big the disk drive, people always manage to fill them up. Although 300K files do not sound too big, a large number of them will fill up a disk drive. We also have people who are exchanging data on floppy disks (360K to 1.44 MB). The microcomputer version of MICRO-IS was developed to eventually replace the mainframe version, which has several files which exceed 40 MB. I have found that in many mainframe-to-microcomputer ports of existing database systems (ORACLE, SAS, etc.), noncompression of data results in requirements for large disk drives. In some cases, you must dedicate your entire hard disk for a database (SAS, ORACLE), depending upon the size of your database. To set a minimal file size at which compression is an issue depends upon the type of machine, type of data, and size of the external storage medium (disk drive) and what the user perceives as the maximum storage requirements for his data. Noncompression of character data can result in massive consumption of storage space, particularly when using fixed-length character fields in which the majority of the character strings stored are less than the maximum length (dBASE III is a prime example). I have found that all users underestimate their storage requirements, and many of our users do not have access to funds to purchase larger disk drives. Granted there is a trade-off between compression and access speed, but it usually depends upon the level and type of compression of data.

21

Applications of Artificial Intelligence to Extracting and Refining Locality Information

MATTHEW MCGRANAGHAN

Introduction

Natural history museums and systematic biology collections contain a great deal of spatially referenced information. The spatial component of this information is often recorded as locality descriptions, rather than as numeric coordinates. Locality descriptions are textual descriptions that indicate where specimens were collected, typically by reference to named places, political jurisdictions, and terrain features. An example is: *Oahu. Palolo Valley. Along stream and up the northeast banks at elevations of 1100–1500 ft.*

Researchers who are interested in standardizing collections databases (TDWG 1989; Croft 1989), or in examining the spatial distributions represented in them, are often stymied by locality descriptions (see Brenan et al. 1975; Allkin and Bisby 1984). The inadequacy of locality descriptions is a well-known theme in the distribution-mapping literature (Henderson 1975; Soper 1975; Adams 1974). Hall (1975) notes that without coordinates, distribution mapping is impossible. The same limitations preclude spatial analysis using geographic information systems.

A geographic information system (GIS) is a computer system to process spatially referenced data. Such data typically pertain to items that have a location as well as other attributes. These two facets of the data, location and attributes, are handled in an integrated manner so that it is equally easy to determine where a given attribute value can be found or what attributes are at

---

Advances in Computer Methods for Systematic Biology: Artificial Intelligence, Databases, Computer Vision, ed. Renaud Fortuner (Baltimore: Johns Hopkins University Press).

a given location. A fully featured GIS supports queries that involve spatial relations such as coincidence, intersection, disjointness, and proximity and could be used to answer a request such as *Show all of the known collection sites that are within Palolo Valley, more than 30 m from any other collection site, less than 50 m from a stream, more than 1 km from a paved road, and at elevations greater than 500 m.* Typically, this type of query involves considerable geometric computation based on the coordinates associated with items in the GIS database.

To make better use of the locational information in collections, it is necessary to convert the locality descriptions to coordinates. This is a complex interpretive task involving natural language understanding, multiple representations of geographic space, and an ability to reason about spatial relations. It requires establishing correspondence between geographic features represented in locality descriptions and the same features recorded in some way that associates the feature with its coordinates. Henderson (1975) suggests that, given sufficiently detailed locality descriptions, it is possible to derive localizations adequate for distribution mapping. McGranaghan and Wester (1988) found that approximately 75 percent of the specimen labels in the Herbarium Pacificum of the Bishop Museum contained data to support localization finer than the political jurisdiction and hamlet. Detailed gazetteers, lists of place names and associated coordinates, and personal travel records (constraining the possible location of a collector at a given time) are of use in converting locality descriptions to coordinates (Henderson 1975; Hall 1975; Soper 1975), as are detailed topographic and habitat data.

The usual method of conversion is to read the locality description, examine maps and gazetteers for places and features referred to in the description, and finally decide on a point or area on a map that seems to match the description. This process involves a considerable amount of searching for things in lists and on maps. This is tedious and error-prone work for humans, and automating it is very appealing. Automating the process is facilitated by numerous collections already having locality descriptions in digital form and by the wide availability of digital cartographic, terrain, and gazetteer data for use in GIS. It requires software that can exert a certain amount of judgment in matching a locality description to a place represented in a GIS database. The need for judgment suggests the application of artificial intelligence (AI). This chapter examines how AI techniques can be applied to these data sets to convert locality descriptions to coordinates. Techniques for pattern matching and knowledge representation are of particular value in deriving useful coordinates from locality descriptions. These techniques, together with an understanding of the form and information contents of locality descriptions and cartographic databases, can provide more useful information from existing collections.

Artificial Intelligence

Artificial intelligence can be characterized as a set of useful programming techniques to make computers behave more intelligently. Two important aspects of AI programs are their methods for representing knowledge and the strategies they use for searching and matching those representations. AI programs tend toward symbolic, rather than numeric, computation. All computation is symbolic (Winograd and Flores 1986), but AI programs tend to use symbols that are more abstract than do typical number-crunching programs. Further, AI programs tend to rely on heuristics, rules of thumb, to guide their progress, rather than strict algorithmic solutions to problems.

Knowledge Representation

Knowledge representation is a fundamental aspect of AI research. The goal in knowledge representation is to provide efficient symbolic representations of things of interest, such that the knowledge they contain can be accessed, reshaped, and used to solve problems. Ascertaining what facts need to be represented, how the symbols representing them will be manipulated and what meaning can be derived from the symbols after they are manipulated are all important issues (Winston 1977).

Frames (Minsky 1975) have received considerable attention in AI research as a knowledge representation technique. With frames, an entity is represented by a collection of labeled slots and values, or fillers, for those slots. The values may be defined explicitly, through inheritance, or by reference to a calculation procedure (Chabris 1989). In converting locality descriptions to coordinates, the knowledge contained in textual locality descriptions and in GIS databases must be compared. Knowledge representations that facilitate this, and that can be built from existing databases, are needed to automate the derivation of coordinates from locality descriptions. The content and form of the knowledge in the locality descriptions and in typical GIS databases are examined below, and suggestions are made on representations that might be appropriate.

Pattern Matching

Locality description interpretation can be thought of as establishing a correspondence, or match, between a representation of a part of the terrain as represented in a locality description and as represented in some database. As we will see below, these representations are drawn from very different sources and created for different purposes, and it is very likely that their representations will not match exactly. AI applications are noted for robust pattern matching, which is the ability to establish matches between knowledge representations that do not match exactly. The patterns to be matched may not represent all of the same facts, or they may conflict about the facts that they do represent. Often a very large number of potential matches must be searched to determine which ones are possible matches and the conditions under which they do match. Heuristic rules can be used to improve the efficiency of such searches by concentrating efforts in directions that are most likely to be successful.

Cognitive and Linguistic Influences on Locality Descriptions

Locality descriptions are produced by human beings, using words that are assumed to be meaningful to others. A locality description is usually written in (or soon after leaving) the field, where the collector sees and navigates among terrain features. It often describes a route for returning to the site. The literature on spatial cognition (see, for instance, Lynch 1964, or Kuipers 1983) suggests that landmarks, areas, edges, routes, and topological relationships among landscape elements are important to people during navigation tasks. Therefore, these are the things that one would expect to find mentioned in locality descriptions. In locality descriptions, one finds reference to both named features (Manoa Stream) and generic terrain features (valley), as well as the relations between these and the site. The proper names for features are most often derived from maps of the collection area. The relations are usually represented by one of a small set of prepositions. Sometimes they are tacitly implied, as when a description begins, *Palolo. Along the . . .* , meaning that the reference is either at a point location or in an areal feature named *Palolo*.

Abstractly, specimen collection sites might be considered as points or areas that are connected to other features by sets of relations. Freeman (1973) has suggested that there are only a dozen possible spatial relations between two objects in two-dimensional images. Several more relations were added by Peuquet (1986, 1988; Peuquet and Zhan 1987) and by Mark and Frank (1991) to account for geographic relations, but there are still relatively few relations to consider. As examples, a site might be a point within an area, or it might be an area that lies some distance in some direction from another object, or it might be found along a linear feature which is itself within an area. The logic of combinations of possible relationships can be formalized (Egenhofer 1989; Egenhofer and Herring 1990). This logic is also intuitively accessible; when we read a location description, even without any knowledge of the particular terrain, we can often tell whether a description makes any sense (i.e., describes a possible situation).

Annette Herskovits (1986) presented an analysis of the spatial relations represented in the English language. She examined three topological prepositions (*at*, *in*, and *on*), each of which is used to describe several different relationships. The different relations are *use types*, in which the basic meaning of a preposition is stretched to convey the meaning intended in an individual instance. For example, Herskovits notes that the word *at* indicates that the basic concept is that one punctiform entity coincides with another and has eight use types. Three examples are: (1) that a spatial entity is *at* some landmark; (2) that an object is *at* the intersection of two lines (e.g., the point where a stream crosses a road); and (3) that an object is *at* some distance from a point, line, or plane. Software that recognizes the use types should be able to "understand" the relations they describe.

The point here is that locality descriptions are not intended to provide an overview of geographic space. The objects and types of relations that are naturally of use for following a path through a landscape are not necessarily the same as those that are of use in providing an overview of it. The overview function requires that anything in the terrain be represented, but does not stress any particular objects or relationships. Objects that are prominent in the navigation task may be inconsequential in the overview. We would therefore expect quite different contents and representations for overview versus path-following functions. The following section describes several different GIS (cartographic overview) data sets to indicate how they differ from the typical content of locality descriptions.

Cartographic Databases

In contrast to the linguistically and cognitively shaped representations found in locality descriptions, the representations used in most GISs are designed to encode overviews of areas. They are often derived directly from maps and tend to focus on geometry, planimetry, and categorizations encoded in map symbols. The prominence of place names and individual terrain features is much lower in most cartographic data sets than in locality descriptions. The typical cartographic data set records points, lines, and areas found on maps. Points are represented as coordinate pairs. Lines are represented as sequences of points, and areas by closed lines (often called polygons). Usually, entities in each of these classes would have attributes attached to them indicating what is symbolized by the point, line, or area on the map. A line, for instance, might have codes indicating it represents a stream, 5 m wide, and with an average flow of 25 m^3/s.

Three data sets are described below, Digital Elevation Models, Digital Line Graphs, and the Geographic Names Information System. Available from the U.S. Geological Survey (USGS 1985,1987), they are representative of the type and quality of digital geographic base data, against which to match locality descriptions. Similar data sets are available from the mapping agencies of other nations, and similar worldwide data sets will soon be available (Mounsey and Tomlinson 1988).

Digital Elevation Models

Digital Elevation Models (DEMs) contain a matrix of elevations to the nearest meter on a regular 30-m square lattice. There are no place or feature names and no explicit indication of where different features are located. Nothing in the data set says *This is a hill*, though it is possible to identify terrain features from the elevations (O'Callaghan and Mark 1984; Frank et al. 1986; Riazanoff et al. 1988). This requires a definition of the feature in terms of elevation configurations. Features are not so easily defined as one might imagine. Like all human mental categories (Lakoff 1987), terrain features do not fall into clean, mutually exclusive categories. Fifty geomorphologists, asked to delineate a hill in an elevation matrix, would likely give fifty (or more) different answers. Knowledge engineering techniques and fuzzy set theory might be of value in developing operational definitions of various terrain features.

Digital Line Graphs

Digital Line Graphs (DLGs) record the cultural and hydrologic features found on USGS topographic maps. These include roads, streams, buildings, power lines, and political boundaries. DLG data do not include place names or topographic configurations. Features are represented as a set of coordinates for points describing their shapes and locations. Each object (set of points) in the database is associated with at least one feature code. The feature codes follow a two-level hierarchy: major and minor codes. The major feature codes coincide with broad categories, such as stream or road. The minor feature codes give more detail, such as the number of lanes, or whether the stream is perennial or ephemeral. This data set was designed primarily for reproducing the nontopographic linework on USGS topographic maps; it can support other, more analytic uses. One shortcoming of DLG is that no proper names are attached to the features represented in the data. This contrasts with locality descriptions in which place and feature names often figure heavily.

Geographic Names Information System

The Geographic Names Information System (GNIS) contains records for the features that are named on USGS topographic maps. GNIS relates feature names to points and categorizes features into sixty-two different classes. The coordinates associated with a name may not refer to the exact location of a feature; in fact, many features are not punctiform. The coordinates of areal and linear features indicate where the name, not the feature, is, and provide a starting point for a spatial search. The feature classes are cognitively salient (e.g., trail, populated place, or valley) but are neither mutually exclusive nor collectively exhaustive of the types of existing terrain features (McGranaghan 1991). In GNIS all of the names are stored in all uppercase letters, without any diacritical marks. This simplifies the content of the database, but it complicates matching place names and increases the possibility of confusing distinct names.

DLG-E: A Data Set with Promise

Test data have not yet been released, but a new object-oriented data set will soon make it easier to use cartographic data sets for more geographic applications. The Enhanced Digital Line Graph (DLG-E) will encode representations of geographic objects with place names and feature types in one data set (Guptill 1991). The DLG-E data model is intended to model the world rather than a map. The design goal is to provide a base which allows one to do more than reconstruct maps. It is intended to allow the user of digital map data to do things that the user of printed maps cannot do. It is anticipated that this data set will reduce preprocessing of data into formats that are better suited to symbolic processing.

A Frame-based Geographic Data Representation

Various authors (Havens and Mackworth 1983; McGranaghan 1985,1991) have suggested that frame-based representations (Minsky 1975), could be very useful for geographic data. The basic idea is to use a frame, a set of labeled slots with values, to represent each geographical object. The set of slots in an object's frame could explicitly record the object's relations to other objects. A collection of frames (like the one below), linked together by their interrelations, could be used to represent a geographic region.

```
(GEOGRAPHIC-OBJECT identifier
(FEATURE-NAME name)
(FEATURE-CLASS class)
(IN identifiers . . . )
(CONTAINS identifiers . . . )
(NEIGHBORS identifiers . . . )
(COORDINATES coordinates . . . ))
```

This type of representation could include other or variable relations. It is not necessary to anticipate all possible future slots when designing the representation of an object in a frame-based system. A map represented as a set of frames would be very similar to a semantic network, and tools for manipulating it could be based on a semantic network environment such as SNePS (Shapiro and Woodmansee 1971; Shapiro and Rapaport 1987).

There are some practical considerations in building a frame-based geographic database. Searches through the database can be much improved by ordering the database by a spatially sensitive scheme (see Preparata and Shamos 1985). Also, because the number of relations between each of the geographic objects in a frame-based database could be quite large, and because most (but not all) interesting relationships would be local, it is likely that a practical implementation of a frame-based GIS would use hybrid processing. Whenever it is possible, simple symbolic processing would be applied to explicitly encoded local relationships. Otherwise, the coordinates of objects could be used to compute spatial relations.

It is easy to envision a dynamic database that, when queried for something which is calculable but has not been previously calculated, creates a slot and stores the result of the query for future use. The Knowledge-Based GIS developed at Santa Barbara (Smith and Peuquet 1985; Smith et al. 1987) made use of such a *calculate once and store* concept. The decision to store or (re)calculate a relation could be based on how frequently it is needed. Frequently needed relations would be added to the explicit store, and infrequently accessed slots could be removed from explicit storage if database size were a major concern.

Pattern Matching with Locality Descriptions

Pattern matching can be applied in locality interpretation for: (1) simple matching of words in locality descriptions and in a gazetteer; (2) recognition of patterns in locality descriptions that might be used to identify the grammatical role, if not the meaning, of the words; and (3) the more complex task of matching representations of the terrain arrangement as found in descriptions and in GIS databases.

Place Name Matching

Place names are very important clues in interpreting locality descriptions. They are symbolic pointers to geographic locations. Through a correspondence table, a gazetteer, place names and coordinates can be associated. Matching place names presents challenges at several levels. It is a conceptually simple matter to go to a gazetteer, find the matching name, and use the corresponding coordinate as a key to begin accessing spatial data. In practice, however, there are several confounding factors, and fairly robust pattern matching is required.

Place names may be accidentally or systematically misspelled. Attempts to match accidentally misspelled place names might use techniques similar to those employed in commercial spelling-checking programs to suggest near matches. These systems often use heuristics based on common typing errors, common spelling errors, and phonetic similarities. More systematic misspellings might be related to translation between languages or transcription from spoken place names. The use of foreign words or local names for features might also fall into this class of problem. Systematic spelling variations, such as abbreviations or suppression of diacritical marks (as in the GNIS data described above) also could be handled through heuristics. Place name matches may be misleading, in that place names change through time and are often used simultaneously for multiple locations. This requires more than robust spelling matchers. Reasoning about temporal relations and locations of collectors at specific times could remove some ambiguities. Gazetteers that include the dates for which a place name was applied to a location (GNIS does), and information about the travels of collectors would be needed. The basis for such temporal reasoning is being formulated (Worboys 1991; Barrera and Al-Taha 1990) though reliable implementations in GIS are not yet available. For the common situation where the same name is used for nested geographical entities (a village within a county) an heuristic like *Use the smaller, more specific entity, unless there is evidence not to* could be applied.

Patterns in Labels

In natural languages, syntax carries a good deal of the meaning. It may be possible to gain information from locality descriptions by noting word order, on the assumption that this will make it easier to identify which words specify places, which specify relations, and which carry little or no information in the description. This would make efforts to match words in the description against the GIS database more efficient. Attempts to modify a program which determines the grammatical parts of simple English sentences to extract more information from locality descriptions showed that the grammar of locality descriptions is not simple (McGranaghan 1989). Prepositions, nouns, and adjectives and other modifiers occur in outwardly jumbled sequences; tracking their structure involves resolving multiple possible interpretations and considerable backtracking.

Geographical Patterns

The purpose in locality interpretation is not to understand the linguistic representation but rather to gather a deeper meaning. What is needed is a match of a locality description against the terrain. This means matching two abstract representations of geographic reality, one conveyed through the textual description with one that is conveyed through a cartographic data set or perhaps a geographic information system database. It was suggested above that a terrain database could be constructed as a set of objects with explicit relations and with explicit coordinates (which could be used to calculate other relations). This could take the form of a set of interconnected frames. Similarly, the relations and objects mentioned in a label could also be represented as a set of frames connected by relations. Finding the location specified by a description becomes a pattern-matching task between abstract representations of a complex system of objects (terrain features) and the relations among them. It should be the case that either of the representations may be partial and that additional relations could be calculated from the base data if they are needed to reduce ambiguity. AI researchers have implemented a number of frame-based and semantic net-based systems that can perform this type of matching. Techniques for matching pieces of frame-based semantic nets are well understood and are described by several of the authors in Cercone and McCalla (1987).

Conclusions

Automated conversion of locality descriptions to coordinates is a desirable goal, and one to which artificial intelligence techniques may be fruitfully applied. Techniques of knowledge representation developed in AI research are appropriate for explicit representation of terrain features, though experience in using this type of geographic database is lacking. Robust pattern-matching techniques which process these representations are available and seem to be appropriate for searching a frame-based geographic database for matches with locality descriptions. Although no commercial GIS currently has these capabilities, they are technologically feasible and in need of demonstration.

There are differences in the types of knowledge contained in locality descriptions and in cartographic databases. These differences are exaggerated by the formats of existing cartographic databases. A frame based geographic knowledge representation promises to reduce the impact of these differences on geographic pattern matching. Research into methods for converting existing cartographic databases and locality descriptions into these representations has begun, but more is needed.

Although general techniques for pattern matching in frame-based representations are available, efficient conversion of locality descriptions to coordinates seems to also depend on cataloging and refining domain-specific heuristics. These should reflect the practices of field collectors and the ecological and geographical principles that govern specimen distributions in order to facilitate search and pattern matching in these databases. This is probably the most difficult part of an automated locality conversion system to develop, and success will depend on the joint efforts of field scientists, geographers, and AI specialists.

Acknowledgments

The author gratefully acknowledges support from the National Science Foundation under grant SES-88-10917 for the National Center for Geographic Information and Analysis for the preparation of this chapter and from Dr. David Mark, organizer of the NATO Advanced Study Institute on Cognitive and Linguistic Aspects of Geographic Space, Las Navas del Marques, July 1990. Discussions at that meeting shaped many of the thoughts presented here.

22

The Use of Geographic Information Systems in Systematic Biology

PAULO A. BUCKUP

Systematic biology is inextricably associated with the task of generating and interpreting geographic information. Decisions about limits and validity of taxonomic species are highly dependent on information about geographic distribution of organisms. Occurrence of significant differences in morphological, biochemical, or behavioral traits of organisms coexisting in the same geographical area (sympatry) is seen as an indication of the existence of reproductive isolation, and thus of complete speciation. Populations exhibiting different allelic and phenotypic frequencies, yet separated by great distances, may require further investigation concerning the nature of this variation before a similar determination can be made. Observed intraspecific variation is regarded as intrapopulation polymorphism if it occurs in a single geographic area, but it is treated as geographic polymorphism or clinal variation if it is correlated with differences in areas of occurrence. At supraspecific levels of analysis, information about species distributions is compared with phylogenetic trees depicting relationships among species and is used to test biogeographic hypotheses.

The Nature of Geographic Information in Systematics

During routine systematic work, geographic information is usually organized into distribution maps. Typically these maps take the form of *spot maps* (Fig. 22.1), where data points representing samples of the organisms are plotted as symbols on a topographic map. Although the final product of systematic work is frequently presented in the form of thematic maps where biological dis-

---

Advances in Computer Methods for Systematic Biology: Artificial Intelligence, Databases, Computer Vision, ed. Renaud Fortuner (Baltimore: Johns Hopkins University Press).

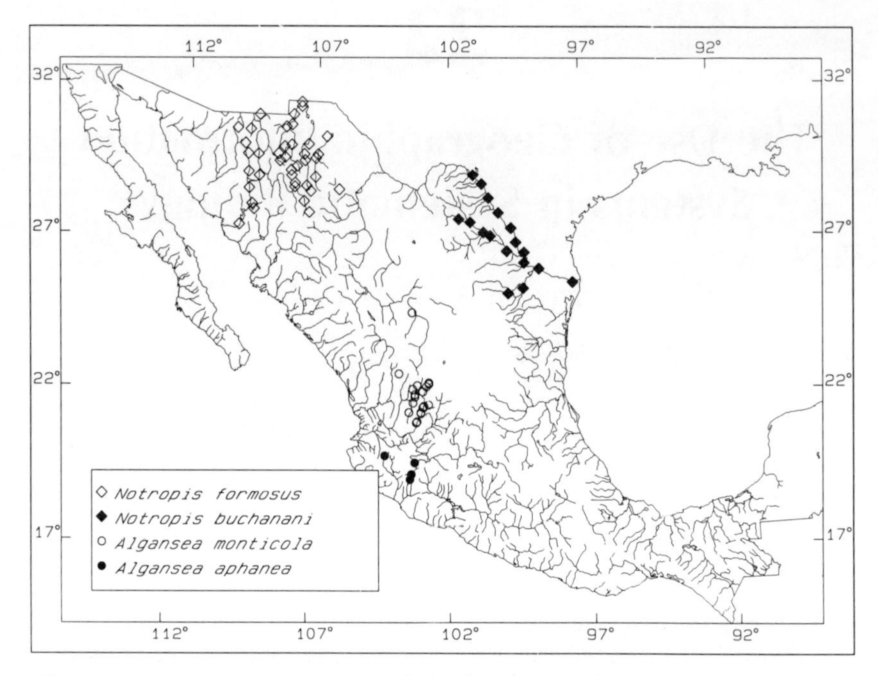

Figure 22.1. Mexican river drainages with the distribution of four species of cyprinid fishes. Produced by FishMap, printed using a 300-dpi laser printer.

tributions are represented by differently shaded areas instead of clusters of points, these maps are ultimately based on data from discrete localized samples. In systematics, the information about the location where specimens were collected or observed is usually referred to as the *specimen locality.* The basic geographical data for systematic studies are individual specimen localities rather than species distributions. The latter are abstractions that result from the systematic analysis of specimen data.

The discrete nature of locality data from systematic collections makes the use of computer-aided automated mapping systems an attractive tool for producing distribution maps. Such systems, usually known as geographic information systems (GISs), can automatically produce distribution maps using locality data stored in computer databases. GISs were first developed as mainframe computer applications with a batch file or command line interface. These initial programs were generally difficult for use by nonexperts because of the noninteractive interface. However, with the proliferation of increasingly powerful microcomputers, typical GIS applications have evolved into user friendly programs with flexible graphic interfaces. GIS applications have also become more independent of output hardware. Originally most GISs required expensive pen plotters, but many current programs are capable of printing maps on common dot-matrix and laser printers. *Desk-top mapping* has become almost as affordable and efficient as desk-top publishing.

Locality data associated with systematic collections traditionally exist in the form of typewritten or handwritten specimen catalogs, but are also being made increasingly available in the form of computerized databases. Several institutions now make data from their systematic collections available through general-purpose database management systems (DBMSs), such as dBASE-compatible systems, or specialized collection management systems (CMSs), such as MUSE and TAXIR. CMSs are special-purpose DBMSs designed specifically for use in curatorial activities associated with the management of systematic collections (generation of specimen labels, management of specimens loans, accession of new materials, etc.)

In systematics, GISs are used mostly for the final, postanalytical phase of the systematic work. Map-generating computer programs are used to prepare final art work for publication. Often, the production of distribution maps is restricted to the use of *paint programs* for desk-top publishing. Even when more sophisticated GISs are used, the data are manually processed prior to use in the GIS itself. I suggest, however, that rather than use a GIS as a means to present the final results of systematic work, systematists should use the GIS as a *source* of systematic data. Ideally, the GIS should be an integral part of the CMS.

GIS Design for Use in Systematic Collections

Assuming that the locality data of a systematic collection are captured in computer format and made available through a CMS, the choice and design of the GIS component of the CMS should take into consideration a number of factors which influence its efficiency and effectiveness. Factors that should be considered include direct data accessibility, program execution speed, memory and storage limitations, user interface, database structure, and remote accessibility. These factors are briefly discussed below.

Direct Access to Data

One characteristic of geographic information in systematic collections is that it is often incomplete and usually stored in a format that is unsuitable for automatic map production. Precise geographic coordinates are seldom included in collection catalogs. Textual descriptions of localities as found in systematic collections are usually inappropriate for direct manipulation through a GIS, because they contain complex spatial relationships (e.g., *three river-miles downstream from bridge on road between . . .*). Administrative units, such as county, district, province, and department, often cannot be assimilated without a human interpreter, either because they have not been explicitly stated in the locality description or because they were not in existence at the time when the specimens were collected. As a result, most systematists end up generating separate map-oriented databases to be used in their GISs. Geographic coordinates for each locality are obtained from atlases and gazetteers and entered into a new GIS-compatible database. This is unfortunate because the upgraded data remain inaccessible to the archival CMS, unless a considerable (from the user's point of view) programming effort is made to integrate the newly generated geographic coordinate data into the existing CMS data files. I suggest that the ideal GIS should be able to directly access the geographic data as they have been originally recorded in systematic collection databases. Moreover, the GIS itself should include specialized functions for upgrading the original locality data, thus freeing the user from having to obtain geographic coordinates manually for later incorporation into a separate GIS database. Therefore, the GIS must be able not only to retrieve the data but also to update the information in the original computerized catalog.

Data Protection

Several commercially available GISs offer some degree of interaction with the most commonly used database file formats, such as the dBASE file format. However, the archival nature of systematic collection databases requires considerable data protection and consistency checking. Geographic coordinates recorded from original field notes should be permanently preserved as a historical record and protected against accidental editing through the GIS graphic interface. A clear distinction must be made between geographic coordinates that are subjectively *inferred* from the original locality data and coordinates that are *exact* translations of these data. The problem is not just writing geographic information into the file, but doing so in a manner that is acceptable for an archival systematic collection database. The system must function as both a GIS and a CMS.

Program Speed

Because species may show considerable variation in geographic range, the GIS should offer fast zooming and panelling capabilities for interactively adjusting the size of the area being plotted. The maps used to plot species distributions often involve very complex objects such as hydrographic systems and relief contours. These graphic objects are defined by a rather large number of individual points when compared with simple polygons such as representations of street blocks, county and state borders, and public utilities, which can be defined by a relatively limited number of points. Rescaling and moving complex images require a large number of computations to relocate each individual point. The time required to perform these computations frequently results in programs that are too slow to be used interactively. Some disciplines, such as engineering, prefer to solve their speed problem by using faster hardware. Mapping programs can be speeded up through the use of math coprocessors to accelerate floating-point calculation of geographic coordinates, sophisticated graphic interface boards, advanced minicomputers, and mainframe computers. However, because hardware resources available to managers and users of systematic collections are usually limited, an emphasis should be placed on adoption of software capable of displaying, zooming, and paneling through distribution maps in a manner fast enough to permit interactive search and retrieval of specimen data. Efficient software with algorithms optimized for speed may be more appealing to systematists than more flexible, but hardware-dependent, programs.

Data Storage Requirements

Although large systematic collections may contain several hundred-thousand specimens, the required storage space for the associated locality data is relatively small by current computer standards. Catalog data from most collections can be adequately stored on a medium-sized microcomputer hard disk (40–100 MB). Because typically only a very small subset (e.g., only a single species out of a few thousand) of the data is accessed in most routine operations, the size of the specimen locality database usually does not present a major performance problem in magnetic disk reading operations. On the other hand, the amount of information necessary to store complex digital maps in magnetic media may be staggering. All or most of the information needs to be accessed for plotting a single map on screen or other output device, so even operations as simple as direct disk reading may become unacceptably slow. Because systematic collections are likely to require a diversity of maps, ranging from local (e.g., county and state maps) to continental and world scales, it is important that systematists adopt GISs that are capable of using highly compressed data formats for storing digital maps.

Because research-oriented mapping programs should emphasize interactive, on-line use of maps, digital map formats that are compact enough to allow data manipulation in RAM are likely to be the most useful in systematic work. To achieve this level of data compression, formats that employ vector representation of geographic objects should be preferred over bit-mapped (raster) graphic representation. If the entire map image can be stored in RAM, the additional processing cost in speed for manipulating complex geographic objects is negligible. In addition to the space-saving benefits, the use of vector technology offers greater size and output-device independence. Using this technique it is possible, for instance, to produce a digital representation of

Mexican river drainages with sufficient detail to permit a twentyfold change in scale without significant loss of resolution (compare Figs. 22.1 and 22.2), while maintaining the size of the map file under 108 K.

User Interface

A major consideration in selecting a GIS is the functionality and ease of use of its user interface. This is true for microcomputer-based applications in general, but is especially important for GIS programs, as they require considerable visual interaction with spatially structured data. Most microcomputer-based GIS programs offer user friendly interfaces including pointing devices and pop-up and pull-down menus. GIS applications for use in routine systematic work should offer ease of use for a broad range of users. Ideally, a GIS linked to a systematic collection should be easy to use for researchers and curators, as well as collection managers and technicians, regardless of their level of computer literacy. Of particular interest for systematists is the possibility of offering visiting scientists on-line geographically oriented interaction with systematic collections. One of the most significant advantages of such interaction is the ability to pinpoint material of special interest to a visiting specialist, even if the material is currently unidentified or misidentified. Using a GIS interface to access a CMS, for example, it would be possible to visually select specimens of a particular genus collected near the type locality of a given species, even if they are not identified as members of that species.

Traditionally, the generation of reports on geographic data associated with systematic collections required considerable data manipulation by an expert resident collection manager. This intervention has been necessary in order to protect the physical medium (e.g., catalog cards) on which the archival records reside. Efficient retrieval of data has also required familiarity with the format and type of information associated with the collection database. Integration of a menu-driven GIS with systematic collection records would greatly enhance their usefulness and cost-effective accessibility for nonresident systematists. Flexible interfaces may allow direct access to the data while still maintaining the physical integrity of the electronic records. Through user friendly interfaces any user of systematic collections might gain easy access to locality data without the need to become familiar with the particular format in which the information is actually stored. As the use of computers becomes more common in systematic collections, we can anticipate increased availability of collection data to the users without the intervention of a collection manager.

Object-oriented Data Structures

Geographic information associated with systematic collections is usually recorded in very disparate formats and frameworks of spatial reference. Collec-

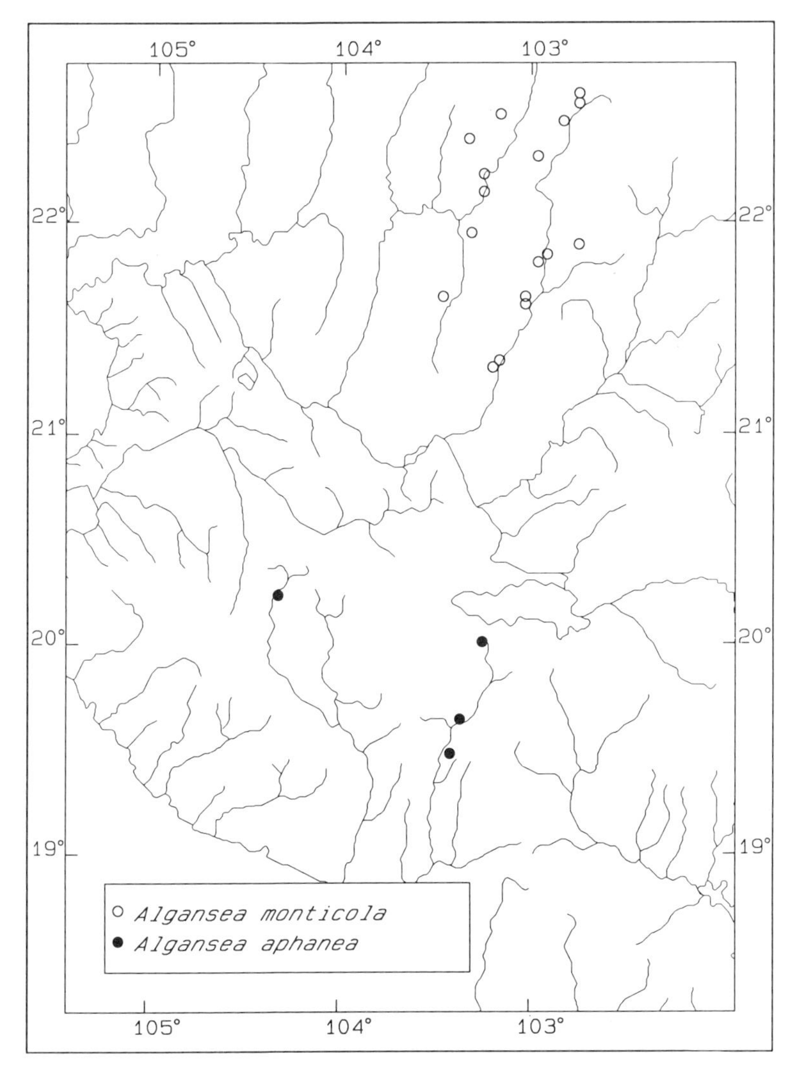

Figure 22.2. Distribution of *Algansea monticola* and *A. aphanea*, produced with the same digitized map used in Figure 22.1.

tion data may include geographic descriptors that range from specific locations described by exact coordinates to more complex entities such as river or coastline segments, or whole areas such as river basins or administrative units. This variability of format requires sophisticated methods for efficient storage and retrieval of the information. Different geographic entities also require different methods for graphic representation and manipulation on screen maps. While a relational DBMS can be used to organize and break down complex pieces of information, it is likely that the resulting database structure (often referred to as the database schema) will be too complex and inefficient to be able to handle the entire range of geographic descriptors found in systematic databases.

Although relatively new and unexplored, object-oriented database management systems (OODBMSs) may offer a valid alternative for managing geographic information in systematic databases. Rather than representing complex entities as a set of relations between simpler but separate pieces of information, an OODBMS stores these entities as self-contained objects. This concept of *data encapsulation* is also used in object-oriented programming languages. The rules (or *methods*) for displaying each geographic entity are directly associated with the corresponding type of object stored in the database. Because of this ability to store information in a variety of formats, object-oriented databases may be more efficient than relational databases when used with GISs designed for systematics.

Remote Access

Computer network communication is becoming a standard feature of the academic research environment. An increasing number of biologists regularly make use of remote computer communications to access information for use in their daily research. Most bibliographic reference services, such as BIOSYS and MEDLINE are available online, and abstracts of published articles can often be downloaded through modems and data lines. Electronic mail and specialty bulletin board systems are becoming an important means of communication among biologists. Molecular databases and resource centers are accessible through computer networks.

A promising area for GIS applications in systematics is the use of remote access to data from systematic collections. Because computerization of systematic collections is a recent phenomenon, not very many museums and research institutions yet offer remote access to their collection databases. It is likely, however, that considerable advances in remote access to systematic collections will be made in the near future. Such developments will have an important impact on several areas of systematic work, such as research in biodiversity and the ever-necessary taxonomic revisions. Remotely accessible collection databases will represent a vast improvement in the infrastructure of systematics, because they will allow systematists to interactively query institutional databases for information on availability of specimens and geographic coverage of their holdings with considerable time and cost efficiency. Integration of a GIS interface to these remotely accessible CMSs will also offer the possibility of immediate visual evaluation of the recovered information.

It can be expected that remote access to collection data will play an increasingly important role as more and more systematic collections become computerized. However, two potential problems distinguish remote use of GISs from other forms of remote data retrieval. These problems will need to be properly addressed in the future. One of the difficulties is that the amount of information necessary to produce digital maps is often unacceptably large for efficient transmission through most modems and computer networks. A solution to this problem would be the use of GIS interfaces that can integrate locally available digital maps with remotely available systematic data.

Incompatibility among graphic device drivers in use by systematists may also impose limitations for real-time access and display of geographic data. This difficulty, however, can be easily eliminated through adoption of systematic databases that support standard query language (SQL) interfaces. Rather than using the host computer to generate the distribution maps, the user should be able to process locally the information obtained from the remote database. In this manner the systematic database can be accessed by virtually any user capable of remote communications, regardless of the characteristics of the user's GIS environment.

Conclusion

Although use of GISs in systematics has been restricted to the final stages of the research process, its use in earlier phases is likely to result in considerable benefits to systematists and users of systematic data. Locality data are an integral part of the basic information associated with specimens used in routine systematic work. As an increasing number of systematic collections are using computer databases to manage locality data associated with their specimens, implementation of collection-oriented GISs will be facilitated by the availability of those data in computer format. Integration of a GIS with a CMS is a promising approach to the management of geographic data associated with systematic collections. The nature of systematic data may require GIS tools and data management strategies that are not common in applications designed for the business and architecture markets. However, as discussed in the previous sections, these requirements can be met within the limits of hardware and human resources available to most systematic collections.

Acknowledgments

The ideas in this chapter benefited from interaction with William L. Fink, Julian M. Humphries, Douglas Nelson, and Scott A. Schaeffer. Robert R. Miller, Mary Anne Rogers, and Kimberly Smith offered suggestions to improve the manuscript. Dr. Miller also made available the species distribution data presented in the figures.

V

Computer Vision and Feature Extraction

23

Introduction to Digital Image Processing and Computer Vision

RAMIN SAMADANI

Introduction

This chapter is written for scientists who need to use or process image data but have little familiarity with the methods involved. The various terms and techniques used are explained below. Then, two examples of the results of successful applications of image processing and computer vision techniques are given to illustrate what is currently possible using the techniques described. These two examples do not include the description of the algorithms used. Readers interested in a detailed understanding of the algorithms can refer to the literature and to the detailed example given in chapter 29 of this volume.

Digital image processing and computer vision are related areas of engineering and computer science research that are almost as old as the computer. The problems encountered in these fields are difficult, but the difficulty is sometimes obscured by the great ease with which human visual perception performs visual tasks. Figure 23.1a shows a portion of a digital image often used by the image-processing community for algorithm testing purposes. We clearly recognize the human eye in the figure. We can answer questions such as, Is the eye open?

This figure is actually a reconstruction of a sequence of about 4000 numbers that represent the light intensity at 4000 points in the image. The larger numbers represent the brighter regions in the image, and the smaller numbers represent the darker regions. Figure 23.1b shows the same information as that shown in Figure 23.1a, but using a different representation. The axes in this plot are the two directions, horizontal and vertical, in Figure 23.1a. In Figure 23.1b the height of the mesh, shown in perspective view, represents the

---

Advances in Computer Methods for Systematic Biology: Artificial Intelligence, Databases, Computer Vision, ed. Renaud Fortuner (Baltimore: Johns Hopkins University Press).

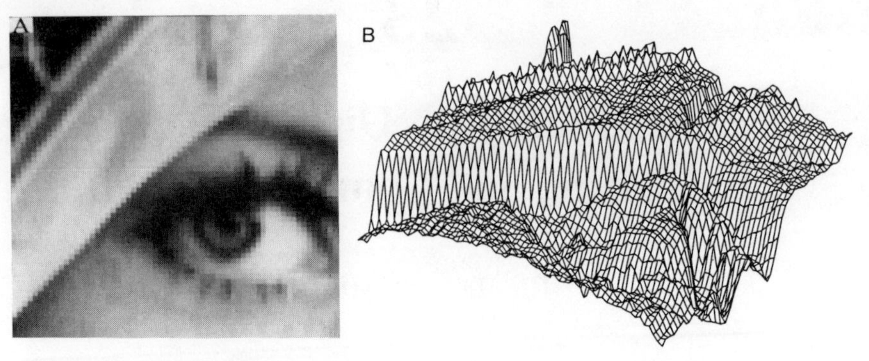

Figure 23.1. Digital image of an eye reconstructed using (a) intensity to represent the numerical data; (b) height of a mesh plot to represent the numerical data.

intensities of the image. Based on the information from Figure 23.1b alone it is much more difficult to answer the question, Is the eye open? In fact, it is even difficult to find the objects such as the eye, the eyebrow, and the eyelash when this representation is used. Questions such as, Where is the eye? are difficult to answer.

Examination of Figure 23.1b gives credence to the fact that there are currently no computer systems that even closely approach the capabilities of our own visual perception. This does not mean, however, that the techniques of digital image processing and computer vision are not useful to scientists. The continuing decrease in costs of computer processing and computer storage has made the use of digital image processing and computer vision affordable for many applications, and this trend is likely to continue or even accelerate in the future. New algorithms have been successful in extracting quantitative information from images for an increasing number of applications.

The success was achieved by setting less ambitious goals than general visual recognition and limiting the types of images acceptable for computer processing. An example of a simple goal that is currently achievable is the storage of images in digital format. Credit card companies store customers' credit slips in digital format, greatly lowering the amount and cost of their paperwork. Changing the representation of an image to digital format is inexpensive and simple. When input images are limited to, for example, images of printed circuit boards, there are image-processing and computer vision techniques currently available for the inspection of defects. There are indeed multimillion dollar companies taking advantage of this market niche. Image processing is particularly useful for this application because humans are not very efficient at repetitive inspection tasks that computers perform without fatigue. A general principle for the successful application of image processing and computer vision can be distilled from the examples above: Use the techniques for solving simple problems that require only limited types of images used as inputs.

Goals and Approaches

Image data can be processed to achieve several practical goals. Some of these goals are: (1) representation of the image data in digital format for easy storage, transmission, retrieval, and processing; (2) improvement of the appearance of the image for display to the user; (3) removal of physical distortions such as misfocus and motion blur that occur during the image acquisition; (4) reduction of the amount of computer storage needed by the images; and (5) extraction of quantitative and qualitative information from the images, including sizes, textures, and shapes of the objects.

Two possible approaches to image processing and computer vision are interactive processing and automatic processing. The choice of approach depends on the amount of data to be analyzed. Interactive processing is the best choice if there are only a few images to be analyzed. Interactive processing allows the very capable human visual system to guide the computer algorithms. For a higher number of images to be analyzed the best approach is automatic processing. Automatic processing algorithms are developed to work without any user supervision. The results of the automatic analysis of the images are then provided to the user.

Several factors influence the success of an automated system. Careful thought should be given to the development time needed for design of the system, performance goals should be set before the system design starts, and the results of the algorithms should be carefully evaluated using statistical techniques. The examples below show the results of two successful applications of automated information extraction from images. Chapter 29 shows a third application in more detail.

From Images to Numbers and Back to Images

Figure 23.2a (image to numbers) shows the steps used for the conversion of an image to a format that can be used by the computer. This figure shows the world viewed by an imaging system that includes a light-sensitive electronic sensor. The output of the electronic sensor is then converted from analog to digital (numerical) information. This information can be represented as a function of two integer-valued variables. This function is known as a digital image. It is stored and processed by computer.

The digital image is the input to image processing and computer vision systems, as shown in Figure 23.2b (numbers to images). Image-processing

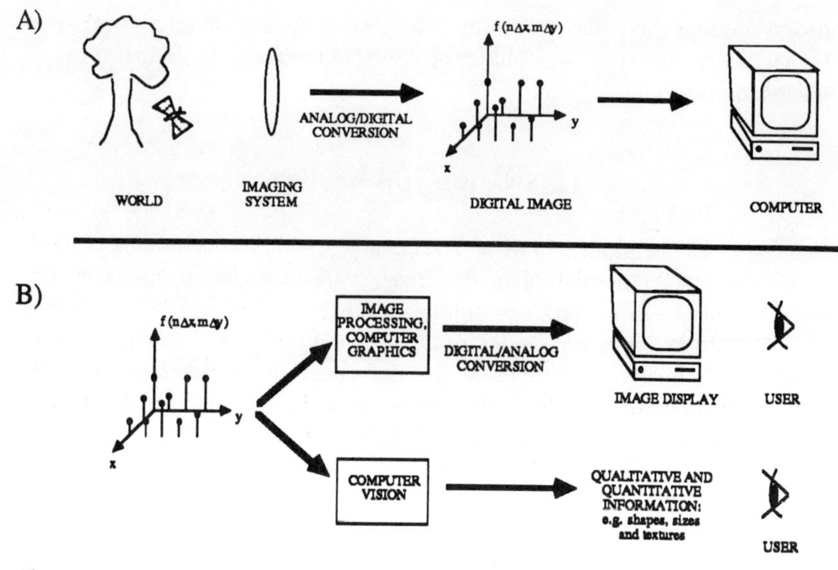

Figure 23.2. (a) Conversion of image data to numerical data. (b) Use of digital image data as the input for image processing and computer vision.

operations simply convert a digital image into another digital image. After processing, the digital image may be converted to an analog electronic signal that controls an image display. The image display converts the electronic signal to radiances on the display screen that are seen by the user. Computer vision systems, on the other hand, convert the digital image information into measurements (e.g., of shapes, sizes, and textures) that are provided to the user.

The Techniques

This section provides a brief view of some of the techniques and terminology used in image processing and computer vision. The goals for each set of techniques are described, together with some of the current approaches to solving the problems. For further studies, the reader is referred to more detailed books on the subject (Rosenfeld and Kak 1982; Jain 1989; Ballard and Brown 1982).

Image Data Compression

Although computer storage costs continue to decrease, digital image data require large amounts of computer storage. The information in most images, however, is highly redundant. Image data compression (sometimes called image coding) reduces the storage requirement for digital images. The techniques for image coding are currently based on statistical information theory (Gallager 1968).

Image data compression is done by *lossless coding* or by *lossy coding*. Lossless coding reduces the storage requirements for a digital image in a manner that allows decoding to exactly recover the original image. Lossy coding, on the other hand, reduces the storage requirements for a digital image in a manner that cannot be decoded to exactly recover the original image.

The decoding process will provide one with an approximation of the original image, and there will always be degradation when using lossy coding techniques. If a certain amount of degradation is acceptable or imperceptible, lossy compression reduces the amount of storage required much more than does lossless compression.

Image Enhancement

Image enhancement improves the subjective appearance of images displayed. This is achieved, for example, by using linear functions of intensity that are the digital equivalents corresponding to the *contrast* and *brightness* settings on a television set. Histogram modification is a nonlinear process useful for images that do not make full use of the intensities available. It can be used to improve the appearance of images acquired under poor lighting conditions. Spatial smoothing, unsharp masking, and band-pass filters are used to make the image appear sharper. Median filters and low-pass filters are sometimes used to reduce the effects of noise. Finally, image magnification and reduction are operations easily performed on digital images and used to allow easier inspection of the images.

Image Restoration

Image restoration is used for removing distortions due to misfocus, motion blur, or atmospheric turbulence. The methods used are based on physical models for the imaging process. Thus, image restoration methods are objective, whereas the image enhancement methods are subjective. Distortions due to the geometry of the imaging systems may also be removed using image restoration methods. Image restoration techniques include linear and nonlinear inverse functions that reduce the effects of the distortion processes.

Segmentation

Segmentation is used for separating an object (or objects) of interest from the rest of the image. Measurements may then be taken on the object of interest. There are two approaches to segmentation: (1) follow the boundary of the object by finding regions in the digital image with large changes in value or (2) coalesce the regions in the image where the values of the digital image are similar.

Operators that enhance the edges in an image are used together with graph-search methods to implement boundary-based segmentation. Region growing and split-merge techniques are used for region-based segmentation. Combinations of boundary-based segmentation and region-based segmentation are also used.

Shape, Size, and Texture

Measuring sizes and shapes of objects from digital images involves discrete geometric methods. Various discrete area measures are used as approximations to the area of arbitrarily shaped objects. If one considers the original image to be a function with the abscissa being the coordinates of the image, and the ordinate being the image intensity, then the various shape representations may be thought of as function expansions using Fourier series. Other function expansions, using polynomials or piecewise polynomials, are also used. Measuring lengths and sizes is also accomplished by discrete approximations to geometric triangulation using two or more images. Extraction of texture information from images also involves decompositions of functions, but using basis functions that are appropriate to texture.

Examples

This section describes the application of image-processing and computer vision techniques to two remote-sensing problems. It is intended to familiarize the readers with what is currently possible using the methods described above. In both applications the goals are to provide useful tools to scientists working with large numbers of images. For this reason, both examples are of automated systems. A description of each scientific problem is given, together with the results of the application of automated computer algorithms to the problem. Readers who wish to know the details of the algorithms should refer to the original publications (Samadani et al. 1990; Vesecky et al. 1988).

Satellite Images of the Earth's Aurora

The aurora polaris is the result of the exchange of energy between the solar wind (the expanding outer atmosphere of the sun) and the Earth's magnetic field. The auroral oval is the result of ionization of the upper-atmospheric gasses by the precipitation of energetic electrons. Images now available from high-altitude polar satellites provide the first global characterization of auroral activity from a single measurement within a single data set. Of particular interest for the understanding of the physical processes of the aurora is the identification of the inner boundary of the auroral oval. From this boundary, useful quantitative parameters can be extracted. For example, the area within the inner boundary varies, and this is thought to be related to the amount of magnetic energy stored in the magnetic field lines that intersect this area.

Figure 23.3a shows a typical Dynamic Explorer (DE-1) satellite image of the polar hemisphere and the aurora. The image is obtained using a photometer sensitive to ultraviolet radiation. The bright crescent shape in the top portion of the figure is due to solar illumination of the Earth. The illuminated ring in the image is the auroral oval.

Until now, the boundaries have been extracted manually. Figure 23.3b shows the results of manual extraction of inner boundaries of the auroral oval for several images. The boundaries are superimposed on the original DE-1 images. Manual extraction of the boundaries is time consuming, and this creates a considerable bottleneck to the analysis of the available data. Over half a million DE-1 images are currently available for analysis. New satellites, soon to be launched, will produce even more images at a greater rate. Detailed analysis of the existing and expected images greatly exceeds the available manpower. Hence, automation is necessary to aid in extracting the information from the images.

Figure 23.3c shows the results of using computer vision (Samadani et al. 1990) to find the boundaries, without user interaction, for the images in Figure 23.3b. Comparison of the manual and automated boundaries shows that the automated technique is successful in reproducing our visual system's capability for this particular task.

Radar Satellite Images of Sea Ice

Sea ice covers about 10 percent of the Northern Hemisphere's oceans, with seasonal fluctuations. The motion of the ice influences the global weather system in ways that are not yet well understood. The location and movement of sea ice also affect arctic off-shore engineering, oceanic transportation, and military operations. In 1978, NASA's SEASAT satellite carried a synthetic aperture radar (SAR) system that could record images of the earth's surface even through clouds, providing a means for continuous observation of sea ice. As the satellite repeatedly observed the same geographic location, a sequence of images of that location was created. Figures 23.4a and 23.4b show two images from the same geographic location gathered by the SEASAT satellite on October 5 and October 8, 1978. The images cover approximately 2500 square kilometers.

Using two images of the same geographic location, a human operator can determine the movement of the ice by matching features. Figure 23.4c shows

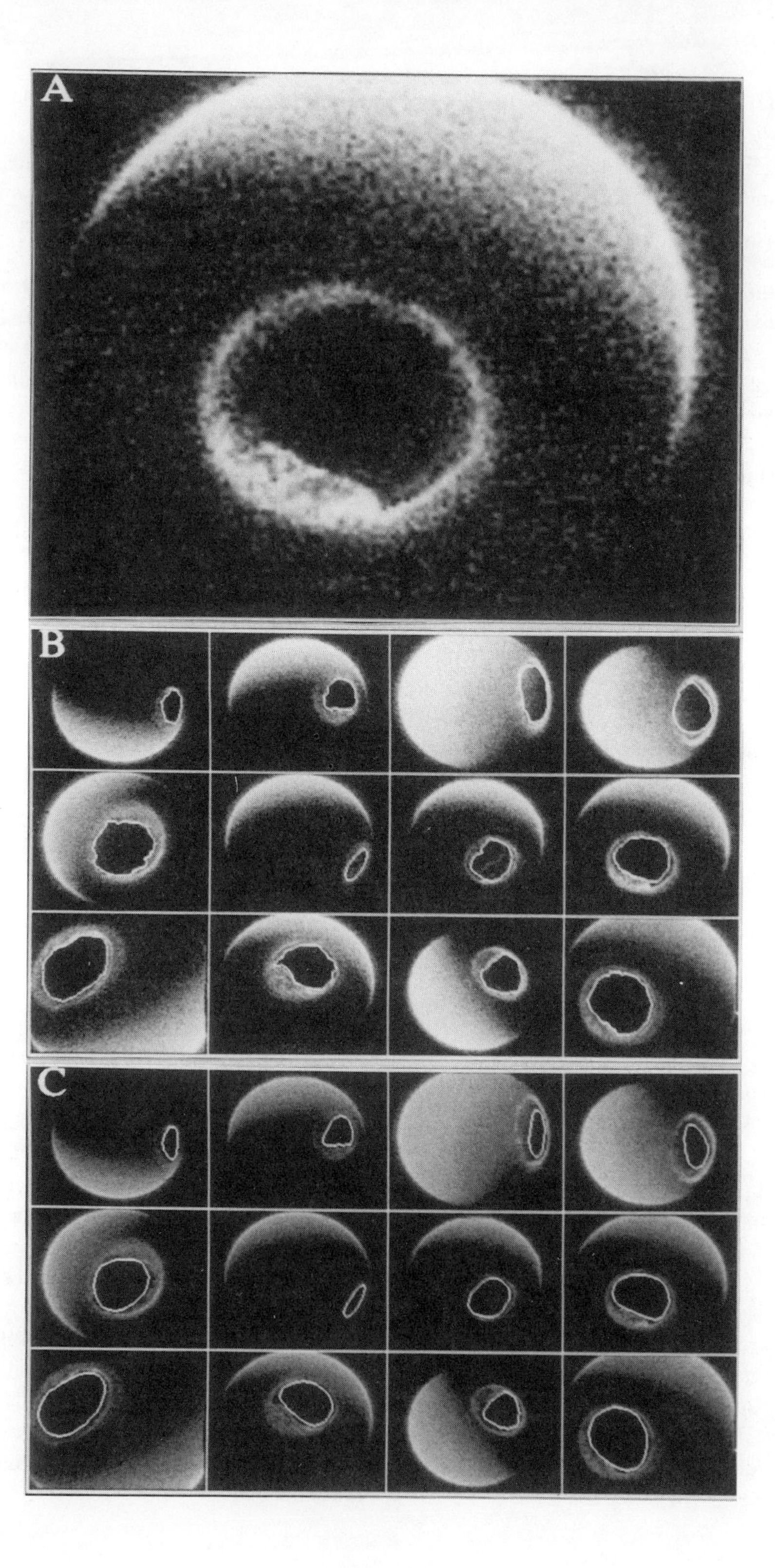
A
B
C

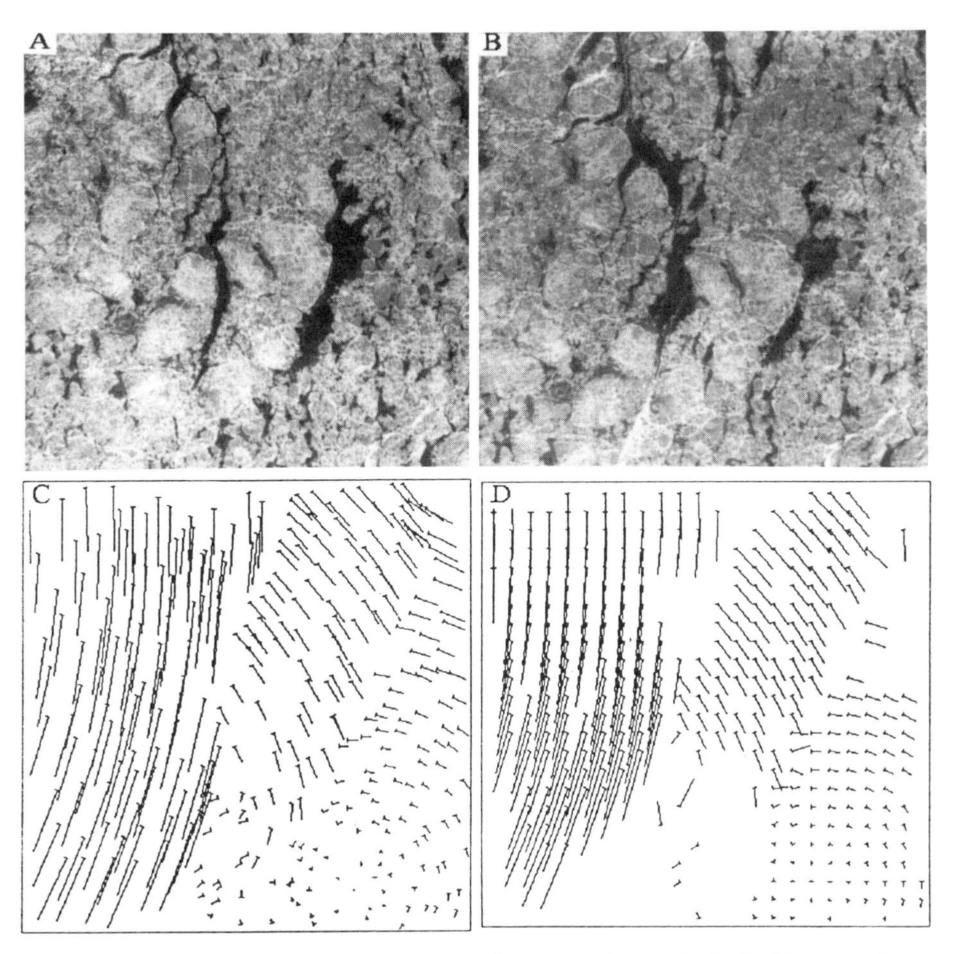

Figure 23.4. SEASAT radar images of sea ice in one location (a) October 5, 1978; (b) October 8, 1978. (c) Manually generated; (d) automatically generated displacement vectors for images (a) and (b).

the vectors manually extracted from the images shown in Figures 23.4a and 23.4b. For the human operator, this task is tedious and time consuming. There are plans to fly SAR systems on new satellites such as ERS-1 of the European Space Agency; Radarsat, constructed jointly by Canada, the United States, and the United Kingdom; and the Japanese JERS-1. Each new SAR system

---

Figure 23.3. (a) DE-1 satellite image showing Earth's auroral oval. (b) Boundaries for twelve DE-1 satellite images of Earth's aurora found by manual measurements; (c) boundaries for the same images found by computerized measurements.

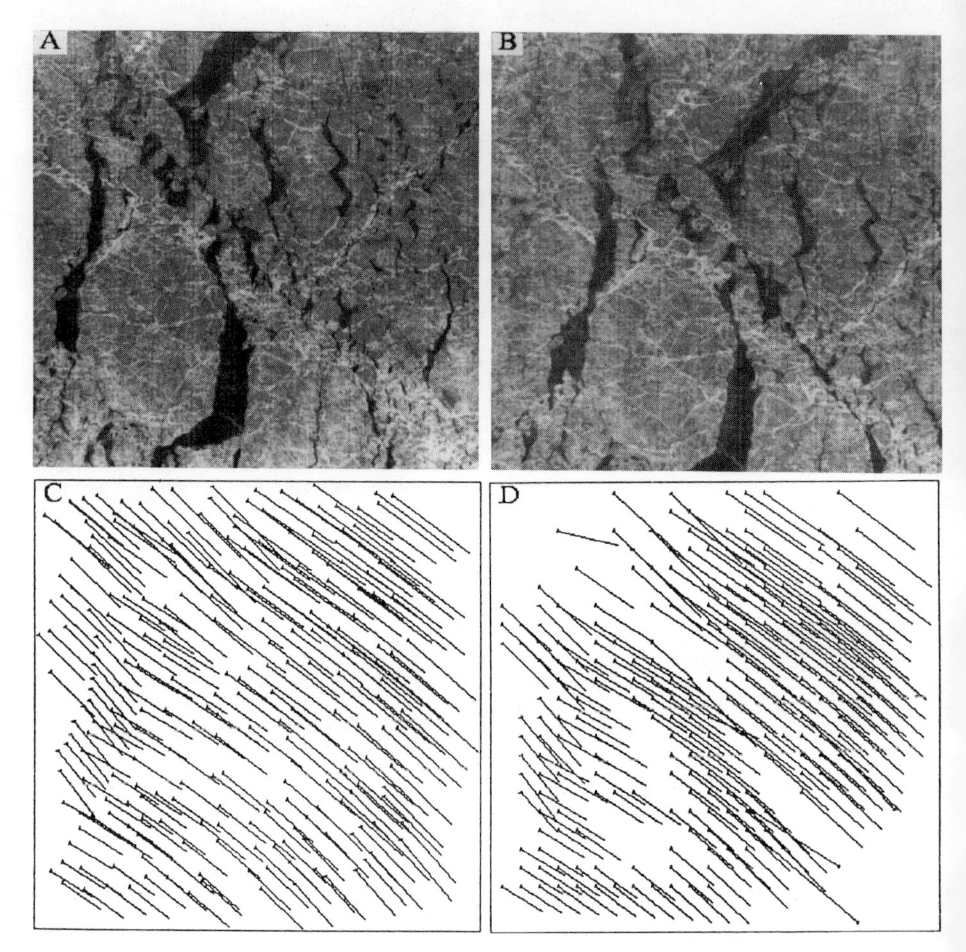

Figure 23.5. (a,b) Second set of SEASAT radar images of sea ice. (c) Manually generated; (d) automatically generated displacement vectors for the pair of images in (a) and (b).

will make available greater amounts of data. To process this vast quantity of data, automated algorithms for motion detection are needed to help relieve the burden on the human operator.

Figure 23.4d shows vectors automatically generated by computer for the images shown in Figures 23.4a and 23.4b. This figure compares favorably with the vectors generated manually (Fig. 23.4c). Figures 23.5a–d and 23.6a–d show other pairs of SAR images with the same comparison of manual and automated displacement vector generation. For this application, the computer system is successful in automating a task that is exceedingly time consuming for humans.

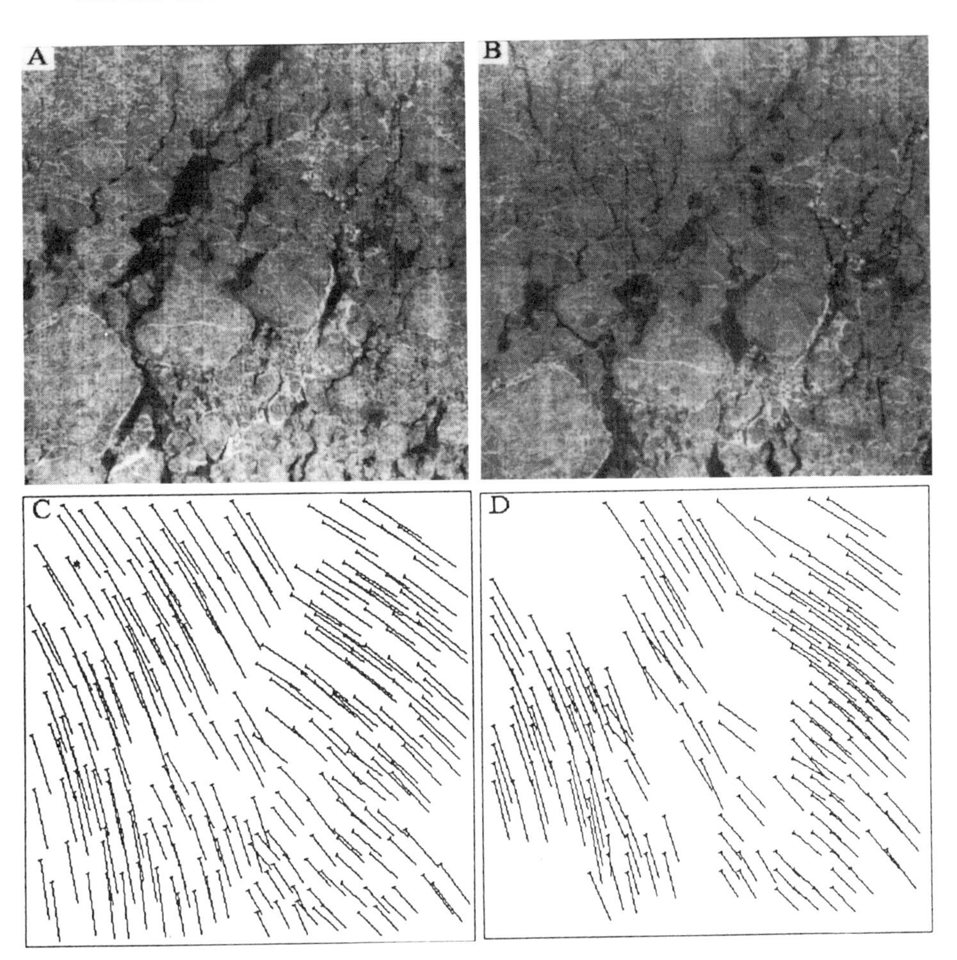

Figure 23.6. (a,b) Third set of SEASAT radar images of sea ice. (c) Manually generated; (d) automatically generated displacement vectors for the pair of images in (a) and (b).

Acknowledgments

The Dynamic Explorer (DE-1) satellite images of the aurora were provided by Drs. L. A. Frank and J. D. Craven of the University of Iowa. The SEASAT images were provided by the Jet Propulsion Laboratory.

Discussion

——: Are you using a model of the aurora for the automatic tracing of the boundaries of the aurora? Is it biased toward a particular kind of configuration that allows it to smooth an arch across some portion of the image?

Samadani: There is a nonlinear least-squares search to find the best parameter starting from some initial guess of the parameters.

——: Why do you need to look at millions of pictures, instead of just a sample? What are the questions to be answered?

Samadani: There is a complicated model for the magnetosphere and ionosphere. The auroral oval is just part of that complicated model. The area of the oval allows the physicists to map the energy distribution in three dimensions. They are interested in correlations with other physical quantities, for example, flux from the sun. They are doing statistical studies about those variations, and they want to analyze as many images as possible.

——: What is the time dependence?

Samadani: For this satellite a new image comes up every twelve minutes.

24

Computer Vision Needs in Systematic Biology

F. JAMES ROHLF

Almost all descriptive data traditionally used in systematics are collected visually. Measurements are taken by visually aligning a structure against a scale; setae, scales, teeth, and other features are counted visually; outlines of structures are assigned to various shape categories by a visual comparison with standards, etc. Thus the potential impact is very large for computer-vision-based automatic data collection systems.

Automation of routine data collection would be expected to make systematists more efficient and allow them to carry out more comprehensive studies than are practical at present. Such automation would also have the advantage of making the definitions of taxonomic characters fully explicit and exactly repeatable by others. The quality of large databases should also be improved since the definitions of the variables would not drift due to operator fatigue or differences in interpretation among personnel during the course of a large study.

The techniques of image analysis are now being used in systematics but only for rather narrow and specialized applications. This chapter discusses the current state of methods for image capture, enhancement, and analysis. The emphasis is on present limitations and problems.

Capturing Information from Images

Several types of devices are used in systematics to obtain descriptive information from organisms. They can be classified according to the type of data they can collect: devices that measure distances between two points, devices that record the position of points relative to some coordinate system, and scanners

---

Advances in Computer Methods for Systematic Biology: Artificial Intelligence, Databases, Computer Vision, ed. Renaud Fortuner (Baltimore: Johns Hopkins University Press).

that record brightness (at one or more wavelengths) over an image. Although only the information produced by the latter device is usually considered the subject of computer vision and image analysis, the former will also be discussed since they still are very useful in practical studies. Fink (1990) gives a useful overview of currently available devices used for data acquisition in systematics.

The resolution or the precision with which measurements can be made with different devices is given some emphasis in the discussion below. Low precision can be a problem in some applications, especially those in which relative levels of variation and covariation are of more interest than comparisons of means of populations (see Rohlf et al. 1982).

Distance-measuring Devices

Rulers and calipers are used to measure lengths of structures and distances between pairs of landmarks on fish, birds, and larger organisms. Digital calipers with interfaces to digital computers are now generally available. Their resolution is typically 0.1 mm, and they can be used to accurately measure lengths from a few centimeters to several meters (using specially constructed digital calipers). Software exists to enter readings from digital calipers directly into standard spreadsheet and word-processing software on microcomputers.

Smaller objects are often measured under a microscope by aligning the structure of interest against a scale in the eyepiece and then either manually recording the distance or speaking into a device able to recognize numbers and a few simple commands. One can also move a specimen (using a calibrated stage that can be digital and under computer control) past a reference mark in the eyepiece.

The above are relatively labor-intensive approaches since the investigator must take an action to record each measurement.

Coordinate Recording Devices

Two-dimensional digitizing devices are very popular. Several types of devices are available. Digitizing tables electromagnetically locate the position of a special cursor on the surface of a digitizing board. Sonic digitizers determine the position of a pointer by measuring the time it takes for sound emitted by the pointer (located very near the tip of the pointer) to reach two or more microphones (located along the periphery of the digitizing area). Coordinates of points on photographs, projected images, and flat objects (such as leaves or the wings of large butterflies) can be digitized and entered into a computer. The length and width of the active area within which coordinates can be recorded range from about 25 cm to more than a meter. This may correspond to a larger or smaller area for the original object if one works from a projected image.

Three-dimensional digitizers are available but are not used as commonly. Sonic digitizers determine the position of a pointer by measuring distances from two sound emitters (located along the pointer) to three or four microphones (located on the periphery of the digitizing volume). Electromagnetic devices also exist that can determine three-dimensional coordinates of a pointer within a volume above a special platform. Optical-mechanical devices allow one to record the coordinates of point of light projected within the visual field of a microscope.

Digitizers usually have a resolution of 0.1 mm. In practice, accuracy is usually limited by one's ability to position the pointing device. Two-dimensional cursors with crosshairs can be positioned very accurately in comparison with a handheld pointer in three dimensions. But as long as the structures of interest take up an appreciable fraction of the visual field, digitizers usually provide excellent precision.

From the x,y (or x,y,z) coordinates of points entered into a computer, various lengths, angles, areas, etc., can be computed. Recording coordinates is more efficient than recording distances since up to $n(n-1)/2$ distances can be derived from a set of n coordinates. One also has the flexibility of being able to derive different suites of measurements from a given set of coordinates. Digitizers can also be used to record a continuous sequence of points to capture the outline of a structure.

Scanning Devices

These devices systematically measure image brightness for an array of points (pixels) across an image. The information recorded may be brightness values (at one or more wavelengths) or just whether or not the brightness was above a specified threshold at each pixel.

The most commonly used input device is an ordinary TV camera interfaced to a digitizing board in a microcomputer. The digitized image usually has a resolution of 480 rows (corresponding to the visible lines on a TV screen in the United States) and 512 or 640 columns (depending upon the digitizer). Lower-resolution digitizers are also used that save only every other row of the image. Still video cameras are an important recent development. They are about the same size as conventional cameras and just as portable. Images of specimens in the field and in museums can be collected and brought back to the lab for analysis and measurement.

Although the length of an object that extends across the entire visual field can be measured with a resolution of about one part in 500, most structures cover only a portion of the field of view and thus can be measured with much less accuracy. One can "zoom in" on different structures, but that requires an accurate registration of images and adjustments for different magnifications. A convenient feature of video systems is that they are interactive—the user can focus, change magnification, etc., and the digitized (and perhaps enhanced) image can be seen directly on a video monitor. There are many software packages developed to facilitate the collection of morphometric data from images (see, e.g., the list in Fink 1990).

Page scanners also can be used to capture images. On microcomputers, driver software usually saves the images as tagged image format files (TIFF) or other standard format image files used in desk-top publishing programs. Depending upon the scanner and the software, the resulting files may represent a black-and-white image at some threshold level of brightness, or it may contain the actual brightness values. Color scanners are also available. Scanners usually have a resolution of 300 dots per inch over about an 8.5- by 11-in. area. This permits accuracy of up to one part in 4000 for a full-scale measurement.

Handheld scanners capture brightness values along only a relatively narrow (10-cm) swath across an image. These devices may be quite useful for certain types of images (especially for elongate structures). Handheld scanners are of interest since they are portable and are relatively inexpensive. The driver software for these devices produces the same types of files as for page scanners and thus requires the same type of software support.

A limitation of scanners is that they arc relatively slow and are not interactive—the image must be captured as an image file before it can be analyzed. Image files can be rather large. A single uncompressed gray-scale TIFF file from a page scanner may require 8.5 MB of storage. In video-based systems the software can directly access the image in a frame buffer, and there is no need to store images as files.

Scanners do not seem to be supported yet by most image measurement packages. The MorphoSys package (Meacham, this volume, chap. 26) can accept an image from a TIFF file, but only if it fits within the dimensions of a video image. The image can be displayed and measurements made interactively as if the image came from a TV camera. Much more software is needed to support the use of scanners in systematics.

Image Storage Needs

Files of video images can be quite large. An uncompressed 480 by 512 video image requires about 246K of storage. Fortunately, these files can usually be greatly compressed. The storage requirements for a compressed image depend upon the complexity of the image and the sophistication of the compression algorithms. At least 50 percent compression can usually be obtained. This implies that about eight images can be stored per megabyte.

Another strategy is to analyze the image at once and save only the x,y

coordinates of landmarks, chain-coded outlines, or other information about the image rather than the brightness values of the actual image. If the appropriate information is saved, one will be able to compute the various features of interest. These files should be much smaller than the original image files. The limitation of this approach is that one must know in advance what kinds of features will be of interest. This can be a problem in taxonomic studies where the later recognition of new species may necessitate the use of new features.

Storage devices are available that should provide sufficient capacity for most applications in systematics. Hard disk drives with up to 2000 MB capacity are available for microcomputers. Optical disk drives are available with up to 1000 MB capacity. Although they are much slower than hard disks, they have the advantage of having removable media, and thus total storage is unlimited. There are also "jukebox" devices that allow automatic selection from among several optical disks. One can also increase the amount of online storage by attaching more than one disk drive to a computer.

The cost of storage has dropped dramatically in recent years. It should soon reach the level where a systematist can afford the cost of storage large enough to hold an image database.

Image Enhancement

One can distinguish two goals for image enhancement. The first is to apply various transformations to the brightness values of an image so as to yield an image that is more pleasing to the human eye or which a human can interpret more quickly and accurately. This is important in semiautomatic image analysis systems where the user must point out objects of interest and make decisions about the locations of landmarks. The second goal is to apply transformations to simplify and improve the accuracy of subsequent automatic computer processing of an image. These two goals may be somewhat in conflict because computer algorithms do not always "see" the image in the same way that a human does.

The application of image enhancement techniques is often fairly straightforward. Most image analysis systems let the user try many different transformations and immediately display the results of each transformation. The user can select the transformation that allows the user to see the structures of interest. Even simple contrast enhancement transformations usually make a large difference in the subjective appearance of an image.

When one has a priori information about what structures should be present in the image, as is almost always the case in systematics, one can try transformations that enhance such aspects of an image. For example, if one knows that the structure of interest has a very simple smooth outline, then edge-preserving smoothing transformations are likely to be useful. Edge and line enhancement transformations are usually useful since landmarks are often located at intersections of lines or at extremal points along curves.

Although the results are sometimes very impressive, image enhancement transformations cannot perform miracles. It is important to start with simple high-quality images that are well illuminated and are as free as possible from distracting elements such as dust, bubbles, stray hairs, and reflections and shadows.

Segmentation of Complex Scenes

The chances of successfully applying image analysis techniques are very much dependent upon selecting the right problem and having clean, non-overlapping images of simple flat objects. It is much easier to work with an image of a single dark leaf against a white background than with an image of a leaf still on the plant. In the former case there is no problem determining which parts of the image corresponds to the object of interest and which parts correspond to background or other uninteresting structures. Unfortunately, such simple images represent only a small fraction of those that need to be processed in systematics. Image segmentation is concerned with the problem of isolating the part of the image that corresponds to the object of interest.

Image segmentation is complicated by a number of factors. (1) There is biological variability in the structures. One cannot depend upon homologous structures having identical geometrical features that can be used to automatically identify landmarks of interest. (2) The structures of interest may not have distinct colors or brightness to allow them to be easily isolated from the rest of the image. (3) Most structures of interest are three-dimensional, so there will be some distortion due to errors in aligning the object or to flattening the object.

For these reasons, considerable time is often spent simplifying the image. For example, structures of interest are physically separated from the rest of the specimen and placed upon a contrasting background. If this is not feasible the segmentation of the image must be done interactively so that the user can point to structures of interest and verify that the computer is able to isolate it from the rest of the scene.

Feature Extraction

Once a structure is isolated in an image it needs to be described. A systematist usually measures its length and width as well as the distance between selected pairs of landmarks. In most cases the selection of characters is rather arbitrary. Characters are often selected because they are easy to measure using conventional techniques.

There are many other types of descriptive features that one can imagine using. Rohlf (this volume, chap. 25) gives an overview of feature extraction methods that have been used in systematics. Most of the methods are very different from those conventionally used in systematics. In many cases the resulting features are coefficients of mathematically convenient functions that have no obvious biological interpretation. They seem quite arbitrary. But there have been a number of studies reporting the usefulness of feature extraction methods in systematics. The use of Fourier coefficients as descriptive features has been particularly popular. Other studies have reported success with elliptic Fourier coefficients, moment invariants, parametric cubic splines, and Bezier curves. Sources of references are given in Rohlf (this volume, chap. 25; 1990a).

Perhaps the only reason for trying to reproduce traditional measurements is in order to be able to compare one's results with those of previous studies.

Conclusion

The once high cost of image analysis systems should soon no longer be a problem for routine applications in systematics. The cost of standard video cameras, digitizers, and computer hardware continues to decrease. The main limitation of such systems is that they provide relatively low-resolution images. While they are sufficient for many applications, higher resolution would be desirable.

The most important current limitations are in the area of automating the recognition and measurement of features from the images. The problem is software, not hardware. Present applications require continuous operator involvement to make sure that the proper structures are recognized and that important landmarks are identified correctly.

Acknowledgments

This work was supported in part by NSF grant BSR-8306004. This chapter is contribution number 790 from the Graduate Program in Ecology and Evolution, State University of New York at Stony Brook.

Discussion

Samadani: How many pictures are you dealing with?

Rohlf: The potential number of images to be processed is very large. Personnel at the U.S. Department of Agriculture have to routinely identify insect, nematode, fungi, and other pests of agriculture; the U.S. Public Health Department has to identify insects, spiders, and other organisms of public health importance. State organizations also do large numbers of identifications each year. Identifications are usually made by technical staff since sufficient numbers of trained taxonomists are usually not available.

——: What is the magnitude of the role of the identification in the USDA or Public Health Service?

Thompson: About 180,000 insects a year at the USDA; that does not include the state organizations.

Diederich: Have you created databases to store this information? To analyze images they have to be in a database.

Rohlf: I have not, but there are now a number of standard commercial packages for image database applications.

Pankhurst: Some protozoologists in the museum thought they could define a fractal representation of shape. I am skeptical, but what do you think?

Rohlf: As an impressive example, Barnsley et al. (1988) have shown that the form of a Black Spleenwort fern could be fit with just twenty parameters. An interesting follow-up would be to use those parameters as features in a taxonomic study and determine the extent to which similar ferns cluster close together in the feature space and how useful they are for identification. One difficulty is that estimating the parameters requires a considerable amount of computation.

Walker: The group at SRI is showing a direct mapping from the fractals you mentioned to Fourier transforms.

Horvitz: Storage and identification efforts should be focused on those features that are most important for taxonomic decision making. If the setae represent a branching factor, just look for the setae as opposed to looking at the whole organism. One could take pictures of structures that are not considered important as a whole and figure out what detail is needed for identification. Is anyone working on that area?

Woolley: Not in a quantitative or rigorous way, but I think that is how people do taxonomy. We would be interested in tools to do that more efficiently.

Rohlf: That is the way identifications are routinely made. There are usually particular structures, for example, the male genitalia of insects, that are especially useful for distinguishing species. That was one of my original motivations for working with mosquito wings. They also have the practical advantages of being relatively flat and easy to observe. They are used for identification but not to the extent that I suspect is possible.

Woolley: We will soon be able to take very rich, full expressions of images and have them accessible to people without having to abstract some set of descriptors. The human eye and brain are capable of integrating an enormous amount of information very rapidly. Richard Pankhurst was saying that botanists like to see pictures. These contain much more information than a few words.

Humphries: Accuracy really is not a concern. I suspect that there are models that can be used to tell us how accurate we have to take measurements to reproduce the results wanted. Resource models would be such an ideal system.

——: But you want your measurement accuracy better than natural variability.

Humphries: That is our immediate answer, but it may not be true. For identification, we may not need the accuracy we think we do. Considering our limited resources, the question is how accurate we need to be.

Allkin: To what extent do you feel visual characters are appropriate for identification?

Rohlf: You can test whether or not a particular set of characters is useful for identification by using known specimens (perhaps using progeny reared in captivity).

Allkin: Would you expect the coefficients of fractals and other functions to be used within an automatic identification system, or would you expect the information to be represented in some other way?

Rohlf: As a biologist I would like to know what the coefficients represent in terms of the development or physiology of the organism. If such an interpretation were possible, I would be tempted to use these explanations as a basis for the generation of characters. This step is not necessary, but it may lead to more stable identification scheme.

——: Has anybody tried fingerprinting mosquitos?

Rohlf: In the field of pattern recognition the term *fingerprinting* refers to having a few features (characters) for which each species would have a unique combination of states. I do not think that would be possible for more than a trivial set of species. In analogy to real fingerprints, one does have small regions with complex structures (such as the male genitalia) where one can see many features at once. The existence of such regions greatly facilitates identification. Unfortunately, it is the female mosquito that bites, and female genitalia are quite simple and not that useful for identification.

25

Feature Extraction in Systematic Biology

F. JAMES ROHLF

This chapter is intended as an overview of computer vision feature extraction methods and their applications in biological systematics. General surveys of image acquisition systems for use in systematics are given by Fink (1990) and by MacLeod (1990). Rohlf (1990a) gives a general overview of image processing and image analysis techniques likely to be useful in systematics. Rohlf (1990c) gives a review (more extensive than in this present chapter) of recent work in morphometrics.

Several terms need to be discussed to help bridge the gap between systematics and computer science. For a systematist the term *character* refers to the various properties used to describe an organism so that it can be compared to other organisms. In pattern recognition these are usually called *features*. In systematics one tries to use biological knowledge to make sure that a particular character means the same thing in different species. For example, if the character is a measurement between two points, then the same measurement in another species must be between the same pair of homologous landmark points.

The notion of a landmark point is important. It is a point of reference that can be used to define an end point of a measurement. It is located where there is a discrete juxtaposition of tissue types, at points of maxima of curvature or other local morphogenetic processes, or at extremal points along a structure (see Bookstein 1990). When points in each of two organisms are both derived from what can be considered the same point in their common ancestor, these points are called homologous landmarks. A practical problem with this definition is, of course, that one does not know the common ancestor—in fact one of the purposes of many studies is to estimate the characteristics of the common ancestor.

Advances in Computer Methods for Systematic Biology: Artificial Intelligence, Databases, Computer Vision, ed. Renaud Fortuner (Baltimore: Johns Hopkins University Press).

An alternative to the expression *homologous landmark* is *operational homology* (see, e.g., Woodger 1945; Jardin 1967). This is closer to the use of features in image analysis. Measurements are operationally homologous if they are well defined in terms of the geometry of the structure. It does not matter whether or not the developmental or evolutionary origin of the tissues is the same at the points at which the measurements are made. If the shape of a structure has changed drastically during evolution then, for example, the two points on the outline that are farthest apart (and therefore used to measure a length) may not correspond to the *same* part of an organism.

The expression *feature space* is used in pattern recognition for the multidimensional space defined by using the various features as variables defining coordinate axes. The usefulness of a feature for some particular application depends on the statistical properties of the points (corresponding to the individual specimens) in this space. It is the multivariate attribute space (A-space, Williams and Dale 1965) in numerical taxonomy (Sneath and Sokal 1973).

The purpose of a systematic study should be considered when selecting features. For identification it is sufficient that the different species (or other taxa of interest) have distributions that overlap as little as possible in the feature space. As Fink (1990) points out, one needs to capture as much information as possible in the hope of getting the information needed. For a phenetic classification, one needs to collect data in a way that results in "similar" species being close together and dissimilar ones far apart in the feature space. It is more difficult to characterize the desired properties of the feature space for a phylogenetic study. The space should be such that an estimated tree (using a particular method) will be close to the true phylogenetic tree (the tree that describes the actual branching pattern of evolution).

Current methods of feature extraction were developed to give convenient, easy to compute, and comprehensive descriptors of an image of an object. They were not developed to be optimal for the purposes listed above (except, perhaps, identification). Applications to systematics have been experimental for this reason. Investigators have tried one or more methods to see empirically how useful they are for some particular purpose—usually identification.

The initial data extracted from images usually include x,y coordinates of geometrically important points (centroids of objects and various extremal points along outlines of structures, such as the pair of points that are farthest apart) and sequences of coordinates of points along the outlines of structures (usually stored compactly as chain codes).

These raw quantities require further processing to transform them into features usable in systematics. Three types of approaches can be distinguished: description of the outlines of structures or whole organisms, descriptions of the relative positions of a specified set of points on the objects (landmarks), and methods for describing an entire gray scale or color image of an object. The discussion given below is for two-dimensional outline and landmark data. Most of the methods listed can be easily generalized to three dimensions. Methods for capture of entire gray scale or color images (e.g., the method of moment invariants; see Rohlf 1990a) are beyond the scope of this chapter.

Outlines

When a structure has few or no landmarks, the conventional types of linear distance measurements used are length (distance between maximally separated pairs of points along an outline) and width (perhaps measured as maximal diameter of the object in a direction orthogonal to the direction at which the length measurement was taken). Such characters capture very little information other than the overall proportions of the object. An alternative is to fit a function to a sequence x,y coordinates of points digitized around an outline. The parameters of the function fitted to these points can then be used as descriptors of the shape of the structure. With enough parameters one can obtain a detailed description of the outline.

Rohlf (1990b) gives a detailed survey of the types of functions that have been used in morphometrics. Simple numerical examples are also presented. An overview of these methods is given below.

Complete Outlines

One needs to distinguish between complete (closed) and partial (open) outline curves. A closed curve can be used to fit the entire outline of a structure or even the complete outline of an organism as seen from a particular direction. An open curve (see below) can be fit to that part of an outline that lies between two landmarks.

There are many examples of the use of complete outlines in systematics. Until recently, the digitized x,y coordinates of points around the outline were usually expressed as polar coordinates relative to a central reference point within the object (either the centroid or a conveniently located landmark). The radii were then adjusted, by interpolation, to correspond to equally spaced angles starting from a reference angle based on a second landmark, usually located on the outline itself. The lengths of these radii were then fit by some function. See Rohlf and Archie (1984) for an example.

It is natural to consider the use of a periodic (trigonometric) function since the starting radius is also the ending radius as one sweeps through radii oriented at angles varying from 0 to 2 π radians. Fourier analysis of the lengths of the radii has been used in many studies. Two mathematically equivalent forms of the parameters of a Fourier analysis are used in morphometric studies. Most studies use the coefficients of the underlying sine and cosine series as the parameters. Others use the harmonic amplitudes and phase angles.

When a second landmark on the outline is not available to provide a unique starting reference angle, the phase angles are usually discarded to yield an orientation-free description of shape (unfortunately this strategy also loses much information about shape). The Fourier coefficients have the mathematically convenient property of being linear (and orthogonal) combinations of the lengths of the radii. Rohlf (1986) describes some of these mathematical properties and the relationships of Fourier analysis to some of the other methods used in morphometrics.

Recent approaches have been more general. A limitation of methods based on polar coordinates is that a given radius may intersect the outline at more than one point when the outline is complex. Zahn and Roskies (1972) proposed using a Fourier analysis of a normalized change in the angle of a tangent to the outline as a function of distance along the outline. This function is also the basis of Lohmann's (1983) eigenshape analysis. Kuhl and Giardina (1982) described an elliptic Fourier analysis that can be used to fit very complex outline contours. Rohlf and Archie (1984) compared several of these approaches and found elliptic Fourier analysis to be particularly useful. Rohlf (1990b) reviews the use of these and other parametric curves to describe complete outlines.

These approaches are all mathematically equivalent in the sense that one can convert from one form to another. As discussed below, they are not equivalent in practical applications to systematics since they result in feature spaces with different relative positions of the points.

Partial Outlines

Open curves can be used to fit part of an outline between two landmarks. For example, one may be interested in a description of that part of the outline of an insect wing that is between the points of intersection of two veins with the margin of the wing. This allows a more localized comparison of structures.

There are many techniques that can be used to describe the shape of a curve between two landmarks. Because one is usually concerned only with the shape of a curve, not its absolute position in space, the points can be rotated so that one end point is at the origin and the other at some point along the x-axis. If the curve is a simple curve it may be possible to fit the points with a polynomial function. If it is more complex a cubic spline function may be useful (see Evans et al. [1985], for a description and an application in systematics).

If the curve is not single valued with respect to the x-axis (i.e., there is at least one x value that corresponds to more than one point on the curve) then parametric functions (the argument of the function is the distance along the curve rather than the distance along the x-axis) will be needed to fit the data. Examples are parametric cubic splines and Bezier curves. Rohlf (1990b) describes these functions and gives numerical examples.

Using Features as Taxonomic Characters

The methods described above can be said to represent "curve-fitting exercises." The functions described above are not based on any particular biological model. The parameters are just mathematical constructs, not direct measures of underlying developmental or evolutionary processes. Therefore, the parameters of these functions should not be expected to have a biological interpretation. However, one may hope that biological information is captured by the parameters and that the feature space they define will contain useful information. One may be able to discover trends and patterns in the feature space and give them interesting biological interpretations.

Since there is a choice of methods, one must ask whether it matters which one is used. Sometimes it does not since the parameters from some methods can be expressed as orthonormal linear transformations of other methods. Usually it does make a difference—the relationship may correspond to an affine transformation or may be nonlinear. The relative importance of this problem depends upon the type of application.

Identification. These methods are used to assign new specimens to existing taxonomic groups with the minimum number of errors (or perhaps with the minimum costs due the consequences of misidentification). There is much less controversy in systematics about the use of different identification schemes than there is about methods for classification and phylogeny estimation. This is undoubtedly because the effectiveness of identification methods can be evaluated empirically using real specimens of assumed known identity (rather than having to depend upon the results of simulation and modeling studies as in phylogeny estimation).

The classical statistical approach to identification is to use Fisher's (1936) linear discriminant function analysis (other methods are described elsewhere in this volume). It is based on the restrictive assumptions of multivariate normality and homogeneity of variance-covariance matrices, but it has the important property of invariance under linear transformations of the feature space. What this means is that it does not matter whether Bezier curves or cubic splines are used to describe a partial outline since the parameters of a Bezier curve can be expressed as a linear combination of the parameters of a cubic spline (Rohlf 1992). On the other hand, sets of features that differ by nonlinear transformations can give different results, and their relative usefulness must be evaluated empirically. For example, Fourier coefficients based on polar coordinates will not give the same results as Fourier coefficients based on tangent angles.

The fact that different sets of features can give different results is not a problem since one can compare different sets of features and select the one that results in the least overlap of the clouds of points (corresponding to the different taxa that one wishes to distinguish) in the feature space.

Phenetic Classification. If your purpose is to construct a phenetic classification (a classification based on the relative degrees of similarity among the taxa [Sneath and Sokal 1973]), then points corresponding to similar species should be close together in the feature space, and species that are dissimilar should be far apart. In this type of application the metric properties of the feature space are very important. The relative proximity of points in the feature space determines whether clusters are present and their membership.

Unlike the problem of identification discussed above, it is difficult to test empirically which method gives the "correct" phenetic classification. One does not have a true classification based on overall similarity with which to compare ones results. About all one can do is to show that certain methods are less stable numerically, exhibit various pathological properties, or are inappropriate under certain models.

The classical techniques of numerical taxonomy (cluster analysis of distances based on standardized characters) are invariant under scale change and orthogonal rotations but not under affine transformations. This means that using the parameters of Bezier curves rather than those of cubic splines, for example, can yield different results. By choosing different features, the relative distances among the species will change and a different impression will be obtained of the relationships among the taxa.

Phylogenetic Inference. While several methods have been proposed for estimating phylogenetic trees from descriptive data, methods based on the principle of parsimony are most popular. Mathematically, the *most parsimonious tree* for a dataset is the Steiner minimal-length tree (Gilbert and Pollak 1968) but based on the Manhattan rather than the Euclidean metric. These trees (there may be more than one for a given dataset) are unrooted evolutionary trees. They are usually rooted based on additional information such as the placement of an out-group species on the tree. Since there is no way to evaluate how accurate a method is for estimating phylogenetic trees from real data, there is considerable controversy about the validity and accuracy of different approaches.

It is well known that the problem of computing Steiner minimal trees belongs to the class of NP-complete problems whose efficient solution appears intractable (Day 1983; Graham and Foulds 1982). I am not aware of any work done on the effects of using types of feature extraction methods on the resulting estimated phylogenies. Clearly there will be important effects since methods based on the Manhattan metric are not even invariant under orthogonal rotations of the feature space. Rohlf et al. (1990) investigated the effects of various aspects of the topology of a phylogenetic tree and of different models for rates of evolutionary change on the accuracy of estimated phylogenies. They did not investigate the effects of transformations of the feature space.

Landmarks

For most organisms it is possible to identify points that are at least operationally homologous across the different organisms under study. These points are often points where structures such as veins, sutures, and plates join or intersect, or points at the locations of geometrically interesting features such as edges, cusps, and spurs. The relative positions of these points are usually recorded as x,y coordinates as seen in some standard viewing plane.

Unless high accuracy is required, coordinates and various types of measurements can now be made very quickly and conveniently using microcomputer-based video image analysis systems. Fink (1990) lists many systems that have been used for morphological measurements (including MorphoSys of Meacham (this volume, chap. 26). Measurement*TV* (developed by Garr Updegraff) is a new program that helps biologists make morphometric measurements. When more accuracy is required, coordinate digitizers are very effective—although completely manual in operation.

Most biological images are very complex scenes. For this reason, image analysis systems used in morphometrics are designed to make measurements under the guidance of an operator. The user points to structures of interest, aligns pointers, etc., in order to specify the coordinates to be recorded or the features to be measured. Completely automatic methods are practical only when the images are very simple. For example, outlines of shells, leaves, bones, etc., can be captured automatically if these objects are placed against a contrasting background with no distracting elements. While the raw coordinates of the landmarks are sometimes used directly as morphometric variables (e.g., Moss and Power 1975; Plowwright and Stevens 1973), they are usually processed in some way. Bookstein (1986) translates, rotates, and scales the coordinates relative to a baseline defined by two landmarks, A and B. After transformation landmark A will have coordinates 0,0 and B 1,0. The coordinates of the other points (relative to A and B) are used as variables in conventional multivariate analyses.

Distance Measurements

The usual use of landmarks has been to define the end points of linear distance measurements between pairs of landmarks. Suites of such variables, recorded for many specimens, are then subjected to various types of multivariate analyses. Strauss and Bookstein (1982) point out that conventional suites of distance measurements tend to sample an organism incompletely and with a high degree of redundancy. The measurements are often in parallel directions and partly overlapping so that an expansion in one part of the organism would affect many of the measurements. They proposed measuring distances in the form of a truss (a networklike system of triangles of measurements resembling a bridge truss).

A limitation of any set of distance measurements is that very little information about the geometrical configuration of the landmarks is preserved. For example, conventional multivariate analyses have no way of taking into account the fact that the distance measurements d_{AB} and d_{AC} share the common landmark, A.

Superimposition Analyses

Sneath (1967) analyzed shape differences in two organisms by superimposing their landmark configurations and then scaling and rotating them until they match as well as possible. A least-squares criterion of fit was used (the sum of squared distances between corresponding landmarks in the two configurations, often called the Procrustes distance). This was a simple and elegant way to compare two configurations of landmarks, especially when they are similar. Differences in shape are shown as residuals. Gower (1971a) further developed this approach and expressed the solution in terms of matrix algebra. Gower (1975) generalized this procedure to handle any number of organisms by computing a best-fitting consensus configuration.

Siegel and Benson (1982) demonstrated that the least-squares fitting criterion has a very undesirable property. There may be residuals at most of the landmarks even if the two configurations differ only in the placement of a single landmark. They proposed a resistant-fit procedure that shows residuals only at the landmarks where there are differences—as long as at least half of the landmarks are stable in position. Rohlf and Slice (1990) generalize this method to handle more than two organisms using an algorithm based on the work of Gower (1975). They also generalized both the least-squares and the resistant-fit procedures to include oblique rotations in the fitting process (affine transformations, to adjust for what Bookstein and Sampson [1987] call uniform shape changes).

One does not expect most organisms to differ only with respect to translation, rotation (perhaps including oblique rotations), and scale. It is usually the residuals that are interesting to interpret. Interpretation is easiest when there are only a few large residuals, that is, when the differences between two organisms can be explained in terms of the change in position of just a few landmarks. Chapman (1990) shows a number of examples. When differences affect many landmarks simultaneously, models based on continuous deformation (see below) may yield a simpler explanation.

This is an active area of research. There have been many applications of this approach. There has also been work on the mathematical properties of the Procrustes distance coefficient (see, e.g., Goodall 1991).

Deformation Analyses

This approach is inspired by the work of Thompson (1917). He presented a series of diagrams to show how an image of one species (the reference image) can be transformed to resemble the image of another (the target image). The required transformation was implied by a set of deformed grid lines representing what would happen if one transformed a simple square grid on the reference image onto the space of the target image. In the deformed grid, regions of local compression are indicated by small cells, and regions of expansion by the larger cells. Curved grid lines imply bending. The deformed grid on the target image was estimated subjectively by Thompson (1917). Bookstein (1978) proposed the use of biorthogonal grid analysis as a way to construct a deformation grid objectively. Bookstein (1989) developed a method based on thin-plate splines needed for the construction of biorthogonal grids.

Thin-Plate Spline. Bookstein (1989) proposed a decomposition of a deformation of a configuration of landmarks into a linear part (an affine transformation) and a nonlinear part. The nonlinear part was modeled in terms of the bending of an infinite uniform thin metal plate. The plate is bent by applying force at the positions of the landmarks for the reference organism until the vertical coordinate of the plate at each landmark in the reference organism is equal to the x and then to the y coordinate of the corresponding landmarks in the target organism. The amount of effort required to bend the plate can be computed as a *bending energy.* This quantity is a function only of the nonlinear part of the deformation. It is not influenced by translation, rotation, or shearing of the plate since these events correspond to a rotation or a tilting of the plate rather than a bending.

The model of the bending of a metal plate can be made into an interpolation formula that transforms coordinates, (x,y), in the reference into coordinates, $(x',y') = [f_x(x,y),f_y(x,y)]$, in the target. The function $f_x(x,y),f_y$ is the height of the metal plate when the x coordinates of the target organism are used, and $f_y(x,y)$ is the height when the y coordinates are used. These functions are easy to compute. While this description (and the discussion of principal warps in the next section) is in terms of two-dimensional configurations of landmarks, the methods have been generalized to work in one- and three-dimensional spaces. An example is shown in Figure 25.1. This figure was generated by TPSPLINE, an IBM PC–compatible program included in the software distributed with Rohlf and Bookstein (1990).

Note that bending energy cannot be used as a measure of morphological distance between two landmark configurations. Bending energy is nonnegative and is zero for a pair of identical configurations. It is not, however, symmetric. The amount of energy required to bend configuration 1 into configuration 2 is not, in general, the same as that required to bend configuration 2 into configuration 1. Bending energy also has the property that it will be larger for displacements in landmarks that are close together than it will be for same size displacements in landmarks that are farther apart. These properties of bending energy do not seem to be reasonable for a measure of developmental or evolutionary "effort."

Principal Warps. Bookstein (1989) also shows that it is possible to decompose the thin-plate spline transformation into a series of components based on the eigenvectors of what he calls a bending energy matrix. Bookstein (1990) calls these eigenvectors the principal warps of the initial landmark configuration. Each principal warp can also be viewed as a thin-plate spline. The eigenvector with the largest eigenvalue corresponds to small-scale displacements of the landmarks (usually a function of the relative placement of the closest pair of landmarks). The eigenvector with the smallest eigenvalue corresponds to the largest-scale deformation.

The coefficients of a thin-plate spline for a particular target configuration can be expressed as a weighted linear combination of the coefficients of the principal warps based on the reference configuration. The overall bending energy also can be decomposed into contributions associated with each principal warp. In this way one is able to attribute the observed differences between two landmark configurations to changes at different scales.

Figure 25.2 shows examples of principal warps for the thin-plate spline shown in Figure 25.1. The magnitude of each warp is shown proportional to its contribution to the overall deformation.

Using the principal warps based on a fixed reference configuration, one can use more than one specimen as a target configuration. For each specimen one can compute the weights needed to combine the principal warps in order to obtain its thin-plate spline transformation and thus define a feature space in which each target configuration corresponds to a point. Various types of multivariate analyses can be used to study variation in this space. It is interesting to note that the results of some multivariate analyses (e.g., linear discriminant function analysis) can be displayed by using the resulting coefficients as coefficients for a thin-plate spline.

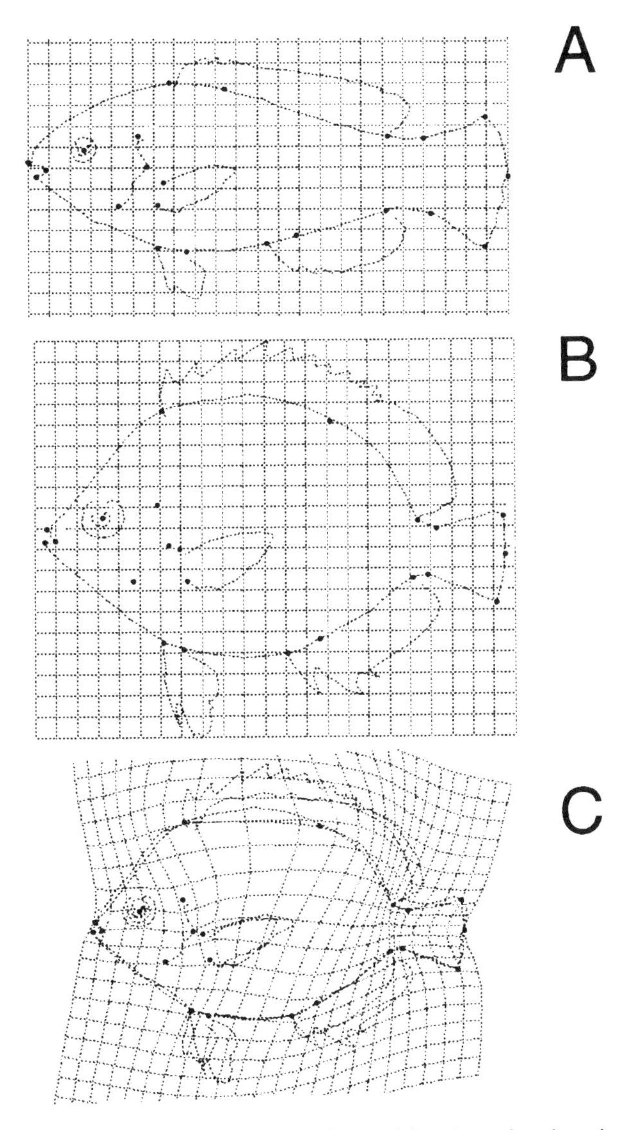

Figure 25.1. Fitting of a thin-plate spline to digitized drawings taken from figures 519 and 520 in Thompson (1917). Reference grids and landmarks were added to the figures. (a) *Scarus* sp. from figure 519. (b) *Pomacanthus* from figure 520. (c) Thin-plate spline transformation of *Scarus* sp. superimposed on drawing of *Pomacanthus* (bending energy is 0.659).

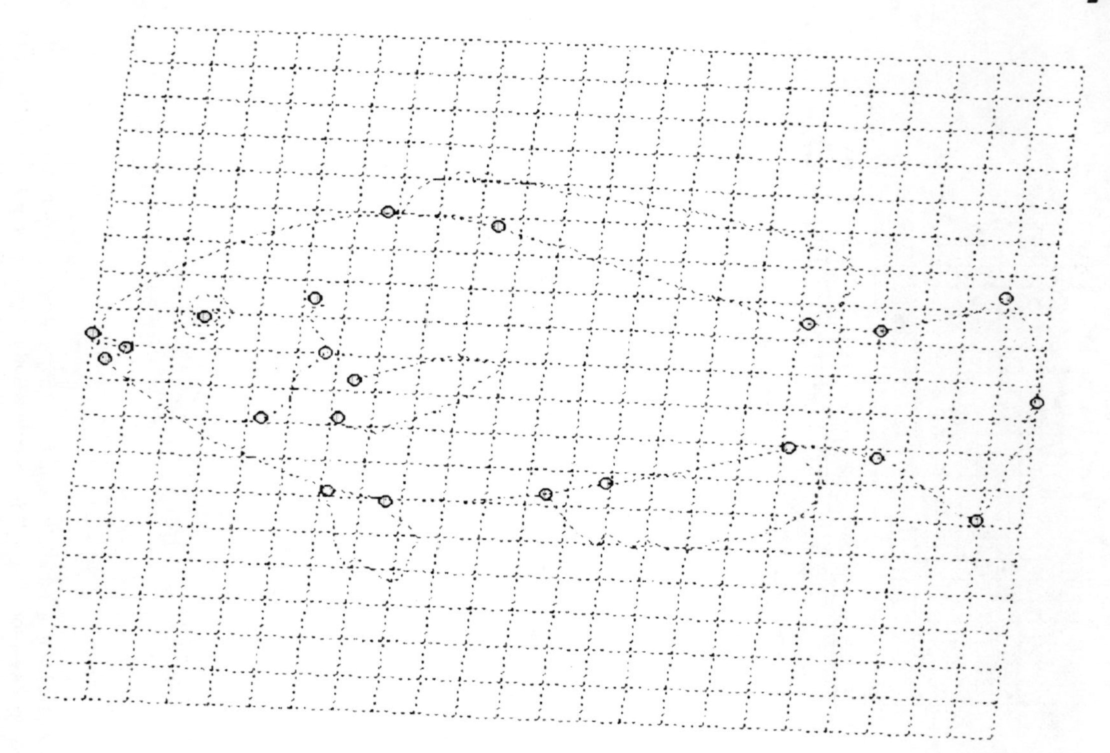

Figure 25.2a. Principal warps of thin-plate spline transformation from Figure 25.1. Affine component.

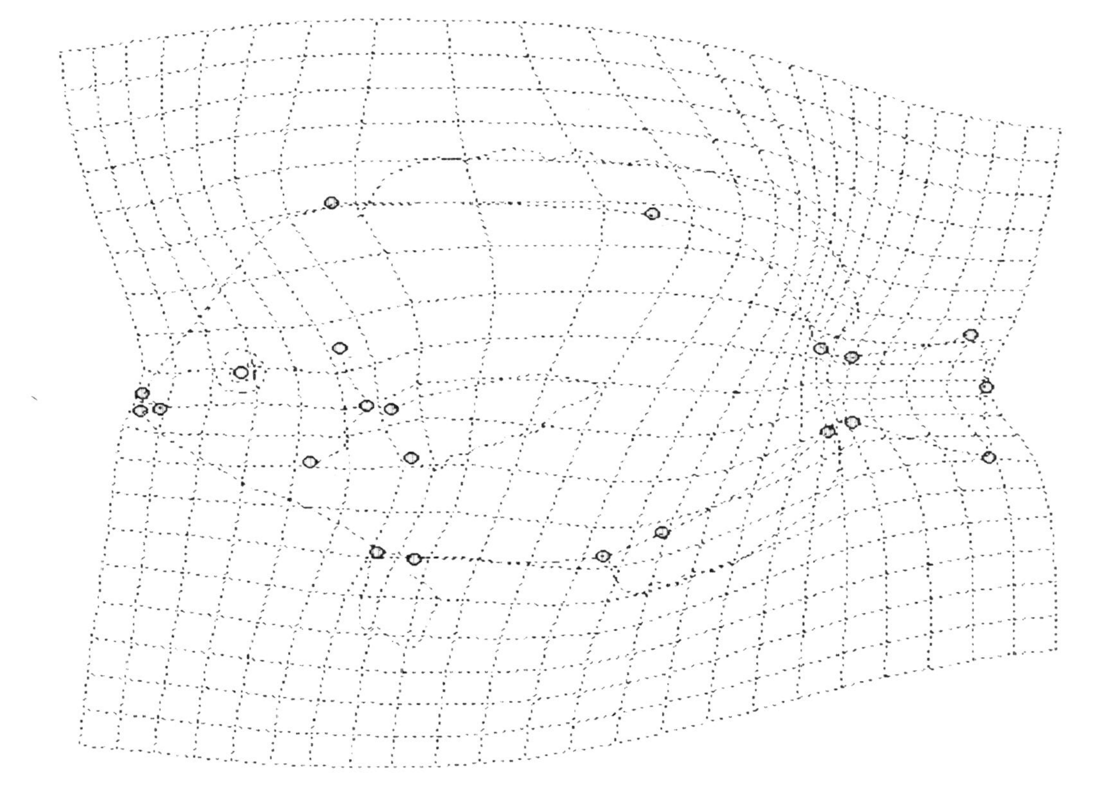

Figure 25.2b. Principal warps of thin-plate spline transformation from Figure 25.1. Nonlinear (strictly inhomogeneous) component.

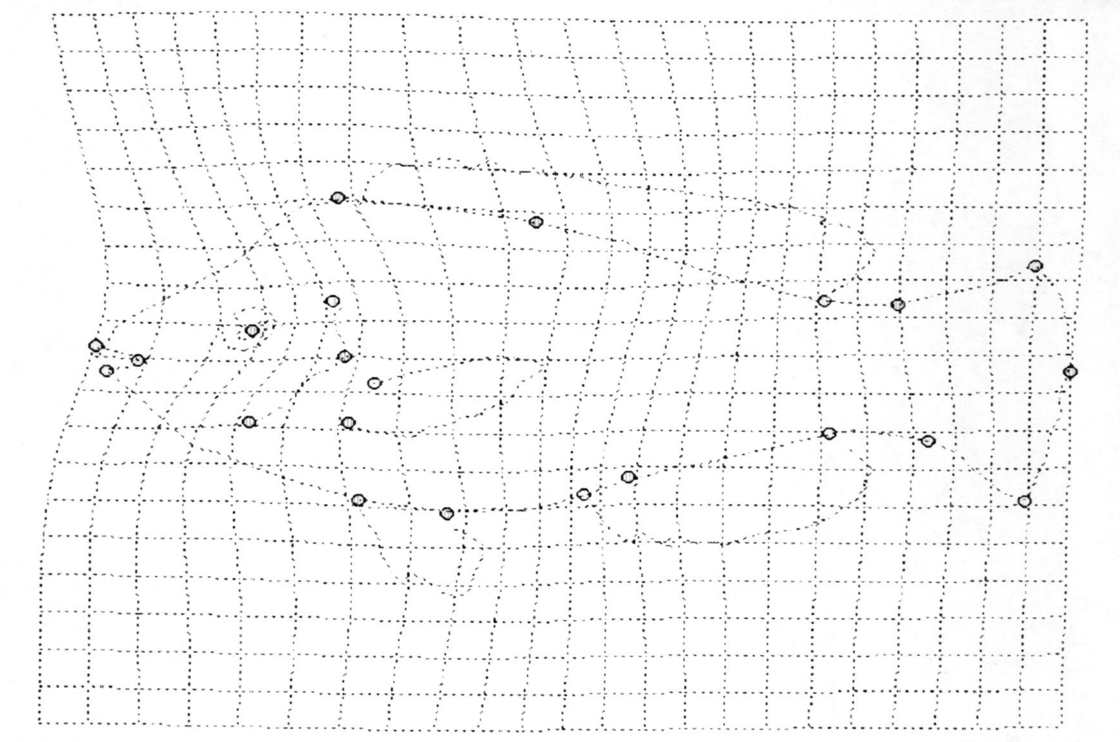

Figure 25.2c. Principal warp with highest bending energy.

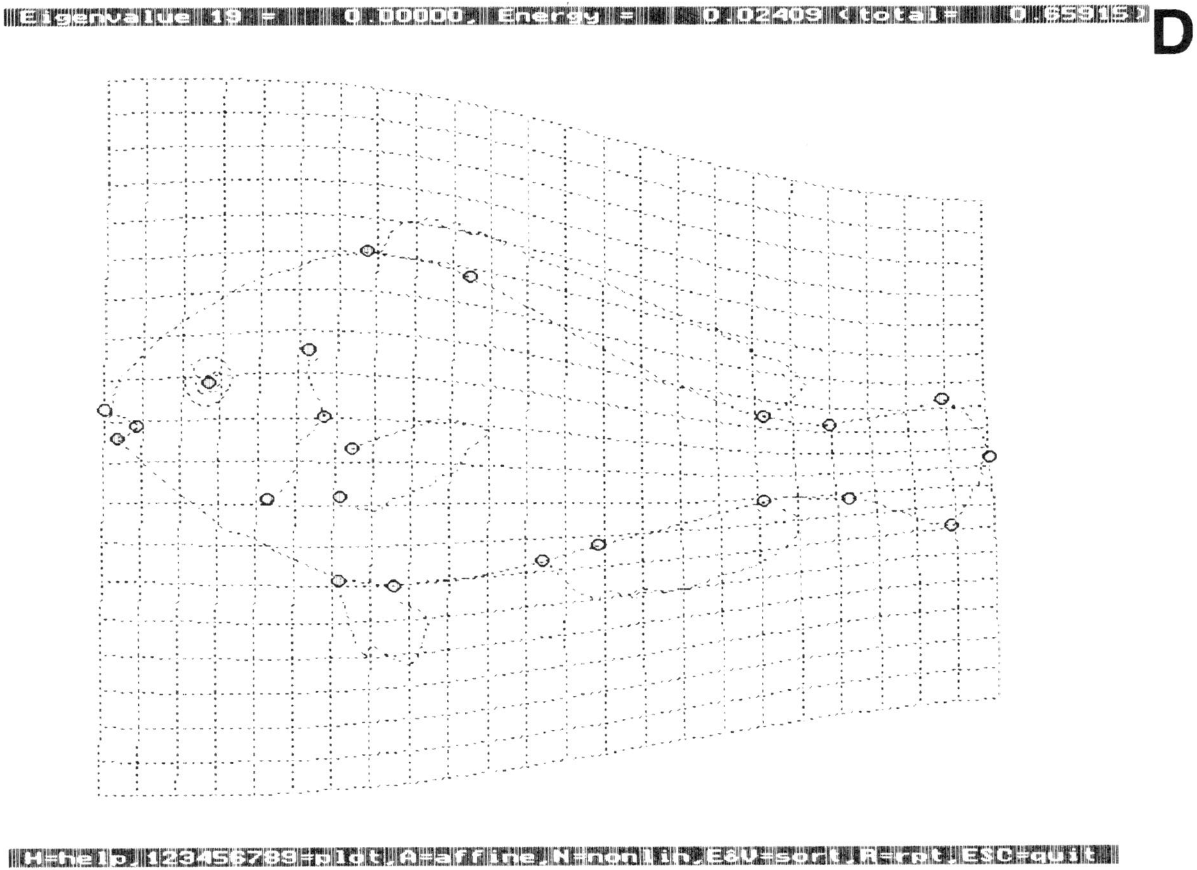

D

Eigenvalue 19 = 0.00000, Energy = 0.02409 (total= 0.65915)

H=help, 123456789=plot, A=affine, N=nonlin, E&U=sort, R=rpt, ESC=quit

Figure 25.2d. Principal warp with the smallest eigenvalue—the component with the largest scale.

The choice of the reference configuration is important. In a study of covariation in a sample from a single population, one source of a reference configuration might be the consensus configuration from a superimposition analysis (see above). In other studies one might use an earlier developmental stage or a hypothesized ancestor as the source of a reference configuration.

Thin-plate splines and principal warps are relatively new and have been used in only a few practical applications. This seems to be a powerful technique. One is now able to capture complex changes in landmark configurations and represent them both as points in a feature space and as continuous deformations in the space of the original specimens. A limitation of the technique is that the thin-plate spline function is based on the properties of a thin metal plate—it is not based on a biological model. Thus one cannot expect each principal warp to correspond to a biological process. One merely hopes that interesting patterns of variation among the landmarks can be discovered by analyzing variation among points in the feature space (e.g., relative warp analysis). The dependence of the results on the selection of a reference configuration is not a problem for its use for identification. It would, however, be a serious problem in any study concerned with classification or phylogeny estimation.

Conclusions

There are many methods than can be used to reduce the coordinates of points along outlines or of morphological landmarks to a small number of descriptive variables. Many of the methods are equivalent in the sense that they capture the same changes in the morphology. They differ, however, in the way in which they represent this information. In particular, different methods of feature extraction do not produce feature spaces that are equivalent in the sense that the same pairs of points are close together or are far apart. This is a serious problem for some applications. Depending upon which method of feature extraction is used one will obtain different results and come to different taxonomic conclusions. This problem is compounded by the fact that we do not yet know enough to be able to select the methods objectively. Of course, this is not just a problem with image analysis. There is an analogous problem when conventional characters are used in a taxonomic study. Different suites of characters usually give somewhat different results. It is usually unclear why one should prefer characters taken from one organ system or one developmental stage rather than another. The usual hope is that if one uses a large number of variables then these discrepancies will tend to average out and one will approximate an overall estimate of dissimilarity among organisms. When working from captured images one can see that adding more data (additional landmarks or more points along an outline) will not have the same beneficial effect. Different feature spaces will give different results. One can even compute the consequences of using different features (e.g., Rohlf 1992)

Despite the theoretical problems, these techniques are quite popular. One often sees titles such as "The Application of Fourier Analysis to . . ." or "The Application of Cubic Splines to . . ." Unfortunately, only rarely is the investigator able to justify why the selected method was used.

Acknowledgments

This work was supported in part by NSF grant BSR-8306004. This chapter is contribution number 789 from the Graduate Program in Ecology and Evolution, State University of New York at Stony Brook.

Discussion

Walker: Is there a rational way of choosing clustering algorithms?

Rohlf: The selection of one clustering method over another implies the acceptance of one definition of "cluster" over another. Users should not, for example, try several different clustering methods in order to find one that gives them the results they expect. Methods should be selected because of the properties that one wants the clusters to possess. This problem does not seem as intractable as the problem of selection of features discussed in the talk. By an empirical study of the data (usually by an ordination analysis) one can often determine whether or not there are any obvious clusters and whether they are round, elongated, curved, etc. This can often suggest what kinds of cluster analyses would be most appropriate.

——: I think that allometric growth is the model that best corresponds to the simple shell growth of gastropods. It can be described in terms of distances measured from point to point.

Rohlf: In the allometric growth model, distances between different pairs of landmarks grow at different rates. This model in itself does not provide enough structure to describe the geometrical relationships among a set of landmarks, and thus it does not seem to lead to a more "natural" approach to feature extraction.

Marcus: The growth models that appeal to me most are those used in conchology because they reflect how the growth of shells occurs. But the corresponding model in the vertebrate skull is very complicated, with growth and resorption etc., between independent bones. That would be a very interesting model it if could be done.

Rohlf: If there is to be a real answer to this problem, I believe that it will come from considerations such as these. The problem is that they are complex and have to be discovered experimentally, not by concocting simple models that seem reasonable based on the geometry of the structure. Ackerly (1990), for example, rejects the geometrically appealing model for gastropod shell growth of Raup (1966).

26

MorphoSys: An Interactive Machine Vision Program for Acquisition of Morphometric Data

CHRISTOPHER A. MEACHAM

The use of modern techniques for morphometric analysis requires capturing large amounts of shape data from specimens. Currently, the most common procedure is to capture data by hand. The specimens must be taken out of collections, manipulated, measured by hand, and these measurements transcribed and manually entered into a computer. This work is labor intensive and routine, but must be done with great care and precision to yield good results. Many of these functions can be performed with current machine vision software and hardware at much greater speed and, many times, improved accuracy. I describe here a software and hardware system that makes the capture of morphometric data much easier. The software is called MorphoSys and was developed in collaboration with Thomas Duncan at the University Herbarium. MorphoSys, besides illustrating the use of current techniques, reveals directions for further development of a unified software environment for morphometric studies.

Image Capture

The most economical means for machine vision at the present time uses image capture based on television technology. Standard television cameras and monitors are relatively inexpensive because of the large numbers produced for the consumer electronics industry and for a wide variety of other purposes. Black-and-white television cameras produce a standard analog video signal with specific characteristics. This standard is RS-170 in North America (NTSC is

Advances in Computer Methods for Systematic Biology: Artificial Intelligence, Databases, Computer Vision, ed. Renaud Fortuner (Baltimore: Johns Hopkins University Press).

the corresponding color standard). Europe and other regions use other video standards. The conversion from analog video signal to a matrix of digitized pixels is performed by a specialized computer board that can be plugged into the bus, or main internal communication path, of a small computer. These boards are often called *frame grabbers* because they can capture and store, in digital form, a frame of video from the incoming analog signal. The most inexpensive boards for digitizing images accept a standard video input although more expensive boards can accept other video signals. This frame grabber accomplishes the basic task of capturing an image from the video source.

MorphoSys runs in a PC environment and uses the PCVISIONplus frame grabber made by Imaging Technology. The PCVISIONplus digitizes the video signal in real time, that is, one entire frame of 640 by 480 pixels every thirtieth of a second (in North America). The PCVISIONplus makes the image available to the software, MorphoSys, by means of the microcomputer bus. MorphoSys can examine and change the contents of pixels, thus locating edges of objects or writing a cursor on the image. The PCVISIONplus also performs the task of reconverting the digitized image, with modifications made by MorphoSys, to an analog television signal for display on a television monitor that is separate from the microcomputer monitor. Many functions of the PCVISIONplus board can be controlled by software. For example, MorphoSys can instruct the board to cease digitizing images after the current image is finished, thus freezing the image in the board buffer. Such freezing of the image is necessary during the brief time that is required to locate and follow the edge of an object. MorphoSys is a real-time machine vision program. It works with a continuously changing image that resides in memory on the frame grabber, rather than analyzing previously captured images. For this reason MorphoSys functionality depends rather heavily on the hardware characteristics of the frame grabber board. Currently, the PCVISIONplus and the related VISION-AT OFG are the only frame grabbers supported, but drivers could be developed for other real-time frame grabber boards that have the same set of basic hardware functions.

MorphoSys Functionality

The goal of MorphoSys is to give the user a tool to manipulate basic shape information, both interactively and automatically. MorphoSys allows the user to capture, edit, archive, recall, display, print, and measure shapes of organisms. MorphoSys has two logically independent functions. The first is feature extraction, which is done in part under interactive user control, but may be augmented automatically. The second is the actual measurement of captured features, which is done automatically under the control of commands supplied by the user. Much flexibility is gained by keeping these two basic functions separate.

Once the extracted features have been digitized and saved on disk, they can be recalled as needed for measurement without recourse to the original specimens. There are a variety of ways of quantifying shape that are used by different methods of morphometric analysis. Some depend on digitizing corresponding landmark points on each shape. Some require an entire outline or contour of a specimen. Both of these kinds of features can be captured and manipulated by MorphoSys. As new techniques for morphometric analysis are developed, users can return to previously captured files to perform different kinds of measurements with very little effort.

Feature Extraction

MorphoSys permits the user to interact with and to control the process of feature extraction. The extracted features are the contours of objects and selected points. When capturing a contour, the user sees a threshold image on the image monitor along with a cursor controlled by a mouse. The user places the cursor near the desired object and presses a button on the mouse. MorphoSys locates the edge of the object and, by examining pixels near the edge, digitizes the contour, typically in one second or less. The user may then accept or reject the contour by clicking the mouse. If the contour needs to be altered, because of damage to the specimen or problems with the image, the user can interactively touch up portions of the outline by using the mouse. Selected points can be captured in different ways, both manually and automatically. To capture selected points that lie on the contour of the object, a user-directed mouse moves the cursor precisely along the outline. Clicking the mouse identifies the current location of the cursor as a selected point. Selected points are indicated on the screen by numbered dots. It is also possible to select points off the contour manually by viewing a gray-scale image and clicking the mouse at the desired location.

A number of contours and selected points can be captured from a single image. Each contour and selected point has a number assigned to it at the time it is captured. These numbers are used in commands to indicate features for processing. Each set of captured features, the contours and selected points, from a single specimen is stored as a single frame. The user will normally capture features from one specimen after another. As each frame is completed it is stored on disk in a plain ASCII text file called a *frame file*. A frame in MorphoSys is not a full image. It is just a skeleton of features extracted from an image. Because of this, a frame takes up much less disk space than the image from which it was extracted. Typically, a frame requires less than two kilobytes of disk space, but frames containing several contours may require three kilobytes or more. Full gray-scale images can also be written to disk and recalled.

Contours are processed internally and saved in the frame file as chain codes, which are simply sequences of digits that indicate the direction of the contour. In MorphoSys, the digits 0 through 3 are used to indicate stepwise moves about the contour in four directions. The digit 0 indicates a step in the positive x direction; 1 a step in the positive y direction; 2 a step in the negative x direction; and 3 a step in the negative y direction. This format allows rapid processing and precise measurement. Points are saved as positions in the x, y coordinate space of the frame grabber.

Table 26.1 shows a single frame, containing one contour and three selected points, written as text. The *F* on the first line indicates the beginning of a frame. This is frame number 1, which has an identifier *517*. The numbers after the colon give the year, month, day, and time that the capture of the frame was begun. The last two numbers of the third line give the number of pixels per millimeter in the x and y directions, respectively, for standardizing measurements. The *C* on the fourth line indicates the beginning of contour 1. Line 5 shows that this contour begins at coordinates (37, −1) and contains

Table 26.1. A MorphoSys Frame, with One Contour and Three Selected Points

```
F 1 '517'
  1.24 0002
  512 480 2.95000 3.64000 'mm'
C 1 : 1988 1 5 13 38 44.53
  37  −1    686
0010001000000010000000000010000000000000003000030003003003003033303030303030300000100100101011112121212
112121212112111121000300300300303030030010010010110111110111211211121112121212122121221221222122222
222222223222222223222222322222212221221212212212212212212212212212212212212212212212212212212122122121
221212212122121212212121221212121221212122121212212212122121221212212122121212212121221212121212212121
212122121212122232212122222322323232323323233233323333233332333333332333333333303323333333033333333
333033333303333033330323033330333333330303303030330303032303303330303303303330303303000333030303
300300300003000000300003000000000000000001000100100101001001010010100100101001001001001001001000010
P 1
  −164  228
P 2
  −68  −11
P 3
  −153  −182
```

686 steps. The steps follow. At the bottom, selected points 1 through 3 are indicated by *P*, and the coordinates of each selected point follow.

Because a MorphoSys frame file is a text file, it can be transmitted by electronic mail easily and can be processed by programs other than MorphoSys. Other programs can be developed to produce files in MorphoSys frame file format.

Measurements

After the shape information is captured in the form of a frame file, the user can go on to measurement. Measurement is controlled by commands entered in a text file by the user (see chapter appendix). The simplest commands determine the most basic parameters of shape: the coordinates of a point, the distance between two points, the angle determined by three points, the length of a contour, and the area enclosed by a contour. The command *D 1 2* requests the distance between selected points 1 and 2. The command *A 1* requests the area enclosed by contour 1. Measurements are written to an ASCII text file that can be read directly by most statistics packages, for instance, SAS.

Besides these basic measurement commands, MorphoSys has a set of commands that calculate new selected points or contours based on geometric relationships among existing selected points or contours. For example, *c 1 4* requests that selected point 4 be placed at the centroid of contour 1. Two selected points can be used to define a line, and other points can be defined from these two lines. The INTERSECTION command, for example, places a new selected point at the intersection of two lines defined by two pairs of points. The command *i 1 2 3 4 5* will place selected point 5 at the intersection of two lines; the first line is defined by selected points 1 and 2, the second line, by points 3 and 4. This kind of command is necessary when the user wishes to calculate the angle between the line 1–2 and the line 3–4, but the point of intersection lies off the screen. The INTERSECTION command can be used to calculate the position of the intersection of the two lines as selected point 5, and selected point 5 can be used as the vertex in calculating the angle even though it may lie far off the screen. The coordinates of a selected point may range from—32,768 to 32,767. The FARTHEST command finds the point on a contour that is farthest from a specified selected point. The command *f 2 1 3* labels as selected point 3 the farthest point on contour 2 from selected point 1.

When the results of a command become available, other commands can use these results as their own operands to perform their function. For example, we might want to measure the length of a skull. We can determine the location of the tip of the snout by finding the point farthest from the centroid of the contour. We can then find the point on the contour farthest from the tip of the snout, and calculate the distance between these two points, which gives us the length of the skull. This would be accomplished by this simple command file:

c 1 1	(Find the centroid of the skull, selected point 1.)
f 1 1 2	(Find the tip of the snout, selected point 2.)
f 1 2 3	(Find the other end of the skull, selected point 3.)
D 2 3	(Calculate the length of the skull.)

An important aspect of the MorphoSys command language is that it is declarative, not procedural. That is, the language specifies what is to be done in terms of abstract geometric relationships, but it does not specify how it is to be done. The ordering of the commands in the command file has no influence on the calculations.

In some cases, for example, with fossil specimens, there may be problems with automatic location of selected points. If the tip of the snout is chipped, selected point 2 may not be located correctly by the commands. MorphoSys supports an editing mode where a frame file is read and a new, edited frame file is written. By making the first pass in this mode, the action of the geometric commands can be monitored by the user. If point 2 is misplaced, the user can intervene with a mouse and move point 2 to its proper location. All of the other geometric commands will be recalculated automatically when point 2 is moved, and point 3 will be moved in response. When the frame is written to the frame file, it will have selected points 1, 2, and 3 saved with it. A second pass on this edited frame file with just the command *D 2 3* will measure all the skulls correctly. In this way, MorphoSys provides an automatic tool for efficient measurement at the same time that it allows the user to control and correct its actions.

Some commands perform complex operations. The CONVEX HULL function produces the contour of the smallest convex polygon that covers the original contour. This function can be used, for example, to quantify the lobing of a dissected leaf. The most recent version in prototype incorporates calculation of elliptical Fourier contours and medial axes of contours. A summary of MorphoSys commands is given in the appendix.

MorphoSys will provide the hardcopy of a frame file on a PostScript capable laser printer (Fig. 26.1). The user can specify the number of rows and columns of frames to place on a page and whether contours are to be filled with shading. If a command file is active, lines representing measurements can be drawn (Fig. 26.2). A hardcopy of the original digitized features is especially useful at later stages of analysis. If a question arises about the validity of particular measurements, it is easy to verify that the selected points were located correctly. If particular specimens are of interest, it is easy to examine them visually.

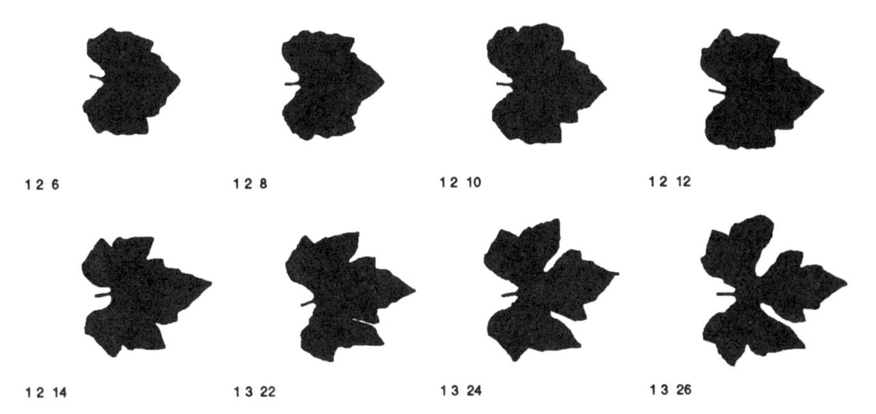

Figure 26.1. Filled cucumber leaf outlines (digitized by Cindi Jones). Frame identifiers indicate treatment, plant, and leaf numbers.

Further Development

Although it is clear that MorphoSys succeeds in many respects, there are many aspects of its use that can be improved.

1. Image enhancement functions are absent, and clear contours can be captured only from opaque specimens with well-defined edges. Stanley Dunn (this volume, chap. 28) discusses techniques that would greatly improve the performance of MorphoSys in this area.

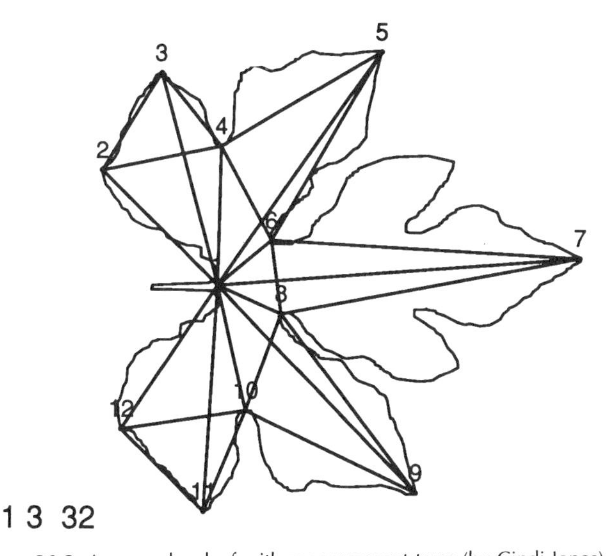

Figure 26.2. A cucumber leaf with measurement truss (by Cindi Jones).

2. The command language is cryptic and error prone. A more structured, intelligible language is needed, perhaps with extensions toward a procedural form. An interactive mechanism for creating a command file by indicating selected points and contours with a mouse would be a great time saver.
3. It is difficult to find and display particular frames in a frame file.
4. Morphosys is limited to two-dimensional projected outlines rather than three-dimensional forms.

One can imagine a unified morphometric data acquisition and analysis package that would include aspects of the current MorphoSys combined, on one hand, with a database management system for frames, images, and measurements and, on the other, with specialized data analysis software. Being able to recover the original shapes in the midst of data analysis would remove some of the "black box" aspects of analyzing data. By combining these tools, it would be possible to retrieve frames on the basis of the analysis and to include shapes graphically in data analysis output (e.g., Rohlf and Archie 1984). Graphical output would ease the interpretation of results by colleagues. Information about shapes is often best communicated by showing the shapes. After all, a picture is worth a thousand words.

Acknowledgments

I thank users of early versions of MorphoSys for their encouragement and advice, especially Howell Daly, Cindi Jones, Stuart Poss, Bruce Riska, F. James Rohlf, and Dennis Slice. The development of MorphoSys was supported by the University of California and by a grant from IBM Corporation.

Appendix. Morphosys Command Summary

The following is a brief description of MorphoSys command functions.

The Measurement Commands

COORDINATES: Calculates the x,y coordinates of a point. The result is a pair of numbers.

DISTANCE: Calculates the distance between two points. The result is a number.

RATIO: Calculates the ratio of the distance between one pair of points to the distance between another pair of points. The result is a number.

PERIMETER: Calculates the perimeter length of a contour. The result is a number.

ANGLE: Calculates an angle based on three points. The result is a number.

AREA: Calculates the area enclosed by a contour. The result is a number.

The Geometric Commands

INTERSECT: Finds the point that is at the intersection of a pair of lines defined by two pairs of selected points. The result is a new selected point.

INTERPOLATE: Finds the point that lies along a line defined by two selected points at a specified proportional distance between them. The result is a new selected point.

PROJECT: Finds the point that is the projection of a selected point onto a line defined by two selected points. The result is a new selected point.

GEOMETRIC POINT: Places a selected point at a position defined on the basis of three existing selected points, an angle, and a proportional distance. The result is a new selected point.

CENTROID: Places a point at the centroid of a contour. The result is a new selected point.

NEAREST: Finds the point on a contour that is nearest to some selected point. The result is a new selected point.

FARTHEST: Finds the point on a contour that is farthest from some selected point. The result is a new selected point.

NEAR BASELINE: Finds the point on a contour that is nearest a baseline defined by two selected points. The result is a new selected point.

FAR BASELINE: Finds the point on a contour that is farthest from a baseline defined by two selected points. The result is a new selected point.

AXIS: Finds the two points on a contour that are farthest apart. The result is a pair of new selected points.

CONTOUR INTERSECTION: Finds the point that lies at the intersection between a line determined by two points and a contour. The result is a new selected point.

SCAN: Similar to the CONTOUR INTERSECTION command above except that the line is specified by a selected point defining an orientation, an angle with respect to the orienting point, and a selected point defining an origin. The result is a new selected point.

LINE CONTOUR: Generates a contour piece that is the straight line segment between two selected points. The result is a new contour.

EXCISE CONTOUR: Produces a new contour that is a piece of an existing contour excised between two selected points. The result is a new contour.

CLIP CONTOUR: Produces a contour that is an existing contour that has been clipped by substituting a straight line segment for the part of the contour that lies between two selected points. The result is a new contour.

JOIN CONTOURS: Joins two contours to produce a new contour. The result is a new contour.

CONVEX HULL: Produces a contour that is the smallest convex polygon that will cover a specified contour. The result is a new contour.

27

Image Processing in Fungal Taxonomy and Identification

GLEN NEWTON AND BRYCE KENDRICK

Fungi are a kingdom of heterotrophic, eukaryotic organisms which may be parasitic, saprophytic, or mutualistic and are important both in nature and in their effect on humans. Fungi are responsible for 70 percent of all crop damage (Kendrick 1985). Millions have died in famines caused by fungi, from the potato blight (*Phytophthora infestans*) in Ireland in the late 1840s to the Great Bengal Famine of 1943, in which *Helminthosporium oryzae* all but destroyed the rice crop (Deacon 1984). Even when the crops have successfully resisted fungal attack in the field, fungi specialized in attacking stored crops often destroy them in storage or contaminate them with dangerous mycotoxins. Fungi infect and destroy most economically important plants (including tobacco, fruit and fruit trees, grain, nuts, lumber, and ornamental trees) as well as cause serious human diseases, some of which can be fatal (bronchopulmonary aspergillosis, aspergilloma, histoplasmosis, candidiasis, mucormycoses) (Deacon 1984; Kendrick 1985).

Fungi have also have a positive side and are important in the production of food (bread, wine, beer, cheeses, miso, tempeh, soy sauce), chemicals (organic acids and enzymes, plant hormones), and pharmaceuticals (alkaloid contractant, antibiotics, etc.). Most higher plants form symbiotic relationships with certain root-colonizing fungi, mycorrhizae, that are necessary for the success and often the survival of the plant (Harley and Smith 1983; Kendrick 1985).

The Identification of Fungi

The reproductive part of a fungus must be available for its identification because its vegetative part (hypha) contains identification characters valid

---

Advances in Computer Methods for Systematic Biology: Artificial Intelligence, Databases, Computer Vision, ed. Renaud Fortuner (Baltimore: Johns Hopkins University Press).

only for higher classification levels (i.e., narrow, septate hyphae separating the division Dikaryomycota [kingdom Eumycota] from the division Zygomycota [kingdom Eumycota], and the protoctistan fungi of kingdom Protoctista).

The life history of fungi includes strategies assuring their dispersal for the colonization of new substrates. Fungi have very specialized reproductive structures that produce spores of many different shapes and sizes. These spores are the agents of dispersal. The spores and the reproductive structures producing them are diagnostic for each individual fungal species. About 70,000 species of fungi have so far been described, with at least twice that many remaining to be discovered (Kendrick 1985).

Despite the large number of taxa, we are still dependent on a small set of diagnostic characters: this makes for difficulties in identification. To illustrate this, we will examine how the identification of an organism from another kingdom might proceed. Here, the characters that differentiate groups are often qualitatively different from one another and do not all revolve around one common theme.

For example, in kingdom Animalia, the branches Radiata and Bilateria are differentiated by radial versus bilateral symmetry; the classes Aves and the Mammalia by feathers versus hair, egg-laying versus live young, etc.; the species *Falco perigrinus* (peregrine falcon) and *Falco tinnunculus* (kestrel) by size, coloration, diet, and mode of flight (Cerny 1984). While moving down the classification tree of the organism, a variety of characters are used at each level, and these characters are usually different between levels. In contrast, although the presence and nature of hyphal septa and wall chemistry can discriminate the divisions Zygomycota and Dikaryomycota, as well as the dikaryomycotan subdivisions Ascomycotina and Basidiomycotina, below this level the reproductive structures are virtually the only diagnostic characters available. This concentration makes for a classification that quickly converges to a very specific set of features.

Further complicating fungal taxonomy and identification is the anamorph-teleomorph dichotomy (Kendrick 1979). Fungi often have a sexual form (teleomorph) and an asexual form (anamorph), which can occupy very different, temporally and spatially distinct niches, as well as tending to have very different reproductive structures and strategies. In many cases, the two different life-forms of a fungus have not yet been connected, and each form is separately described as an independent species. If it is a teleomorph (sexual form), it is classified and placed in its proper subdivision (i.e., Basidiomycotina); however, if it is an anamorph (asexual form) and no connection can be made to the appropriate teleomorph, the fungus is classified into an Anamorph-Class and form-genus (i.e., Anamorph-Class Hyphomycetes, genus *Penicillium*), pseudo-taxa created for convenience (Carmichael et al. 1980, Kendrick 1980). This is an extra level of complexity that does not exist for other groups of organisms.

With the taxonomic difficulties in mind, as well as the success and ubiquity of this group, it can be appreciated that an automated system for identification that could carry out a preliminary screening of unknown organisms would have significant practical value. It could reduce the number of routine identifications that must be carried out by professional mycologists and would free them for more much-needed basic taxonomic research.

Although several biomedical image-processing systems have already been developed (red blood cell identification [Bacus and Weens 1977; Zajicek and Shohat 1983], blood vessel detection [Stevenson et al. 1987], neuron boundary locating [Selfridge 1986], ventricle boundary detection [Duncan 1987], brain tissue categorization [DeLeo et al. 1985], etc.), a general spore-identification system would differ in having to deal with the much greater range in sizes, shapes, colors, textures, and complexities of fungal spores. Building a model-based feature extraction system for anything but a trivial number of spore types would be too generalized to successfully deal with the problem. This is why a more generalized approach is seen as more effective, though not necessarily more easily designed and implemented.

An Image-processing System for Identification

Research at the University of Waterloo

The primary goals of the computer taxonomy and identification research in the Mycology Lab at the University of Waterloo are to: (1) examine the feasibility of the construction of an automated or semiautomated expert system or pattern recognition system using image processing for the identification of fungi from fungal spores and to build a prototype, and (2) examine the methods used and made available by the automated identification with the hope that some of the lower-level identification methodologies (i.e., feature extraction, quantitative shape analysis) can be used directly by fungal taxonomists. This chapter examines some of the methods being developed for the segmentation of fungal spore contours (outlines) and internal structure in raster images of spores.

Hardware

All research and development have been in C, on a Sun 386i with 8 MB of memory, 91 MB local disk, 500 MB remotely mounted disk, and an 8-bit 1024 by 768 color display. Images (512 by 482 pixels) are grabbed using an AT&T Targa M8 frame grabber, connected to a standard NTSC video camera mounted on a Nikon phase-contrast compound microscope (phase-contrast microscopy is often necessary due to the transparent or hyaline nature of many fungal structures).

Software for the display of images in the SunView and X11 windowing environments, numerous filters, and image format converters have also been developed. The image format used is the IM image format (Paeth 1986), developed at the Computer Graphics Lab, University of Waterloo.

Image Processing

The first step in almost every image-processing system, both biological and digital, is the separation of the area of interest from the rest of the image, called segmentation. As the area of interest can be different things depending on the observer's interests, segmentation is an extremely context-sensitive process. So, to take an engineering view and say that segmentation is the separation of signal from noise would be very naive: one person's noise is another person's data.

Segmentation is generally classified into global (region-oriented) and local (pixel-oriented) methods. Local methods examine simple low-level features and build them up into more complex structures. Global methods depend on larger-scale features or collections of features to discriminate the various regions in an image. The method chosen to segment the fungal spore images is a local method, using an edge detector to measure the likelihood that a pixel lies on a contour that separates the area of interest from other areas. This information is then used to construct contours using linking and graph searching methods.

Segmentation. The goal of segmentation of images of fungal spores is to extract the outline or contour of the spore, so that the size and shape (using various methods) of the spore might be measured. The internal structures of spores are also important, and they too have to be segmented. The size and shape as well as the relations with the spore contours and other internal features must be measured.

The quality of the images grabbed from the video camera by the frame grabber of the subject on the stage of the microscope is quite poor (i.e., noisy, poor contrast), and so a certain amount of preprocessing is needed. After experimenting with a number of techniques including smoothing, sharpening, contrast enhancing filters, etc., we found that transforming several grabbed images of the same subject by their principal components (Gonzalez and Wintz 1987) and using the transformed image corresponding to the largest eigenvalue produced an image that contained less noise and performed better in subsequent processing than the single images.

Filtering. The edge detector is then applied to every pixel in the transformed image returning a value reflecting the amount of contrast in the neighborhood of the pixel. This local contrast is associated with the shift in gray scale when one moves from one region in the image to another. The edge detector used here is a discretized Sobel edge detector (Gonzalez and Wintz 1987). This detector uses four Sobel masks, which return an edge strength and associated edge direction. If the mask returns a negative strength, this indicates a negative direction, producing a total of eight directions at 45° intervals. The largest edge strength and its direction are taken to be representative for each pixel.

The edge detector produces information as to how close a pixel is to a discontinuity in gray levels indicating a crossing of a boundary separating two regions in the image. As one approaches a discontinuity, the edge detector produces increasing edge strengths until the edge strength is maximized when the pixel being examined is on the discontinuity (or at least closer than any other pixel), and then starts decreasing as one continues into the next region (Fig. 27.1). The ideal edge point is therefore at the apex of a curve perpendicular to the edge direction and greater than its neighbors perpendicular to the edge direction. The points that do not meet this criterion are discarded. Contours also have a certain amount of smoothness associated with them, so edge pixel candidates that are not relatively continuous (±45°) with their neighbors in the line of the edge direction are also discarded, as are points having an edge strength below a certain threshold, about 50 (found empirically) for spore images.

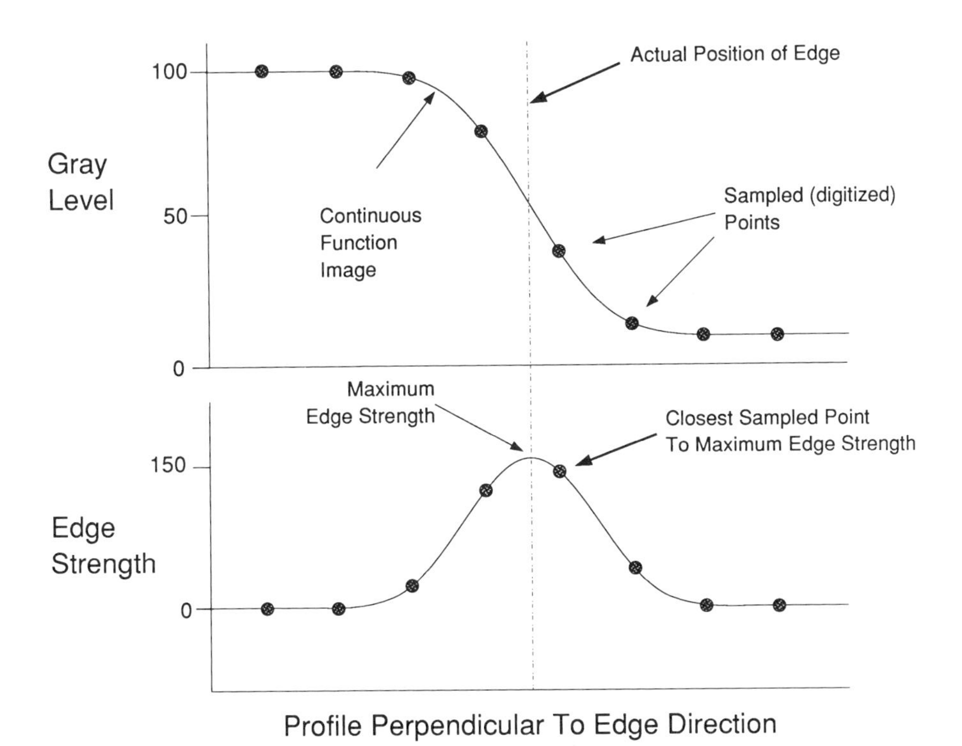

Figure 27.1. Edge magnitude behavior orthogonal to the edge direction.

This filtering process produces, almost exclusively, points that are part of some contour. These points form contiguous groups which can be categorized as follows: (1) groups that completely describe a contour, with no unlinked branches (branches are where a single line contour splits and continues in two independent lines); (2) groups that completely describe a contour, with some dangling branches (dangling branches do not close, but just end abruptly); (3) groups that completely describe a contour, with one or more dangling branches; (4) groups that form an open curve which has two end points and no branches of any sort; and (5) groups that form open curves with more than two end points and one or more branches. Some of the points retained by the filtering process are derived from noise or belong to features that, in spite of their sharp contrast, are not part of some interesting contour or internal spore structure. Conversely, some points discarded by the filtering process are part of a structure of interest. It is therefore necessary to link together the unclosed contour groups, while avoiding the groups produced by various unwanted features or noise.

Contour Linking Using Heuristic Search. The linking of the unclosed contours is performed using the classic artificial intelligence graph-searching methodology (Nilsson 1980; Winston 1984). First, the unclosed contours are rated for use in the searching. This rating is based on the average edge strength, the smoothness of the contour, the number of branches in the contour, the edge strength of the end points of the contours, and the consistency of the gray levels in the pixels of the contour. Then, the search begins by selecting the most likely contour (the starting contour) and its best (strongest-edge strength) end point. A local area around this end point is searched for the end points of other contours, and a short list (less than six) of these points is made. This is a list of candidate goal points, that is, end points of contours that are likely to be part of the complete contour with the starting contour. A search attempting to link the starting contour best end point to each of the candidate goal points is performed. If any searches are successful, the best one is chosen, and the two contours (the starting contour and the contour belonging to the candidate goal point) and the searched path are merged. If the new contour to be merged with the starting contour has more than two end points, the search is continued from the best end point of the contour. The algorithm continues until the contour is closed or a search fails. If this happens the search continues from the other end point(s).

The heuristics of the searches used in the linking of the unclosed contours are dependent on the method for estimating the path total cost (the cost $g(n)$ of the path up to a certain point, n, and the estimated cost $h(n)$ [heuristic function] of the path from n to the goal point G [Winston 1984; Poole and Goebel 1988]). A number of different search strategies for the linking of the unclosed contours have been implemented and evaluated. They are: depth-first search, breadth-first search, hill climbing, best-first search, branch and bound, and

A* (Winston 1984; Poole and Goebel 1988). The path cost *g(n)* up to *n* has been calculated using edge strength, path smoothness, gray-level consistency, and combinations and functions of these three. The estimate of the cost *h(n)* from *n* to goal point *G* is calculated using Euclidian distance and the square of the Euclidian distance.

The search strategies have been applied to several images of fungal spores, and the results can be found in Table 27.1 (note that depth-first and breadth-first do not use either *g(n)* or *h(n)*, best-first and hill climbing use *h(n)*, branch and bound uses *g(n)*, and A* uses both.) In Table 27.1, data pooled from four images are used to compare the different search strategies by the number of nodes searched, the success rate in closing contours, and (subjectively) how well the closed contours matched the human perception of the subject. Not too surprisingly (A* is supposed to find the optimum path; see Nilsson [1980, pp. 81–84] for the admissibility of A*), A* was the most effective in space (the number of nodes searched), as well as success in closing contours. Figure 27.2 shows an outline of the entire image-processing, segmentation, and heuristic search processes. Figure 27.3 shows a digitized *Bipolaris* sp. spore and the extracted contour produced using the A* search strategy and edge strength, path smoothness, and the square of the Euclidean distance as heuristic. Other types of spores, septate (having internal divisions) and aseptate, more pigmented and more hyaline, and having greater shape complexity have been segmented with a high degree of success and accuracy using this method.

Conclusions

The extraction of spore outline and internal structures is seen as a first step in the automation of fungal spore identification. Manual measurement of size

Table 27.1. Effectiveness of Search Strategies
Greater variation is not categorized as a successfully closed contour

Search strategy	Number of nodes searched	Success rate	Quality
Width first	9987	78%	Fair[a]
Depth first	2165	82%	Good
Best first	1044	87%	Excellent
Branch and bound	1057	87%	Good
Hill climbing	1032	87%	Good
A*	986	94%	Excellent

[a]Fair quality: limited variation from the contour of not more than three pixels.

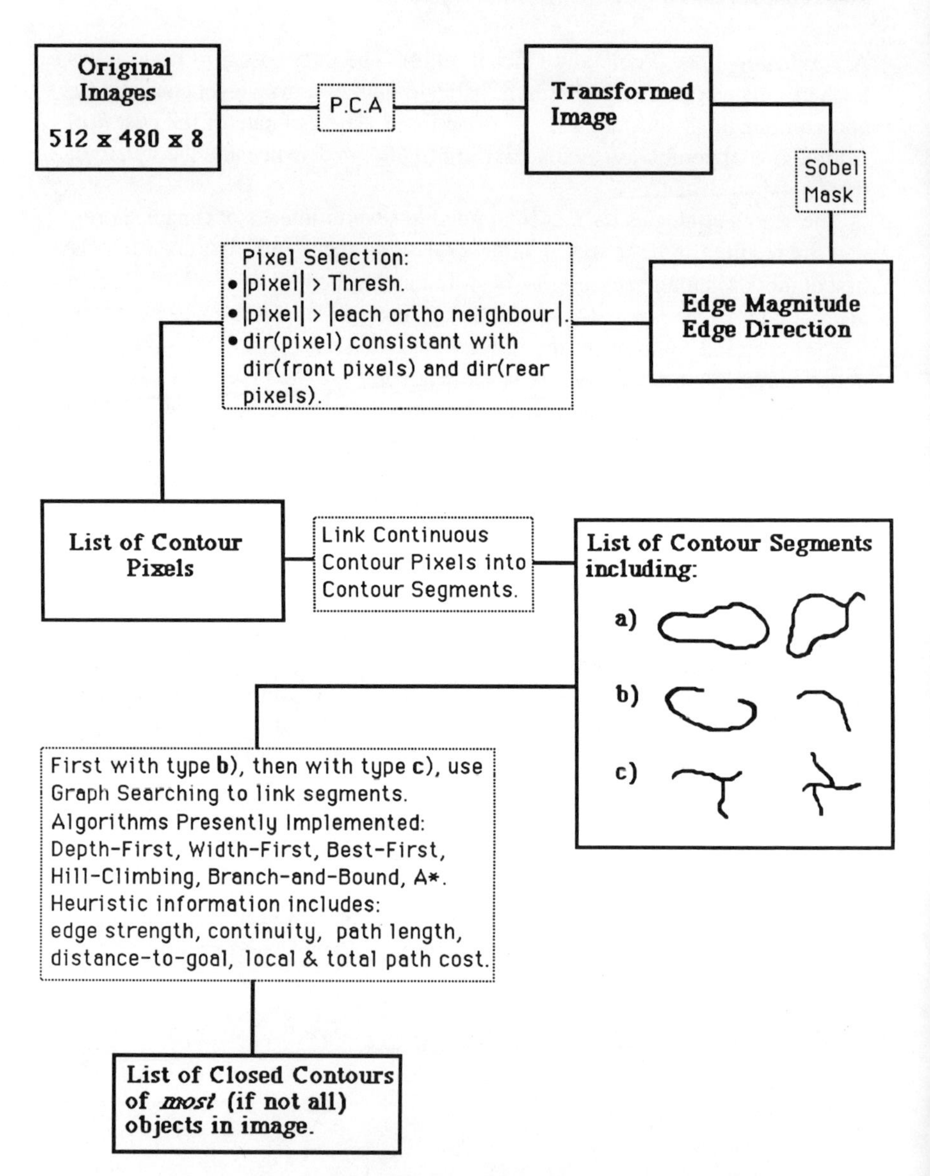

Figure 27.2. Algorithm outline.

and shape is quite tedious, and many fungal taxonomists, when describing a new species, usually give a range of lengths and widths derived from the measurement of less than twenty spores. This is not a significantly useful sample, and the automatic measurement of size and shape will allow for the

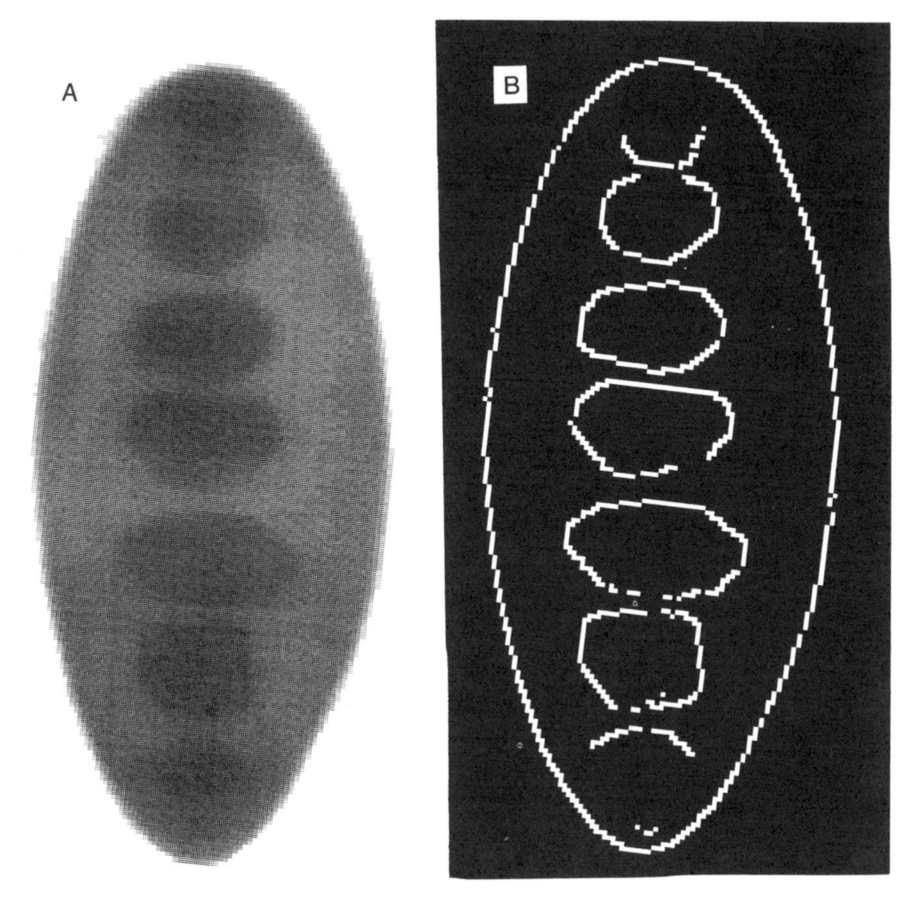

Figure 27.3. (a) Image (200 × 160 pixels) of *Bipolaris* sp. spore. (b) Contours of Figure 27.3a extracted with A*.

processing of large numbers of samples to build a statistically sound representation of these values.

This is seen as being important, as the subsequent automatic identification will depend on probabilistic reasoning and decision-making methods. So will the correlation of qualitative shape descriptions used by fungal taxonomists with various shape descriptors from the literature in an attempt to attach some quantitative meaning to these labels. These are the steps now being undertaken in our research.

In the discussions at the ARTISYST meeting, it became evident that some of the biologists held the opinion that probability need not be considered in expert identification systems and that the emphasis should be placed on the development of deterministic systems and methodology. This, it was said, was because of dependency on the type specimen. This may well be a problem for extinct organisms where only one or two samples of a taxa may exist, but it is not the case for extant species, where the definition of a taxon is constructed from the careful examination of as many samples as possible, and the building, in the mind of an expert, of a probabilistic representation of features and distributions (Atkinson and Gammerman 1987; Petrini and Rusca 1989). The type specimen is in no way definitive for a taxon. Although the type specimen is important for the nomenclature and identity of a taxon, the description should rely on statistical principles. Uncertainty and variation are the stuff of biological diversity, and any system of identification must take them into account.

Acknowledgments

The authors would like to thank D. Newton and L. Patten for assisting in some of the software development and debugging, as well as the PAMI Lab, University of Waterloo, for use of their hardware early in the project. This work is possible due to the open minds at the National Engineering and Science Research Council of Canada through operating and equipment grants.

Discussion

Walker: Have you built all of the codes yourself, or are you using off-the-shelf codes for any of the processing?

Newton: No, I built all of the code myself. We wanted to use Lisp, but we could not afford a commercial Lisp compiler for the Sun. So we had to write all the code in C, but we were thinking of using one of the buggy semi–public domain Common Lisp.

——: Why did you want to use Lisp? C is an obvious language to use for this.

Newton: Well, there were a lot of other things. Implementing searches is much easier in Lisp. For the low-level image processing, C was used, but FORTRAN could just as easily have been used. For anything built on the image-processing part of this system, like some sort of knowledge base that we are interested in, something like Lisp or CLOS (Common LISP Object System—the AI power of Lisp and the object-oriented power—and more—of Smalltalk combined) will work much better.

28

Image Analysis in Systematic Biology: Models of Expected Structure

STANLEY DUNN

The difficulty with human interpretation of images of natural and biological data is the wide variability in the criteria and rules used to identify the structures present. A good example of this occurs in arthropod hematology (the study of blood cells of arthropods), where imprecise feature description has led to disagreement among researchers over nomenclature and classification of the forms and functions of the hemocytes (blood cells) and their organelles (constituent elements).

Arthropods (lobster, shrimp, insects, spiders, and others) have six kinds of hemocytes or blood cells: the prohemocyte, plasmatocyte, granulocyte/coagulocyte, spherulocyte, adipohemocyte, and oenocytoid. Two of these, the granulocyte (GR) and the plasmatocyte (PL), provide cellular immunity (phagocytosis and encapsulation) to these animals. In recognition of the ability of the GR and PL to discriminate between self-tissues and non-self-tissues, Gupta (1986) coined the term *immunocyte* for these hemocytes.

One of the nagging problems in arthropod hematology is the lack of reliable, error-free criteria to correctly identify and classify these various types of hemocytes. Often, the hemocytes are classified by their morphological characteristics, staining reactions and functions, and it is not uncommon to discover that the same hemocyte has been differently or incorrectly identified by two or more authors. Poor control of staining and use of imprecise, qualitative morphological features have inevitably resulted in confusing terminology (Gupta 1986), and it has not been possible to consistently attribute the same immunologic function to the same hemocyte types. In addition, there are long-standing controversies regarding such hemocytes as coagulocytes,

---

Advances in Computer Methods for Systematic Biology: Artificial Intelligence, Databases, Computer Vision, ed. Renaud Fortuner (Baltimore: Johns Hopkins University Press).

the so-called crystal cells, adipohemocytes, and others that are found in arthropods.

Computer image analysis has a role in solving this problem because the computer embodies a precise set of detection criteria and the performance of the feature detection does not change with time. The features can be identified in an unbiased way; the blood cells can be classified according to the features extracted, and agreed-upon rules can be used to classify the hemocytes.

Computer-based identification of blood cells is routine in clinical laboratories, and we have used established image-processing methods (Ballard and Brown 1982; Rosenfeld and Kak 1982) to determine a robust set of image features (morphology and stain characteristics) to recognize some of these cells. The features describe mainly the size, shape, textural characteristics (of the stain), and external boundary continuity of cell organelles like mitochondria, granules, and vacuoles. The GR is found in all major arthropod taxa (Gupta 1979) and is their main immunocyte (Gupta 1986).

The long-term goal of this research program is to develop a computer vision system to visualize complete three-dimensional images of both internal and external features of arthropod blood cells by light, transmission, and scanning electron microscopy. By manipulating such three-dimensional images, we could study the role of such organelles as marginal microtubule bands, which are crucial in maintaining the flat and discoid nature of these cells and increase sevenfold in number in an activated state (Han and Gupta 1988). Some preliminary results of computer-assisted identification of arthropod blood cells were presented in Dunn and Gupta (1988). The examples shown in this chapter are of two-dimensional segmentations of serial sections; these could be used as a first step for successful three-dimensional analysis. With a proper segmentation of a consecutive series of microscope slices and a good identification of the objects present in these slices, one can proceed to a three-dimensional reconstruction of the internal and external structures of both the cells and their organelles.

This chapter outlines an automatic image interpretation (see below) system and its applications in the interpretation of transmission electron micrographs (TEM) of arthropod hemocytes. The features of the hemocytes are detected using mathematical models derived from qualitative models found in the literature. These experiments yielded promising results, supporting our belief that this image interpretation paradigm and knowledge representation scheme can be applied to other image-processing tasks in other natural and/or biological domains. We present results in cell discrimination by a human expert, based on the computer results. Our long-term goal is to automate this process and produce a computer-based cell identification and classification system.

The purpose of this chapter is threefold: first, to introduce the reader to, and provide a general overview of, model-based image processing and image understanding; second, to provide a summary of how computer image processing can be used in insect hematology and other systematic biology applications; and third, to act as a reference to literature in this area. The aim is to provide material from which one can assess the state of the art, identify research opportunities, and learn about the established image processing operations.

General Considerations

The Task of Image Interpretation

Image interpretation constitutes the extraction of useful, accurate, and reliable information about the objects that are present in a scene captured in a given image and the use of this information in understanding the scene. The first major task in image understanding systems is to segment the image into meaningful entities that are characterized in a way that is suitable for machine recognition. For this task to be accomplished, the intermediate-level tasks that must be performed should be well planned. A good segmentation plan requires knowledge of the image formation process, knowledge of the domain of analysis, and some qualitative description of the objects that are present in the image. The translation of domain and image formation knowledge and the qualitative object descriptions into a quantitative object model to guide the segmentation process constitutes what we call *segmentation with models of expected structure*. The interpretation paradigm is based on using object models that describe the expected presentation of those objects likely to be present in the images. The user inputs the image, information about the image formation process, and a description of the domain, and the image is segmented to yield regions that most closely resemble the models of the expected structure.

Computer Vision

Image processing and interpretation (Ballard and Brown 1982; Russ 1990; Marr 1982; Rosenfeld and Kak 1982; Nevatia 1982; Serra 1982; Pavlidis 1982) is an engineering discipline of gathering pictures from one or more sources, performing a series of operations, and interpreting the results to answer a question. These images can be obtained from television cameras in normal room lighting, infrared cameras, magnetic resonance imaging, or ultrasound, as well as electron microscopy. The applications are just as numerous, ranging from medicine to industrial inspection to autonomous navigation. What all these environments and imaging systems have in common is the processing between image acquisition and answer extraction. The field of computer (or machine) vision has evolved to become the study of general processing algorithms and recognition strategies without regard for the particular domain or image acquisition method. This has met with moderate success; the difficulty is the experimental verification that comes from the use of these general methods in military, industrial, and medical applications. At the moment one of the greatest stumbling blocks is the lack of computational power available to complete the desired task in a reasonable amount of time. Inspection and navigation tasks require response in a second or less from algorithms that typically have taken minutes or hours to complete on a single-processor computer. Thus, there is a strong motivation to study how to perform image processing on advanced computers. The emphasis in this chapter is on the development of working systems; details on computer architectures for image analysis can be found in Reeves (1984).

Image Processing

Image-processing operations are those performed on digitized pictures and are designed to detect the presence of features of interest. These features in the image should correspond to physical characteristics of interest, such as intensity highlights, repeated patterns, and edges. Regions of little to no intensity variation are identified by performing image processing to remove intensity variations due to noise.

Operations on pictures that fall into this category are characterized as two-dimensional signal-processing operations. These may be expressed as a convolution operation, as functions of the global intensity distribution, or as nonlinear filtering operations in two dimensions. Figure 28.1a is a sample transmission electron micrograph of a hemocyte from a horseshoe crab with a number of organelles. Figures 28.1b to 28.1d show some sample image-processing operations on Figure 28.1a. Figure 28.1b is the result of a convolution operation to extract edges, Figure 28.1c is a thresholding operation to extract the organelles using the global intensity distribution, and Figure 28.1d is a noise removal process based on a nonlinear filtering operation.

In all cases, however, the operations map an image to an image. The result of an image-processing operation is an image of the same size (the border elements may be handled differently for reasons that will be discussed later). The computational complexity of these operations is proportional to the number of picture elements (pixels) multiplied by the size of the local neighborhood used in the operation (which may range from 1 to 225, typically). Images are typically rectangular and range from 64 by 64 (65,536 elements) to 2048 by 2048 (4,194,304 elements). As the images increase in size there is a greater demand for more computational resources. On a typical single-processor computer the image is processed in a raster fashion: row by row from top to bottom. As the size increases the execution time increases in proportion to the square of one of the dimensions. To curb this growth in execution time, either more processors or special-purpose hardware must be added.

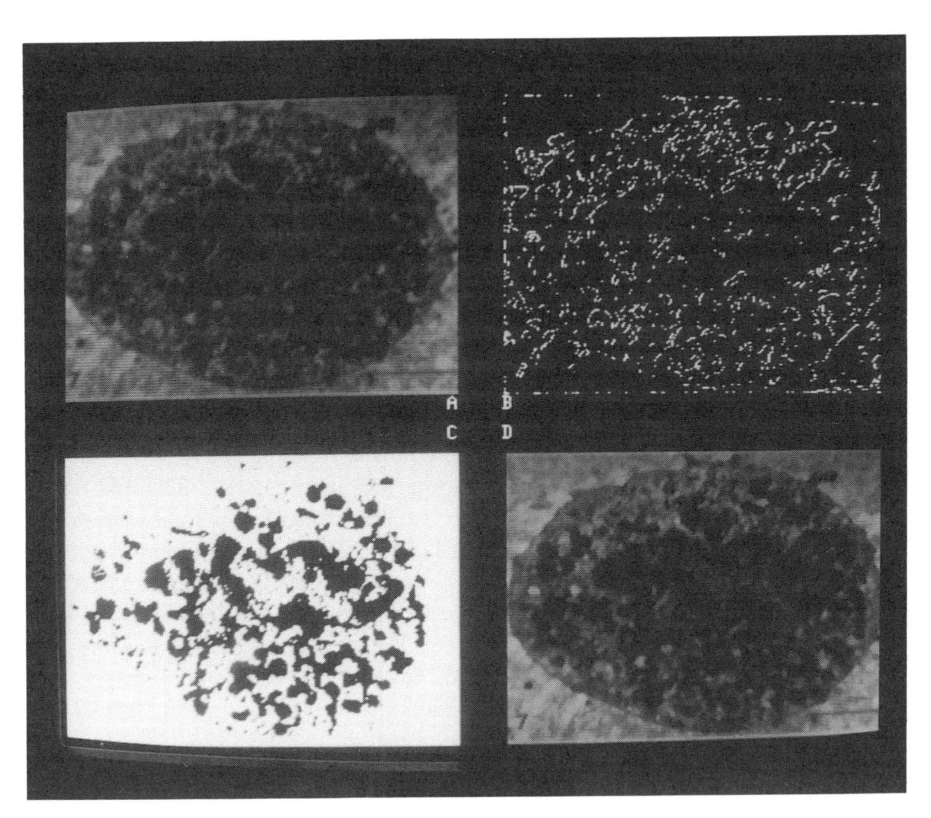

Figure 28.1. Image-processing operations on a transmission electron micrograph of a hemocyte from a horseshoe crab.

Image Understanding: Assimilating Features of Pixels

Operations on image data that fall into this category differ from image processing in that these are operations on pixel features to synthesize features of neighborhoods and regions. These include the problems of determining motion from a sequence of images, computing surface shape, and assigning labels to these segmented regions. Image-understanding processes require a great deal of computation, either numeric or symbolic. While the first two items in the list require mathematical computation, they involve typical optimization procedures and are computationally expensive. The optimization methods can be as elementary as finding the least-squares solution for a set of parameters (Ballard and Brown 1982) or involve surface fitting using a regularization algorithm (minimization with smoothness constraints; see Horn 1986) or one of its alternatives. The last item in the list, assigning labels to regions, is an example that requires symbolic computation, that is, computation with character and symbols and not necessarily numbers.

Image-understanding tasks differ from image processing in that much of the intermediate data and final results are symbolic rather than numeric. As is true of many symbolic processes, the data structures are typically linked lists and graphs (Ballard and Brown 1982) instead of arrays. The number of regions in an image cannot, in general, be fixed, and consequently data structures that can grow dynamically are used as in most artificial intelligence applications. While in image processing the operations are image to image, in image understanding there are image-to-list transforms and operations. These lists can be arbitrarily long, and there can be arbitrarily complex sublists that represent image properties and regions, knowledge about the world, or knowledge about how to process an image. As expected, the complexity of the image-understanding operations are dependent on the complexity and contents of the image.

There are three major computational problems limiting advances in research and applications of image processing and understanding. The first is the sheer volume of the data required to complete a task. It is not uncommon for images to be 480 rows by 512 columns or 512 rows by 512 columns. Each pixel is typically 8- to 16-bit integer data; some applications, such as nuclear magnetic resonance (NMR) spectroscopy, are using 32-bit floating-point numbers. Although each individual image by itself may not be a great deal of raw data, the application may require that they be processed very rapidly. Inspection tasks where an image-processing system may be placed on an assembly line or an autonomous vehicle may require that an image be processed in a very small fraction of a second. In medical imaging, the volume of data comes from attempts to build three-dimensional reconstructions of internal organs of the body. To obtain relatively accurate reconstructions, the individual images correspond to thin sections, and consequently there may be several hundred (typically 256) images in a reconstruction.

The second problem is a consequence of dealing with spatial data: Many computer vision algorithms including local processing and interpretation require some knowledge of local properties (either numeric or symbolic) to complete the operation. Finally, computer vision problems are unique in that they require both numeric and symbolic computation.

Computer Image Analysis in Cytology

The study of insect blood cells is expected to provide insight into the structure and function of human blood cell structure and function. The problem of automatically classifying human blood cell images has been studied for over twenty years; however, there has been little work in developing comprehensive research tools that would allow the researcher to study the complete internal and external structure of the cell as well as automatically classify the cell structure.

One of the earliest efforts in automatic cell classification is the work of Prewitt and Mendelsohn (1966). A group of twenty-two blood cells were prepared with a special staining procedure, digitized, and processed. The cells were classified into four categories. Young (1969) and Bacus (1971) extended this classification to all six major categories of normal cells. They used an image enhancement technique, the Whitening Transformation (Bacus 1976), assumed a bivariate distribution of image pixels, and, using two filtered monochromatic images (of different colors) from the same cell, produced a density image (black and white) and a color image of the cell. They then used these results to segment the images, obtaining a better segmentation than with simple gray-level histogram thresholding.

Bacus et al. (1976) extended the area to classify over seventeen types of blood cells. Red blood cells could be analyzed, and these authors developed methods to identify subcategories of normal and abnormal blood cells. They proposed an error measure for assessing the quality of their results (Yasnoff et al. 1977, 1979) that quantified the disagreement between the "true" and test segmentations, and the errors in pixel classification. The measure, which was a function of the proportion of misclassified pixels and the spatial distribution of pixels within a given class, was image-domain independent, image-size independent, and easy to compute. With this measure, one could compare various segmentation algorithms and correct or modify some of the heuristic techniques. Mui et al. (1978) later proposed a feature selection and statistical pattern recognition approach to the automatic classification problem.

The work of Bacus et al. (1976) led to the development of the leukocyte automatic recognition computer (LARC) to classify leukocytes in human blood specimens. Bacus (1979) gave a comprehensive historical analysis of the development of automated differential systems, and Trobaugh and Bacus (1979) examined the design and performance of the LARC. Clinical evaluations of this system and other similar systems have confirmed the theoretical results and also exposed some weaknesses both in the physics of the design (Bacus et al. 1980) and in the image analysis and classification algorithms.

Little progress has been made on the problem of automatic identification since then. The early heuristic research efforts of the middle 1960s evolved into some interesting theoretical results in the 1970s. In the same period, there was remarkable progress in the fields of computer vision, pattern recognition, computer graphics, and artificial intelligence (reviewed by Horn 1986). In addition, more sophisticated electron, scanning confocal, and phase contrast microscopes have been developed that produce images with such high resolution that we can now make a comprehensive analysis of the internal and external structures of specimens, thus increasing the chances for a more accurate identification and classification.

Model-based Image Analysis

General Considerations

The objective of an image interpretation system is to construct an explicit description of the scene captured in a given image. To do this, a computer image interpretation system must solve two problems. First, the system must extract an explicit description of the scene from the image and, second, it must decide what interpretative techniques should be used to complete the vision task (arrive at an understanding of the scene). The first major task in any image-understanding system is thus to segment the scene into meaningful entities, which are given a characterization suitable for machine recognition. This segmentation is the result of the image-processing steps. The task of image understanding becomes much easier if the first task is done well.

A good segmentation scheme is the key to a good image interpretation system. Constructing meaningful entities from iconic data (pixel raster representation with gray levels) poses problems that are generally linked to the type of objects to be identified. Restraining segmentation algorithms to a particular class of images may simplify the process, but the ability to use the system for another class of images is lost and is a very serious handicap of such a solution. The complex human visual system first extracts reliable intrinsic information from the input image and successively processes the extracted information with some domain knowledge input as the interpretation progresses from the retinal (acquisition) to the cognitive (interpretative) levels (Cohen and Sherman 1983; Hubel and Wiesel 1977). A near-optimal segmentation scheme should adopt this approach. Many solutions have been proposed for the segmentation problem (Ballard and Brown 1982; Marr 1982; Rosenfeld and Kak 1982), but few utilize the formal mechanism of reasoning that is apparent in the biological visual system. A few of these techniques have been quite successful in a variety of specific instances, but do relatively poorly in the general case. Computer vision systems built to analyze scenes of man-made objects are often successful because most objects in the scene can be easily modeled to a high degree of accuracy. When it comes to natural objects with large variations in structure, edge-based or region-based segmentation algorithms (Haralick and Shapiro 1985) present serious inadequacies.

The most general image segmentation scheme that would be suitable for images of both man-made and natural objects would be data driven. That is, it would use only the raw image data and make no reference to the domain of analysis. Such an ideal system has been found to be impossible to implement even in the most complex systems. It has been shown experimentally (Dunn et al. 1990) that the best such a computer system can achieve is a preliminary grouping of pixels and initial partitioning of regions. Obtaining a satisfactory final segmentation requires domain-specific knowledge. The efficient organization and optimal insertion of this knowledge in the control processes and tasks to be performed are the key to building a good image-understanding system.

However, data-driven algorithms can be used to build an initial low-level segmentation module that will serve as a basis for a general image interpretation scheme (Dunn and Shemlon in press). Such a module would act as an input processor that accepts iconic data from any domain and produces the initial partitioning into primitive elements and their abstract description for use by later processes in the interpretation hierarchy. At the intermediate and higher levels, the processes could be more specialized and domain specific. The complex problems of scene analysis require a combination of regular image segmentation techniques and artificial intelligence (AI) techniques for reasoning, planning, and learning to be applied, with a proper stratification of domain knowledge use (Barry et al. 1988; Suetens and Oosterlinck 1987).

Hemocyte Modeling

Several examples of the use of models in image segmentation and interpretation can be cited (Ejiri et al. 1989; Mitchell and Gillies 1989; Suetens et al. 1989). In most of these examples, model-driven image interpretation tasks have been limited to high-level matching and recognition. The use of models in their abstract symbolic form needs to be included in the segmentation process to permit greater control of the lower-level processes, to allow greater flexibility, and, most important, to act as an aid or filter in reducing the amount of noise in the output of the low-level operators. Perhaps the greatest challenge in building image interpretation systems for images of biological structure is discriminating between biological variability and noise in the output of low-level image operators.

In most cases where the use of models has been attempted, only elementary cues like the curvature of edges and gray-level similarity are used (Pavlidis 1986). Higher and more sophisticated characterizations of models of scene objects need to be exploited. Reynolds et al. (1984) propose a flexible intermediate-level representation for multiresolution segmentation using multiresolution models. They propose foveation (focus of attention) as a means of speeding the process. The system is, however, limited by the fact that it uses model descriptions that are suitable only for man-made objects. In the same line, Suetens et al. (1989) propose a system with hierarchical instantiation of models. They implement matching and interpretation as integrated processes. The system has the inconvenience that knowledge representation is still primitive (at the iconic level). In their work, Kapouleas and Kulikowski (1988) describe a system with solid models for three-dimensional vision. They propose a very good example of the use of expected biological structure in three dimensions. With this system, they are able to segment and interpret magnetic resonance imaging (MRI) images of the human brain. They start with an initial low-level segmentation, which they successively refine with increased use of domain knowledge.

In the system described in this chapter, models are used in both the segmentation and interpretation processes. A segmentation is successively refined as we go from coarser to more complete model descriptions. In this process candidates are successively dropped that do not fit the description of expected objects or from which we are least likely to build the expected structures (Dunn et al. 1990; Dunn and Shemlon in press). The process consists of determining the goodness of fit with one object model parameter at a time and dropping the worst fits. In this way, biological variability does not have to be explicitly modeled. Rather, candidates that are least likely to build the expected structures are assumed to be outside the limits of possible biological variability for the given expected structure and are dropped.

An Example of a Hemocyte Image Analysis System

Global System Description

The organization of the image models, domain models, and image-processing tasks is shown in Figure 28.2. The digitized image is processed in three parallel paths. The first path is quantized thresholding, to extract features to be matched against the models of expected structure, extract principal boundaries (forbidden boundaries, boundaries that regions cannot cross) for use in assessing the quality of the initial segmentation, and define control cues for region growing.

The second path is quantized smoothed histogram equalization, to extract cues for computing texture parameters to be used in comparing image object texture features. A histogram of the original image is filtered to obtain a smooth curve, and histogram equalization is done around the maxima and minima of the filtered histogram (see Dunn and Shemlon [in press] for details).

The last path is data-driven segmentation where edge and image intensity information is used to generate an initial fine segmentation without any image type or domain-knowledge use. First, the image is filtered. After edge detection, pixels within this filtered image are grouped using a modified version of the connected components algorithm (see Dunn and Shemlon [in press] for details). Information about regions resulting from this initial segmentation is generated in the form of a global database (Liang 1989). After this initial segmentation, domain-independent data-driven region-growing algorithms are used to obtain a finer segmentation. The algorithms use two segmentations of the given scene (a coarser and a finer segmentation) to obtain an intermediate segmentation that is more refined than the coarser segmentation and devoid of the unnecessary detail in the finer segmentation (see Dunn et al. [1990] and Dunn and Shemlon [in press] for details).

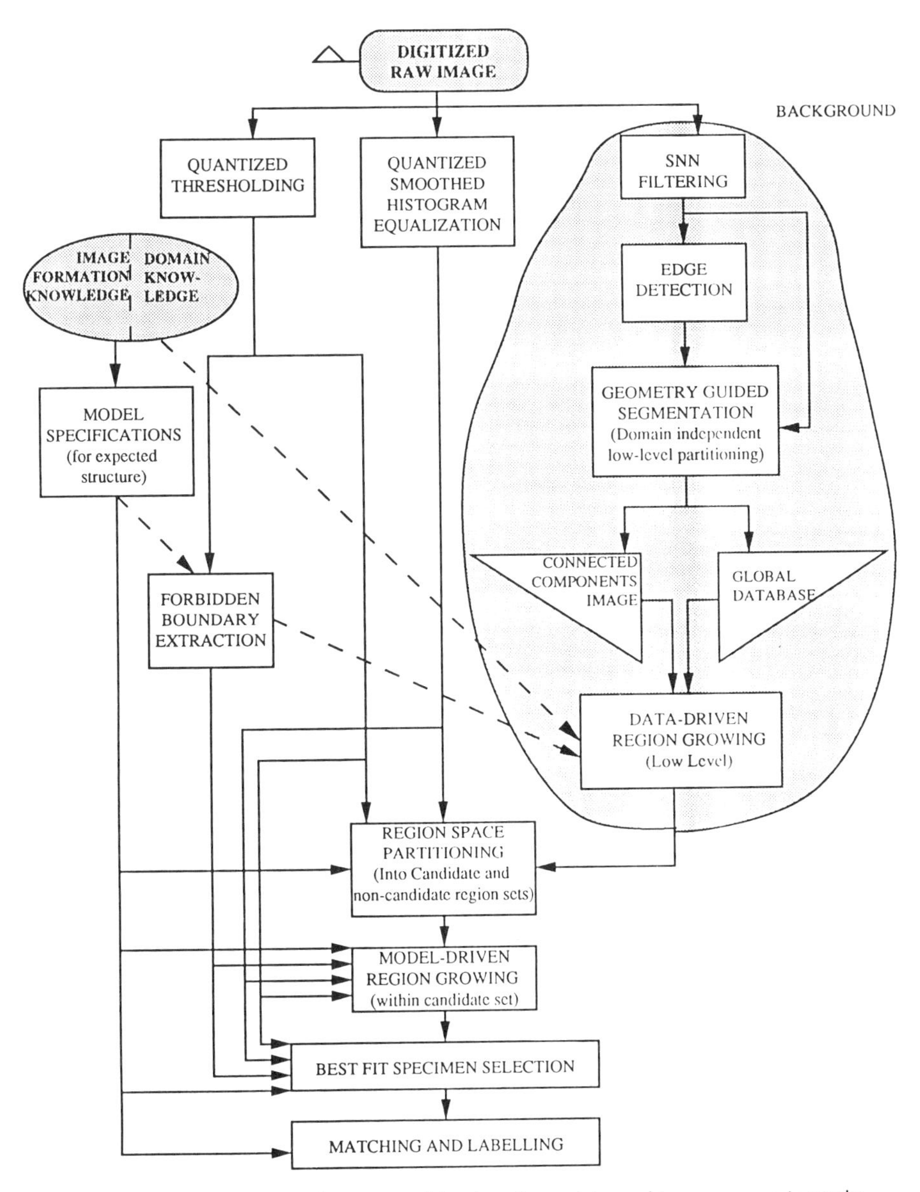

Figure 28.2. Organization of image models, domain models, and image-processing tasks.

The results of these three parallel paths are then combined in model-driven region growing where domain knowledge and information about the image formation process are used to guide the merging of small regions to form candidates that most closely match one (or possibly more) of the models of expected structures. Knowledge of the domain of analysis (e.g., insect hematology) and of the image formation process (e.g., transmission electron microscopy) is exploited in the construction of quantitative object descriptions that characterize structures expected to be present in such images, from qualitative object model specifications.

After region growing, the best fits in the final segmentation (final regions whose characteristics best match model specifications) are selected as specimens. The other regions in the final set are compared to these best fits, and the closest matches are selected and labeled along with the specimens as the targeted objects.

The whole segmentation process from raw data to object extraction is automatic. No human intervention is needed for any threshold selection. All thresholds are selected automatically using knowledge about the data being analyzed. The system needs only knowledge of the domain (insect hemocytes), the objects we are looking for (granules, mitochondria, or other organelle), and information on the image formation process (TEM).

The segmentation process is speeded up by using prior knowledge of the final characteristics of the regions we are building. The initial segmentation region space can be split into two classes: one composed of likely candidates (from which the desired structures can be grown) and the second constituting the background. This reduction in the search space increases the probability of obtaining a near-optimal final segmentation of desired objects. Merging rules, grouping criteria, model specifications, and search space constraints are successively enhanced as the region growing progresses (more complete model descriptions and stronger constraints are used) until a final set of most likely candidate regions is obtained. As the segmentation progresses, the size of the candidate region space shrinks (since many regions are merged into one). This will ensure that the time complexity of the algorithm is not exponential in the number of regions.

Model Specifications

Regions are grown to fit structures that we expect to find in the image. If the final segmentation step produces regions that match the expected description, they are extracted, and the remaining regions are considered either to belong to another object's candidate region space or to be part of background.

Quantitative object descriptions result from combining qualitative object models in a given domain (arthropod hematology in this case) and image formation models. These quantitative parameters describe object appearance in the image. For the moment, we shall concentrate on the form of the object descriptions and how they are used. Object descriptions take the form of mathematical expressions or statistical parameters that characterize image regions. Regions are characterized by their shape and intensity parameters.

The brightness of a region is measured by comparing its average gray level to a computed threshold. The region's position in the image is given by its centroid. The chain-code encoding of the region's bounding contour is used to extract information on the region's concavity. It can also be used to characterize the fact that a region is lobed or protruded or has a smooth outline. The contour profile obtained from the chain code can be used as a measure of shape irregularity. A more explicit characterization of shape, using moments and other descriptors to compute geometric invariants, is given below. Parameters for characterizing texture are also examined.

The visible regions in the image are characterized by their shape and intensity parameters. We characterize shape using geometric invariants for various two-dimensional shapes. Texture is characterized using statistical parameters computed from the gray-scale distribution of pixels within a region (Dunn et al. 1990; Dunn and Shemlon in press). We have used moments, areas, and perimeters as shape descriptors to compute some geometric invariants for simple two-dimensional shapes that model some of the objects we expect to find in the images being analyzed. As an example, the following two-dimensional shapes can be quantitatively modeled as:

$$\text{round,} \quad \frac{P^2 M_{xx} M_{yy}}{A^2} = 1;$$

$$\text{elliptical,} \quad \frac{A^2(M_{xx}^2 + M_{yy}^2)}{2PM_{xx}^2 M_{yy}^2} \approx 1;$$

$$\text{square,} \quad \frac{3P^2 M_{xx} M_{yy}}{4A^2} = 1;$$

and

$$\text{rectangular,} \quad \frac{P^2}{8[6(M_{xx}^2 + M_{yy}^2) + A]} = 1,$$

where A = area, P = perimeter, M_{xx} = square root of the normalized second moment about the x axis, and M_{yy} = square root of the normalized second moment about the y axis.

We characterize texture using very simple statistical parameters. The average gray-level of a region (AGL), its standard deviation (SD), the percentage of pixels with gray-level value greater than 2 standard deviations above the mean (TP2), between 1 and 2 standard deviations above the mean (TP1), and less than 1 standard deviation above the mean (TP0) are all used to characterize a region's texture. These texture parameters are computed from a texture-enhanced version of the original image that is obtained using the quantized smoothed histogram equalization and quantized thresholding techniques (see Fig. 28.2).

Creating Quantitative Models of Organelles

The algorithms described above were used to process and interpret images of transmission electron micrographs of insect hemocytes. The following are some of the features we use to describe regions in our segmentation paradigm: shape: triangular, rectangular, round, elliptical, trapezoidal, or irregular; position: specification of location or bounding window; relative size: very small, small, average, large, or very large (relative to the sizes of similar objects in the image); relative texture: highly, medium, or nontextured (relative to the textures of similar objects in the image); intensity specification: very dark, dark, medium, bright, or very bright. For example, the following organelles often present in TEM images of insect hemocytes (Gupta 1979; Gupta 1986) could be qualitatively modeled as follows:

Granules: dark or very dark, round or elliptical in shape, average to large in size, relatively nontextured, located everywhere except the nucleus;
Mitochondria: dark, round, or elliptical in shape, average to large in size, textured, located everywhere except the nucleus.

These qualitative descriptions were converted into quantitative measures that are usable by the segmentation system. The descriptions of *dark* or *light* are translated into criteria that compare the average gray-level of image regions to gray-level thresholds obtained by quantized thresholding. Qualitative descriptions of shape are translated into quantitative measures using the shape parameters and two-dimensional geometric invariants. The classification *textured* or *nontextured* is quantitatively expressed using the texture parameters that are computed from the quantized smoothed histogram equalized image.

With these quantitative object descriptions, a TEM serial section is segmented using the following model-based, region-growing procedure. All dark regions in the primary segmentation constitute the initial class of candidates that can be grown to nuclear chromatin, mitochondria, granules, or nucleoli. We can use the shape constraint to extract a subset of most likely granules, mitochondria, or nucleoli. Since granules are generally nontextured, a subset of most likely granules can be extracted leaving another subset of most likely nucleoli and mitochondria. From the subset of most likely granules (if nonempty), the best fits will be chosen to act as granule specimens. The subset of most likely nucleoli and most likely mitochondria is extracted using texture discrimination with parameters defined above; the best fits in these two subsets also become specimens. Other objects present in the image are matched to the selected specimens to generate the final class of objects that best fit the models of the structure (organelle) we are looking for.

Experiments and Results

These algorithms were used to process and interpret TEM serial sections of insect hemocytes. Figure 28.3A shows one of the original TEM images used in these experiments. The images were first processed to eliminate noise and enhance edge features. The filtering algorithm also eliminated some of the fine texture in the granules, mitochondria, and chromatin. To avoid this inconvenience, the filtered images were used only for edge detection and data-driven segmentation. The parameters used in model-driven region growing, specimen selection, and matching were computed from images obtained by quantized smoothed histogram equalization and quantized thresholding.

As we can see in Figure 28.3b, all dark regions are extracted as candidates to be merged to form granules, nucleoli, chromatin, and mitochondria. Figure 28.3c shows the targets around which primary region growing will be done. Here the criterion used to select these regions is size since we know our expected structures are larger than the average region produced by the early

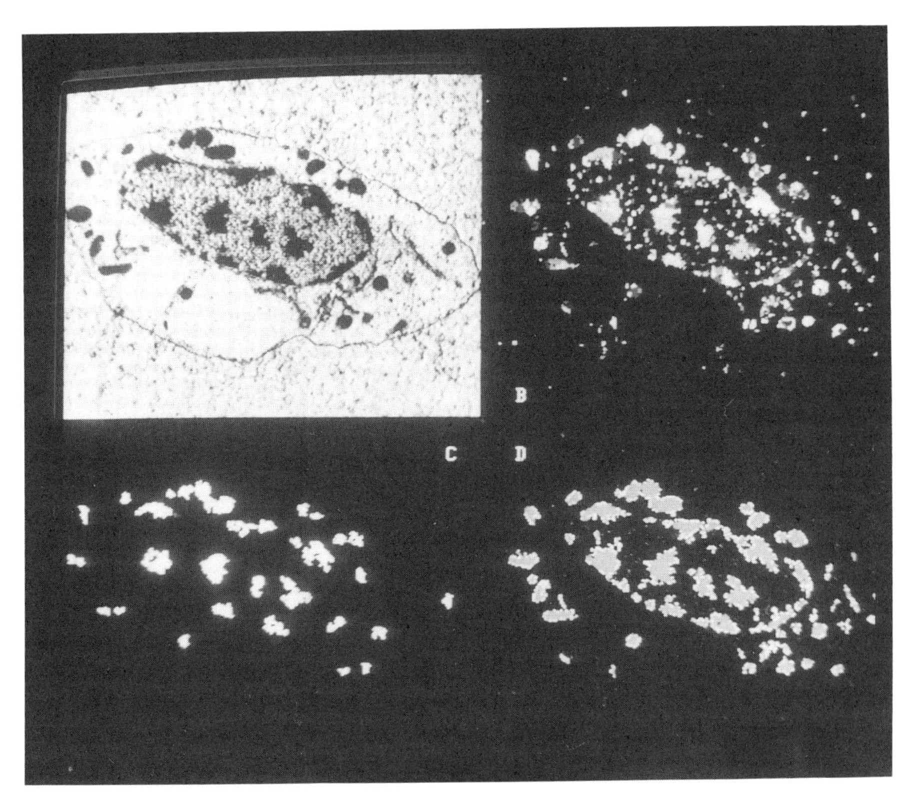

Figure 28.3. Interpretation of the transmission electron micrograph of an insect hemocyte.

segmentation process. Figure 28.3d shows the final segmentation. Using the computed shape parameters, we evaluated the various invariant expressions for each shape for a given region and expressed the degree of approximation of the region's shape to the given shape as a percentile. This percentile is the ratio of the computed value to the expected value. Figure 28.4 shows some sample regions extracted from the granulocyte shown in Figure 28.3a; the letters in the figure denote specific regions whose descriptors are shown in Table 28.1. Table 28.1 shows the computed shape descriptors (area A, perimeter P, and moments M_{xx} and M_{yy}) and the geometric invariants, which express how close the given regions approximate circles, ellipses, squares, and rectangles. The most likely candidates for granules and mitochondria are those closest in shape to ellipses or circles, for example, regions a, e, g, i, j, and k. We remark here that due to digitization errors a Cartesian plane approximates figures with straight edges better than it does for curves. It is therefore no surprise that the computed geometric invariants should be better for squares and rectangles than for circles and ellipses. In Figure 28.5, we show preliminary experimental results of using organelle labeling to differentiate types of insect hemocytes. Based on the granules and mitochondria extracted compared with the number of organelle present, we can label a granulocyte in Figure 28.5a, a plasmatocyte in Figure 28.5b, a partial cross section of a spherulocyte in Figure 28.5c, and a coagulocyte (considered to be granulocytes by Gupta [in press]) in Figure 28.5d.

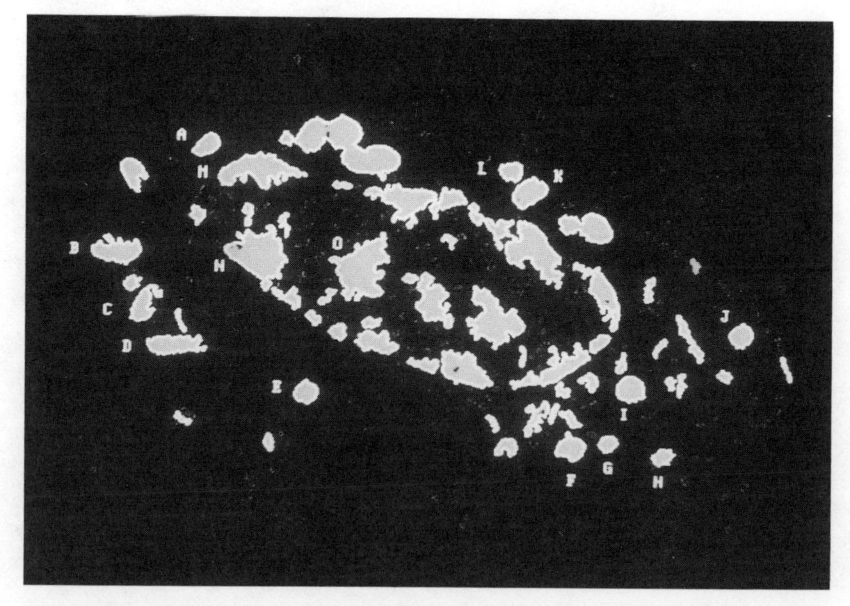

Figure 28.4. Sample regions extracted from the granulocyte in Figure 28.3a.

Table 28.1. Shape Descriptors and Values of Geometric Invariants for Regions in Figure 28.4

					Fit (%)			
Region	A	P	M_{xx}	M_{yy}	Circle	Ellipse	Square	Rectangle
a	185.50	58.18	4.44	4.33	53	53	71	98
b	432.00	131.54	4.98	8.78	25	29	33	48
c	265.00	112.57	7.85	4.52	16	18	21	48
d	385.00	117.40	3.76	9.97	29	43	38	62
e	192.00	55.94	4.51	3.89	67	68	89	97
f	240.00	73.60	4.61	5.00	46	46	62	76
g	117.00	41.80	3.48	3.18	71	71	94	87
h	150.50	58.53	4.16	4.18	38	38	51	84
i	275.50	74.53	5.23	4.84	54	54	72	84
j	216.00	58.77	4.72	4.15	69	70	92	95
k	352.50	78.33	5.75	5.87	60	60	80	99
l	168.00	62.77	4.27	3.90	43	43	57	75
m	713.00	218.45	6.29	13.60	12	16	17	35
n	989.00	209.28	10.64	9.35	22	23	30	40
o	1014.00	213.28	12.22	8.66	21	23	28	42

Cell discrimination at this stage is not yet automatic. The decisions are still made by a human expert, but they are based on the computer results. There is work in progress to automate this process.

Analysis and Discussion

To judge the accuracy of our segmentation, we present in Table 28.2 a summarized analysis of the experimental results on nine images. In these experiments we were interested in extracting only granules and mitochondria. The segmentation is in error if: objects present in the scene are not identified either because they have been merged with close-by nonobjects or because they fell outside the range of close matches (e.g., a granule that is too close to the nucleus could be merged to nuclear chromatin or a lightly stained granule would not be dark enough to be considered a close match); objects are identified but incorrectly labeled (e.g., identifying a granule and labeling it as mitochondrion) either because two close-by objects (of different types) were incorrectly merged together or because similarities among different objects introduced errors; nonobjects are incorrectly identified and incorrectly labeled (e.g., labeling nuclear chromatin, which is not one of the objects we are searching for, as either granule or mitochondrion).

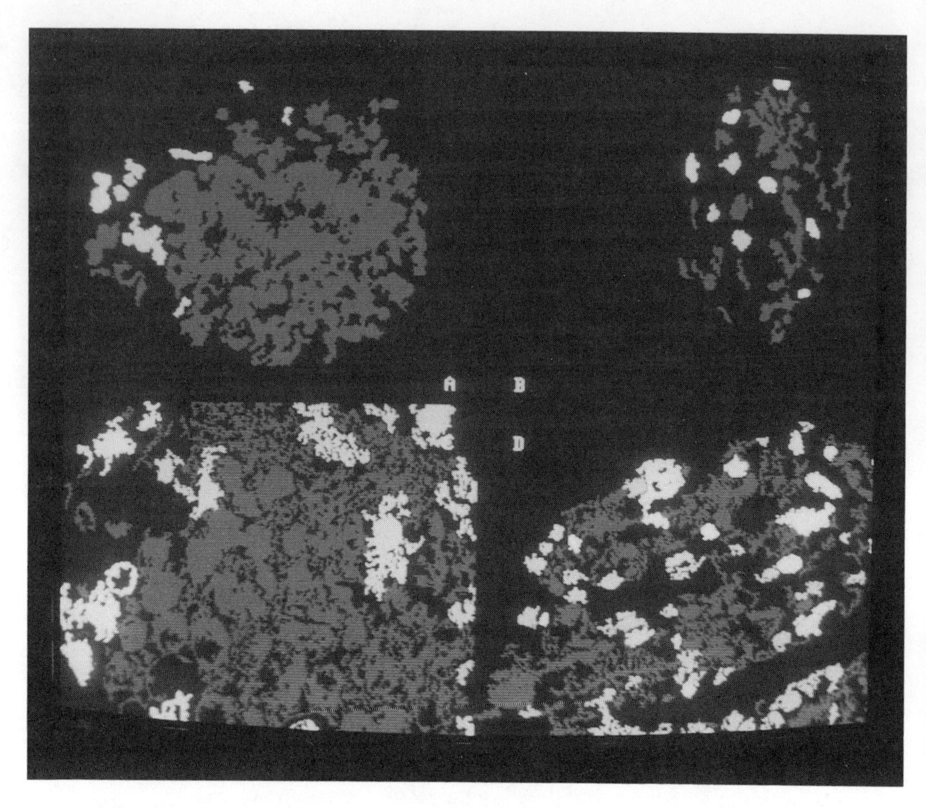

Figure 28.5. Use of organelle labeling to differentiate types of insect hemocytes.

Globally, we have an object extraction and identification success rate of about 58 percent for organelle in insect hemocytes (55 percent for granules and 62 percent for mitochondria). In the specimen extraction phase, the success rate is near 100 percent, as all the specimen choices are almost always perfect. In all our experiments, only one case of wrong specimen choice was reported. The image in this case was very unevenly stained. About 41 percent of objects present in insect hemocytes were not extracted (44 percent for granules and 38 percent for mitochondria). In Table 28.2, *other* represents nucleoli or dark nuclear chromatin, two structures that can easily be confused with either granules or mitochondria. About 5 percent of the granules identified and 6 percent of the mitochondria identified were false positives.

The mislabeling rate was quite low. Only about 1 percent of the granules were mislabeled as mitochondria. This demonstrates the reliability of our texture measures in texture discrimination. Less than 1 percent of regions with very little or no texture were misclassified as textured regions. There is no

Table 28.2. Three-Way Scatter Matrix for Error Analysis

Real	Classification		
	Granules	Mitochondria	Other
Granules	12.17 (55%)	0.17 (1%)	9.67 (44%)
Mitochondria	0.00 (0%)	10.17 (62%)	6.17 (38%)
Other	1.17 (5%)	1.33 (6%)	19.83 (89%)

misclassification of textured regions (mitochondria) as nontextured regions (granules). However, such results cannot be guaranteed if the images are noise contaminated or if the serial sections are unevenly stained.

In the matching of other objects in the scene to the selected specimens, various errors can be introduced for the reasons cited above. Errors can also be introduced by other discrepancies. In cases of bad iconic data (uneven staining, for example, in Figure 28.5c), the system can fail to produce specimens for a scene containing expected structure. Such bad data can also introduce errors like identifying parts of a single object as two different objects or identifying just parts of objects. In cases of very noisy images (Figs. 28.5a and 28.5b), many objects present in the scene can be missed, and the probability for mislabeling and identifying nonobjects increases.

Summary and Conclusions

In biology and medicine, one major problem often encountered in building computer systems to analyze and interpret images of natural scenes is the difficulty in characterizing objects present in these scenes. The objects often appear in such varied forms that it seems impossible to design a single complete model to represent a given object and all its variations. The main problems to be solved in the model design phase are how to represent deformation and how to establish parameters for shape and textural structure. The system must be able to recognize similar objects in the scene in spite of variations in size, position, or orientation. For example, granules in a TEM serial section come in various sizes, shapes, and orientations and can be anywhere in the image except in the nucleus. The system must also allow for variations in the textural characteristics of the same object. For example, the mitochondria in a TEM serial section do not necessarily have exactly the same texture, gray level, or shape.

The image analysis system described in this chapter uses models of expected structure. These are computer models that represent what the user expects to see in the microscope sections. The objects are characterized by quantitative parameters computed from the image based on qualitative descriptions of these objects that are found in the biological literature. The processes that determine permissible observed variations in these parameters are described in Dunn et al. (1990) and Dunn and Shemlon (in press). Dunn and Shemlon (in press) explain how to build these models for objects other than arthropod hemocyte organelle, including human anatomy and how it appears in radiographs and magnetic resonance imagery.

Future work will consist of building structural, functional, geometric, and morphological model descriptions for all organelles. At the moment our model descriptions have not been optimized, so the identification is not perfect and the success in classification is quite low, as can be seen from the results. In future work, we hope to optimize both the model descriptions and the implementation of the extraction and labeling algorithms, leading to better classifications.

Some of the crucial problems that remain to be solved include the modeling of complex objects that can have a completely different morphology depending on how the data are recorded. A good example is the golgi complex. In TEM images of insect hemocyte cross sections, this structure presents itself like microtubules or microfilalae, depending on whether the cell is cut transversely or longitudinally. Another example is the nucleus. Some of the errors cited in the analysis of the results can be corrected by optimizing the model descriptions. It is difficult to model all the forms of granules in a cell undergoing cell division. Other organelles in such a cell would also have structures that are distortions of the normal when the cell is undergoing cell division. However, near-optimal model descriptions must be developed. With experimentation, we can also optimize the matching algorithms by learning which parameters should be relaxed and which should be prioritized for particular data. This can help reduce the number of objects that are missed, wrongly labeled, or wrongly extracted.

The problem of poor data recording also has not been completely addressed. Uneven staining of microscope sections for a TEM cross section can produce images in which known structure can very easily be mislabeled. Regular methods for normalizing intensity have proved not to be very satisfactory solutions to this problem. Images obtained using equipment with lower resolution than that of a TEM need to be processed with a different image formation model. The resolution and complexity of the object description and image data must correspond for such a system to be guaranteed to give good results. As an example, the descriptions of granules and mitochondria would change if one uses light microscopy in place of electron microscopy. The object model remains the same, but as one changes the image formation process the object description must change. This, however, does not require a change in the organization of the segmentation system, since the paradigm is not changed. Our experiments with micrographs, radiographs, and aerial images support this assertion.

In the future, we shall consider the problems of reconstruction and interpretation of three-dimensional structures from a series of well-segmented, consecutive two-dimensional serial sections. From the three-dimensional volume and surface data we can synthesize an arbitrary section containing the objects of interest. The long-term goal of this research is to produce a computer-based cell identification and classification system.

Acknowledgments

The support of NIH research resource grants RR-7058 and RR-2483 is gratefully acknowledged. The micrographs used in this chapter were produced by Dr. S. S. Han, when he spent his sabbatical leave in A. P. Gupta's laboratory, and we are very grateful for his permission to use them. The help of Mr. Tajen Liang and Mr. Stephen Shemlon in implementing the programs is gratefully acknowledged.

29

The Automatic Design of Low-Level Image-processing Operators Using Classification and Regression Trees

RAMIN SAMADANI

Image-processing algorithms proceed in hierarchical steps from the image data to a desired result. Low-level image-processing operators are mathematical operations that convert an image into another image. These operations are often local, taking as input only a small region of the input image, and they provide as output a value at the center of the input region. These local operations will be referred to below as *masks*. Low-level operators are often used as a first step toward enhancing or extracting information from images. Edge detectors, for example, highlight the pixels that mark the boundaries of an object. The edge detector output image may then be used to delineate the boundary of an object for further processing such as shape determination or object recognition.

Low-level image processing operators are currently designed manually. The operators typically require the setting of parameters to tune their performance. These parameters are often set by slow trial and error, and new operators are necessary for each new class of image to be analyzed. This is an expensive design process. This chapter discusses an automatic method for designing low-level image-processing operators. For the technique discussed here, the user "trains" the computer to find pixels similar to ones chosen by the user. The technique, in effect, resembles supervised classification. It is based on automatically designed classification and regression trees (CART).

The solution presented in this chapter proceeds as follows: (1) The user selects candidates for local measures to be used to design the image operator.

Advances in Computer Methods for Systematic Biology: Artificial Intelligence, Databases, Computer Vision, ed. Renaud Fortuner (Baltimore: Johns Hopkins University Press).

This set of local measures or features is extracted for each pixel in the image. (2) Based on these features and training pixels with known (user-trained) operator response, a pruned decision tree is automatically constructed. (3) The pruned decision tree is used as a low-level image operator (a classification or regression tree consisting of local masks) to be used to process new images.

Pruned Decision Trees

Swain and Hauska (1977) gave an early example of the use of decision trees for remote-sensing classification. In remote sensing, classification means identification of each pixel in an image according to predefined criteria such as land cover class (vegetation, urban, etc.).

Among the advantages of decision tree classifiers are efficiency, the ability to use various types of features, including categorical features, and the ability to classify even when some data are missing (Breiman et al. 1984). Decision trees were originally designed manually, and this prevented their widespread use. Recently automated design techniques for decision trees have been developed (Breiman et al. 1984; Chou 1988; Smyth 1988).

This chapter reports preliminary results of the application of one such technique, optimally pruned trees, to the design of image operators. Already, pruned trees have been used for image data compression (Riskin et al. 1989), and nonpruned trees have been used to develop efficient image edge detectors (Smyth 1988).

The trees used in this chapter are designed using the technique described by Breiman et al. (1984). A brief review of the technique follows, but the reader will find detailed information and examples in the above reference. The construction of the tree is in two stages: (1) build a large balanced tree using the training set and (2) optimally prune the tree using cross-validation with the internal training set to estimate the true misclassification rate of the tree. Note that a variation of the above procedure allows the design of regression trees rather than classification trees, but the results in this chapter use only classification trees. The operators designed below thus have their thresholds automatically determined.

The first step of building the tree is not globally optimal, but uses a greedy-tree-growing algorithm. The greedy tree growing chooses a new tree based on locally optimal decision criteria. This greedy tree growing is not guaranteed to be globally optimal, but it is computationally much more efficient. For the greedy tree growing, the questions for the ordered variables are of the form, *Is feature* n *smaller than value* m? This type of question applies to the numerical features commonly used in computer image processing. The question of this form that best splits the training set is used to add new nodes to the current tree. The training data are then split and sent down to the appropriate terminal nodes, and the node creation process is repeated. The successively larger trees thus created have smaller apparent misclassification rates when one considers only the training data.

The very large tree that is built using the greedy growing may perform poorly when presented with new data since the nodes near the leaves were formed using much smaller training samples and may be due to statistical fluctuations. Another possibility is that some of the nodes are due to decisions made with features that do not have good predictive ability for the current problem. An optimal pruning algorithm is used, together with cross-validation, to generate a decreasing nested sequence of trees. At the large end the sequence includes the fully grown tree, and at the small end the sequence includes the root node. For each of the trees in the sequence the true error rate is estimated using cross-validation. The tree with the smallest estimate for the true error rate is chosen. This pruning technique precludes the design of a tree overly specific to the training data.

The constructed tree defines the low-level image-processing operator. Its use on new images is straightforward. The classification proceeds at each pixel by traversing the tree from the root node to a terminal node. The label of the terminal node will determine the output of the operator. The pruned tree design described above offers two additional advantages. First, it is trained by examples, mitigating the need for manual knowledge transfer from an expert to a computer. Second, the pruned decision tree operation is itself easily inspected and understood.

Examples of Applications

This section shows two examples of the use of pruned decision trees on computer-simulated images. The two examples are finding *pluses* and *exes* in speckled images. Very noisy computer-generated images like the one shown in Figure 29.1a are used as input to the decision tree. This speckled image was obtained from the true test image in Figure 29.1b by generating random numbers with a gamma distribution of order 4 at each pixel. The mean of the distribution at each pixel is the true pixel value shown in Figure 29.1b. These characteristics are used to simulate radar images that are created with coherent illumination. Figure 29.2 shows the features extracted from the images and used as input for the design of the pruned tree. The figure shows the numerical weights (blanks represent a weight of 0) for 13 masks that are applied to the input image. The masks shown consist of the pixel intensity values in the 3 by 3 neighborhood of a pixel and the four line-finding masks *vert, horz, pdiag*, and *ndiag*.

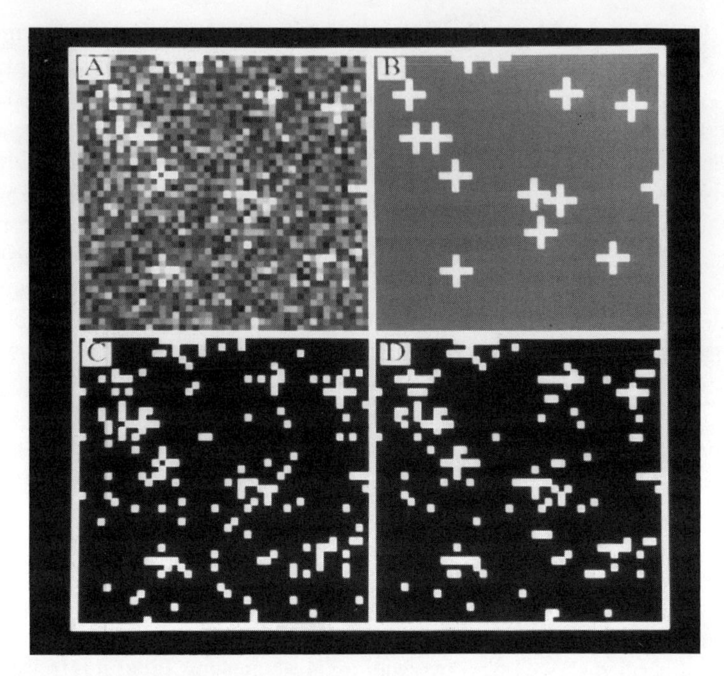

Figure 29.1. (a) Speckled test image. (b) True test image. (c) Output of a threshold at 122. (d) Output of the pruned tree of Figure 29.3.

First Example

The first example of the application of the decision tree methodology is that of finding *pluses*. A training image is used to build a pruned decision tree using the CART software package from California Statistical Software, Inc. Three test images are then used to test the performance of the pruned tree classifier. The results for one of the three test images are shown in Figure 29.1. The image B is the true classification of pixels for the input. It shows the output of the pruned tree classifier. For comparison, the input image thresholded at the value 122 is shown in Figure 29.1c. This threshold value results in the same number of missed object pixels as the pruned tree classifier. Visual inspection shows that the decision tree output is somewhat more acceptable than the image thresholded at the optimum value. The cumulative number of misclassified pixels for the pruned tree and for thresholding are compared using the results for all three test images. The threshold is selected for each image to have about the same number of missed-object pixels as does the pruned tree. The total number of missed-object pixels for the pruned tree classifier is 102 and for the threshold is 101. The pruned tree incorrectly classified 243 back-

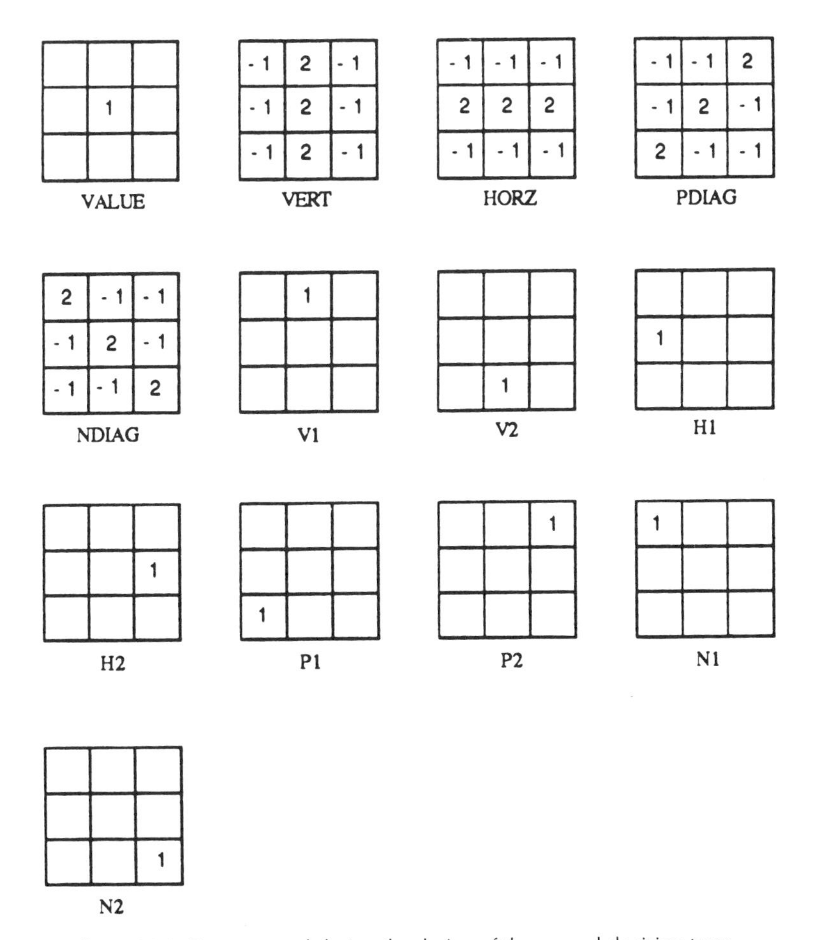

Figure 29.2. Features used during the design of the pruned decision trees.

ground pixels as object pixels, and the thresholding incorrectly classified 309 background pixels as object pixels.

The decision tree automatically constructed for this demonstration is shown in Figure 29.3. Near each internal node of the tree is the question that is asked at that node. If the answer to the question is *yes*, one follows the descendant node to the left. If the answer is *no*, one follows the descendant node to the right. The terminal nodes show the predicted classifications. The root node of the tree is a pixel intensity threshold. The next internal node bases its decision on the *horz* line-finding mask. The action of the decision tree offers a simply understood alternative to thresholding: threshold with a higher intensity value first and then threshold the output of the *horz* convolution. This procedure

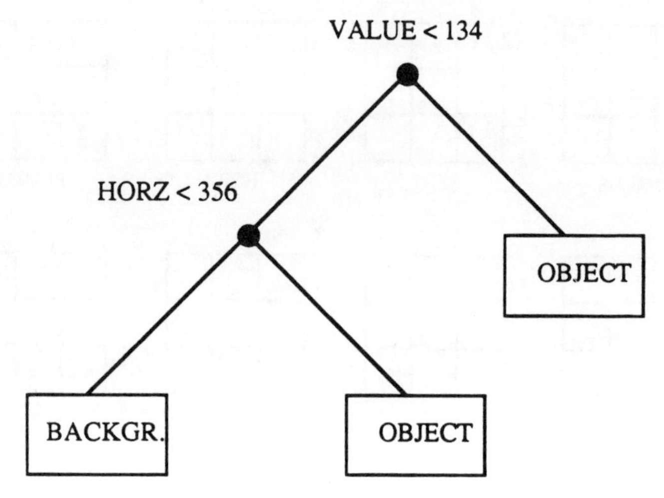

Figure 29.3. Pruned decision tree for finding *pluses* in speckled images.

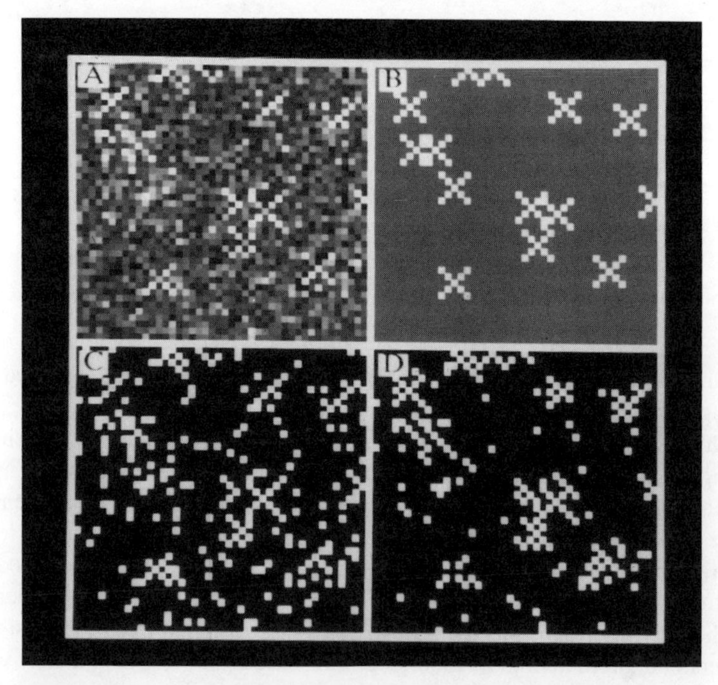

Figure 29.4. (a) Speckled test image. (b) True test image. (c) Output of a threshold at 107. (d) Output of the pruned tree of Figure 29.5.

performs better than thresholding alone. Finally, only two features out of the thirteen are used to classify the image.

Second Example

The same procedure was used to design a pruned tree to find *exes* in speckled images. The results for one of the three test images are shown in Figure 29.4. For comparison, the input image thresholded at 107 is shown on the bottom left. The tree classifier performs much better than simple thresholding. The number of missed object pixels for three test images both for the pruned tree classifier and for thresholding is 69. The pruned tree incorrectly classified 267 background pixels as object pixels, and thresholding incorrectly classified 515 background pixels as object pixels. The decision tree that was constructed for this demonstration is shown in Figure 29.5, using features from Figure 29.2. This tree has more nodes than the one used to find *pluses*, but it is similar to it

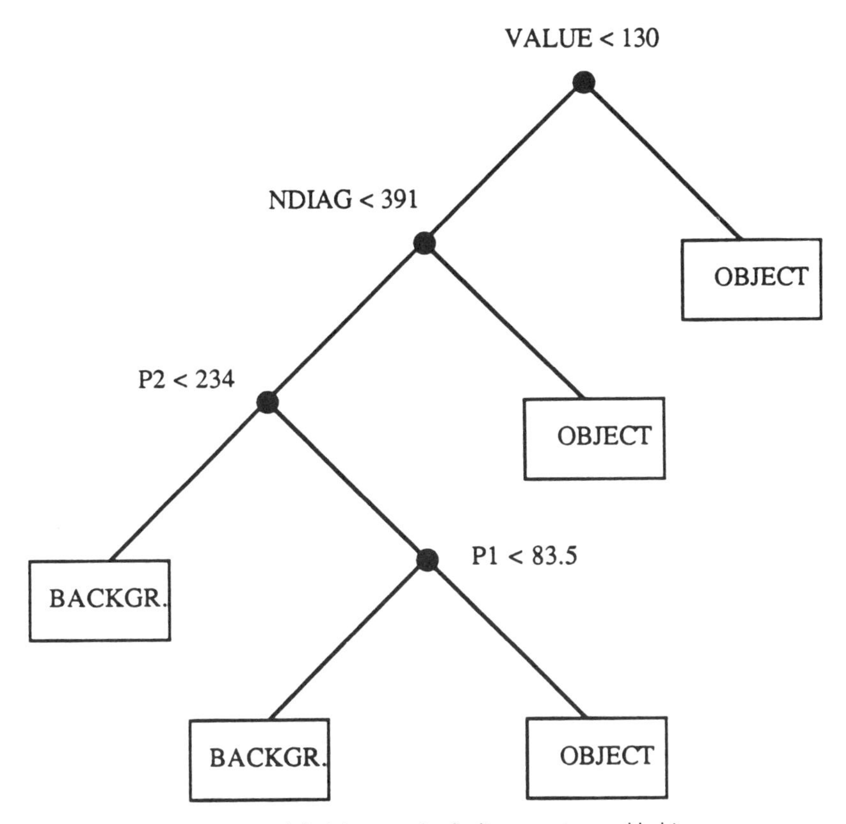

Figure 29.5. Pruned decision tree for finding exes in speckled images.

in structure. The root node still has an intensity threshold. This is again followed by a local line-finding mask. In this case the mask is *ndiag*. For this tree two neighboring pixel intensity tests using features *p1* and *p2* have been added.

Application to Radar Images

This section describes the application of the decision trees to finding pressure ridges in satellite synthetic aperture radar (SAR) images of sea ice. Figure 29.6 shows a SEASAT image of sea ice. The dark regions are water or thin new ice, and the brighter textured regions are older ice, built over several years. The brightest curvilinear features in the image are believed to be pressure ridges. Polar oceanographers are interested in locating pressure ridges since they characterize the internal stresses of the ice.

Figure 29.6 is part of the training set used to design an operator for extracting pressure ridges. The user has selected examples of pixels that belong to a pressure ridge and examples of pixels that do not belong to a pressure ridge. These pixels are highlighted with plus signs in the image. Using this image and other training images, the user selected seventy-six pixels that were classified either as *ridge* or *background*. These values were used to build a

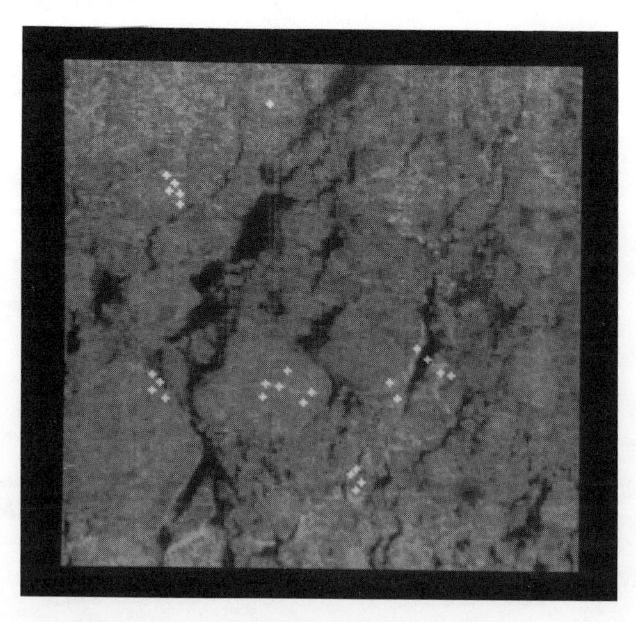

Figure 29.6. SEASAT satellite radar image of sea ice.

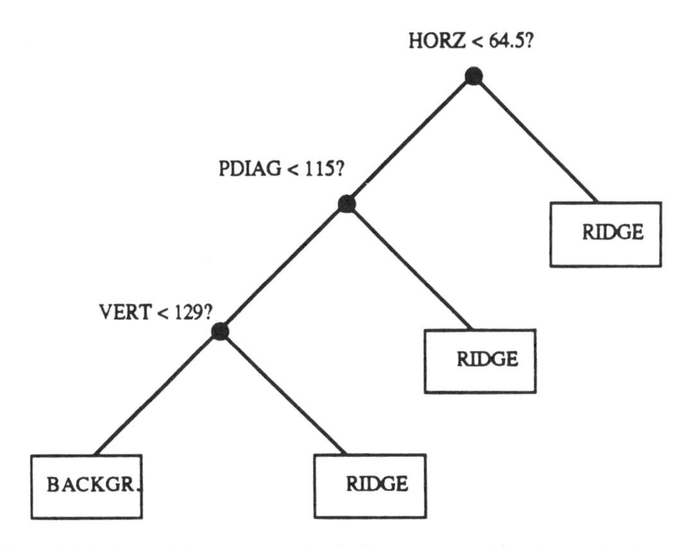

Figure 29.7. Pruned decision tree for finding pressure ridges in sea ice images.

pruned decision tree for finding pressure ridges in the radar images. The tree that results from this process is shown in Figure 29.7. If the answers to the questions at each node are *no*, the right path is taken.

Figure 29.8 shows a sequence of steps in the application of the decision tree pressure-ridge-finding operator. Figure 29.8a shows the input image. Figure 29.8b shows the output of the image-processing operator. The pressure ridges are successfully located in this image, but many spurious features are also found. These spurious features have been removed in Figure 29.8d by deleting short features. Figure 29.8c shows the pressure ridges found with this algorithm superimposed on the original image. These results are better than previously developed competing techniques (Bovik 1988; Samadani and Vesecky 1990; Vesecky et al. 1990).

Conclusions

Optimally pruned decision trees can be used to automatically design low-level image-processing operators, using training images, and these operators can subsequently be used to detect features in previously unexamined test images. The pruning process precludes the construction of trees that are too specific to the training set and thus improves their performance on new images. Furthermore, the pruned tree design process selects features that are relevant for finding the objects of interest. For instance, the features *value* and *horz* are selected for finding pluses, and the features *value*, *ndiag*, *p1*, and *p2* are

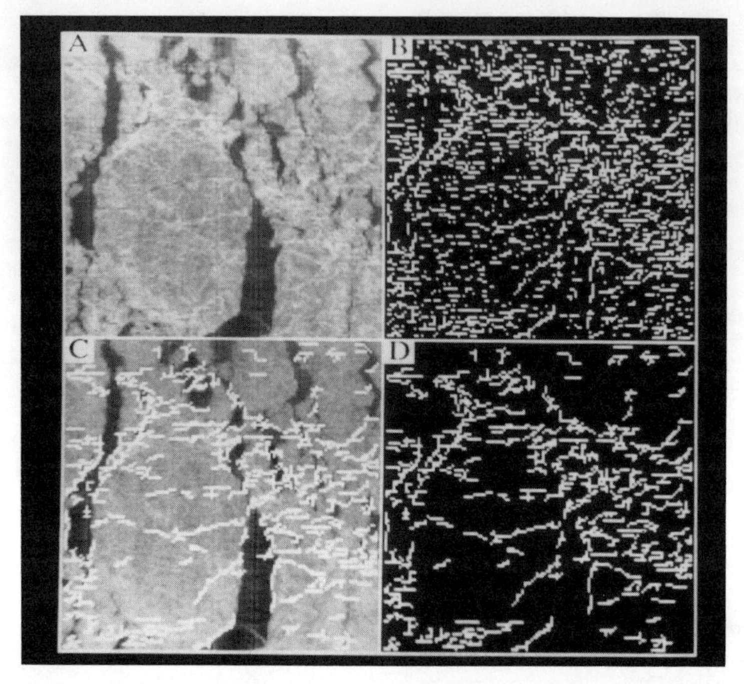

Figure 29.8. Applying the decision tree for extracting pressure ridges. (a) Input. (b) Output of the decision tree operator. (c) Results overlaid on the input. (d) Results of (b) after removal of short segments.

selected for finding exes. The operations of the pruned trees are simple to understand, are efficient, and do not require the troublesome process of expertise transfer needed by expert systems. Finally, with the computer simulations described above, for a given false miss rate, the trees resulted in lower false alarm rates than simple thresholding.

The application of the trees to SAR radar images of sea ice successfully located pressure ridges. The technique described in this chapter was demonstrated on an application of current interest to the author. The technique is, however, general, and it should apply as well to some applications of interest to systematic biologists.

Discussion

——: What is the definition of the term *greedy*?

Samadani: At each step of the tree-growing process you try to do the best you can; you are greedy. In other words, you take locally optimum steps. But in the overall scheme of things, taking those steps may not be the best thing to do if you look ahead and consider the total situation.

VI

Conclusions

30

Directions for Computing Research in Systematic Biology

PATRICK A. D. POWELL

The biological sciences are making increasing use of computers for basic scientific research in their disciplines. Many of the discussions and presentations of computer systems for systematics presented at the ARTISYST workshop show a deep and sophisticated understanding of the mathematical aspects of computing as well as software systems. However, there appears to be a large amount of duplicated effort by systematics researchers in developing software and hardware systems that either already exist or are being undertaken by researchers in engineering or computer science. For example, the problem of generating and porting graphical user interfaces to hardware and software systems has been a research and development concern of the computing community for many years. There are ongoing efforts in computer science to develop tools and methodologies to standardize the types of interfaces, as well as the methodologies and software to support them.

In this chapter I discuss areas of overlapping interest to both computer science and systematic biology and discuss methods to involve computer scientists in the development of research software and methods in these areas. These concerns include workstations, networks, the development of scientific databases, and information storage and distribution. I will try to identify common areas of interest, as well as areas that I see as promising for joint research and development.

Workstations

An emphasis has been placed on the development of computer workstations for systematics. Yet, differing usage of this term has resulted in some confu-

Advances in Computer Methods for Systematic Biology: Artificial Intelligence, Databases, Computer Vision, ed. Renaud Fortuner (Baltimore: Johns Hopkins University Press).

sion. Within the computer science community a workstation is a powerful single-user system with a graphical user interface. The use of the term *expert workstation* (Diederich and Milton, this volume, chap. 7) refers more to the set of all services (programs) that are needed to be run on a particular workstation. In case any confusion remains I will briefly address the issues of computer hardware and operating systems, before addressing the application programming issues.

For a computer scientist, a powerful computer workstation usually refers to a small (desk-top) Reduced Instruction Set Computer (RISC)–based system that provides local support (i.e., using the microprocessor) for interactive computing. The workstation is usually connected to other workstations or central computing facilities by a high-speed local area network, which in turn is connected to a wide area network. The user interface provided by the workstation is usually a high-resolution (1024 by 1024 pixels) monochrome or color monitor, together with any necessary mouse, keyboard, or user-level input-output (IO) devices. Most workstation support software provides a windowing environment such as X Windows or Sunview. Workstations are available from many high-end computer manufacturers, such as Sun Microsystems, Data General, DEC, IBM, and NCR, to name only a few.

The computational and graphical capabilities of currently available workstations cover an enormous range. Based on presentations of programs and systems during the ARTISYST workshop, it would appear that a workstation able to support most programs and applications is widely available at reasonable prices. For example, a RISC-based workstation with a 10 MIPS (million of instructions per second) execution rate, a 200-MB hard disk drive, a high-resolution monochrome graphics monitor, and 8 MB of memory is available for a little over $3000 to educational institutions, and the cost would be lower if large numbers were purchased. Prices for color systems are slightly higher (Sun Microsystems, University Pricing Structure, January 1991, for SUN SPARC Workstations, and Data General Corporation, List Prices, February 1991. University and research discounts applied; specific prices are available on request).

An alternative to a workstation would be to use a personal computer, such as those available from IBM or Apple Computers. However, in order to provide the same capabilities as a workstation, a PC requires substantial investment in time, network interfaces, and applications software upgrades. Also, most workstations use variants of the UNIX operating system, which provides support for multiprocessing and a standard set of IO and networking facilities. Most personal computers use a single-user DOS system and have primitive IO support. Finally, workstation software is now available that emulates an IBM or Apple PC, usually at satisfactory speeds for most simple applications.

In addition to workstations that are characterized by their interactive graphics capabilities, many researchers have found that they needed powerful computing engines for numerical processing (number crunching) or other computationally intensive (compute bound) tasks. Traditionally this type of computation has been done using centralized computing facilities. However, due to the development of high-performance microprocessors, these central computing facilities are rapidly becoming obsolete. While much of the application software developed for these systems will continue to be used, it is clear that developing new software for these systems will be difficult to justify, and porting old software to new systems will be an ongoing effort.

From these comments, it appears that the computer hardware needed to support the current set of application programs is within the reach of a well-funded researcher. The difficulty in convincing software developers to use a workstation rather than a personal computer for development and use has decreased as workstation costs have decreased. Finally, the multiprocessing, graphics, and network capabilities that are fundamental to workstations are proving necessary for advanced programs and research.

Databases and File Systems

Although the current database methodologies developed by computer scientists are able to handle existing financial and commercial databases, the growing concensus is that greatly improved methods will have to be developed to deal with large volumes of scientific data. For example, the Human Genome Project will be generating information that must be stored in a database. This database will need to be updated, queried, distributed, and managed in ways that differ radically from the current commercial transaction-oriented databases. Problems in data distribution, database update, database consistency, and information retrieval will need to be solved in order to access this information. The area of large-scale scientific databases will be very active until these problems are solved. I will briefly mention three areas that need to be considered when dealing with databases: physical storage of data, database management systems, and information retrieval.

Storage Media

The type of media used to store database information is strongly dependent on the size of the database (usually measured in bytes), the number and types of database updates, and the number and types of database queries. If a database is small (under 10 MB), then it can be stored on a magnetic disk drive. However, as both the number and sizes of databases grow, the choice of storage media becomes a major consideration.

For static databases, that is, ones with relatively few updates and of moderate size (10–650 MB), optical disk technology can be very cost effective. There are three types of optical disk systems: CD-ROM (compact disc read only memory), WORM (write once read many), and EMO (erasable magnetic-optical) drives. The CD-ROM drives use the same media and recording techniques as the audio CDs, and many CD-ROM drives can also be used as audio CD players. Each diskette holds about 600 MB of information. The CD-ROM drives are available in the $500 range, and the disks cost about $20 each in quantities of 10,000 units. The WORM drives record information by burning or melting part of the disk media using a high-powered laser, and then use a low-powered laser to read the information back. The WORM drive costs about $3500, and each 650-MB disk costs about $200. The EMO drives use various techniques to read and write information on a magnetic media. The drives cost about $4000, and the disks about $350.

The major benefit of the optical disk technology is the ability to remove and replace media in a simple manner. The CD-ROM is currently the most cost-effective manner to distribute large volumes of information to many users, but has the disadvantage of having a slow access rate. While the WORM and EMO drives have faster access times, they are substantially slower than a magnetic disk drive. The removable media have been exploited by creating a "jukebox" that has a mechanism for automatic insertion and removal of up to 128 different disks. This makes it possible to construct a data library that has slow but automatic access to large volumes of information.

For databases that need frequent update or rapid access, the magnetic disk drive is still the most effective medium. Individual disk drives with 1.2 GB capacity will soon be available for less than $4 000. Including power supplies, cabinets, cables, and other miscellaneous items, a 10-GB file system for a workstation would cost less than $40,000.

Database Management Systems

Most of the currently available database management systems (DBMSs) have been created for the management of transaction or commercially oriented information. These databases are based on a relational data model, which is basically table and entry oriented. Adapting the relational model to represent information of a hierarchical, recursive, and object-oriented nature has proved to be more difficult than first assumed. The design and development of new types of DBMSs for scientific information are of great interest to computer scientists. A joint effort of biologists, who have the requirements and the data, and computer scientists, who have the computer and data management experience, should be of mutual benefit.

Information Retrieval

One of the major problems to be faced when dealing with databases is finding the desired information. This can be partitioned into the following problems: specifying the questions to be put to the database management systems, understanding the answers that are returned, and locating the database that contains the desired information. Each of these problems is an ongoing research problem for computer scientists. The design of new database management systems will have to take into consideration the types of questions that will be asked. If the designers are not aware of database use, the results may be unsatisfactory for the users. Presentation of information will have to be done in a manner understandable to the questioner. This will require research in user interfaces and information presentation, a field of great interest to software engineers.

The problem of locating a database that contains the desired information is referred to as the *federated database problem*. If information is scattered through several databases, then these will first have to be physically located, the information extracted from each of them, and finally the information combined. If there are a large number of individual databases, each with a unique coding and database management system, the problem will rapidly become insurmountable. It is important for the implementors and designers of databases to consider these issues before implementating yet another format and database system. Again, there is considerable interest in the computer science community in these areas.

Image Processing

Several of the speakers at the ARTISYST workshop showed overwhelming need for some easy, cost-effective, compact, portable, robust, and reliable method to handle and collect visual data. The range and volume of data that needed to be recorded and analyzed were tremendous. The printed text of classical works of systematics, including diagrams, drawings, and other information, would be useful for many purposes. The need to record details of plant, insect, and animal specimens, in laboratory, conservatory, and natural surroundings, has been discussed at length, and the problems of specimen storage and deteriorating displays could be addressed by recording the specimens before they (literally) rot to bits.

Image processing is also important for the automatic recognition and classification of specimens. Automating chores such as counting and identification of insect specimens would have tremendous impacts on workers in the Department of Agriculture, where they need to process literally thousands of insects per month. This would also have an impact for researchers in population studies, allowing them to perform many more samples than are currently feasible. Unfortunately, while the goals are visible, the desire obvious, and the need apparent, the technology, engineering, and research are only in the initial stages. Much of the work in image studies is directed toward quality control, robotics, and pattern analysis of handwriting, fingerprints, and reconnaissance photographs. Although a joint effort of biologists and vision researchers should be strongly encouraged, many of the problems with image processing have proved more difficult to solve than was initially assumed.

In order to store and capture visual information, we are going to need extensive research and development in the areas of image capture (cameras), storage, and retrieval. Our available technology for capturing high-resolution video images is still tremendously primitive. Until we are able to provide electronic image resolution to the limits of 35-mm film, we will continue to have poor-quality images. This area of research is also of interest to the medical community, which is trying to deal with the problems of long- and short-term storage of medical records. The need to record and store X-ray images, tomographies, and other information is becoming increasingly important. Image storage must overcome the huge data requirements by developing new methods for both data storage and image data compression. Finally, we need to develop methods to perform image analysis that are more versatile and robust than those currently available. These methods will have to be incorporated into tools and applications for biological researchers, whose definitions of adequate performance and recognition may differ substantially from those of the computer science and engineering research community.

One of the interesting developments during the last decade has been the development of high-resolution (400 dots per inch) document scanners and low-resolution (100 dots per inch) handheld, portable document scanners. Advances in this area may provide an answer to the problems of image capture.

Graphical User Interfaces

Graphical user interfaces for application programs and general use are becoming extremely sophisticated as well as more complex. This, in turn, has driven users to demand an integrated graphical presentation environment, where all of the tools and their interfaces have a standardized user interface. User interface standardization has become necessary as the number of application programs used by individuals has greatly increased and the graphics user environment has become widely available. Rather than discuss the benefits of one windowing system over another, I will focus on the areas of development that I believe will have a substantial impact in the near future.

The most important advance for user interfaces will be the development of a standardized and complete set of help or assistance functions. I feel that the major difference between most user-developed software and commercially developed software is the level and sophistication of the user help and documentation. One of the areas of interest to researchers in software engineering would be to study the needs of users for help and online diagnostic functions and then develop automatic or semiautomatic ways to generate user assistance.

A second development in user interfaces is the ability to tune the level and type of interface. For example, users can be classified as novice, casual, familiar, and expert. The type of interface, prompting, and presentation of information could then be tuned to the level of expertise of the user. Tuning could also be done depending on the use of the software. For example, a multipurpose database or taxonomical query system could be used for many different purposes, such as finding the single best match or finding how a particular specimen differs from a model. While the same software could be used for both problems, the presentation of information would depend on the problem area.

A standardized set of tools for creating user interfaces would ease the problems of providing a common set of functionality and uniform methods of presenting information. This area is of interest to software engineering researchers.

Finally, one of the most difficult parts of constructing a user interface is to make it portable and extensible. In order to be portable, a proprietary graphics or windowing support system cannot be used. At the current time there are few, if any, portable, easy to use, and general-purpose graphical interface systems readily available to the application developer. However, the X Window System appears to be at least a partial answer to this problem. Many graphic system suppliers are now providing support for X Windows, either at the application library or at the full distributed network level. In passing, it should be noted that most workstations have X Window System support and user interfaces.

Networks

One of the most important developments in the computer science community in the last decade has been the availability and ease of access to wide area and local area networks. This has caused a revolution in computer science research. For example, Lynn Conway (1981) has related how in 1978 and 1979 a VLSI (very large scale integrated circuit) design course was carried out at the Massachusetts Institute of Technology and the California Institute of Technology. She emphasized that many of the projects were successful only because of the fast communication and fast file transfers available using the

ARPANET wide area network. Using this initial experience, the National Science Foundation (NSF) established MOSIS, a facility to fabricate integrated circuits for research and teaching purposes. Over one hundred universities who are offering VLSI design courses use the INTERNET, the successor to the ARPANET, to obtain design information and transfer designs.

The super-computing community has also made heavy use of wide and local area networks to carry out research activities. Chemists, physicists, civil engineers, and others are using computers on the other side of the continent in real time. As a result of the growing use of the INTERNET for research activities, the NSF has undertaken to upgrade and improve the backbone of the network. Within the decade they hope to have 1-MB data links connecting the major users (i.e., generators of large amounts of traffic) on the INTERNET. This will make fast communication more readily available and inspire new uses of the network.

One of the novel uses of a high-speed network would be to reduce the amount of duplicate information stored at each site on a network. For example, let us assume that a high-speed disk drive has an average access time to fetch a record of 16 ms and that it transfers data at a rate of 1 MB (or 8 megabits) per second. To transfer 1000 records of 1000 bytes (for a total of 1 MB of information) in sequential order on the disk drive would take approximately 1.016 s. If you have two workstations connected by a local area network with a 1-MB/s transfer rate, you would need a total of 2.016 s to transfer the data to the other workstation. However, it is possible to overlap the disk transfers and network transfers so in fact the total time to transfer information would take only very little more than 1.016 s.

These observations have led to the development of diskless workstations that have no disk drives and that use a high-speed local area network to access information from file server which has disk drives. This means that there need be only a single physical copy of a file or database and that workstations can access portions of it as they need it. However, if there are too many workstations trying to access the same files or to use the network at the same time, performance will suffer. Thus, most workstations usually have a small (200 MB) local drive on which is stored frequently accessed files and which is used for temporary working space.

Clearly the idea can be extended to providing access to large databases using the same simple technique. Only a single physical copy would have to exist, and users would access it over a high-speed network. In order for this to be feasible, the total number of users and accesses to the database would have to be small enough so the network did not become saturated and the server could respond to requests in a reasonable amount of time.

If the biological community wants to get access to the large amount of information currently stored in databases, it is clear that using the INTERNET will be the most effective way to proceed. Unfortunately, this requires an investment in installing both a local area network for your local facility and establishing a link to one or more sites already on the INTERNET. The good news is that most university computer science and electrical engineering departments have already made this investment, and it is usually possible to get access to their facility. The bad news is that you will have to make serious financial commitments to running cables and an ongoing commitment to monitoring your network for problems.

Distributed Processing

Once a network and high-performance workstations with local disk drives are in place, high-performance parallel processing and distributed data bases can be investigated. For example, suppose that we wish to search a database of textual information for a particular pattern and discover that the only way to do the search is to exhaustively examine all items in the database. Surprisingly, this is not an uncommon event when searching protein or DNA sequence information databases.

If all of the information is stored on a single disk drive, then the maximum search rate is limited by the lower of the disk data transfer rate or the processor throughput. For example, if the disk can transfer 1 MB/s, but the processor can examine only 100K/s, the processor is the bottleneck. If we have a 100-MB database and the processor was the bottleneck at 100K/s, it would take at least 1000 s to search the database; if the disk was the bottleneck at 1 MB/s, it would take 100 s to search the database. Now let us assume that we have one hundred processors with local disks, each with 1 MB of the database. If we could broadcast a simultaneous request to all of the processors to search their portions of the database and the disk transfer was the bottleneck, it would take a only 1 s for each processor to perform its search. Thus, we could effectively speed up our search times by adding more disk drives and partitioning the database further.

If we provide redundancy by having duplicates of portions of the database stored on several processors, then, while the searching process could take longer, we could lose several processors and still perform the search. Thus, we would enhance our reliability and accessibility to the information at only a slight cost.

System Development

Software design and development has become easier and less expensive, but this has been balanced by a growing need for more complex systems with higher performance. Thus, we are faced with a problem in software development: Who will want to, or even be able to, fund the development of software and systems that are desired by users? Also, who will support and extend these systems and repair defects and errors when they are found?

One solution has been to use research funds to develop and maintain software. As several persons have pointed out during the ARTISYST workshop, developing better forms of existing software is not computer science research. However, developing methods and tools to do such development usually is computer science.

A second solution has been to commercialize software and sell it to the users. With a high enough demand and large enough user community this could be a feasible method to fund development. One barrier to this is the small number of users and limited funds available for software purchase in the systematics research community.

Another method, which is used in other research fields, has been for a funding agency or other agency to fund the development of commercial software. This has the benefits of a centralized development and maintenance activity for the user community at low cost to individual users. This appears to be the most fruitful method for future software development.

Conclusions

If we put all of the material I have outlined together, we arrive at the following picture of the workstation of the future. It will be a high-performance color graphic workstation, with a network connection to perhaps thousands of other workstations. It will have a high-capacity laser disk for local storage of high-volume material. It will have a high-speed magnetic disk for temporary storage as well, with access to data bank servers across a local area network. On the network will be high-performance compute servers that will be able to be accessed for computationally intensive tasks. Finally, all the systems on the network will contribute a small amount of their computing resources to a common pool for parallel processing and database searches. The workstation may have attached a high-resolution camera for image capture. A major funding agency, it is hoped, will be developing software for use on this system, with the systematist providing input on the types and kinds of systems required.

There are many areas yet to be explored. For example, sound classification and voice recognition systems are of interest not only for user interfaces, but also for the analysis of insect and bird songs. Portable, low-power, and rugged versions of these workstations would also be useful. Programs for the general public, which will run on personal computers, should also be explored. Finally, the problem of finding the funding and qualified staff to establish, develop, and maintain such complex systems still needs to be studied.

Appendix A: Workshop Notes

Bayes' Rule, Belief Networks, and Discriminant Analysis

Walker: Several people have asked about Bayes' rule, Eric Horvitz has mentioned belief networks, and some people have asked about the relationship of discriminant analysis to neural nets and to nonparametric methods of identification. I am going to explain these various points.

The Concept of Feature Space and Bayes' Rule

If you want to develop a classifier to distinguish apples and oranges, you can use color and weight as features. In Figure A.1 we have a scattergram of points representing apples and oranges, with their color and weight. This scattergram represents a feature space.

To develop a program that will distinguish apples from oranges, you can start with a training set with known apples and oranges for which you have determined the color and weight. Using this training set, you can draw various lines in the feature space for separating apples from oranges. You can draw a straight line as in Figure A.1, or a curve, or some other way of partitioning that feature space.

The formal, mathematical way of deciding which group an unknown specimen belongs to is to use Bayes' rule. Using the medical terminology, Bayes' rule gives the probability of having a disease, given some evidence:

$$P(\text{Disease}|\text{Evidence}) = P(\text{Evidence}|\text{Disease}) \cdot P(\text{Disease}) / P(\text{Evidence}) .$$

This probability P(Disease/Evidence) is equal to the probability of observing this evidence given the disease, P(Evidence|Disease), times the prior probability of the disease, P(Disease), divided by probability of the evidence, P(Evidence).

You can get the prior probability of observing the disease by interviewing an expert and asking him to give an estimate the proportion of the population that has the disease. For medical diagnosis or specimen identification, you may limit the estimation of prior probabilities to the geographic area where you are working. You can also get this prior probability from a large patient

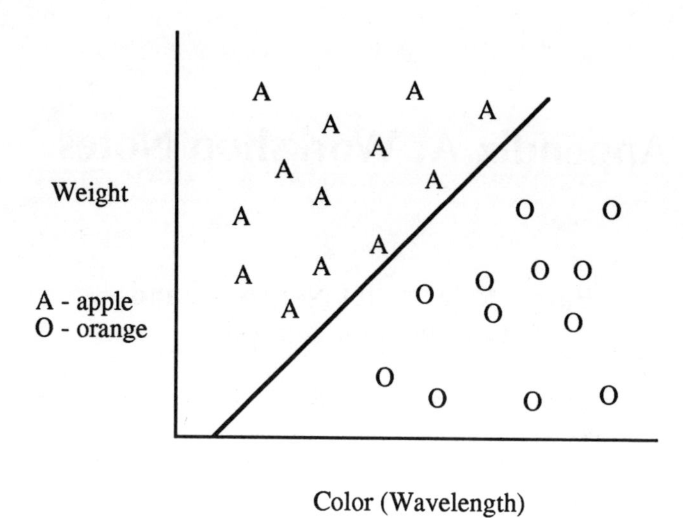

Figure A.1. Scattergram of apples and oranges, in a feature space defined by the features' weight and color.

database by assuming that whatever you are going to see in the future occurs with the same frequency as it does in this database. You can get the probability of seeing a particular piece of evidence from the same database. The conditional probability of evidence, given the disease, is an estimate of the chances that the particular piece of evidence will show up among all your patients who have that disease. You can find this by doing a frequency analysis from the database or by several other methods.

A high prior probability of disease increases the probability that a particular patient has the disease, regardless of the values of the other probabilities in the Bayes' formula. This means that the probability of having systemic lupus, a moderately rare disease, is low compared to the high probability of having a very common disease such as a cold.

By analogy, if I have to identify a nematode from California, the prior probability that it belongs to a species known only from Manchuria is low. Strictly speaking, the prior probability should apply for the entire universe, but in practice you know what area you are looking at, and you can pretend that the whole universe is restricted to that area. If you are considering the whole universe, the prior probability that you give for California then becomes a conditional probability; that is, you condition the probability on the fact that you are in California. It will work out as long as you calculate all the other probabilities using the same restriction. If you cannot give any estimate at all of the prior probabilities, you could make the approximation that all prior probabilities are equal.

Dallwitz: Another consideration is the probability that a particular species is being presented for identification. Houseflies are very common, but they are never presented for identification because everyone recognizes them.

Walker: Yes, this is absolutely correct. Generally speaking, sometimes the prior probabilities really matter and people will use them, and sometimes they won't. It depends on the domain, on the type of problem you are solving, whether you are an expert, etc. To estimate a probability, you can use all types of data, qualitative or quantitative characters, mass function, density function, real value variables or binary variables, and so on.

Pankhurst: Many taxonomists are reluctant to use probability, and I wonder whether these methods work as well as or worse than keys.

Walker: The question is important because it motivates whether you bother paying attention to the rest of my explanations. I can answer very quickly. The simple answer is that, analytically, there is no way of knowing ahead of time which discriminant method is going to work best for any data set. The only exception is if your data are normally and identically distributed except for relative location in feature space. In that very special and very rare circumstance, linear discriminant analysis is going to work better than anything else. But outside of that, all bets are off.

Diederich: Can you compute the probabilities if you have a database of species rather than a database of specimens?

Walker: You really want specimens, as subsequently you will try to identify specimens. If you have only data about species, you have to ask the expert. If everything else fails, if you only know a species from a single specimen, and you still want to apply Bayes' rule, you can always say that the probability of seeing the evidence given that species is 1.0 for all its characteristics, and if you do so you end up with a key. If you don't know prior probabilities, you say that they're all equal. I would like you to come away with the idea that all the classification methods eventually depend on Bayes' rule and they're just variations on Bayes' rule in terms of what assumptions you make or what data are available.

Conditional Independence and Belief Networks

If you have several pieces of evidence, for example if a patient says that he has a fever and a temperature of 101, he has not given you two independent pieces of information and you should not count that evidence twice. If on the other hand he says that he has a fever and he has a family history of cancer, you will probably be willing to treat these two facts as being independent and weigh the evidence twice. When people talk about *idiot Bayes*, they mean that you are making the simplest possible assumption; that is, all your pieces of evidence are independent of all the other evidence. That is the whole idea of conditional dependence or conditional independence. In Bayes' rule, you actually look for the conditional probability of evidence 1 and evidence 2, given the disease. The expression *conditionally independent* means that these two pieces of evidence are independent, conditioned on having the disease or not having the disease.

Belief networks are one of hundreds of methods people have proposed to try to deal with conditional independence. For purposes of explanation here, I want to treat them as statistical methods, although they sometimes are called knowledge bases because they are underneath an expert system. You may have something that looks like an expert system shell, and deep in its heart is a belief network, but you wouldn't know that. Outwardly, it looks like a diagnostic program such as MYCIN or any rule-based expert system.

Fortuner: How do you define independent characters? Would you say that their coefficient of correlation is equal to zero?

Walker: Two pieces of evidence are independent if the joint probability of observing these two pieces of evidences equals the probability of observing the first one times the probability of observing the second one:

$$P(A \text{ and } B) = P(A)\,P(B)\ .$$

If there's a 0.5 chance that I'm going to have a fever and a 0.5 chance that I'm going to have a temperature of 101, this joint probability should be 0.5 times 0.5, which equals 0.25. We know that's nonsense and that these two pieces of evidence are not independent. If you have multiple variables it is not enough to look at them pairwise, although you may be forced to do it for lack of data. But, yes, the correlation should be zero for real-valued variables. It may not be exactly zero because of sampling error. You have to do a chi-square test to see if it is significantly different from zero.

Discriminant Analysis

R. A. Fisher (1936), working on agricultural and medical problems, wanted to distinguish two or more groups. He found that, by making two assumptions about the data, there is an analytical solution to the problem of finding the best possible line to separate the two groups in feature space. This is Fisher's linear discriminant analysis, which you find in all standard statistic packages: SAS, BMDP, SPSS, etc. The specific assumptions are that: (1) both groups are normally distributed and (2) they have identical covariance matrices. This means that they have exactly the same shape in feature space and they differ only in their mean location.

How is linear discriminant analysis related to neural network methods? I bet that if Fisher had had computers, he would have used neural nets. Neural networks use a heuristic search to find the dividing line. This means sticking a line someplace in feature space and asking if it separates the two groups well; then move the line, ask again, and keep moving it around until you find a good dividing line. Search methods give you many ways of doing it. Neural nets use many variants of the basic method of back-propagation and least-squares error.

Now, to come back to the question from Richard Pankhurst, how do you know if this method is any better or worse than linear discriminant analysis or using keys? The answer depends on the criteria you use. One very good criterion is, Can I run it on a computer that I own or can afford? Other very good criteria are, Do I have any data to estimate the numbers? Can I build it in a system that the users are willing to use? Is my domain expert willing to give me the numbers I need? Those are all very important issues, separate from the mathematical form of the solution, that may in fact be of overriding importance.

If you draw a dividing line in feature space based on assumptions about the distribution of the groups you are using a parametric solution. Linear discriminant analysis is parametric because you assume the normal distribution. You can use other algorithms such as the k-nearest neighbor algorithm. To identify a new point using the k-nearest neighbor algorithm for apples and your oranges, you just look at that point's neighborhood in feature space and ask, "What's the nearest thing to it?" You assign the new point to the same class (apple or orange) as its nearest neighbor. You can also take the average of the three nearest neighbors, or the average of the three nearest neighbors weighted by how close they are to the unknown.

A slightly different version uses Parzen kernel density estimation. Conceptually, it is almost exactly the same thing as the k-nearest neighbor but, instead of just counting the nearest neighbors, you define a region, a circle, or some other shape, and you do a weighted average of the points that fall within or close to that region. In practice, that region is often defined as being something like a cone in the sense that the points nearest to the peak of the cone get weighted the most. The height of the cone is proportional to the weight given to each point. You can think of the cone as a graph of the weighting function.

Identification Methods

Pankhurst: The expert systems people have a very impressive methodology there but it's too general. They use probabilities; they have rule systems which deal with fuzzy thinking by experts . . . You are not going to get probability data from most biologists. We can't get it for you and we don't need to, we can do better than that. We don't need rules, we can't give you probabilities, but we can provide data matrices.

Dallwitz: Most taxonomic works do not require the probabilistic approach, and most taxonomists work in a deterministic way, saying that a species either has or doesn't have particular character states. They fall back on things like discriminant analysis only for groups which cannot be separated by deterministic means. It is important to have systems that will enable the taxonomists to record data in a deterministic manner. There are fairly mature and successful deterministic identification methods, and there are also purely probabilistic medical identification systems. Ordinary taxonomists would not use medical diagnosis type of programs. The data are not of that type. Taxonomists are not interested in probabilities; they don't understand them; they don't want them. But there are cases where taxonomists are not able to use deterministic methods. There are species groups that cannot be separated by those methods, and discriminant analysis or other methods may be needed. It would be better if the different methods could be integrated so we could use whichever is necessary.

Pankhurst: We need methods that operate off data matrices. We don't see that they are really out there. If you want original expert system research, following what I said before, what you want is nonprobabilistic, non-rule-based systems, since this makes the problem harder, something like matrix-based expert systems to make the problem easier.

Fortuner: I believe that biological data are fundamentally probabilistic, whether we record them as such or not. Probabilistic identification should not be systematically ruled out. It may be useful in some circumstances, and rule-based expert systems may be useful in other circumstances. On the other hand, some of the approaches advocated by computer scientists may have requirements that are not met in nature. For example, Mike Walker said earlier that independence of characters means that their correlation should be zero. In biological specimens there is no such thing as two absolutely uncorrelated characters.

Walker: Let's say that we want whatever identification method works best, and it may be probabilistic or ruled based or matrix based. Another need is for systems that integrate both the probabilistic and the deterministic approaches. We also need systems that make better use of the metadata, such as the uncertainty about the data.

Classification and Identification Data

Allkin: We are assuming that we have a set of descriptive information about the taxa and that we are going to produce a diagnostic tool. Now, systematists fit descriptive data into data matrices or databases to do classification rather than identification. This raises a number of questions, and I wonder whether computer science might help. One is, How do you select sets of characters for identification or for classification? They are not the same thing. For classification, you need characters that are shared by several taxa so you can group these taxa into a higher-level taxon. For identification, you want sets of characters that are unique for each taxon. The second question is, How do you manipulate the information from a database and represent it in different ways for different purposes? For classification, you want your data to be very precise and detailed. You are prepared to work at your microscope for hours to capture that particular fine measurement or to draw shapes in a very detailed way, whereas, for a diagnostic tool, the information has to be much more obvious. Rather than a precise quantitative measurement, for example, you may want just a quick range. We need a method of converting, redefining the same information.

The final problem is that not only you need different information, but sometimes the same data need to be presented differently depending on what you want to do. For my grandmother, a poinsettia has big red flowers, but an expert knows that the flowers are white and minute and hidden away and that those big red things are leaves.

Systematists, largely through lack of experience and lack of contact with computer scientists, or maybe because of contacts with certified computer salesmen, assume that once they have their data in the machine, they can do everything with it. They can do a classification, they can produce an identification tool, and then they can write a book. They must understand that they need different sorts of information for doing different things. We should be able to rephrase information in a knowledge base constructed for classification to make it usable for diagnostic purposes. We also want to exchange the knowledge that may be germane to other activities.

White: What you really need are tools able to handle your data and adapt them to different uses depending on what you want to do.

The ASN.1 Data-Exchange Standard

Karp: Many database developers expect that the database they produce will be used outside of their organization. For example, the National Library of Medicine (NLM) is developing a new database of DNA and protein sequences (called the GenInfo Backbone) for use by the molecular biology community. A number of problems arise in attempting to exchange data among different organizations, particularly if we assume that those organizations use different hardware platforms and different database management systems (DBMSs). We must have a means of describing the structure of a large database in abstract terms so that users of the database will understand all of the elements of that database and the relationships of those elements to one another. We must also have a means of describing the exchange format of the database with great precision so that, on any hardware platform, a program can automatically determine the boundaries between individual data elements and can convert the machine representation of any data element in the exchange format into the preferred internal machine representation used on that hardware platform.

This section provides a brief introduction to a data exchange standard called Abstract Syntax Notation One (ASN.1) that has been approved both by the International Standards Organization (ISO) and by the American National Standards Institute (ANSI) (ISO 1987a, 1987b; Gaudette et al. 1989). The NLM is using ASN.1 to encode the GenInfo Backbone for widespread distribution, and other developers of biological databases are also beginning to adopt ASN.1. Developers of systematics databases should consider using ASN.1 as a standard for distribution of their databases.

ASN.1 Datatype Specification Language

The ASN.1 standard consists of two primary components. The first is a language for describing abstract data types, which I call the datatype specification language. This language allows us to define new datatypes by composing primitive ASN.1 datatypes. Database developers can use it to produce an abstract description of what data are contained by a given database. For example, Table A.1 shows an ASN.1 definition of a datatype called Compound in a database of metabolic compounds. It also contains definitions of datatypes referenced by the Compound datatype. The Compound datatype is defined as a set of fields, such as a field called synonyms that is of type String-Set. At the bottom of Table A.1, we see that the String-Set datatype is in turn defined as a set of VisibleStrings (ASCII strings). We also see a field within Compound called charge, which is of type INTEGER, and which is not required to be present in the description of a given compound, and thus is marked OPTIONAL. The structure field is fairly complex because it contains a collection (this time a sequence, where the order of the fields must be preserved) of subfields, each of which is another complex datatype. This definition spells out very clearly what the structure of the Compound datatype is. Accompanying documentation is of course required to specify the meaning of these fields, for example, the fact that the field "kingdoms" lists the animal

Table A.1. ASN.1 Specifications of Datatypes Used within a Database of Chemical Compounds

```
Compound ::= SET {
    unique-id VisibleString,
    synonyms String-Set,
    systematic-name VisibleString OPTIONAL,
    boeman-quadrants Integer-Set OPTIONAL,
    chemical-formula VisibleString OPTIONAL,
    molecular-weight REAL OPTIONAL,
    charge INTEGER OPTIONAL,
    structure SEQUENCE {
        atoms String-Sequence,
        atom-charges Charge-Sequence OPTIONAL,
        bonds Bond-Sequence OPTIONAL,
        display-coords-2d XY-Sequence OPTIONAL
    },
    acid-base-variants String-Set OPTIONAL,
    kingdoms String-Set OPTIONAL,
    cas-registry-numbers String-Set OPTIONAL,
    mesh-ids String-Set OPTIONAL,
    roots String-Set OPTIONAL,
    sources String-Set,
    reviewed VisibleString OPTIONAL,
    templates string-Set OPTIONAL,
    all-templates String-Set OPTIONAL
  },
Bond-Sequence ::= SEQUENCE OF
                  SEQUENCE {
                    atom1 INTEGER,
                    atom2 INTEGER,
                    blood-type INTEGER }
XY-Sequence ::= SEQUENCE OF
                SEQUENCE {
                  x REAL,
                  y REAL }
Charge Sequence ::= SEQUENCE OF
                    SEQUENCE {
                      atom INTEGER,
                      charge INTEGER }
String-Set ::= SET OF VisibleString
String-Sequence ::= SEQUENCE OF VisibleString
Integer Set ::= SET OF INTEGER
```

kingdoms in which a given metabolic compound is found.

Systematists would use ASN.1 to define datatypes that are of interest within systematics and that correspond to datatypes present in systematics databases. For example, to encode a museum database we might define a datatype called Specimen, and to encode a taxonomic database we might define a datatype called Taxon. As a data exchange standard, ASN.1 provides an approved way for a particular museum (for example) to clearly describe their Specimen datatype and to distribute their specimen database to outside users. But the definition by one museum of a Specimen datatype does not necessarily imply that this definition of Specimen is an approved standard within the systematics community—other museums might use ASN.1 to describe different Specimen datatypes. However, should the systematics community agree on a standardized definition of a Specimen, the ASN.1 standard could be used to describe it.

ASN.1 Encoding Rules

The second component of the ASN.1 standard is a set of encoding rules that describe how to encode instances of any ASN.1 datatype into a particular machine representation. Although the definition in Table A.1 specifies the abstract definition of a compound very clearly, it says nothing about how to encode the data describing a particular compound as (for example) an ASCII file or a binary file. The ASN.1 encoding rules do just that. The binary encoding rules (ISO 1987b) describe how to encode instances of ASN.1 datatypes into binary byte streams with sufficient precision that the binary encoding can be decoded on any hardware platform. The ASCII encoding rules describe how to encode instances of ASN.1 datatypes into ASCII files. The binary encoding rules are more of an official standard than are the ASCII encoding rules, and the binary rules produce more compact representations of datatype instances. ASCII encodings are much easier for humans to read and are therefore more suitable for exposition and for program development.

Table A.2 shows an ASCII encoding of an instance of the Compound datatype that describes the compound L-proline. The ASCII encoding of every datatype begins with the name of the datatype followed by the "::=" characters. The encoding of the data then depends on the definition of the datatype. The Compound type is a set; to encode a set we encode each of its elements, separating the elements with commas, and enclosing all of the elements between the "{" and "}" characters. The first element of the set for proline is the unique-id field. We encode such a field by printing the name of the field and then its value. In this case, the value is the VisibleString *pro*, which we encode by printing *pro* between double quotation marks. The next field is the synonyms field; we first print the name of the field and next print its value. Here the value is of type String-Set, which in turn is a set of Vis-

Table A.2. The ASCII ASN.1 encoding of an instance (L-proline of the Compound datatype)

```
Compound ::=
{unique-id "pro",
 synonyms {"L-proline", "prol", "proline", "pro"},
 boeman-quadrants  {7},
 cas-registry-numbers  {"147-85-3"},
 molecular-weight  {11512999, 10, −5},
 chemical-formula "N1 C5 O2 H9",
 structure
  {atoms  {"C", "N", "C", "O", "O", "C", "C", "C", "H"},
   bonds
     { {atom1 8, atom2 6, bond-type 1},
       {atom1 3, atom2 9, bond-type 1},
       {atom1 8, atom2 7, bond-type 1},
       {atom1 7, atom2 3, bond-type 1},
       {atom1 6, atom2 2, bond-type 1},
       {atom1 5, atom2 1, bond-type 1},
       {atom1 4, atom2 1, bond-type 2},
       {atom1 3, atom2 1, bond-type 1},
       {atom1 2, atom2 3, bond-type 1} },
   display-coords-2d
   { {x {77, 10, −2},                y {154, 10, −2} },
     {x {−215, 10, −2},              y {188, 10, −2} },
     {x {−65, 10, −2}                y {188, 10, −} },
     {x {−26500002, 10, −7},         y {39, 10, −2} },
     {x {162, 10, −2}                y {−15, 10, −1} },
     {x {248, 10, −2},               y {99, 10, −2} },
     {x {−17, 10, −2},               y {4, 10, −1} },
     {x {−14, 10, −1},               y {−5, 10, −1} },
     {x {12799999, 10, −7},          y {−5, 10, −2} } } },
 kingdoms  {"all"},
 mesh-ids  {"D12.125.72.401.623", "D011392"},
 roots  {"proline"},
 sources  {"ihcsdb", "mimavro", "boeman"},
 templates  {"amino-acids"},
 all templates {"amino-acids", "all-amino-acids"}
```

ibleStrings; we again encode this set by printing its elements between braces and separating them with commas. Integers are encoded as we might expect, but real numbers are encoded as triplets of integers: a mantissa, a base, and an exponent from which the actual real number can be computed. White space in the ASCII encoding is ignored and is used only to make the representation more readable.

Conclusions

A researcher might do several things when given a database encoded in ASN.1 form (such as Karp's Compound knowledge base, which contains 1000 metabolic compounds). He might write a program that translates the ASN.1 form of the data into a form with which he is more familiar, such as a relational database, or an object-oriented database. This process would involve defining relations in the relational database that correspond to the type definitions in Table A.1, determining how to translate any Compound datatype into the associated relational schema, and then performing this translation for the entire dataset. He could then execute SQL queries with respect to the resulting relational Compound database. Since ASN.1 is not a database management system, it has no notion of a query language. Another possibility is to write programs that manipulate the ASN.1 data directly to solve a particular problem. Such programs might parse an ASCII ASN.1 encoding directly, or they might call on existing software for manipulating databases in ASN.1 form, such as Karp's Path-Manipulation Package (Karp 1990), or the Free Value Tool (Gaudette 1989), or the ISO Development Environment (Rose 1990).

For example, the Path-Manipulation Package will parse an ASN.1 datatype into an internal form, and then return elements of that dataset to user programs via a set of C subroutine calls. Users refer to desired elements using path expressions that uniquely identify elements of a datatype. The function call

```
ASN_GetInteger(compound,"cas-registry-numbers[0]")
```

would return the first Chemical Abstracts Service registry number from the compound data in Table A.2. In this call, the variable compound would point to an internal form of the compound that was created by a parser that is part of the Path Manipulation Package.

Another possible use for ASN.1 is as a common format for communicating computer programs. For example, the molecular biology community has developed a large number of software tools for performing sequence analysis, such as global sequence alignment, multiple sequence alignment, and identification of promoter regions. Complex computations may require the integration of a number of these different tools. Such a task would be very time consuming now because these tools all expect inputs and produce outputs in different formats, and often those formats are not machine parsable. Therefore, it is difficult to use the output of one program as the input to another program. Use of ASN.1 to describe program input and output data would greatly facilitate the interconnection of diverse software tools.

It should be acknowledged that many ideas for the use of ASN.1 at the NLM came from James Ostell.

Pankhurst: Is ASN.1 in any way different from a C structure?

Karp: The abstract datatypes that can be defined in ASN.1 bear a number of similarities to the datatypes that can be defined in the C programming language. The fundamental difference between the C language and ASN.1 is that the definition of the C language does not include the binary or the ASCII encoding rules that provide a well-defined specification of how to encode instances of those datatypes into binary and ASCII files. Of course, every implementation of the C language does in fact encode instances of C datatypes in some form. However, every implementation is free to choose a different form because those encodings are not part of the definition of the C language—for good reason: the C language is not a data exchange standard so every implementor should be free to choose the representations that are most efficient for a given hardware platform. The fact that the ASN.1 standard specifies encoding rules guarantees that only one encoding of an ASN.1 datatype can exist, thus allowing the exchange of data among different organizations.

Diederich: If you dump the data from an object-oriented database, you are going to lose object identity. That could be a serious problem because it is a very powerful notion within object-oriented databases. Every object has an internal numerical identity. Other objects will then refer to it by that identity. When I dump my data in this format, I lose the fact that other objects are pointing to this one.

Karp: The ASN.1 datatype that represents a given object would then have to include an explicit representation of the object identifier.

Diederich: But you may have to include such an identifier for every attribute, because every attribute, down to the very lowest level, may be an object with an identifier. You don't want to create manually such a large number of identifiers, so you have to build an automatic identifier creation mechanism.

Karp: Yes, a computer tool for translating object-oriented databases into ASN.1 might be more useful if it automatically generated object identifier fields in the datatype for every object class.

——: How can you add more information to an ASN.1 datatype definition?

Karp: You can alter the ASN.1 datatype definitions at will to add or delete fields as needed. Such changes may, of course, require changes in software that manipulates those datatypes.

Humphries: Would the field-name information in Table A.2 be repeated for every compound in the database?

Karp: Yes, in the ASCII encoding the information is repeated—that encoding is not particularly compact. However, no field-name information is used in the binary representation.

Humphries: How are primary key relationships maintained? If you generated an ASN.1 representation of a relational database it's not clear how you keep the key relationships so that someone can reconstruct a relational database from the ASN.1 representation.

Karp: ASN.1 is not specifically designed to facilitate the encoding of any particular data model, such as the relational model. Therefore we can expect that ASN.1 may have trouble capturing intricacies of that model, such as information about which fields of a datatype represent the primary key and which fields of a datatype should be used to construct indices. To represent all aspects of a relational database we may have to add special fields to the ASN.1 datatype definitions to capture this information.

On the one hand, we can view the fact that ASN.1 is not closely aligned with any data model as a deficiency since it means that, if we are concerned only with that data model, ASN.1 may have difficulties in capturing subtle aspects of that model. If, for example, I have a relational database that I wish to distribute to a number of users, all of whom plan to access the data in relational form, it may be easier to distribute the data in a form that has been tailored for the relational data model, such as SQL table definitions and delimited ASCII dumps of each table. Diederich raised essentially the same point in his question where he pointed out that ASN.1 may have difficulties in capturing subtle aspects of the object-oriented data model. But if, on the other hand, we wish to make very few assumptions about the form in which users will utilize the data, ASN.1 becomes a better choice. Imagine that I develop my compound database in relational form, but I know that some end users will translate it to an object-oriented form, others will translate it to a knowledge-base form, and still others prefer not to take the time to translate it and not to spend the money to buy an expensive DBMS—and therefore prefer to use it in whatever form I distribute it. In this case it is not particularly helpful to capture information about what fields constitute the primary keys since the concept of a primary key is defined only within the relational data model and not within the frame knowledge-base model.

Computer Products

ALICE Species Checklist Database Management System

Allkin: ALICE is a database system for biologists wishing to create annotated species diversity databases. ALICE combines intelligent data capture and efficient storage with an easily used interface and does not require previous computer experience. An ALICE database may contain nomenclature, synonyms, common names, geographical distribution, status as native or introduced, short notes or long text entries, and descriptors for habitats, uses, and any number of user-defined characters. Any fact can be cross-indexed to a citation list. A bibliographic reference file is built automatically by the program.

The names of genera, species, subspecies, and varieties can be included, together with their authorities, and any of the data associated with a taxon may be accessed using its accepted name or any of its synonyms. ALICE also provides for homonyms, names which have been misapplied or used in more than one sense, and names of uncertain status. Geographical distributions can be described by areas arranged in three hierarchical levels defined by the user, such as continents, countries, and regions within countries. This hierarchy is explicitly stored within the database and used to detect inconsistencies.

The ALICE software consists of a suite of associated data management programs which work together to create and support an ALICE database and provide a flexible and easy-to-use query mechanism. Some of these programs were demonstrated. There are flexible design facilities for the creation of various kinds of reports, and subsets of the data can be exported in a range of formats, such as XDF, DELTA, and dBASE, suitable for entry into many other programs.

ALICE requires an IBM PC XT- or AT-compatible or PS/2 running DOS 3 or later, with 512K or preferably 640K RAM and at least 3 MB free hard disk space. No graphics are used.

ARBO Image Shape and Pattern Description Package

White: ARBO is a suite of programs to allow the automatic measurement of shape descriptors for object outlines and surface patterns. It has been developed for two-dimensional biological applications, including leaf and seed shape and insect wing patterns. Design principles are that minimal operator intervention be required during the measurement process and that the measurement algorithms be generally applicable to various object types without requiring reprogramming for each new study. For these reasons, the use of landmarks is limited, and ARBO is complementary to the approach taken by Christopher Meacham's software (see Meacham, this volume, chap. 26).

For shape measurement, the outline (or "boundary" or "perimeter") of an image is traced automatically, and further computations are based solely on the outline, which can be stored in a very compact form for archival purposes. Four main approaches to shape measurement have been implemented so far. These are distances between a few automatically located landmarks, chain-code statistics, moment invariants, and elliptic Fourier coefficients. Normally, the matrix of measurements of a selected suite for a given set of objects is used later for multivariate analysis, frequently canonical discriminant analysis to distinguish between different samples of objects. The use of ARBO represents an enormous saving of effort compared with the traditional manual measurement of many characters on the hundreds of specimens required by such analyses.

For pattern measurements, only the moment invariant technique is currently available. This treats an image as a two-dimensional distribution of data values (image brightnesses) which can be described, rather sketchily at present, in seven coefficients that are independent of the orientation of the image. Both the outline and pattern measuring processes were demonstrated, using seed shapes and butterfly wing patterns as experimental material.

The present version of ARBO requires any IBM PC XT- or AT-compatible or PS/2 running DOS 3 or later. A math coprocessor chip is desirable for speed but is not essential. A memory level of 512K RAM is sufficient, and all common graphics hardware (CGA, Hercules, EGA, ATT 6300, 3270 PC, MCGA, VGA, 8514/A) is supported. Images have to be acquired; at present this is by video camera connected to a PC via specially constructed or Matrox PIP-512 interface boards. The image input routine has to be modified for other video camera interfaces, but there are plans to read PCX or TIFF images to permit the use of document scanners.

The XDF Data Exchange Format and the XGPM Program

White and Allkin: XDF (the Exchange Data Format) is a medium for the exchange of data between different database projects and application programs. It has been adopted as a standard by TDWG, the IUBS Commission on Taxonomic Databases. Data sets prepared in XDF consist of one or more text files that contain only normal printable characters, and thus can be independent of the hardware, software, data structures, or character coding schemes used in any particular project. XDF is an extensible data description language with its own syntax and vocabulary. Special procedures required to manage certain data sets or to import data into certain programs can be specified within XDF. These manipulations can include reorganizations of the data tables contained in the XDF data set.

XDF is designed for use primarily with multivariate taxonomic data sets including structured text and numeric data, but is also extensible to new classes of biological data. To limit the deleterious effects of uncoordinated extensions, three facilities are available:

1. Extensions can be described within an XDF data set, or reference made to an external file containing them;
2. Specialist exchange formats can be defined and agreed upon for particular projects or application areas; and
3. Provision is made for default definitions of the common core elements of systematic data sets.

In the demonstration session slides were shown to summarize the features of XDF, and then one way to support these features was demonstrated, using a macroprocessor program (XGPM). XGPM is being actively developed as a means to enable users and application developers to gain access to the capabilities of XDF without the need for major program development. Forms of implementation now being tested include three ways to provide these functions to an application program or DBMS:

1. A library of functions to be linked into the program;
2. An independent coresident module containing the same functions to be called from it; and
3. An entirely separate stand-alone program to perform manipulations of data files before or after running it.

XGPM requires an IBM PC XT- or AT-compatible or PS/2 running DOS 3 or later. Memory of 512K RAM is sufficient. No graphics are used. XGPM is written in portable C, and versions could be produced if required for other types of computer.

Commercially Available Object-oriented DBMSs

Winslett: Listed below are the commercially available object-oriented DBMSs, as of fall 1990. These systems are typically intended for use in computer-aided engineering (the design of engineering artifacts, such as cars and computer chips) or for office automation (the 1990s version of business data-processing needs). These systems are not intended for use in scientific data management, which means that they will have some of the needed features, but others (such as handling of uncertainty, checking of complicated integrity constraints, unifying data from radically different databases) would have to be programmed in.

Distributed Object Management Environment by Dome Software Corporation,
Gemstone by ServioLogic,

Gbase by Graphael,
Object Store by Object Design,
Ontos (formerly VBASE) by Ontologic,
Orion/Itasca by Xidak,
Statice by Symbolics, and
Versant Introductory Package by Versant

Frame Representation Systems

Karp: The following list summarizes the availability of both commercial and noncommercial FRSs. Note that most noncommercial systems are provided as-is with no user support.

Commercial Systems

ProKappa, from IntelliCorp Inc., Mountain View, California. Implemented in C. Hardware: SUN and other UNIX workstations.

Nexpert Object, from Neuron Data, Palo Alto, California. Implemented in C. Hardware: Macintosh and UNIX workstations.

Noncommercial Systems

THEO, from Carnegie-Mellon University. Implemented in Common Lisp. Available for nonprofit research, with no right to redistribute. No fee. Potential users must complete a licensing agreement form. Hardware: SUN workstations, IBM RT. Contact: Professor Tom Mitchell, School of Computer Science, Carnegie-Mellon University, Pittsburgh, PA 15213, Tom.Mitchell@cs.cmu.edu

Parmenides, from Carnegie-Mellon University. Implemented in Common Lisp. Available for nonprofit research, with no right to redistribute. No fee, unless it is to be used for commercial purposes. Potential users must complete a licensing agreement form. Hardware and software requirements: Parmenides runs on most dialects of Common Lisp. Contact: Peter Shell School of Computer Science, Carnegie-Mellon University, Pittsburgh, PA 15213 pshell@cs.cmu.edu

Framekit, from Carnegie-Mellon University. Implemented in Common Lisp. Available for nonprofit research only, with no right to redistribute. No fee if distributed for nonprofit research. Potential users must complete a licensing agreement. Hardware and software requirements: Should run in any Common Lisp. Contact: Dr. Eric Nyberg Center for Machine Translation, Carnegie-Mellon University, Pittsburgh, PA 15213 ehn+@cs.cmu.edu

FROBS, from University of Utah. Publicly available. No fee and no licensing but may not be used for commercial applications. Hardware and software requirements: Common Lisp. Contact: Dr. Robert Kessler, Computer Science Department, University of Utah, Salt Lake City, UT 84112 kessler@cs.utah.edu

Samadani: CART is a software package from California Statistical Software, Inc., Lafayette, California.

Annotated Bibliography

The following works have been selected as particularly interesting for gaining a general understanding of various subjects. The reasons for their selection are given in brackets following each reference.

Artificial Intelligence

Luger, G. F., and Stubblefield, W. A. 1989. *Artificial intelligence and the design of expert systems*. Redwood City, Calif.: Benjamin Cummings Publishing Company. [A good basic text on AI and expert systems. Includes introductory material on some advanced topics as well.]

Winston, P. 1977. *Artificial intelligence*. Reading, Mass.: Addison-Wesley. [A good introduction to AI.]

General Technical Interest

IEEE Spectrum Magazine. [Very high-level and easy to read articles on general engineering issues. This journal includes a calendar section of conferences and announcements.]

IEEE Computer Magazine. [Most articles present material of a current and advanced technological nature. Very good review articles.]

Electronics Design Magazine. [Commercial magazine that usually has good articles on state-of-the-art electronics. Interesting advertisements. Conference dates are usually listed.]

Workstations

IEEE Micro Magazine. [Microprocessor hardware and software issues.]

WESCON Conference. [Latest technology in computer hardware. Usually held in Fall. See *Spectrum Magazine* for dates.]

COMDEX Conference. [Personal computer hardware, peripherals. Usually held in Spring. See *Electronics Design* for date.]

Networks

Tanenbaum, A. S. 1989. *Computer networks* (2d ed.), Englewood Cliffs, N.J.: Prentice-Hall. [Very easy to read book on networks.]

Comer, D. 1991. *Internet working with TCP/IP* (2d ed.). Englewood Cliffs, N.J.: Prentice-Hall. [A review and expository treatment of networking, strongly tied to the NSFNET, ARPANET, and INTERNET network facilities.]

Stephens, W. R., 1990. *UNIX network programming*. Englewood Cliffs, N.J.: Prentice-Hall. [Discusses how to write network programs. Very good introduction to the facilities provided by different flavors of UNIX operating systems.]

LAN Magazine. [Mainly commercial articles, but the advertisements are interesting.]

Neuro Computing

Anderson, J. A., and E. Rosenfield (eds.) 1988. *Neurocomputing: Foundations of research*. Cambridge, Mass.: M.I.T. Press. [Recommended as a source for anyone who is interested in understanding the history of neurocomputing. This anthology contains most of the classic papers on neurocomputing through 1986.]

Caudill, M. 1990. Neural network primer. *AI Expert*. [This special edition of the *AI Expert* journal contains reprints of a number of articles that previously appeared in the journal. It provides a nice introduction to neural networks. Copies may be ordered from *AI Expert*, 1990, Miller Freeman Publications, 500 Howard St., San Francisco, Calif. 94105.]

Rumelhart, D., and McClelland, J. (eds.) 1986. *Parallel distributed processing: Explorations in the microstructure of cognition. Vol 1: Foundations.* Cambridge, Mass.: M.I.T. Press.

McClelland, J., and Rumelhart, D. (eds.) 1986. *Parallel distributed processing: explorations in the microstructure of cognition. Vol. 2: Psychological and biological models*. Cambridge, Mass.: M.I.T. Press.

McClelland, J., and Rumelhart, D. (eds.) 1988. *Explorations in parallel distributed processing: A handbook of models, programs, and exercises.* Cambridge, Mass.: M.I.T. Press. [These three volumes provide systematic and detailed discussion of a number of learning paradigms, transfer functions, and theoretical arguments in neural networks. The 1988 edition contains disks that have neural network paradigms that one can operate on a PC. The text has exercises and a good discussion of how the various networks operate.]

Wasserman, P. 1989. *Neural computing theory and practice*. New York: Van Nostrand Reinhold. [A good introduction to the topic of neural networks; Wasserman provides a cogent treatment of a wide range of important topics without overwhelming the reader with all of the mathematical details.]

Systematics

Hawksworth, D. L. (ed.) 1988. *Prospects in systematics*. Syst. Assoc., vol. 36. Oxford: Clarendon Press. [Important volume covering many aspects of systematics from theory to practice.]

International Code of Botanical Nomenclature. 1988. Konigstein: Koeltz Scientific Books. [This is the latest revision of the rules of botanical nomenclature, hammered out by the editorial committee of the International Association for Plant Taxonomy and voted on by the IAPT membership at the 1987 International Botanical Congress in Berlin. For nontaxonomists, the rules are probably most remarkable for their level of detail and legal air. Settling on the correct name for a taxonomic concept is much more involved than one might think. What hath Linnaeus wrought?]

International Code of Zoological Nomenclature. 1985. Berkeley and Los Angeles: University of California Press. [The corresponding volume for the rules of zoological nomenclature.]

Patterson, C. 1982. Cladistics and classification. *New Scientist* 94:303–306. [Clearest well-written short summary of cladistics.]

Schoch, R. M. 1986. *Phylogeny reconstruction in paleontology*. New York: Van Nostrand Reinhold Co. [Textbook treatment covering more than title implies.]

Sneath, P. H. A., and R. R. Sokal. 1973. *Numerical taxonomy.* San Francisco: W. H. Freeman. [Still the most comprehensive overview of numerical taxonomy; see Felsenstein 1983 (ed.) for more up-to-date material.]

Stuessy, T. F. 1990. *Plant taxonomy: The systematic evaluation of comparative data*. New York: Columbia University Press. [A nice, comprehensive description of modern techniques, concepts, and theories of classification in botanical systematics. The author also discusses the classes of data used for botanical classification in considerable depth.]

Swofford, D. L., and G. J. Olsen. 1990. Phylogeny reconstruction. In: Hillis, D. M., and C. Moritz (eds.) *Molecular systematics*. Sunderland, Mass.: Sinauer Assoc.: 411–501. [Clear overview of algorithms and numerical methods. Well-balanced introduction.]

Wiley, E. O. 1981. *Phylogenetics: The theory and practice of phylogenetic systematics*. New York: John Wiley and Sons. [Most recent comprehensive textbook on methodology.]

Wilson, E. O. 1988 (ed.) *Biodiversity*. Washington, D.C.: National Academy Press. [Many aspects of subject in short articles.]

Important journals for discussion of systematic methodology and theory: *Systematic Biology* (formerly *Systematic Zoology*); *Plant Systematics and Evolution*; *Taxon*; *Symposia of the Systematics Association*.

Systematic Trees and Phylogenetic Inference

Dunn, G., and B. S. Everitt. 1982. *An introduction to mathematical taxonomy*. Cambridge: Cambridge University Press. [Good brief overview of methods.]

Felsenstein, J. 1984. The statistical approach to inferring evolutionary trees and what it tells us about parsimony and compatibility. In Duncan, T., and Stuessy, T. F. (eds.), *Cladistics: Perspectives on the Reconstruction of Evolutionary History*. New York: Columbia University Press. pp. 169–191. [One of several worthwhile papers by this author; see References for others.]

Felsenstein, J. 1988. Phylogenies from molecular sequences: Inference and reliability. *Ann. Rev. Genet.* 22: 521–565. [This is perhaps the standard overview of the methods available for inferring trees from sequences as well as for estimating their reliability. Some background knowledge is assumed.]

Penny, D., Hendy, M. D., Zimmer, E. A., and Hamby, R. K. 1990. Trees from Sequences: Panacea or Pandora's Box. *Australian Systematic Botany* 3: 21–38. [This paper gives a general overview for nonspecialists of both the potential and problems of sequence data in classification. Some extensions to existing methods are included.]

Swofford, D. L., and Olsen, G. J. 1990. In *Molecular Systematics*. Hillis, D. M., and Moritz, C. (eds.), Sunderland, Mass.: Sinauer Assoc. [These authors give an introduction to methods of tree inference that assumes little prior knowledge of the field.]

Identification

Delta Newsletter. Editor: Dr. Robert D. Webster, USDA/ARS/SMBL, Bldg. 265, BARC–East, Beltsville, MD 20705, USA. [Designed to promote communication among scientists developing and applying computer technology in the collection, storage, analysis, and presentation of taxonomic data for descriptions, keys, interactive identification, and information retrieval.]

Fortuner, R. (ed.), *Nematode identification and expert-system technology.* New York: Plenum. [Contains discussions on the process and methods of identification and the description of an identification domain.]

Pankhurst R. J. (ed.) 1975. *Biological identification with computers.* London: Academic Press. [Proceedings of a meeting held in Cambridge, England, in September 1973. Contains papers on theory, data, programs, and traditional methods of identification, and a classified bibliography of computers and identification.]

Pankhurst, R. J. 1978. *Biological identification: The principles and practice of identification methods in biology.* London: Edward Arnold, [A textbook on identification, soon to be replaced by an updated work.]

Databases

Allkin, R. 1980. *Some difficulties of computer stored taxonomic descriptions.* Xalapa, Mexico: Instituto Nacional de Investigaciones sobre Recursos Bióticos. [One of the earliest attempts to summarize some of the problems raised by attempts to construct systematic databases.]

Allkin, R., and Bisby, F. A. (eds.) 1984. *Databases in systematics.* London: Academic Press. [Proceedings of an international symposium held in Southampton, England, in December 1982. Contains papers on nomenclatural, biogeographic, curatorial, and descriptive databases.]

Allkin, R., and Bisby, F. A. 1988. The structure of monographic databases. *Taxon* 37: 756–763. [Argues against the storage of descriptive data as unstructured text and in favor of structured database design; some of the advantages of structured databases are described, especially the ability to support different views of the same data for different users.]

Date, B. J. 1990. *An introduction to database systems* (5th ed.). New York: Addison-Wesley. [The classical introduction to databases. Very easy to read, with the emphasis on classical business-oriented databases.]

Fikes, R., and Kehler, T. 1985. The role of frame-based representation in reasoning. *Comm. Assoc. Computing Machinery* 28: 904–920. [Overview of frame representation systems.]

Finin, T. 1986. Understanding frame languages. *AI Expert* 44–50. [Overview of frame representation systems.]

Korth, H. F., and Silberschatz, A. 1991. *Database system concepts*, 2d ed. McGraw-Hill. [This is the best text on databases for undergraduate computer science majors. It concentrates on the relational model and contains a chapter on object-oriented approaches, including an introduction to object-oriented concepts. It also includes many references to additional articles on object-oriented databases.]

McGranaghan, M., and Wester, L. 1988. Prototyping an herbarium collection mapping system. *Technical Papers: 1988 ACSM-ASPRS Annual Convention: GIS* 5: 232–238. [Introduction to the problems of using label data in particular to determine geographic coordinates.]

Pankhurst, R. J. 1988. Database design for monographs and floras. *Taxon* 37: 733–746. [A view of systematic database design.]

Ullman, J. D. 1982. *Principles of database systems*. 2d ed. Rockville, Md.: Computer Science Press. [Basic overview of relational DBMSs.]

Valduriez, P. 1987. Join indices. *Assoc. Computing Machinery, Trans. Database Systems* 12: 218–246. [Although this is a highly technical description of operations within the relational model, it is accessible to biologists who would like to know more about data structures that would support the management of taxonomic concepts and classification information. The paper discusses the use relations (i.e., records in a relational database system) to link nodes that can be conceptually represented in a directed graph structure.]

Books not written for a computer science audience unfortunately tend to be how-to-do-it manuals oriented toward a particular DBMS product. Books of this nature are available in every technical bookstore. Currently available books on object-oriented DBMS are written for DBMS specialists, and so are even more impenetrable to the non-computer-scientist.

Computer Vision

Levine, M. D. 1985. *Vision in man and machine*. New York: McGraw-Hill. [Includes an especially good chapter on shape measurement with an extensive bibliography.]

Ballard, D. H., and C. M. Brown 1982. *Computer vision*. Englewood Cliffs, N.J.: Prentice-Hall. [Covers many aspecs of computer vision including low-level image processing to high-level symbolic processing. Intermediate level.]

Jain, A. K. 1989. *Fundamentals of digital image processing*. Englewood Cliffs, N.J.: Prentice-Hall. [Covers almost all of the aspects of image processing except for computer vision. Somewhat mathematical, but self-contained. Intermediate level.]

Niblack, W. 1986. *An introduction to digital image processing*. Englewood Cliffs, N.J.: Prentice-Hall. [A very readable, basic introduction to the techniques. Only basic mathematical background is required.]

Pratt, W. K. 1978. *Digital image processing*. New York: Wiley. [The second edition of this classic book will be printed very soon. Very complete. Advanced mathematical preparation is helpful in reading this book.]

Rosenfeld, A., and Kak A. C. 1981. *Digital picture processing*, vols. 1 and 2. New York: Academic Press. [Covers many aspects of computer vision. Written by the first researchers on this subject. Intermediate level.]

White, R. J., and Prentice, H. C. 1988. Comparison of shape description methods for biological outlines. In: Bock, H. H. (ed.) *Classification and related methods of data analysis*. Amsterdam: Elsevier (North Holland): 395–402. [Landmark, chain-code, moment invariant and elliptic Fourier approaches to automatic shape measurement are tested using computer-traced outlines of three different sets of plant organs. The ability of canonical discriminant analysis to differentiate groups within each test set is assessed using each measurement suite.]

White, R. J., Prentice, H. C., and Verwijst, T. 1988. Automated image acquisition and morphometric description. *Can. J. Bot.* 66: 450–459. [Landmark, chain-code, moment invariant, and elliptic Fourier approaches to the automatic shape measurement of computer-traced outlines of birch tree leaves are used to reveal the nature of shape variation between two species and even between different parts of the same tree.]

Appendix B: Report to the National Science Foundation

The ARTISYST Workshop, *ART*ificial *I*ntelligence and modern computer methods for *SYST*ematic studies in biology, September 9–14, 1990, Napa, California.

Introduction

The science of systematics makes heavy use of computers. Modern systematic methods (phenetics, cladistics) could not be carried out without them. Activities such as construction of phylogenetic trees, data entry, descriptive statistics and statistical discrimination of taxa, literature searches, identification aids, have come to rely heavily on programs, database systems, etc. Most available software, however, relies on traditional methods (e.g., conventional database management systems, and algorithmic methods). The resources offered by modern computing techniques such as artificial intelligence, advanced database management systems, image analysis, modern graphics interfaces, integrated workstations, etc., have been little used although they have been applied in other fields for similar applications.

To assess the way in which modern computing methods might be more fully used in systematic biology, to encourage collaboration among computer scientists and systematists, and to recommend how to proceed with future research support, the Systematic Biology Program of the National Science Foundation (NSF) sponsored a workshop on Artificial Intelligence, Expert Systems, and Modern Computer Methods in Systematic Biology (ARTISYST workshop). The workshop was held September 9 to 14, 1990, in Napa, California, with forty-three biologists and computer scientists (listed at the end of this report).

The workshop provided a general review of important state-of-the-art computing methods and of some of the possibilities they offer for applications in systematics. It offered systematists the opportunity to make productive contacts with computer scientists potentially interested in these applications. Recommendations stemming from this interaction provide a yardstick for funding of computer-related proposals to systematic biology during the next few years.

Five areas in systematic biology that look particularly promising for modern computing applications were selected by the workshop organizing committee:

1. Phylogenetic inference and mapping characters onto tree topologies;
2. Expert systems for identification, and identification methods;
3. Literature data extraction and geographical data;
4. Machine vision and feature extraction; and
5. Expert workstations for systematists.

For the fifth area, the participants interested in each of the first four areas were asked to study the characteristics of scientific workstations that would be most appropriate for their area and for systematics in general. Computer scientists working on tool-based expert workstations presented results and guidelines to the participants to suggest this as an alternative to building traditional expert systems for use in systematics.

In the remaining four problem areas, various approaches in artificial intelligence, expert systems, knowledge and database systems, knowledge representation, uncertainty, image processing, and conventional computer science and statistical methods were discussed. Emphasis was placed on current software support and its shortcomings in the selected areas, and on important research directions. Participants demonstrated working examples of current software tools.

During the workshop, participants in both the computer science and the systematic biology communities established potentially productive working relationships. Specific steps will be taken to continue the interaction among the group of participants and to expand the group. Besides general sessions where overviews of systematics and modern computing methods were presented, the participants were divided into smaller groups in which representatives from the more specialized areas met and discussed the practical realization of projects in these areas, as well as general research directions. The conclusions and recommendations of these groups are presented below.

Phylogenetic Inference and Mapping Characters onto Tree Topologies

Overview

Phylogenetic inference, that is, inferring evolutionary relationships among organisms, is usually considered an essential part of systematic studies. Many systematists are dissatisfied by at least three factors that they feel are interfering with the practice of phylogenetic inference: (1) imperfectly developed analytical methods; (2) difficulties and delays in the implementation of new ideas and methods; and (3) inadequate information available to users about the assumptions, strengths, and weaknesses of the different approaches.

Methodology

There is a need for a better understanding of the current methods, their robustness and their assumptions, depending on the complexity of the data. The conclusions of such studies could then be communicated to the general users by computerized tools that would allow choice of the best methods and algorithms, given the nature of the data and choice of model and assumptions.

Robustness of Methods. Little is known about the robustness of methods when the biological mechanisms of the evolutionary process deviate from the model and, conversely, how far it is possible to deviate from the assumptions of each method before it ceases to be applicable. Model assumptions may include: homogeneous or heterogeneous rates of change, both of characters at different sites, and along edges (internodes); lack of independence between characters; degree of internal nodes (binary or polytomous); branch length measures; and specification of ancestral states.

Model and Data Sensitivity Analysis. The more complex the data, the more sophisticated the models for treating the data can become. Yet, it is not clear how far the number of parameters of the models can be increased before any additional increase becomes superfluous. The points to be considered include: choosing between simple and complex models for different types of data; sufficiency of data for the chosen model/hypotheses (number of characters, signal/noise ratio); effect of variation of different sets for same taxa; effect of variation of model assumptions; relative plausibility of alternate trees; and confidence limits on sub-tree edges and trees.

Developments in Phylogenetic Theory. Better methods are needed for the estimation of ancestral information and tree and sub-tree comparison (using both graphical and metric indices). Theory and algorithms with realistic biological assumptions need to be developed and implemented in computer programs. The mechanisms currently assumed are too simple from a biological point of view, but more realistic mechanisms become untractable. The power of exact algorithms for larger numbers of taxa should be increased to up to thirty taxa from the current limit of about twenty. Including more taxa would allow better comprehension of biodiversity.

Tool Needs

Computer tools should allow systematists to explore different methods and their options and allow an enriched display of the different types of trees. Systematists could then compare different trees, or compare trees built with different methods, through graphic or other comparison methods. These tools should be method, model, and hardware independent, portable, and widely available. New methods should be quickly added into existing packages for easy distribution.

A systematic workstation should be presented as a package of tools. It should be developed as a shell into which the users can insert additional algorithms. The workstation also should provide for standard data interchange formats, as discussed by the Database Group.

The computer tools should allow the display of various kinds of data on trees from various sources. The data to be displayed could be text, numbers, graphics or images, and include uncertainty measures of tree structure.

Data structures and displays should permit cycles (reticulation, anastomoses). Graphics could be used to summarize relationships within large sets of trees. Trees should be displayed so that the user can visually compare local differences between trees. Display of trees embedded in three-dimensional spaces (e.g., principal components ordination space) and display of trees at different levels of detail (for large trees) also should be available.

There is a need for interactive editing of trees (branch swapping, reordering) and output of publication-quality trees and other graphics. The display representation must remain under user control. Finally, there need to be standard formats for coding and exchanging trees.

Implementation of the systematic workstation and the tools listed above represent a significant amount of work, yet this effort would be worthwhile for a more efficient use of the power of phylogenetic inference methods.

Recommendation:

1. That NSF support the development of computer software for the workstation proposed above.

User Understanding of Methods

Provided with a multiplicity of methods, a systematist needs guidance, not so much with the choice of methods as with the implications of this decision. While developing decision-making systems, assumptions built into phylogenetic methods need to be made explicit. Thus, the construction of decision tools may help to explain many "hidden" assumptions that are involved in our routine analytical decisions.

Choice of Models and Assumptions. Systematists often assume differing models in building phylogenetic trees. Decision analysis techniques can be used to understand the probability of phylogenetic hypotheses given different models, and which models may be most useful given differing types of data. The effective use of probabilistic techniques for the assignment of a measure of knowledge about alternate hypotheses is complex. It requires the assessment of several classes of data and the construction of models that capture the relationships between genetic processes and observed data. However, there is opportunity for taking the initial steps to apply formal decision-theoretic analyses to supplement more commonly used statistical approaches.

Interactive Exploratory Tools. It would be useful to have the ability to generate multiple sets of results, each based on different assumptions (e.g., parsimony versus maximum likelihood, ontogeny versus outgroup for determination of polarity, etc.). Computer-based decision tools could help systematists with the analysis of the models and assumptions by providing an easy-to-use data sensitivity analysis. Such tools might allow a biologist to inspect the implications of alternate sets of assumptions, or to compare quickly the results of the different analyses. The development of useful data sensitivity tools would greatly benefit from the specification of parameters in models that can be modified.

Choice of Algorithms. Given a model, there exist different algorithms for the construction of trees. Knowledge about the relative efficiency of alternate algorithms for processing different types of data, and more generally for solving different problems, is not readily accessible to all systematists. Yet, the capabilities exist to collect and collate existing knowledge concerning the various trade-offs between optimality (or resolution) and computation time for different problems. It could be productive to integrate algorithmic, statistical, and expert knowledge about the performance of alternate algorithms (given different models and data sets), and to make that knowledge available to biologists.

Many approaches to collecting and diffusing such knowledge exist, from a simple printed manual on the use of selected algorithms to more automated approach, such as a rule-based production system that would capture the expertise of experienced individuals. Clearly these systems need sophisticated explanation facilities to fulfill the task of advising. Backward chaining may not be able to give the kind of explanations needed. Other approaches such as the tool approach should be considered. Given the interest and expertise in this area in computer science and the possibility for significant methodological advance in phylogenetic systematics, this appears to be a very promising area for funding.

The performance of different algorithms, given different data sets, is typically uncertain. Given the uncertainty about the usefulness of alternate methods in different settings, an automated decision-theoretic system could be constructed as an embedded metalevel tool in a larger (traditional) package of tree-processing algorithms. Such a metalevel reasoner could dynamically discriminate the preferred approach given the best knowledge available (e.g., probability distributions about the performance of alternate algorithms).

Uncertainty in Data and Assumptions. To date, almost all results in systematic biology have been reported without an indication of the degree of uncertainty of the results or the explanatory power of models. Where uncertainty has been reported, the results have been expressed as gross confidence intervals. It would be useful to represent qualitative and more detailed probability knowledge (as it becomes available), and to propagate such uncertainty explicitly through the solution process.

Probabilistic representation and propagation of uncertainty provide a theoretically sound method for ensuring that conclusions are reported with no more accuracy than is embodied in the initial data or models. Techniques developed in the decision-science community (originally for modeling noisy sensors) can be applied to the problem of most appropriately drawing conclusions from noisy or uncertain data. The decision-science approaches allow the uncertainty or reliability of the data to be effectively integrated into the conclusions of the analyses.

Many investigators have resisted using a priori probabilities of alternate models primarily because of the subjective nature of these models. However, there is no need to attribute firm Bayesian a priori probabilities to scientific analyses; we can assume the prior probabilities to be variables, and observe the results of different assumptions of probability values.

Data Collection and Experimentation. Given the level of uncertainty associated with our evolutionary knowledge base and the costs associated with the acquisition of alternate data sets, we need methods of evaluating the efficacy of allocating limited resources (time, money, etc.) to specimen and knowledge acquisition.

Recent advances in the automation of decision-theoretic inference offer a solution to evaluating and improving the data collection problem. This approach involves incorporating knowledge about the uncertainty in alternate hypotheses, the power of alternate data or experiments to discriminate among those hypotheses, and the costs of performing different experiments or of collecting alternate classes of data, as has been implemented for providing computer-based acceptance in other arenas, such as medicine and business, but not yet for phylogenetic analysis applications. The theory for value-of-information computation is straightforward. Yet, modeling the data-collection problem for complex problems in biology will undoubtedly require a challenging modeling and assessment task. For example, we need to collect metaknowledge about the uncertainty in different hypotheses, the costs of data acquisition for different data types, and the discriminatory value of different classes of data.

Implementations of data-collection and experimentation advisory systems may be useful for streamlining and optimizing the use of limited resources. Expert systems should have the capacity to advise researchers on the likely productivity of different experimental designs, given the researcher's own knowledge about the uncertainties, and the costs of different tests. Given a specific research question, the system can help a researcher with questions concerning the best new data to collect, the availability and quality of important specimens and other data, and the likely outcomes of alternate experimental designs. A researcher may query the expert system for the availability of specimens, species, and data, the relative costs in acquiring new specimens and data, and which data types have the most probable discriminatory powers, given these availability and costs, in meeting the initial goals under specified resource constraints.

Recommendations:

2. That NSF support studies on the theoretical basis of current methods and their improvement.
3. That cooperation between developers of algorithms, models, and programs be encouraged to provide systematics users with suites of programs representing diverse philosophical and methodological viewpoints for both inferring and analyzing trees.
4. That NSF fund projects to investigate the feasibility of incorporating decision-theoretic tools into phylogenetic analysis.
5. That NSF fund the development of computerized bases of systematics information, as well as computer science research that is needed to support these expert decision-guidance systems. NSF should support the construction of these information bases, and the implementation of a system for accessing this information from decision-analytic tools.

DNA Sequencing

Phylogenetic systematists are primarily interested in the evolutionary history of particular groups of organisms on which they are experts. The tools that they are trained to use include computational methods of phylogenetic inference and those necessary to acquire character data. Useful characters in phylogenetic inference include those from morphology, behavior, ecology, ontogeny, and molecular data. Many systematists do not have the facilities to learn DNA sequencing techniques but are interested in analyzing the data.

Systematists need access to automated sequencing laboratories from which these data could be acquired at a cost that is lower than that involved in establishing personal sequencing labs.

Recommendation:

6. That NSF support automated sequencing laboratories that can answer sequencing requests from systematists.

Expert Systems, Expert Workstations and Other Tools for Identification

Overview

Many critical issues in biology today, such as the characterization and preservation of biological diversity and the evaluation of the effects of climatic change, involve the collection and identification of a large quantity and diversity of organisms. Timely and accurate identification of specimens is the foundation for all research in systematics, ecology, and most other biological disciplines. Demand for taxonomic services currently overwhelms the scarce resources of the systematics community. One of the best ways to use the expertise of taxonomists, both within thc taxonomic community and by other end users, is through the development of more efficient computer-based identification systems. The development of these identification systems provides problems of both practical and theoretical interest to both the systematics and the computing communities.

On the practical side, systematic and ecological studies will benefit through the production of identification systems for large and diverse taxa. Such studies would be aided by more efficient mechanisms to process and analyze the massive and complex data involved.

On the theoretical side, the process of identification will benefit from some concepts developed in computer science (e.g., the principles of accessing very large databases, the use of probabilistic methods, alternate means of data and knowledge representation, the development of integrated workstations, etc.).

Finally, the ability of the scientific community to put tools into the hands of the public that yield the answer to that basic question, What is it? would go a long way to sustain public curiosity and interest in science.

Recommendations:

7. That NSF encourage proposals to develop computer-based identification systems.

8. That NSF encourage the development of special-purpose software for taxonomic identification by specific groups. Also, the development of computer-aided identification devices that might enhance the general understanding of the activity and its importance should be embraced. In particular, identification systems that can be used by biologists without detailed systematic expertise, as well as proposals promising theoretical advances or novel approaches in computer science or systematics should be solicited.

Methodology

Current computerized identification systems built and used by biologists make use of only a few characters and are often based on the dichotomous principle, either directly or indirectly. Existing systems address the identification of small groups and miss the complexity of identification in large groups (orders or higher levels). The small number of characters used and the somewhat rigid identification strategies are a drawback to more comprehensive, flexible, reliable, and easy to use identification systems. Also, most existing systems make only limited use of modern interface methodology to aid users, researchers, and developers.

Some promising approaches were discussed and demonstrated during the workshop. The probabilistic approach used in some areas of expert system as applied to medical domains may be seen as an alternative to the deterministic approach. Important issues remain regarding the efficacy of this approach in biological domains, and some participants questioned the need or naturalness of using probabilities in this domain. Combining deterministic and probabilistic reasoning might be an interesting solution, both in terms of usefulness to systematics and in terms of interesting research for computer scientists. It was felt that more research is needed in this area.

Recommendations:

9. That NSF support research in deterministic identification systems.
10. That NSF support research in probabilistic systems and in combining deterministic and probabilistic approaches.
11. That NSF support building biological databases with more character data thus giving a more realistic representation of species. This also is addressed in the report on data and knowledge base support.

Tool Needs

As identification systems use more sophisticated paradigms, user interfaces will take on an increasingly important role in a comprehensive identification system. The various types of interactive activities for the display of systematic information in identification workstations are just beginning to be defined. Clearly, further studies are needed, particularly on the effective use of image data, digitized images, and on flexible interfaces allowing different use of the system depending on the level of expertise and personal preferences of the end-user. Development of a good user interface typically takes 50 to 80 percent of the total system development, and user interface development should become a major part of future studies. It seems important to fund tools and methodologies aimed at reducing this massive investment in time and effort on individual systems.

Another problem is that there is no efficient method for the transfer of systematic knowledge. Common data standards do, however, promote efficient data sharing. There is a pressing need for the use and improvement of standard data formats, such as DELTA (as one example). Existing standards must evolve, or new standards be developed, to meet new circumstances and requirements; for example, standards must be able to handle image data, they must allow for metaknowledge about the data, etc. Common data standards will enlarge the number of potential software users. This might entice outside developers and industry to provide tools for use by the scientific community. Still, generic data standards may not be rich enough to support certain applications, and the investigators should have the freedom to use their own standards, provided they explain why existing standards cannot be used.

Existing systems do not provide an integrated set of tools needed to improve the sharing of information among researchers and between applications. Such tools might make use of modular programming, customizable user interfaces, interface "shells," text-to-database conversion tools, more efficient means to handle graphical data, and so on.

Recommendations:

12. That proposals seeking funds to build computerized identification systems be judged in part on the interface proposed for the system. While interfaces are not a direct component of any identification paradigm, systems will only be useful if they are easy to use, flexible in that users with different degrees of expertise can use them quickly and efficiently, and consistent with user expertise. Thus, research into identification system architectures should pay careful attention to the interface capabilities.
13. That development and use of identification systems consider current methods of encoding information in existing systems, particularly when there is a format that has gained acceptance in several systems.
14. That awareness of data standards by the investigators be one criterion used in the evaluation of proposals. However, data standards should not be rigidly enforced.

15. That NSF encourage the use of electronic communications networks by systematic biologists for data sharing and collaboration.
16. That NSF proposals for the development of new tools be evaluated in part for uniformity, generality, portability, and extensibility of the computer code and data storage mechanisms.

Databases, Literature Extraction, and Geographical Information Systems

Overview

Meaningful organization and management of organism data have always been central to systematic biology. To fulfill this mission better, the discipline needs to compile, manage, and distribute taxonomic data in an efficient manner. Currently, the bulk of classification, taxon, specimen, character, and phylogenetic data is not in digital form. The limited data in computer information systems are not easily accessible via networks, are not available in standardized formats, and are not well known to discipline scientists. There has not been a clear determination of what is common and different in the various branches of systematics, and scientists tend to think their problems are unique and cannot be fitted to a common standard.

Future Objectives

Computer science advances are needed in both database systems and knowledge representation systems. Collaboration between computer scientists and systematists is necessary to ensure that computer science solutions are brought to the problems in systematics and that state-of-the-art techniques are further advanced to yield general solutions to information management problems in systematics. In addition, collaboration is necessary to ensure that computer scientists solve problems that are of real value to systematists.

From a long-term perspective, information systems for systematics should be able to represent complex biological information. This includes handling multiple taxonomic hierarchies, data at several levels of integration (individual specimens, populations, species, and higher categories), complex structure of biological characters (redundant, related, derived, fuzzy, summary, dependent, etc.), multimedia information, multiple versions, and partial information. Information systems also should allow logical inference over information bases. They should simplify modification of the information schema for rapid prototyping, detect inconsistent data through the use of integrity constraints, provide specialized query languages, simplify integration of multiple heterogeneous information bases, and support shared access by multiple scientists to very large bases of information. No existing database or knowledge representation system provides all these capabilities.

An example of such an information system might be an information base that represents in its structure the genealogical relationships among all known taxa. To date, there have been no efforts to construct information bases using the data or the structural information from phylogenetic hypotheses. Such an information base would be large, containing information contributed by systematists, developmental biologists, functional morphologists, ecologists, and others. They may want to access and contribute to this information base simultaneously. This information base could be interfaced to databases that are currently only used for species identification; they contain some characters that also would be useful in the area of phylogenetic analysis. Biologists outside systematics do not have easy access to character data, nor is their comparative data accessible to systematists for the generation of evolutionary hypotheses. Similarly, some phylogenetic information will be useful for identification. Given the diversity of models of evolution and differing views of classification, there is a need within systematics to represent alternate relationships within a single information base. Also, it is essential to represent uncertainty in our knowledge of genealogical relationships and in the assumptions on which a taxonomic hierarchy is based.

Specific Problems and Opportunities

Modern advances in computer science are typically slow to be communicated to other disciplines, and discussions at the workshop suggest that systematic biology is no exception. While many systematists use computers for some aspects of their work, they may not be aware of some basic computer science research issues that might help them in handling their huge data needs. Indeed, basic knowledge about computers might give a false sense of security that is harmful in the long term if it leads to a failure to support the development of industrial-strength systems. Many biologists lack knowledge not only of recent computer-science research, but of well-established computer technology that is in use by their colleagues, such as of commercial database systems and international data standards.

Effective research use of systematics data requires information methodology that does not currently exist. Biological data require non-conventional data structures, non-conventional query mechanisms and highly specialized processing. The dispersed nature of systematics data sources, systematics collections, and systematics research requires careful attention to enterprise-wide data integration. Data networks are the appropriate structure, but wide-scale interchange of information for systematics requires communications and data standards that are currently underdeveloped and underapplied. The systematics subdisciplines have different data requirements, which constrain the specification of those standards.

Existing specimen data include geographic components that vary greatly in resolution and in precision (lacking precise and consistent use of geographical descriptors).

Any database, once populated with the relevant information at the time of its first implementation, will quickly become obsolete unless new information that will become available after that event is entered into the system. Yet, funding for database maintenance is notoriously difficult to obtain.

Recommendations:

17. That NSF support an inventory of the databases that have been implemented in the various disciplines in systematics, including their content and form. Investigations of the database requirements of the various disciplines will allow the definition of what is common and different among them.
18. That NSF support an assessment of the adequacy of existing data standards. Some participants suggested organizing a conference to discuss the results of this assessment and to formulate improved data standards. These data standards should then be communicated to the community for approval, use, and further enhancements. Later, NSF should support the need to develop new standards as future requirements dictate.
19. That NSF support studies and implementation of improved methods for the interchange of data.
20. That NSF support investigations into new methods for scanning and conversion of published textual materials and other automated data capture methods; actively promote the future collection of data in electronic form; and encourage collaboration with researchers in areas such as expert systems to aid interpreting and formatting of data, or enforcing the use of standard nomenclature.
21. That NSF support research directed toward the construction of more sophisticated biological data and knowledge base management systems. Their capabilities must include support for multiple taxonomic hierarchies, multiple versions, multiple views, multimedia information, rapid prototyping, and inference.
22. That NSF support the integration of existing and future biological databases into a network easily accessible by researchers. With the development of systematics databases at organizations dispersed around the world, research is also needed on access methods for the future transparent, easy, and open access to these sources of systematic information.
23. That NSF support research into automatic methods such as natural language understanding for the conversion of geographic data into standardized formats suitable for proper geographical information systems (GIS) interpretation. Also geographical data standards (e.g., geographical coordinates or other standards) for future specimen collections are required.

24. That NSF fund equipment, networking, and personnel for the design, construction, and long-term maintenance of databases that include various types of systematics data.

Machine Vision and Feature Extraction Applied to Systematics

Definitions

The following terms are defined for this report. Image processing comprises operations on an image to produce another image. Image analysis comprises operations on an image to produce a set of local descriptors, which may be numerical or qualitative. Image understanding (also called scene analysis and image interpretation) comprises extraction of objects, their descriptions, and their interrelationships. Pattern recognition is a set of quantitative or logical methods for identification (called "classification" in the pattern recognition literature) or clustering of data. Machine vision (or computer vision) is the application of automatic image processing, analysis, and understanding and pattern recognition.

A feature is a human- or machine-recognizable characteristic, and feature extraction is a function from images to sets of features. Low-level processing involves pixel-level operations; mid-level processing involves region-level operations; high-level processing involves object-level operations. Bottom-up implies processing from low-level to high-level; top-down implies processing from high-level to low-level.

Systematics is defined here, for the benefit of persons interested in computer vision, as an area where descriptive features (called "characters" by systematists) are used for identification and classification. A large number of features need to be extracted from a large number of specimens. The goal is to be able to determine patterns in nature.

Overview

The primary reason for using computer vision in systematic biology is that there is more material that requires visual processing than can be processed by the available systematic biologists. Demand outstrips the resources.

Advantages of computer-based image analysis and image understanding systems, besides the ability to process many more images, include the following: reproducible feature extraction; well-defined features; improved precision and accuracy of feature recognition; the ability to archive and retrieve images; and reduced loss of information caused by degradation or destruction of specimens.

Several factors hinder the use of computer vision in systematics. Current systems are often too expensive for general use, particularly those that offer sophisticated capabilities. Current methodology is inadequate for many problems of biological interest; there is a lack of good methods for fully automatic image processing and analysis of typical biological materials. Current systems are capable of image processing and analysis only under extremely well-controlled image acquisition conditions. There is a low level of awareness among biologists of the capabilities of existing systems to perform useful and cost-effective tasks. There are few collaborations between systematic biologists and computer vision specialists, and there is little funding to support collaborations.

Needs and Opportunities

There is a distinction between biological features and computable features. Many features in biological specimens that are used for identification or classification cannot be automatically extracted with current techniques. One important direction for research is to bring closer together these two largely disparate classes of features, either by implementing biological feature extraction by computers, or by using different biological features that would be easier for the computer to recognize while still acceptable for systematics.

Image archiving and retrieval systems for biological data are not readily available. Such systems would reduce the loss of information caused over the years by degradation of specimens stored in museums and provide faster access to specimen data.

Software is needed for modeling image formation and the physical imaging process used for biological images. This should allow correction for optical distortion in light microscopy or correction for uneven staining in electron microscopy. Software should provide the ability to redress distortions in specimens. For example, some features may be seen at an angle, which shorten the distances to be measured. Microscopic specimens may have been flattened on slide preparations, and their true shape needs to be reconstructed. Fossils are often distorted, missing parts have to be reconstructed, and tools are needed for recreating and interpreting the once-living organism and its environment.

Inexpensive systems are needed that allow capture and use of full-color imagery. The capture of color imagery is simple, but requires additional storage and processing power. The advantage of using color information for image analysis is that many biological specimens are easily and naturally distinguished in terms of color. Having only gray-scale images artificially increases the difficulty in distinguishing the objects.

Fully automatic processing systems are needed. The amount of available data requires the automated processing and extraction of information from the images. This extraction should be reviewed by the user. Even if only 80 percent of the data analysis can be automated, that would be a great improvement over the currently used manual methods, which are time consuming and preclude analysis of most of the data.

Image-processing packages should allow a combination of low-level processing (filters) and high-level, knowledge-based processing (model-driven analysis). The processing components should include models with knowledge of biological structure and allow for biological variability.

Inexpensive hardware is needed to allow image capture at higher resolution. This may include lighting and optical equipment considerations.

There is a need for basic research in both computer science and computer vision for the following vision applications in systematic biology:

1. Shape descriptors for biological shapes to account for variability. The necessary components include three-dimensional image capture and understanding and models for deformation description. New methods are needed for describing biological shape and variability. New matching algorithms and techniques are necessary for matching models of deformation to observed biological objects.
2. Learning, inference, and techniques for user-trainable algorithms. For example, adaptive techniques that learn to find features after being trained by examples from the users.
3. Measuring change of form through time, deformation rate. Research on motion analysis is related to this problem and may provide useful new ideas. There are, however, additional characteristics that appear in biology that have not been addressed by the current research in motion.

Recommendations:

25. That NSF encourage computer vision applications and research in systematic biology. Such research currently falls through the cracks because there is no program that explicitly supports it.
26. That NSF encourage research on models and languages for shape descriptors that allow irregular, variable, and deformable objects.
27. That NSF encourage research on adaptive, learning algorithms applicable to image analysis problems in systematic biology.
28. That NSF encourage research on the preparation of samples and image capture methods that will help computer vision applications in systematic biology.
29. That research proposals and publications describing new techniques include validations of the techniques with different data sets for broad applicability in systematics. To support this, we recommend creation of a data repository of images, descriptions, annotated images, and sources of further information to allow evaluation and comparison of new algorithms.

30. That research proposals and publications involving development of new methodologies include a statement of experimental design and statistical analyses. Many proposals and publications lack such statements.
31. That an online index be developed and maintained for describing computer vision techniques, including features and algorithms. The index should contain descriptions and citations of the literature.

Miscellaneous Issues

During the workshop and the discussions of the various groups, several issues were considered that were seen as cutting across the boundaries of all areas.

Interdisciplinary Projects

The interdisciplinary nature of the tasks we are considering, uniting two complex areas in computer science and systematics, poses some unique problems. Due to differences in language among disciplines, the scientists involved in collaborative projects may have unique difficulties in communicating. Additionally, the projects they submit to NSF offer unique problems for reviewers.

Difficulties in Communications. There is a need to improve interdisciplinary communication between computer scientists and systematists in problems of interest. There is a lack of awareness of computer science opportunities in biosystematics. Computer scientists and biologists are not sufficiently familiar with each other's disciplines and need to collaborate effectively in areas of mutual interest. Computer scientists need to attend biological meetings to understand biological specialties and problems, and vice versa.

Recommendations:

32. That NSF support opportunities for cross-fertilization through various means, including additional inter-disciplinary workshops, summer institutes, post-doctoral training, fellowships, and inter-disciplinary courses for undergraduates, e.g., in bio-informatics programs.
33. That NSF provide travel funds to enable collaboration and to enable computer scientists to attend systematic biology conferences and vice versa.

Unique Needs of Collaborative Projects. Collaborators in computer science and biology are often physically separated, either in different buildings on the same campus or in different institutions. This can place an added burden on collaboration if some hardware and software resides only at one site. Collaborative projects will have unique needs such as duplicate equipment purchases.

Recommendation:

34. That NSF support the greater investment in hardware, software, maintenance, and logistics that will be needed for collaborative research.

Preparation of Collaborative Projects for Funding. The complexity and interdisciplinary nature of the projects involved cause extra difficulties during the preparation of the grant request. It is in the best interest of all persons and organizations involved that this situation be acknowledged and the preparation of the grant request be aided in substantive ways.

Recommendations:

35. That the investigators proposing complex interdisciplinary research proposals be encouraged to submit pre-proposals to the relevant programs.
36. That pre-proposals be examined by computer scientists and biologists and their comments be communicated to the investigators.
37. That, after review of the actual grant application, grant applicants have the opportunity to respond to the comments made by the reviewers before final decision by the panels.

NSF Grant Review Process. While any interdisciplinary work will face inherent difficulties, discussions at the workshop centered on several potential problems unique perhaps to this particular collaboration. The general desires of computer scientists to do computer science as opposed to applications means that the desired collaborations must be made attractive. The existence of competing moneyed areas such as medicine and engineering puts an extra burden on funding agencies if systematic biology is to compete successfully for the limited number of computer scientists who might become collaborators. While limited funding in biology seems a stark reality for the near future that we must simply accept, it was felt that NSF Systematic Biology should be particularly careful in the review process. Shared anecdotes suggested that this has not always been the case in the past, and ill-informed comments from reviewers may jeopardize further collaboration. Ideally, the reviewers should have knowledge in both fields, but the number of such scientists is low.

Recommendations:

38. That interdisciplinary proposals such as those for systematic tools and identification systems be reviewed by joint teams [pairs] of computer

scientists and systematists. These teams should ideally produce a single joint review.

39. That NSF study the feasibility of review panels for joint proposals with both computer scientists and systematists. The computer scientists in the joint panel should have an appreciation for biological problems gained through some experience in that field.

Need for Broader Systems

Software development elements should be generic rather than just for that one project, to avoid duplication of effort and the waste of the scarce resources currently available within the systematic community.

Recommendation:

40. That proposers of software for very narrow applications justify carefully the difficulty of generalizing it to broader applications, as well as the benefit of carrying out the research in the restricted domain.

Nomenclature

Many authors find it difficult to follow the rules of the International Codes of Nomenclature (both zoological and botanical nomenclature), and particularly the grammatical rules that govern the formation of Latin binomina. To solve this problem, one suggestion is to do away with the Codes and the requirements that biological names be written in Latin. An alternative to this drastic solution would be to set up rule-based expert systems, for ensuring that the rules of the Codes of Nomenclature are properly followed, and for assisting authors with the few Latin grammar rules that are used in the formation of Latin binomina.

Recommendation:

41. That NSF support funding for the implementation of expert systems for International Codes nomenclatural rules and Latin grammatical rules.

Request for More Workshops

There was strong support for more workshops on the same general subject. The objective of the ARTISYST workshop (to provide a review and guidelines to NSF for future funding of collaborative projects in modern computing techniques applied to systematics) prevented the participants from going past the point of problem identification. Future workshops could take off from that point and begin to investigate problem solutions.

Regular professional meetings could provide an appropriate place for some of these workshops, but it may be necessary to request NSF support for special workshops on specific subjects. The problems surrounding the design and implementation of databases for systematic biology were singled out as specifically in need for such a meeting.

Recommendation:

42. That NSF fund a workshop on biological databases.

Conclusions

Benefits

During the workshop, fruitful contacts were established between the participants from both disciplines. Actual collaborative projects were started for no less than six different groups of people, and it is hoped that others will follow suit. The workshop should result in the implementation of sophisticated systems that will be very beneficial to systematists.

It is expected that computer science will benefit from ideas and methods developed in systematics. The serendipity that will result from collaborative efforts between systematists and computer scientists looms large. For example, systematists will contribute to computer science theory, and the complexity of morphological data will prove to be challenging if computer scientists try to develop better database management systems. The connections that might emerge cannot be predicted, but they are welcomed.

The interest in this workshop indicates that systematic biology is trying to establish standards that would be adopted by its various fields. Hardware and software needed by the systematic community and the larger biological community are not being developed, partially because of a perceived lack of sufficient market. The support requested from NSF for broader systems may act as an incentive for commercial software developers to start investing their time and efforts in what should become a large market, if consolidated.

ARTISYST Network

At different times during the workshop the important issue of communication arose. Researchers were already establishing networks and were making provisions to share data and establish closer contacts. The computer science community has used electronic communication as a primary means of communication for nearly a decade now. We are taking specific steps to begin a moderated network, initially populated by the ARTISYST participants, which should help maintain the valuable scientific contacts established during the workshop.

Recommendation:

43. That NSF take a more active role in promoting electronic communication of all forms in the biological community.

NSF Endorsement

Recommendation:

44. That NSF circulate its response to these recommendations to the workshop participants and to other interested parties and that the NSF include a summary of those recommendations that it accepts in its instructions to grant application reviewers.

ARTISYST Workshop Participants

Bob Allkin, Royal Botanic Gardens, Kew, England, United Kingdom
Jim Archie, California State University, Long Beach, California
James H. Beach, Harvard University, Cambridge, Massachusetts
Paulo A. Buckup, The University of Michigan, Ann Arbor, Michigan
Ann Budd, University of Iowa, Iowa City, Iowa
Diane M. Calabrese, Independent scholar, Columbia, Missouri
Peter Cheeseman, NASA, Ames Research Center, California
Michael J. Dallwitz, CSIRO, Canberra, Australia
James Diederich, University of California, Davis, California
Stanley M. Dunn, The State University of New Jersey, Rutgers, Piscataway, New Jersey
Linda S. Ford, American Museum of Natural History, Central Park, New York, New York
Renaud Fortuner, California Department of Food and Agriculture, Sacramento, California
Dan Gusfield, University of California, Davis, California
Michael D. Hendy, Massey University, New Zealand
Steven L. Heydon, University of California, Davis, California
Eric Horvitz, Stanford University School of Medicine, California
Julian M. Humphries, Cornell University, Ithaca, New York
H. Joel Jeffrey, Northern Illinois University, De Kalb, Illinois
Peter D. Karp, National Institutes of Health, Bethesda, Maryland
Paula M. Mabee, Dalhousie University, Halifax, Nova Scotia, Canada
Matt McGranaghan, University of Hawaii, Honolulu, Hawaii
Leslie F. Marcus, American Museum of Natural History, New York, New York
Richard L. Mayden, University of Alabama, Tuscaloosa, Alabama

Christopher A. Meacham, University of California, Berkeley, California
Jack Milton, University of California, Davis, California
Glen Newton, University of Waterloo, Ontario, Canada
Richard J. Pankhurst, British Museum, London, England, United Kingdom
Alain Pavé, Université Claude Bernard, Lyon, France
David Penny, Massey University, New Zealand
David A. Portyrata, ARC, Atlantic Research Corp., Rockville, Maryland
Patrick A. D. Powell, University of Minnesota, Minneapolis, Minnesota
Ann F. Rhoads, Morris Arboretum, University of Pennsylvania, Philadelphia, Pennsylvania
F. James Rohlf, State University of New York, Stony Brook, New York
Ramin Samadani, Stanford University, STAR Lab, Stanford, California
Richard E. Strauss, University of Arizona, Tucson, Arizona
F. Christian Thompson, USDA, Smithsonian Institution, Washington, D.C.
Linda Trueb, The University of Kansas, Lawrence, Kansas
Michael G. Walker, Stanford University School of Medicine, California
Richard J. White, The University, Southampton, England, United Kingdom
Marianne Winslett, University of Illinois, Urbana-Champaign, Illinois
Jim B. Woolley, Texas A&M University, College Station, Texas
Robert A. Zerwekh, Northern Illinois University, De Kalb, Illinois

Glossary

affine transformation: a transformation of a character space that preserves parallel lines but not necessarily the angles between lines that are not parallel.
algorithm: a series of steps or instructions by which a problem can be solved.
allometric growth: growth of an organ at a rate different from that of another organ to which it is compared.
alpha taxonomy: *See* taxonomy.
analog: a continuously varying quantity, like a voltage level or brightness level, that has not been quantified in discrete units (digitized). *See also* digital.
artificial intelligence: the exploration of computer-based methods for solving tasks that have traditionally been solved by people. Such tasks include complex logical inference, diagnosis, visual recognition, comprehension of natural language, game playing, explanation, and planning.
ASCII: American Standard Code for Information Interchange.
ASCII file: a file in which all of the bytes (apart from end-of-record and end-of-file markers) are meant to be interpreted as printable symbols according to the ASCII coding scheme.
associative memory: content-addressable memory.
attribute: 1: a property of an object. 2: a field (column) within a relation (table) in the relational data model. 3: an instance variable of an object. 4: a character (biology).
backward chaining: chaining together of a set of rules that creates an inference path, or chain, from causative agents to observed symptoms. *See also* forward chaining.
belief network: a system that allows people to express qualitative, in addition to quantitative, knowledge about beliefs, preferences, and decisions. It is a directed acyclic graph (DAG) that contains nodes representing propositions (e.g., hypotheses and observations) and arcs representing probabilistic dependencies among nodes. *See also* influence diagram.
beta taxonomy: *See* taxonomy.
binary character: having only two distinct and mutually exclusive states. Also called two-state character.
binomen (pl. **binomina**): the combination of a generic name and a specific name that together constitute the scientific name of a species, for example, the scientific name of Man, *Home sapiens*, is formed with the genus name *Homo* and the species name *sapiens*. Also called binomial or binominal name.
binomial nomenclature: the system of nomenclature whereby a species is denoted by

---

Some of the definitions in this glossary are quotes from Futuyma 1986; Pankhurst 1975; Sneath and Sokal 1973; and the International Code of Zoological Nomenclature 1985. Other definitions were proposed by the authors of this book.

a combination of two names, a genus name and a species name. Interpolated names (subgenus name, subspecies name, etc), if present, are not counted as words of a binomen.

biological diversity: the study of the greatest number of taxa, observed in, or collected from, the greatest number of sampled habitats. The totality of plant and animal species found in a given place.

biological species: groups of organisms that are reproductively isolated from other such groups. *See also* species.

bitmap: 1: a method for storing or processing graphics and monochrome images where image intensities are converted to different densities of dots in the bitmap. 2: a bitmap image.

Boolean expression: a logical relation involving only true/false statements and the operators *and, or*, and *not*.

branch and bound: search technique where paths through a search tree are pruned if subgoals generated by those paths will be less valuable than the best approach discovered so far.

button: an area on a computer screen that may be "pushed" (usually by clicking a mouse) to start an action or set a context.

causal networks: a data structure used in artificial intelligence in which causal relations are represented as a connected graph. Vertices in the graph represent events, and edges in the graph represent causal links between events.

CD-ROM: Compact Disc Read Only Memory. A disc that contains digitally encoded information readable by a computer. The disc cannot be modified (read only).

chain code: a method of coding a path, usually in two dimensions, as a sequence of unit steps that are represented by digits. For instance, if the digit 0 represents "one step east," 1 represents "one step north," etc, the chain code 00112233 would trace a square, two steps on a side, in a counterclockwise direction. Some chain-code conventions also permit diagonal moves and thus have eight directions for steps.

character: 1: a characteristic such as *color of petals*, *shape of leaf*, or *length of body* that can be used to differentiate two objects (biology). Also called property, feature, or attribute. It is often composed of two parts, the name of an organ or organ part, and a character state or value as in *body elongate*. 2 a: a symbol (as a letter or number) that represents information. b: a representation of such a character that may be accepted by a computer.

character independence: in evolutionary theory, state changes in two or more characters that are uncorrelated across evolutionary time; in phylogenetic inference, changes in two or more characters that are assumed to be uncorrelated from node to node along the branches of trees. *See also* dependent characters.

character state: the values taken by qualitative characters, such as *pink* or *elliptic*. *See also* character value.

character types: characters may be qualitative or quantitative. Quantitative characters can be ordinal, for example, *body short, medium, long*, or nominal, for example, *oviduct type disk or tube*. Quantitative characters can be integers, for example, *number of petals* or real numbers, for example, *height of stem*. Characters can also be continuous, discrete, or binary.

character value: the numerical expression attached to a quantitative character in a specimen, a population, or a taxon. *See also* character state.

characteristic: a feature or property of an object, whether or not it can be used to distinguish this object.

cladistics: a taxonomic theory that holds that only shared derived characters states should be used to determine common ancestry.

cladogram: a treelike diagram or dendrogram that indicates ancestor-descendant relationships. A branching treelike diagram that reflects at each node or branching point the branches or sister taxa, that is, those taxa believed to be most similar by common ancestry.

class: 1: a description of a group of objects having common properties and procedures. 2: a group of such objects. Each object in a DBMS belongs to one or more classes.

class hierarchy: a directed acyclic graph of classes, based on their similarity to one another. The children of a class in the hierarchy differ from their parent in that they can contain additional attributes, define additional procedures or functions not present in the parent, override default values specified in the parent, and redefine procedures and functions that are defined in the parent.

classification: 1 a: the delimitation, ranking, and ordering of organisms and taxa into groups, usually hierarchical, reflecting their relationships (biological definition). b: the end product of the process of classification, for example, a classification of plants. 2 a: the assignment of unidentified objects to the correct group or class once a classification has been constructed (computer science and statistics usage, i.e., identification). b: the operation that automatically computes the proper position of a frame in a taxonomic hierarchy of frames.

collection: the biological specimens gathered in a museum for study, comparison, or exhibition.

collection management system (CMS): a special-purpose DBMS designed for use in curatorial activities associated with the management of systematic collections; a typical CMS includes functions for generation of specimen labels, management of loans of specimens, accession of new materials into the master catalog, updating and documentation of taxonomic information associated with the specimens, etc.

computational complexity: the time required for a computation of an algorithm can be described by a function $f(n)$ which is proportional to the number of computational steps required. If $f(n)$ is bounded by a polynomial function of n then the algorithm is classed as polynomial. Such algorithms are considered as efficient. In contrast, some belong to a class of nonpolynomial algorithms which are considered as inefficient and are therefore limited to smaller data sets. For example, parsimony algorithms for n taxa are of exponential order and therefore nonpolynomial. *See also* NP-complete problems.

computer vision: machine vision.

conditional probability: probability of an observation A, given the presence of hypothesis H.

consistency index: a measure used to evaluate the number of times that a suite of characters change between states when superimposed on a phylogenetic tree calculated by comparing the observed number of changes to the minimum possible number for the suite of characters.

content-addressable memory: a memory system with the ability to recall a complete memory based on partial descriptions of the memory; also known as an associative memory.

convergence: evolution of similar features in unrelated organisms, from different antecedent features or by following different pathways. *See also* parallelism.

corruption: events such as disk hardware failures and operating-system failures that render an information base wholly or partially unreadable by the computer. The information base might be destroyed or left in an inconsistent state.

data format: the arrangement or layout of data, especially in a file prepared for input to some application program. Files containing structured descriptive data typically have records whose fields are either of fixed width and position, or are "free-format" with variable length and separated by some specific delimiter character. An explanation of the nature of each piece of data is also required, but may be implicit or held in some other form (such as on paper) for simple formats.

data language: in taxonomy, the way descriptions are written. The data language can be coded, as in the DELTA system, or it can be a restricted (specialized) set of the natural language.

data management: in systematic biology, activities undertaken by systematists when they are managing their collections, undertaking their taxonomic research and disseminating their results to the scientific community.

data matrix: a machine-readable table with numerical or symbolic values of characters for specimens, populations, or taxa.

data primitive: the basic conceptual unit of an information system; a data primitive determines how information is organized within that system.

database: a collection of interrelated data that is intended for use in more than one application and is maintained separately from those applications. It is managed by a database management system.

database management system (DBMS): a program or suite of programs to assist in the construction and maintenance of a database and in data entry, storage, update, and retrieval.

database structure: the design of the database tables used to hold the data in a database, including their records, fields, logical relationships, and meaning. *See also* schema.

DBMS: database management system.

decision tree: a branched structure that allows decisions to be made by sequentially asking *yes-no* questions.

dendrogram: a treelike diagram representing the relationships of organisms. *See also* cladogram, phenogram.

dependent character: characters that are connected by logical rules. For example, if character 1 (presence of an organ) equals *organ absent*, then all the characters that describe the properties of this organ are impossible. *See also* character independence.

determinism: 1: a philosophical doctrine that holds that occurrences in nature are determined by antecedent causes. 2: the belief that the observation of a given character state or value is enough to reject the hypothesis that the organism belongs or does not belong to a particular species. *See also* probabilism.

diagnosis: identification.

dichotomous key: an identification aid composed of a succession of paired choices between alternatives of specified characters. Each choice leads to another character or to the name of a taxon. It is a decision tree written on paper in a formal way and traced out manually by a human user. There are two forms of key, corresponding to whether the decision tree is traced out first from top to bottom, and then left to right (parallel type) or else in the opposite order (bracketed type).

digital: 1: a quantity that is represented as a number or discrete units. 2: calculation by number or discrete units. *See also* analog.

distributed database: a database in which different parts of the information reside on different computers that are connected by networks. Users may or may not be able to tell that a given information base is distributed over several machines (locational transparency). *See also* replicated database.

distribution map: in systematic biology, a geographic map depicting the area of occurrence of a taxonomic group of organisms, usually a species.

dot map: a type of distribution map in which the occurrence of an organism is represented by discrete symbols, such as squares, triangles, and circles, which correspond to separate records of captured specimens or field observations. Sometimes called spot map.

Euclidean metric: a space in which distances are measured as straight-line distances. In two dimensions, the distance is the square root of the sum of the distances in the x and y dimensions. *See also* Manhattan metric.

expert system: computer program that could draw conclusions by performing logical inference on a large knowledge base of information acquired from experts. It includes a rule base of condition-action rules and a separate inference engine that find inferences by forward or backward chaining.

expert system shell: an expert system in which all domain-specific knowledge is represented explicitly as rules, rather than as code in the inference engine, thus allowing the replacement of those rules for a different task.

expert workstation: an integrated set of tools designed to support activities of an expert user within a domain of interest: a tool-based system. It may also be utilized by nonexperts in certain domains. The main conceptual components are: presentation (interface), representation (data and knowledge base), and rules, with primary emphasis on the first two. *See also* software tools, workstation.

export format: transfer format.

facet: a facet is a property of a slot, just as a slot is a property of a frame. A character.

feature: 1: a character (systematic biology) 2: mathematical operations that take as input a small area in an image and provide as output a numerical value (computer vision). For example, edges are features that are defined by image intensity gradients.

feature extraction: a function from images to sets of features such as, computing the strength of the edges in the image.

format: the layout specification for data in a file. *See also* free format; data format; transfer format.

forward chaining: chaining together of a set of rules that creates an inference path, or chain, from observations in the world to possible causes. *See also* backward chaining.

frame: the frame knowledge representation system data primitive; it represents a single concept or object and describes properties or attributes of that object using a collection of slots. Each frame contains several classes of knowledge that experts have identified as being useful in a given problem.

frame knowledge representation system (FRS): an implementation of the frame data model; it manages a computer database that is structured as a hierarchy of frames.

free format: a data format in which data elements do not have to appear in fixed positions within records.

FRS: frame knowledge representation system.

function: the desired effect of an operation (carried by the software alone) or an action (involving the user). Operations and actions are low-level components of a piece of software. The function of operation or action is to achieve something. For example, *search* is a function in a text editor.

functionality: The set of all the functions included in a piece of software or in a section of this piece of software.

gamma taxonomy: *See* taxonomy.

gazetteer: a list of place names and the geographic coordinates (e.g., latitude and longitude) associated with each place.

generalization: the ability to apply learned concepts or distinctions to new cases.

geographic coordinates: a pair of magnitudes, usually longitude and latitude expressed in degrees, minutes, and seconds of arc, used to define the position of a point on the surface of the Earth. In contrast with Cartesian or projection coordinates which are used to position points in maps plotted on paper or computer displays, geographic coordinates are defined in terms of actual geographic landmarks.

geographic information systems (GIS): a class of computer system (hardware, software and data) which supports collection, storage, manipulation and query of spatially referenced data. It usually includes a graphic interface for displaying geographic maps and optionally tools for acquisition or validation of data.

GIS: geographic information systems.

graceful degradation: gradual loss of performance in response to a decrease in resources. 1 a: for a system, gradually losing the ability to process information. b: suffering gradual or partial loss of accuracy or speed due to system damage. c: ability to make a best-guess with partial information. 2 a: for an identification system, ability to use an incomplete set of characters. b: robustness of the system when fed erroneous data, related to the number of characters in which a specimen may differ from a taxon before it is considered to not belong to the taxon. c: gradual loss of the ability of the system to correctly identify an unknown specimen as an increasing number of errors are made in the description of this unknown specimen, for example, diminished accuracy of the probability distribution generated by a Bayesian system, or of the conclusions of a production-rule system, or of the coefficient of similarity of a matching system. d: ability to tell the user when a case is outside the system experience. e: ability to inform the user that the system does not have much confidence in its answer.

greedy tree algorithm: an algorithm that constructs a tree by sequentially adding objects. An agglomerative algorithm, using steps that are only locally optimal.

heuristic: a rule of thumb. 1: providing aid in the solution of a problem but otherwise unjustified or incapable of justification. 2: a heuristic method or procedure.

hill-climbing algorithm: an optimization procedure for maximizing a function (called an objective function, merit function, or optimality criterion) of one or more independent variables by a process of iterative refinement from a set of initial values. If the values of the function with respect to the variables can be portrayed as a geometric landscape, then the initial values compose a starting point, and the algorithm can be pictured as climbing the landscape toward the nearest extremum (optimal value), which might be local or global. Such algorithms may be heuristic (direction-set methods) or may use derivatives (gradients) to find the shortest path.

holophyletic taxon: a monophyletic group that includes all the descendants of the most recent common ancestor of the group. *See also* paraphyletic taxon.

homology: similarity between character states in two or more taxa that is due to inheritance from a common ancestor rather than to independent acquisition (*see also* homoplasy). Two concepts are prevalent in systematic biology: (1) the operational concept, in which homologous morphological structures are identified by similarity in position, shape, and structure; and (2) the taxic concept, in which homologous character states are defined to be those that diagnose monophyletic groups (i.e., taxa restricted to all of the descendants of a common ancestor). The former is independent of the evolutionary conclusions sought, while the latter depends critically on those conclusions.

homonym: In biological nomenclature, a homonym is created when the same scientific name is given to two distinct taxa, that is, taxa that are based on two independent type concepts. Homonyms are relatively uncommon and are usually produced when an author is unaware of the previous use of a name for a different taxon.

homoplasy: 1: an instance of character change on a phylogenetic tree that is inferred to be a reversal in state or a parallel or convergent change because the change occurs elsewhere on the tree. 2: any instance of character state change on a phylogenetic tree above the minimum possible necessary to account for the different states of a character.

homoplasy-excess ratio: a measure used to evaluate the number of times that a suite of characters change between states when superimposed on a phylogenetic tree. It is calculated by comparing the observed number of changes to the mean number of changes observed for a set of comparable data that reflect only random character change.

icon: a pictorial representation of a system entity.

identification: the allocation or assignment of additional unidentified objects to the correct class once a classification has been established. 1: the process of assigning a new object to one of the groups in a classification. 2: the assignment of a taxonomic name to an individual organism. Essentially synonymous with the computer science definition of classification. Also called diagnosis.

image acquisition: sensor and electronic system that converts energy (such as light or electron beam) into a pictorial form that is then converted by digitization into a format that can be processed by computer.

image analysis: operations on an image to produce a set of local descriptors, which may be numerical or qualitative.

image data compression: removal of the redundancy in image data to allow for lower storage space requirements.

image enhancement: the operation that improves the quality of an image for viewing by the user.

image processing: an operation that transforms an image into another image that is more suitable for viewing or for further processing.

image restoration: the removal of physically understood degradations caused by the image acquisition system.

image understanding: extraction of objects, their descriptions, and their interrelationships. Also called scene analysis and image interpretation.

inference: the procedure (reasoning) used to derive new beliefs from old.

inference engine: the reasoning mechanism in an expert system. *See also* backward chaining, forward chaining.

influence diagram: a system similar to a belief network but containing several types of nodes.

inheritance: the transmission of structural or behavioral characteristics from one frame or object in a hierarchy to its children or subclasses.

interactive key: an interactive computer program to assist with identification.

judgment space: an *n*-dimensional vector space for simulating a specific human judgment skill.

key: 1: an identification aid. 2: specifically, a dichotomous key. *See also* interactive key.

knowledge base: a collection of information about expertise within the domain of interest, possibly including data, represented in such a way (e.g., by rules, frames, semantic nets) that the expertise is explicitly available for use by the application software.

LAN: local area network.

landmark : an anatomical reference point from which morphometric measurements can be taken.

landscape: the assemblage of physiographic and biotic features on a region of the Earth's surface as might be seen from an airplane *See also* terrain.

likelihood: the probability of the joint occurrence of a set of independent events under a specific hypothesis. The product of the conditional probabilities of each event given the hypothesis.

local area network (LAN): a computer communication network used to interconnect computer systems in a relatively small area, usually a building or floor of a building. *See also* wide area network.

locality: a place on the Earth's surface. *See also* type locality; locality data.

locality data: in systematic biology, the information associated with the collection of an organism, including a description of the site of capture (described in terms of geographic coordinates, administrative units, or topographic landmarks), collector, date of collection, environmental conditions, etc.

machine vision: the application of automatic image processing, analysis, and understanding and pattern recognition. Also called computer vision.

Manhattan metric : a space in which distances are measured according to the distance one would walk along streets in a city such as Manhattan, rather than as straight-line distances. In two dimensions, the distance is the sum of the distances in the x and y dimensions, as opposed to the square root of that sum in the Euclidean metric. Also called city bloc metric and L1 norm. *See also* Euclidean metric.

maps: *See* distribution map; spot map.

matching: the comparison of the states or values of characters in unknown specimens to the corresponding states or values in the various taxa to which these unknown specimens are being compared.

menu: *See* pixel menu, pop-up menu.

minimum-length tree: a tree (hypothesis) that requires the minimum possible number of character-state changes to explain the distribution of the character states among the taxa.

monophyletic taxon: a group of species descended from a single (*stem*) species that includes all the species descended from that stem species. 1: a group originating from a single ancestor. 2: a group that contains the common ancestor of a group and all of the descendants of that common ancestor. *See also* holophyletic taxon, paraphyletic taxon, polyphyletic taxon.

multiple-entry key: a key where the user is free to choose the character to be used at each step of the identification process. *See also* polyclave.

network: a set of several computers and terminals electronically linked for sharing data and software. *See also* local area networks; wide area networks.

neural network: an information-processing structure that is modeled after the massively parallel structure of the brain; it has many relatively simple individual processing elements that are highly interconnected. It computes by strengthening or weakening the interconnections in response to input, rather than in the traditional way of storing information in memory cells and operating on it. Also called connectionist systems, neurocomputers, and artificial neural systems.

nomenclature: 1: a system of names and provisions for their formation and use. 2: the study of the relationships between published names of organisms and between those names and the taxa they represent. Codified separately for botanical and zoological nomenclature (International Code of Botanical Nomenclature, International Code of Zoological Nomenclature).

nomenclatural synonyms: duplicate names that represent taxa that are based on the same type concept or the same type specimen (botany). Also referred to as homotypic synonyms. They commonly occur when a species is transferred from one genus to another in a new classification. The binomial of the species changes (usually just the generic name), and the original binomial then is considered a nomenclatural synonym of the new name. Both names refer to the same type specimen. Nomenclatural synonyms also result when two or more classifications interpret a particular concept at different levels in the taxonomic hierarchy. For example, black maple is considered to be a species [*Acer nigrum* Michaux f.] by many taxonomists, a subspecies of sugar maple by some [*Acer saccharum* subsp. *nigrum* (Michaux f.) Desmarais], and a variety of sugar maple by still others [*Acer saccharum* var. *nigrum* (Michaux f.) Britton]. All three names are based on the same type specimen. The correct name depends on the rank at which the taxon is recognized; the other two names become nomenclatural synonyms. These rules do not apply in zoological nomenclature.

NP-complete problem (nondeterministic polynomial problem): a computational problem that is equivalent to a class of problems for which no efficient polynomially bounded solution has been determined. Thus, the time required to solve the problem grows more than polynomially in the number of objects given in the problem. *See also* computational complexity.

numerical taxonomy: the grouping by numerical methods of taxonomic units into taxa on the basis of their character states. *See also* phenetics.

object: a conceptual unit of the real world often represented in object-oriented programming languages or databases as a set of values (which may be other objects) for named attributes and a collection of procedures which can be applied to it.

object-orientation: 1: In software development, a method based on the objects identified in an application and the operations performed on them as opposed to a functional or procedural decomposition of the application. A relatively new approach to software development thought by many to be superior to procedural approaches. 2: A term applied to any programming language or database system that uses objects as the central conceptual framework in addition to supporting features such as object classes, hierarchies, structural and procedural inheritance, and identity independent of the objects' content.

object-oriented language: a computer language for creating and manipulating objects.

object-oriented database: a database in which objects are stored. Built via an object-oriented database management system, which supports operations for manipulating the objects.

objective synonyms: two names based on the same type, as, for example, two genus names based on the same type species (zoology).

ontogeny: the history of the development of an individual organism.

operational taxonomic unit (OTU): used in numerical taxonomy studies to indicate the units being studied: individual organisms, species, or groups of organisms at any taxonomic level of the same rank.

orthogonal rotation: a rigid rotation of two or higher dimensional space. Angles between lines will be unchanged by this transformation.

orthonormal linear transformation: a transformation that corresponds to a rigid rotation of a character space. Angles between lines will be unchanged by this transformation.

OTU: operational taxonomic unit.

pane: an individualized part of a window.

paradigm: an outstandingly clear or typical example or archetype.

parallel processing: performing simultaneous computations in a number of processing units; in contrast to sequential processing where computations must be performed one after another in one processor.

parallelism: evolution of similar features independently in related taxa, by similar modifications of the same developmental pathways. *See also* convergence.

paraphyletic taxon: a monophyletic group that does not contain all of the descendants of the most recent common ancestor of that group. *See also* holophyletic taxon.

parsimony: the principle of invoking the minimal number of evolutionary changes to infer phylogenetic relationships.

pattern recognition: a set of quantitative or logical methods for identification (called classification in the pattern recognition literature) or clustering of data.

phenetics: a taxonomical theory that bases its classification on overall similarity of taxa. Similar species are combined to form genera, similar genera to form families, and so on.

phenetic species: morphologically similar populations in a definite geographic area and distinct from other populations assigned to different species. *See also* species.

phenogram: a branching-tree diagram, or dendrogram, that indicates at each branching level the level similarity or difference among the objects at that level. The phenogram was not designed to depict hypotheses of shared ancestry.

phylogeny: the genealogy of a group of taxa.

phylogenetic relationships: the pattern of shared common ancestry among organisms.

phylogenetic species: groups of organisms with a shared history that can be diagnosed or differentiated from other such species. *See also* species.

phylogenetic systematics: *See* cladistics.

pixel: a picture element 1: the smallest unit of a digital representation of a picture. 2: any of the discrete numbers that represent a digital image.

pixel menu: a pop-up menu whose list of possible selections depends on the position of the cursor at, or close to, a particular pixel in a picture rather than on what pane it is in.

planning research: study of common sense reasoning.

polyclave: a multiaccess identification device. *See also* multiple-entry key.

polyphyletic taxon: a group whose most recent common ancestor is not cladistically a member of that group. A group that shares more than one common ancestor.

population: a group of specimens belonging to the same species, found in a single location.

pop-up menu: a menu of possible selections that appears at the cursor when a mouse button is pressed. The item in the menu depends on what pane the cursor is in.

prior probability: the prevalence of a state of the world, conditioned only on prior assumptions or hypotheses.

probabilism: 1: a philosophical doctrine that held that certainty is impossible and that probability suffices to govern belief and action. 2: the belief that the observation of a given character state or value only allows to attribute a certain probability to the hypothesis that the organism belongs or does not belong to a particular species. *See also* determinism.

production system: *See* expert system.

property: character.

pruned decision tree: a decision tree that has some of its nodes removed in order to improve its performance with new data.

qualitative character: *See* character type.

quantitative character: *See* character type.

query: a question about the data in a database, phrased in a special data access language supplied for that purpose by a database management system.

rank: the level of a taxon in the classification hierarchy, species, genus, family, order, class, phylum, etc, including extra levels formed by the prefixes *sub* and *super*: subfamily, superspecies, etc.

relation: a collection of tuples (rows). A relation (table) is the relational DBMS data primitive.

relational data base management system: an implementation of the relational data model; it manages a computer database that is structured as a collection of relations.

replicated database: a database whose records are duplicated and stored on multiple computer systems. The replication allows individual computer systems to fail without preventing access to the database information through another machine. *See also* distributed database.

retention index: a measure used to evaluate the number of times that a suite of characters change between states when superimposed on a phylogenetic tree. It is calculated by comparing the observed number of changes to the number of changes observed on a tree with no structure where all branches radiate from a central root.

rule-based system: expert system.

schema: the description of the structure of the data in the database, including information on the organization of the data into relations or classes, the fields and attributes of those relations or classes, and constraints on what data values are legal for particular fields and attributes. *See also* database structure.

segmentation: the separation of an image into individual objects.

semantic net: a method of knowledge representation usually represented graphically as a collection of nodes and edges with different kinds of meaning attached to them.

semihierarchical database: hierarchical database which is limited to parent-child record relationship in which there is only one child level. There are no child-subchild relationships in the database.

sequential processing: performing one computation after another in one processor. *See also* parallel processing.

shell: expert system shell.

similarity coefficient: a measure of the resemblance between two objects (two OTU), particularly between an unknown specimen and a taxon to which it is being compared. The similarity can be estimated by scoring the similarity of each character as it appears in the unknown specimen and in each taxon, or by various measures of taxonomic distances.

slot: a component of a frame that represents a property or attribute of that frame, or a relation between that frame and another frame.

software tools: a system of discrete programs that communicate or interact with each other to form an expert workstation. Each tool is designed to fulfill a small, well-defined related set of tasks within the application area.

species: *See* biological species; phenetic species; phylogenetic species.

specimen: an individual organism.

spot map: dot map.

state: character state.

stem species: *See* monophyletic taxon.

synonym: Each of two or more scientific names of the same rank used to denote the same taxon. They can be of several types, for example, nomenclatural synonym and taxonomic synonym.

systematic biology: systematics.

systematics: the scientific study of the kinds and diversity of organisms and of any and all relationships among them. 1: the scientific discipline which is concerned with the discovery and identification of the diversity of living and fossil organisms. 2: the process of taxonomy, or a wider form of information management in which other biological information, not itself used to form the classification, is organized according to a taxonomic hierarchy. Also called systematic biology.

systematic data: biological data used in systematic studies, and likely to involve a taxonomic hierarchy and descriptive data with variability, missing values, character dependence, and characters of different types such as multistate, quantitative, and binary characters.

systematic database: a database containing systematic data, including descriptive, taxonomic, nomenclatural, bibliographic, biogeographic, and curatorial data.

taxa: plural of taxon.

taxon (plural taxa): a class of related organisms, defined in terms of their shared characteristics, for example, *Mus musculus* (house mice), Poaceae (grasses).

taxonomic hierarchy: a specific more or less widely held opinion of the hierarchical classification of a group of organisms, or the ordered set of levels (ranks) used in such a hierarchy.

taxonomic database management system: a DBMS designed specifically for handling systematically organized biological data for use by taxonomists or other biologists without detailed knowledge of the database structure; requires systematic database structures, taxonomically intelligent algorithms and an interface that can be used by biologists. In many applications, the users will determine the set of descriptors to be included and will want to be able to define or redefine them and their properties.

taxonomic synonyms: duplicate names for taxa that are based on different type concepts. At the species level (and below) taxonomic synonyms are created when two or more existing concepts, each based on different type specimens, are merged into a single taxon. The names for the species that no longer exist in the revised classification become taxonomic synonyms of the name for the new, more broadly defined taxon. Taxonomic synonyms occur at higher ranks when taxa based on distinct type concepts are "lumped" into a single taxon. For example, if a small, segregate genus is merged into a genus that is based on an older type species concept, the name of the genus being "sunk" becomes a taxonomic synonym of the name of the larger revised taxon. Also called heterotypic synonyms.

taxonomic view: a summary of a taxonomist's current classification with reference to previously held nomenclatural systems, including the relationships between published names and between those names and the taxa they represent.

taxonomic work-bench: a set of reliable and easy to use tools suitable for a taxonomist's or other biologist's work-bench data management system or expert workstation.

taxonomy: the theoretical study of classification, including its bases, principles, procedures, and rules. 1: the theory and practice of classifying biological diversity. 2: sometimes used as a synonym for systematics. Alpha taxonomy is concerned with the attribution of names to species and publication of their descriptions; beta taxonomy with the formation of larger groups nested in a hierarchy and described in the form of a hierarchical classification; and gamma taxonomy with the study of the biological aspects of taxa including intraspecific population studies, speciation or how new species arise, and evolutionary rates and trends.

terrain: landscape, but with somewhat more emphasis on physiographic features.

tool: software tool.

transfer format: a format for the export of files of data to be read by other programs. These export formats range from simple free-field and fixed-field record formats like SDF and dBASE to complex specifications involving the detailed syntax and interpretation of data exchange files such as DELTA and XDF.

truss: a networklike system of triangles of measurements of a biological specimen resembling a bridge truss. The apexes of the triangles are called landmarks.

tuple: a sequence of items, analogous to a row, which is a sequence of columns.

two-state character: binary character.

type: or *nomenclatural type* is that element to which the name of a taxon is permanently attached. In botany, at the level of species and below, a specimen is designated to represent the name or *type* of a taxonomic concept. The type specimen is not necessarily the most biologically typical or representative element of the taxon.

It serves as a point of reference for taxonomic research in that it ties a name and a concept to a particular, archived specimen. At the levels of genus and above, type concepts consist of a designated taxonomic entity at a level below. For example, a *type species* is usually the species the original author had in mind when establishing the genus. In cases where a historical author did not indicate a type for a genus, a type species is subsequently designated. The type system of biological nomenclature was designed to promote the consistent and conservative application of scientific names to taxonomic concepts.

type locality: locality where the name-bearing type (holotype, lectotype, neotype, syntypes) was originally collected.

user's view: the user's concept of the arrangement of the information stored in a database that differs from the actual internal structure of the database. The complexity of the data and therefore of the database design may be such that direct data entry and retrieval would be a highly complex process involving a deep understanding of the design of the database. A special-purpose database management system for biologists acts as an interface to the underlying database structure and DBMS, translates a user's requests into commands appropriate to this structure and DBMS, and converts the results of retrievals back into the user's view of the data.

value: character value.

visual recognition: the process of identifying objects in arbitrary scenes.

wide area network: a computer network connecting systems relatively far apart. For example, Internet. *See also* local area network.

window: an area of the computer screen with a border, an interior, and sometimes a label. It may be divided into several panes.

workstation: a high-performance computer, usually used by an individual. *See also* expert workstation.

References

Abbott, L. A., Bisby, F. A., and Rogers, D. J. 1985. *Taxonomic analysis in biology.* New York: Columbia University Press.

Ackerly, S. C. 1990. Using growth functions to identify homologous landmarks on mollusc shells. In: Rohlf, F. J., and Bookstein, F. L. (eds.), *Proceedings of the Michigan Morphometrics Workshop*. Special Publ. No. 2, Ann Arbor: Univ. Michigan Museum Zool.: 339–344.

Adams, R. P. 1974. Computer graphic plotting and mapping of data in systematics. *Taxon* 23(1): 53–70.

Agrawal, R., Borgida, A., and Jagadish, H. 1989. Efficient management of transitive relations in large data and knowledge bases. *Proc. Assoc. Computing Machinery Intern. Conf. Management Data* 18: 253–262.

Aho, A., Hopcroft, J., and Ullman, J. 1983. *Data structures and algorithms.* Menlo Park, Calif.: Addison-Wesley.

Allison, L., Wallace C. S., and Yee, C. N. 1990. Inductive inference over macromolecules. *Annual Rev. Genetics* 22: 512–565.

Allkin, R. 1984. Handling taxonomic descriptions by computer. In: Allkin, R., and Bisby, F. A. (eds.), *Databases in systematics.* Syst. Assoc., vol. 26. London and Orlando: Academic Press: 263–278.

——— 1988. Taxonomically intelligent database programs. In: Hawksworth, D. L. (ed.), *Prospects in systematics.* Syst. Assoc., vol. 36. Oxford: Oxford University Press: 315–331.

Allkin, R., and Bisby, F. A. (eds.) 1984. *Databases in systematics.* Syst. Assoc., vol. 26. New York: Academic Press.

Allkin, R., and White, R. J. 1988. Data management models for biological classification. In: Bock, H. H. (ed.), *Classification and related methods of data analysis.* Amsterdam: Elsevier: 653–660.

——— 1989. XDF: A language for the definition and exchange of biological data sets, description and manual. Las Palmas: presented at the meeting of the TDWG (IUBS Commission).

Allkin, R., White, R. J., and Winfield, P. J. 1992. Handling the structure of biological data. In: Witten, M. (ed.), Mathematical models in medicine. *Mathematical and Computer Modelling* (special issue).

Allkin, R., and Winfield, P. J. 1989. *ALICE user manual, version 2.* Kew, England: Royal Botanic Gardens.

——— 1990. ALICE: The taxonomic and nomenclatural core module. *DELTA Newsletter* 5: 1–3.

——— 1992. Software development strategies for global plant information systems.

In: Bisby, F. A., Russell, G. F., and Pankhurst, R. J., (eds.), *Designs for a global plant species information system*. Oxford: Oxford University Press.

Amarel, S. 1968. On representations of problems of reasoning about actions. *Machine Intelligence* 3: 131–171.

Andreassen, S., Woldbye, M., Falck, B., and Andersen, S. 1987. MUNIN: A causal probabilistic network for interpretation of electromyographic findings. In: *Proc. Tenth Intern. Joint Conf. on Artificial Intelligence, Milan, Italy*. San Mateo, Calif.: Morgan Kaufmann: 366–372.

Andrews, H. C. 1970. *Computer techniques in image processing*. New York: Academic Press.

Anonymous 1985. *International code of zoological nomenclature*. 3d ed. London: International Trust for Zoological Nomenclature and British Museum (Natural History).

——— 1988a. *IUCN Red list of threatened animals*. The IUCN Conservation Monitoring Centre. Cambridge, U.K.: IUCN.

——— 1988b. *CLASSIC—Générateur de systèmes experts en classification et en diagnostic*, Manuel de l'utilisateur, version 2.2. Gentilly, France: ILOG.

——— 1989. *Zoological record search guide*. Philadelphia: BIOSIS.

——— 1990a. How many species are there on earth? *Biometric Bull.* 7: 25.

——— 1990b. *Zoological Record*. 126 (Section 12): i–vii.

Archie, J. W. 1989a. A randomization test for the presence of phylogenetic information in systematic data. *Syst. Zool.* 38: 239–252.

——— 1989b. Homoplasy excess ratios: New indices for measuring levels of homoplasy in phylogenetic systematics and a critique of the consistency index. *Syst. Zool.* 38:253–269.

——— 1989c. Phylogenies of plant families: A demonstration of randomness in DNA sequence data derived from proteins. *Evolution* 43:1796–1800.

——— 1990. Homoplasy excess statistics and retention indices: A reply to Farris. *Syst. Zool.* 39: 169–174.

Ashlock, P. D. 1974. The uses of cladistics. *Ann. Rev. Ecol. & Syst.* 5: 81–99.

Atkinson, W. D., and Gammerman, A. 1987. An application of expert system technology to biological identification. *Taxon* 34: 705–714.

Ayala, F. J. 1978. The mechanisms of evolution. *Scientific American* 239(3): 56–69.

Bacus, J. W. 1971. *An automated classification of the peripheral blood leukocytes by means of digital image processing*. Ph.D. Thesis, University of Illinois Medical Center, Chicago.

——— 1976. A whitening transformation for two-color blood cell images. *Pattern Recognition* 8: 53–60.

——— 1979. The development of automated differential systems. College of American Pathologists. *Differential Leucocyte Counting* 1: 95–117.

Bacus, J. W. , Belanger, M. G., Aggarwal, R. K., and Trobaugh, F. E. 1976. Image processing for automated erythrocyte classification. *J. Histochem. Cytochem.* 24: 195.

Bacus, J. W., Watt, S., and Trobaugh, F. E. 1980. Clinical evaluation of a new electrical impedance instrument for counting platelets in whole blood. *J. Clinical Path.* 73: 655–663.

Bacus, J. W., and Weens, J. H. 1977. An automated method of differential blood cell classification with application to the diagnosis of anaemia. *J. Histochem. Cytochem.* 25: 614–632.

Ballard, D. H., and Brown, C. M. 1982. *Computer vision.* Englewood Cliffs, N.J.: Prentice-Hall.

Barnsley, M. F., Ervin, V., Hardin, D., and Lancaster, J. 1986. Solution of an inverse problem for fractals and other sets. *Proc. Natl. Acad. Sci. USA*, 83: 1975–1977.

Barr, A., and Feigenbaum, E. 1982. *The handbook of artificial intelligence,* vol. 2. San Mateo, Calif.: William Kaufmann.

Barrera, R., and Al-Taha, K. 1990. Models in temporal knowledge representation and temporal DBMS. NCGIA Technical Paper 90-8.

Barry, M., Cyrluk, D., Kapur, D., Mundy, J., and Nguyen, V. 1988. A multi-level geometric reasoning system for vision. *Artificial Intelligence* 37: 291–332.

Barsalou, T. 1989. An object-based architecture for biomedical expert database systems. In: *Proc. Twelfth Symp. Computer Applications in Medical Care, Washington, D.C.* Los Angeles, Calif.: IEEE Computer Society Press.

Beaman, J. H. 1990. *Revision of* Hieracium *(Asteraceae) in Mexico and Central America.* Ann Arbor, Mich., American Society of Plant Taxonomists, Systematic Botany Monographs No. 29.

Beaman, J. H., and Regalado, Jr., J. C. 1989. Development and management of a microcomputer specimen-oriented database for the flora of Mount Kinabalu. *Taxon* 38: 27–42.

Berliner, H. J. 1980. Backgammon program beats world champion. *Artificial Intelligence,* 14: 205–220.

Binford, T., Levitt, T., and Mann, W. 1989. Bayesian inference in model-based machine vision. In: Kanal, L., Lemmer, J., and Levitt, T. (eds.), *Uncertainty in Artificial Intelligence 3*. New York: North Holland: 73–96.

Bisby, F. A. 1984. Information services in taxonomy. In: Allkin, R. and Bisby, F. A. (eds.), *Databases in systematics*. Syst. Assoc., vol. 26. New York: Academic Press: 17–33.

Bookstein, F. L. 1978. *The Measurement of biological shape and shape change. Lecture notes in biomathematics*, vol. 24, Berlin: Springer-Verlag.

——— 1986. Size and shape spaces for landmark data in two dimensions (with discussion and rejoinder). *Stat. Sci.* 1: 181–242.

——— 1989. Principal warps: Thin-plate splines and the decomposition of deformations. *IEEE Trans. Pattern Analysis Machine Intelligence* 11: 567–585.

——— 1990. Introduction to methods for landmark data. In: Rohlf, F. J., and Bookstein, F. L. (eds.) *Proceedings of the Michigan Morphometrics Workshop*. Special Publ. No. 2, Ann Arbor, Univ. Michigan Museum Zool.: 215–225.

Bookstein, F. L., and Sampson, P. 1987. Statistical models for geometric components of shape change. *Proc. Section Statist. Graphics, 1987 Ann. Meet. Am. Statist. Assoc.*: 18–27.

Bovik, A. C. 1988. On detecting edges in speckle imagery. *IEEE Trans. Acoustics Speech Signal Processing*, 36: 1618–1627.

Brachman, R. J. 1988. The basics of knowledge representation and reasoning. *AT&T Technical Journal* 67:7–24.

Brachman, R. J., and Levesque, H. 1986. On knowledge base management systems. In: Brodie, M. L., and Mylopoulos, J. (eds.), *The knowledge level of a KBMS*. New York: Springer-Verlag: 9–12.

Breiman, L., Friedman, J. H., Olshen, R. A., and Stone, C. J. 1984. *Classification and regression trees*. Monterey, Calif.: Wadsworth and Brooks.

Bremer, K. 1988. The limits of amino acid sequence data in Angiosperm phylogenetic reconstruction. *Evolution* 42: 795–803.

Brenan, J. P. M. 1975. EDP in major herbaria. In: Brenan, J. P. M, Ross, R., and Williams, J. T. (eds.), *Computers in botanical collections*. New York: Plenum Press: 9–16.

Brenan, J. P. M., Ross, R., and Williams, J. T. (eds.) 1975. *Computers in botanical collections*. New York: Plenum Press.

Brinkley, J. F. 1985. Knowledge-driven ultrasonic three-dimensional organ modelling. *IEEE Trans. Pattern Analysis Machine Intelligence* 7(4): 431–441.

Brodie, M. L. 1988. Future intelligent information systems: AI and database technologies working together. In: Mylopoulos, J., and Brodie, M. L. (eds.), *Readings in artificial intelligence and databases*. San Mateo, Calif.: Morgan Kaufmann: 623–642.

Brodie, M. L., and Manola, F. 1988. Database management: A survey. In: Mylopoulos, J., and Brodie, M. L. (eds.). *Readings in artificial intelligence and databases*. San Mateo, Calif.: Morgan Kaufmann: 10–34.

Brodie, M. L., and Mylopoulos, J. 1986. *On knowledge base management systems*. New York: Springer-Verlag.

Brown, W. 1990. A new tree of life takes root. *New Scientist,* 11 August: 30.

Buchanan, B. G. 1982. Research on expert systems. In: Hayes, J., and Michie, D., (eds.), *Machine intelligence,* vol. 10. Chichester, England: Ellis Howard: 269–299.

Buchanan, B. G., and Shortliffe, E. (eds.) 1984. *Rule-based expert systems: The MYCIN experiments of the Stanford Heuristic Programming Project*. Reading, Mass.: Addison-Wesley.

CACM 1990. Special issue on object-oriented design. *Comm. ACM* 33, No. 9, September.

Callomon, J. H. 1981. Dimorphism in ammonoids. In: House, M. R., and Senior, J. R. (eds.), *The Ammonoidea*. Syst. Assoc., vol. 18. New York: Academic Press: 257–273.

Camin, J. H., and Sokal, R. R. 1965. A method for deducing branching sequences in phylogeny. *Evolution* 19: 311–326.

Carmichael, J. W., Kendrick, W. B., Conners I. L., and Sigler. L. 1980. *Genera of Hyphomycetes*. Edmonton: University of Alberta Press.

Cavalli-Sforza, L. L., and Edwards, A. W. F. 1967. Phylogenetic analysis: models and estimation procedures. *Am. J. Human Genetics* 19: 233–257.

Cavender, J. 1978. Taxonomy with confidence. *Math. Biosc*. 40: 271–280.

Celko, J. 1989. Parts explosion in SQL. *DBMS* (November): 40–43.

Cercone, N., and McCalla, G. (eds.) 1987. *The knowledge frontier*. New York: Springer-Verlag.

Cerny, W. 1984. *A field guide in colour to birds*. London: Octopus Books.

Chabris, C. F. 1989. *Artificial intelligence and Turbo C*. Homewood, Ill.: Dow Jones-Irwin.

Chailloux, J., Devin M., Dupont F., Hullot J.-M., Serpette B., and Vuillemin J. 1986. *Le-Lisp de l'INRIA, Version 15.2, Le Manuel de Référence,* 2d ed. Rocquencourt: INRIA.

Chapman, R. E. 1990. Conventional Procrustes approaches. In: Rohlf, F. J., and Bookstein, F. L. (eds.), *Proceedings of the Michigan Morphometrics Workshop.* Special Publ. No. 2, Ann Arbor: Univ. Michigan Museum Zool.: 251–267.

Charniak, E., and Goldman, R. 1989. A semantics for probabilistic quantifier-free first-order languages, with particular application to story understanding. In: *Proc. Eleventh Intern. Joint Conf. on Artificial Intelligence*, Detroit.

Cheeseman, P., and Kanefsky, R. 1990. Evolutionary tree reconstruction. In: *Minimum message length encoding.* AAAI Spring Symposium, Stanford, Calif.: Stanford University.

Chernoff, B. 1986. Systematics and long-range ecologic research. In: Kim, K. C., and Knutson, L. (eds.), *Foundations for a national biological survey.* Lawrence, Kans.: Assoc. Syst. Coll.

Chou, P. A. 1988. *Applications of information theory to pattern recognition and the design of decision trees and trellises.* Ph.D. Thesis, Stanford University.

Clifford, H. T., Rogers, R. W., and Dettman, M. E. 1990. Where now for taxonomy? *Nature* 346: 602.

Codd, E. F. 1970. A relational model of data for large shared data banks. *Comm. ACM* 13(6): 377–387.

Cohen, B. 1977. The mechanical discovery of certain problem properties. *Artificial Intelligence* 8: 119–131.

Cohen, D. H., and Sherman, S. M. 1983. The nervous system. In: Berne, R. M., and Levy, M. N. (eds.), *Physiology.* St. Louis: C. V. Mosby Company: 5–21 and 69–356.

Cohen, P., and Feigenbaum, E. 1982. *The handbook of artificial intelligence*, vol. 2. San Mateo, Calif.: William Kaufmann.

Colless, D. H. 1970. The phenogram as an estimate of phylogeny. *Syst. Zool.* 19: 352–362.

Conway, L. 1981. The MPC adventures: Experiences with the generation of VLSI design and implementation methodologies. *Invited Lecture, 2d CaliforniaLTECH Conf. on VLSI.* Palo Alto, Calif.: VLSI-81–2, XEROX Palo Alto Research Center.

COSMOS 1987. *Advanced Revelation.* Wash., D.C., and New York: Bellevue.

Cox, R. 1946. Probability, frequency and reasonable expectation. *Am. J. Phys.* 14: 1–13.

Cracraft, J. 1989. Speciation and its ontology: The empirical consequences of alternative species concepts for understanding patterns and processes of differentiation. In: Otte, D., and Endler, J. A. (eds.), *Speciation and its consequences.* Sunderland, Mass.: Sinauer Assoc.: 28–59.

Croft, J. R. 1989. Discussion paper and data dictionary. HISPID (Herbarium Information Standards and Protocols for the Interchange of Data) Workshop, Canberra, Australia.

Crosby, M. R., and Magill, R. E. 1988. *TROPICOS: A botanical database system at the Missouri Botanical Garden.* St. Louis, Missouri, Botanical Garden.

Dallwitz, M. J. 1974. A flexible computer program for generating diagnostic keys. *Syst. Zool.* 23: 50–57.

——— 1980a. A general system for coding taxonomic descriptions. *Taxon* 29: 41–6.
——— 1980b. *User's guide to the DELTA system. A general system for encoding taxonomic descriptions*. 1st ed. CSIRO Division of Entomology, Canberra, Australia.
——— 1984. Automatic typesetting of computer-generated keys and descriptions. In: Allkin, R., and Bisby, F. A. (eds.), *Databases in systematics*. Syst. Assoc., vol. 26. New York: Academic Press: 279–290.
——— 1989a. Diagnostic descriptions from INTKEY and CONFOR. *DELTA Newsletter* 3: 8–13.
——— 1989b. Diagnostic descriptions for groups of taxa. *DELTA Newsletter* 4: 8–13.
Dallwitz, M. J., and Paine, T. A. 1986. *User's guide to the DELTA system—a general system for processing taxonomic descriptions*. 3d ed. CSIRO Div. Entomol., Camberra, Australia, Report 13.
Dallwitz, M. J., and Zurcher, E. J. 1988. *User's guide to TYPSET. A computer typesetting program*. 2d ed. CSIRO Div. Entomol., Camberra, Australia, Report 18.
Dantzig, G. B. 1963. *Linear programming and extensions*. Princeton, N.J.: Princeton University Press.
Dantzig, G. B., and Fulkerson, D. R. 1954. Minimizing the number of tankers to meet a fixed schedule. *Naval Research Logistics Quarterly* 1: 217–222.
Dantzig, G. B., Fulkerson, D. R., and Johnson, S. M. 1954. Solution of a large-scale traveling-salesman problem. *Operations Research* 2: 393–410.
Date, C. J. 1983. *An introduction to database systems*, Reading, Mass.: Addison-Wesley.
——— 1990. *An introduction to database systems*, vol. I. 5th ed. Reading, Mass.: Addison-Wesley.
Davies, R. A. G., and Randall, R. M. 1989. Historical and geographical patterns in eggshell thickness of African fish eagles *Haliaeetus vocifer* in relation to pesticide use in Southern Africa. In: Meyburg, B. U., and Chancellor, R. D. (eds.), *Raptors in the modern world*. Berlin: WWGBP: 501–513.
Davis, R. 1982. Consultation, knowledge acquisition, and instruction. In: Szolovits, P. (ed.), *Artificial intelligence in medicine*. Boulder, Colo.: Westview Press: 57–78.
Davis, R., Buchanan, B., and Shortliffe, E. 1977. Production rules as a representation for a knowledge-based consultation program. *Artificial Intelligence* 8: 15–45.
Day, W. H. E. 1983. Computationally difficult parsimony problems in phylogenetic systematics. *J. Theoret. Biol*. 103: 429–438.
Deacon, J. W. 1984. *Introduction to modern mycology*. 2d ed. Oxford: Blackwell Scientific Publ.
Dean, T., and Boddy, M. 1988. An analysis of time-dependent planning. *Proc. AAAI-88 Seventh Natl. Conf. Artificial Intelligence, American Association for Artificial Intelligence*: 49–54.
Debry, R. W., and Slade, N. A. 1985. Cladistic analysis of restriction endonuclease cleavage maps within a maximum-likelihood framework. *Syst. Zool*. 34: 21–34.
DeLeo, J. M., Schwartz, M., Creasey, H., Cutler, N., and Rapoport, S. I. 1985. Computer-assisted categorization of brain computerized tomography pixels into cerebrospinal fluid, white matter and gray matter. *Computers Biomed. Res*. 18: 79–88.
Dextre Clarke, S. G. 1988. The use and future of bibliographic database systems. In:

Hawksworth, D. L. (ed.), *Prospects in systematics*. Syst. Assoc., vol. 36. Oxford: Clarendon Press: 305–314.

Diederich, J. R., and Milton, J. 1991. Creating domain specific metadata for scientific data and knowledge bases. *IEEE Trans. Knowledge Data Engineering* 3(4): 421–434.

——— 1987. Experimental prototyping in Smalltalk. *IEEE Software* 4(3): 50–64.

Donoghue, M. J. 1989. Phylogenies and the analysis of evolutionary sequences, with examples from seed plants. *Evolution* 43: 1137–1156.

Doucet, M. E. 1989. Identification of nematodes in areas where nematology is little developed. In: Fortuner, R. (ed.), *Nematode identification and expert-system technology*. New York: Plenum Press.

Doyle, J., 1979. A truth maintenance system. *Artificial Intelligence* 12: 231–272.

Dreyfus, H. L., and Dreyfus, S. E. 1986. *Mind over machine*. New York: Free Press, Macmillan.

Duncan, J. S. 1987. Knowledge directed left ventricular boundary detection in equilibrium radionucliotide angiography. *IEEE Trans. Medical Imaging* 6: 325–336.

Duncan, T., and Meacham C. A. 1986. *MEKA Version 1.1. A general purpose multiple-entry key algorithm*. Berkeley: University of California, Herbarium.

Dunn, G., and Everitt B. S. 1982. *An introduction to mathematical taxonomy*. Cambridge University Press.

Dunn, S. M., and Gupta, A. P. 1988. Computer imaging of arthropod immonocytes. *18th International Congress of Entomology, Vancouver.*

Dunn, S. M., Shemlon, S., and Gupta, A. P. 1990. Interpreting transmission electron micrographs of insect hemocytes. *J. Computer Assisted Microscopy* 2: 133–160.

Dunn, S. M., and Shemlon, S. In press. Expectation driven image segmentation. *Visual communications and image representation*.

Edwards, A. W. F., and Cavalli-Sforza, L. L. 1964. Reconstruction of evolutionary trees. In: Heywood, V. H., and McNeill, J. (eds.), *Phenetic and phylogenetic classification*, London: System. Assoc. Publ. No. 6.: 67–76.

Egenhofer, M. 1989. A formal definition of binary topological relationships. In: Litwin, W., and Schek, H.-J. (eds.), *Third Intern. Conf. Foundations of Data Organization and Algorithms (FODO), Paris, France, Lecture Notes in Computer Science,* vol. 367. New York: Springer-Verlag: 457–472.

Egenhofer, M., and Herring, J. 1990. A mathematical framework for the definition of topological relationships. In: Brassel, K., and Kishimoto, H. (eds.), *Fourth Intern. Symp. Spatial Data Handling*. Zurich, Switzerland.: 803–813.

Ejiri, M., Yoda, H., Sakou, H., and Sakamoto, Y. 1989. Knowledge-directed inspection for complex multilayered patterns. *Machine Vision and Applications* 2: 155–166.

Eldredge, N., and Cracraft, J. 1980. *Phylogenetic patterns and the evolutionary process*. New York: Columbia University Press.

Endler, J. A. 1989. Conceptual and other problems in speciation. In: Otte, D., and Endler, J. A. (eds.), *Speciation and its consequences*. Sunderland, Mass.: Sinauer Assoc.: 625–648.

Erwin, T. L. 1988. The tropical forest canopy. The heart of biotic diversity. In: Wilson, E. O. (ed.), *Biodiversity*. Washington, D.C.: Natl. Acad. Sci.: 123–129.

Estabrook, G. F. 1972. Cladistic methodology: A discussion of the theoretical basis for the induction of evolutionary history. *Ann. Rev. Ecol. Syst.* 3: 427–456.

Euzenat, J., and Rechenmann, F. 1987. Maintenance de la vérité dans les systèmes à base de connaissances centrées-objet. *Sixième Congrès Reconnaissance des Formes et Intelligence Artificielle, Antibes, novembre 1987.*

Evans, D. G., Schweitzer, P. N., and Hanna, M. S. 1985. Parametric cubic splines and geological shape descriptions. *Math. Geol.* 17: 611–624.

Faith, D. P. 1992. Cladistic permutation tests for monophyly and nonmonophyly. *Syst. Zool.* (in press).

Faith, D. P., and Cranston, P. S. 1991. Could a cladogram this short have arisen by chance alone? On permutation tests for cladistic structure. *Cladistics* 7: 1–28.

Farris, J. S. 1970. Methods for computing Wagner trees. *Zoology* 19: 21–34.

——— 1972. Estimating phylogenetic trees from distance matrices. *Amer. Nat.* 106: 645–668.

——— 1977. Phylogenetic analysis under Dollo's Law. *Syst. Zool.* 26: 77–88.

——— 1981. Distance data in phylogenetic analysis. In: Funk, V. A., and Brooks, D. R. (eds.), *Advances in cladistics: Proceedings of the first meeting of the Willi Hennig Society*, Bronx, New York Botanical Garden: 3–23.

——— 1983. The logical basis of phylogenetic analysis. In: Platnick, N. I., and Funk, V. A. (eds.), *Advances in cladistics: Proceedings of the second meeting of the Willi Hennig Society.* New York: Columbia University Press 2: 7–36.

——— 1986. Distances and cladistics. *Cladistics* 2: 144–157.

——— 1988. *Hennig86.* Port Jefferson, N.Y.

——— 1989. The retention index and the rescaled consistency index. *Cladistics* 5: 417–419.

Farris, J. S., Kluge, A. G., and Eckhardt, M. J. 1970. A numerical approach to phylogenetic systematics. *Syst. Zool.* 19: 172–189.

Faught, W. S. 1986. Applications of AI in engineering. *IEEE Computer* 19(7): 17–27.

Feigenbaum, E. 1964. *Computers and thought.* New York: McGraw-Hill.

Felsenstein, J. 1973. Maximum likelihood and minimum-steps methods for estimating evolutionary trees from data on discrete characters. *Syst. Zool.* 22: 240–249.

——— 1978. Cases in which parsimony or compatibility methods will be positively misleading. *Syst. Zool.* 27: 401–410.

——— 1981a. Evolutionary trees from gene frequencies and quantitative characters: Finding maximum likelihood estimates. *Evolution* 35: 1229–1242.

——— 1981b. A likelihood approach to character weighting and what it tells us about parsimony and compatibility. *Biol. J. Linnean Soc.* 16: 183–196.

——— 1982. Numerical methods for inferring evolutionary trees. *Quarterly Rev. Biol.* 57: 379–404.

——— 1983. Methods for inferring phylogenies: A statistical view. In: Felsenstein, J. (ed.), *Numerical taxonomy.* Berlin: Springer-Verlag: 315–334.

——— 1984a. The statistical approach to inferring evolutionary trees and what it tells us about parsimony and compatibility. In: Duncan, T., and Stuessy, T. F. (eds.), *Cladistics: Perspectives on the reconstruction of evolutionary history.* New York: Columbia University Press: 169–191.

——— 1984b. Distance methods for inferring phylogenies: A justification. *Evolution* 38: 16–24.

Hendy, M. D., and Penny, D. 1982. Branch and bound algorithms to determine minimal evolutionary trees. *Mathematical Biosciences* 59: 277–290.

——— 1989. A framework for the quantitative study of evolutionary trees. *Syst. Zool.* 38: 297–309.

——— 1992. Spectral analysis of phylogenetic data. *J. Class.* (in press).

Hennig, W. 1950. *Grundzüge einer Theorie der phylogenetischen Systematik*. Berlin: Deutscher Zentralverlag.

——— 1966. *Phylogenetic Systematics*. Trans. D. D. Davis and R. Zangerl. Urbana: University of Illinois Press.

——— 1979. *Phylogenetic Systematics*. Trans. D. D. Davis and R. Zangerl, with a foreword by D. E. Rosen, G. Nelson, and C. Patterson. Urbana: University of Illinois Press.

Henrion, M., Breese, J., and Horvitz, E. 1991. Decision analysis and expert systems. *AI Magazine* 12(4): 64–91.

Herskovits, A. 1986. *Language and spatial cognition*. Cambridge University Press.

Heywood, V. H. 1988. The structure of systematics. In: Hawksworth, D. L. (ed.), *Prospects in systematics*. Syst. Assoc., vol. 36. Oxford: Clarendon Press: 44–56.

Hoagland, K. E. 1990. Hearings on biological diversity bill point to need for redefinition of the relationship between systematists and the Nature Conservancy. *Assoc. Syst. Coll. Newsletter* 18: 3–4.

Horn, B. K. P. 1986. *Robot vision*. Cambridge, Mass.: M.I.T. Press.

Horvitz, E. 1987. Reasoning about beliefs and actions under computational resource constraints. *Proc. Third Workshop on Uncertainty in Artificial Intelligence, Seattle, Wash.*

——— 1988. Reasoning under varying and uncertain resource constraints. In: *Proc. AAAI-88 Seventh National Conference on Artificial Intelligence, Minneapolis, Minn.* San Mateo, Calif.: Morgan Kaufmann: 111–116.

——— 1990. *Computation and action under bounded resources*. Ph.D. Thesis, Stanford University, Stanford, Calif.

Horvitz, E., Breese, J., and Henrion, M. 1988. Decision theory in expert systems and artificial intelligence. *Intern. J. Approximate Reasoning* 2: 247–302.

Horvitz, E., Cooper, G., and Heckerman, D. 1989a. Reflection and action under scarce resources: Theoretical principles and empirical study. In: *Proc. Eleventh Intern. Joint Conf. Artificial Intelligence*: 1121–1127.

Horvitz, E., and Heckerman, D. 1986. The inconsistent use of measures of certainty in artificial intelligence research. In: Kanal, L., and Lemmer, J. (eds.), *Uncertainty in artificial intelligence*. New York: North Holland: 137–151.

Horvitz, E., Heckerman, D., and Langlotz, C. 1986. A framework for comparing alternative formalisms for plausible reasoning. In: *Proc. AAAI-86 Fifth Nation. Conf. Artificial Intelligence, Philadelphia, Penn.* San Mateo, Calif.: Morgan Kaufmann: 210–214.

Horvitz, E., Heckerman, D., Ng, K., and Nathwani, B. 1989b. Heuristic abstraction in the decision-theoretic Pathfinder system. In: *Proc. Thirteenth Symp. Computer Applications in Medical Care, Washington, D.C.* Los Angeles, Calif.: IEEE Computer Society Press: 178–182.

Horvitz, E., and Rutledge, G. 1991. Time-dependent utility and action under uncertainty. In: *Proc. Seventh Conf. on Uncertainty in Artificial Intelligence, Los Angeles, Calif.*: San Mateo, Calif.: Morgan Kaufmann.

Howard, R. 1966. Decision analysis: Applied decision theory. In: Hertz, D., and Melese, J. (eds.), *Proc. Fourth Intern. Conf. Operational Research*: 55–77.

——— 1968. The foundations of decision analysis. *IEEE Trans. Systems Sc. Cybernetics* 4: 211–219.

——— 1989. Knowledge maps. *Management Science* 35(8): 903–922.

Howard, R., and Matheson, J. 1981. Influence diagrams. In: Howard, R., and Matheson, J. (eds.), *Readings on the principles and applications of decision analysis*, vol. 2. Menlo Park, Calif.: Strategic Decisions Group: 721–762.

Hubel, D. H., and Wiesel, T. N. 1977. Functional architecture of macaque monkey visual cortex. *Proc. Phil. Trans. Royal Soc.* 198: 1–59.

Hughes, N. F. 1989. *Fossils as information: New recording and stratal correlation techniques*. Cambridge University Press.

Humphries, J. M., Biolsi, D., and Beck, R. 1990. *MUSE tutorial and reference manual*. Ithaca, N.Y.: The MUSE Project.

ISO 1987a. *Information processing systems, open systems interconnection, specification of Abstract Syntax Notation One (ASN.1)*. Switzerland: International Standards Organization ISO-8824: 1–50.

——— 1987b. *Information processing systems, open systems interconnection, specification of basic encoding rules for Abstract Syntax Notation One (ASN.1)*. Switzerland: International Standards Organization, ISO-8825: 1–15.

Iversen, G. R. 1984. Bayesian statistical inference. Beverly Hills, London, New Delhi: Sage Publications.

Jain, A. K. 1989. *Fundamentals of digital image processing*. Englewood Cliffs, N.J.: Prentice-Hall.

Janvier, P. 1984. Cladistics: Theorem, purpose and evolutionary implications. In: J. W. Pollard (ed.), *Evolutionary theory: Paths into the future*. New York: Wiley: 39–75.

Jardin, N. 1967. The concept of homology in biology. *Brit. J. Phil. Sci.* 18: 125–139.

Jeffrey, H. J. 1991. *LIBRARY: The expert librarian*. Expert systems with applications, Maxwell House, Fairview Park, Elmsford, N.Y.: Pergamon Press.

Jersy, J., and Smith, T. 1988. A fundamental division in the *Alu* family of repeated sequences. *Proc. Natl. Acad. Sc.* 85: 4775–4778.

Johannes, J. D. 1977. *Automatic thyroid diagnosis via simulation of physician judgment*. Ph.D. Dissertation, Systems and Information Science Dept., Vanderbilt University, Nashville, Tenn.

Judd, W. S. 1979. Generic relationships in the Andromedeae (Ericaceae). *J. Arnold Arboretum* 60: 477–503.

Jukes, T. H., and Cantor, C. R. 1969. Evolution of protein molecules. In: Munro, H. N. (ed.), *Mammalian protein metabolism*. New York: Academic Press: 21–132.

Kapouleas, I., and Kulikowski, C. A. 1988. A model based system for the interpretation of MR human brain scans. *Proc. SPIE Medical Imaging II Conf.*

Karp, P. D. 1988. *A process-oriented model of bacterial gene regulation*. Technical Report KSL-90-75, Knowledge Systems Laboratory, Stanford University, Stanford, Calif.

——— 1990. *The ASN.1 printfile parser and path-manipulation package*. Bethesda, Md., National Center for Biotechnology Information, National Library of Medicine.

Keeney, R., and Raiffa, H. 1976. *Decisions with multiple objectives: Preferences and value tradeoffs*. New York: Wiley.

Kendrick, B. (ed.) 1979. *The whole fungus*. Vols. 1 and 2. Ottawa: National Museums of Canada.

——— 1980. The generic concept in Hyphomycetes: A reappraisal. *Mycotaxon* 11: 339–364.

——— 1985. *The fifth kingdom*. Waterloo, Ontario: Mycologue Publications.

Kiff, L. F. 1989. DDE and the California condor *Gymnogyps californianus:* The end of the story? In: Meyburg, B.-U., and Chancellor, R. D. (eds.), *Raptors in the modern world*. Berlin: WWGBP: 501–513.

Kim, W., and Lochovsky, F. (eds.) 1988. *Object-oriented concepts, applications, and database systems*. Reading, Mass.: Addison-Wesley.

Kimura, M. 1980. A simple method for estimating evolutionary rate of base substitutions through comparative studies of nucleotide sequences. *J. Molecular Evolution* 16: 111–120.

Kishino, H., and Hasegawa, M. 1989. Evaluation of the maximum likelihood estimate of the evolutionary tree topologies from DNA sequence data, and the branching order in Hominoidea. *J. Molecular Evolution* 29: 170–179.

Kluge, A. G., and Farris, J. S. 1969. Quantitative phyletics and the evolution of anurans. *Syst. Zool.* 18: 1–32.

Kock K.-H., Duhamel, G., and Hureau J.-C., 1985. BIOMASS (Biological Investigations of Marine Antarctic Systems and Stocks): Biology and status of exploited antarctic fish stocks: A review. *Biomass Scientific Series*, vol. 6.

Kohonen, T. 1988. *Self-organization and associative memory.* 2d ed. Berlin: Springer-Verlag.

Korf, R. 1980. Towards a model of representation change. *Artificial Intelligence* 14(1): 41–78.

Kuhl, F. P., and Giardina, C. R, 1982. Elliptic Fourier features of a closed contour. *Computer Graphics and Image Processing* 18: 236–258.

Kuipers, B. 1983. The cognitive map: Could it have been any other way? In: Pick, H. L., and Acredolo, L. P. (eds.), *Spatial orientation: Theory, research and application*. New York: Plenum Press.

Lake, J. A. 1987. Rate-independent technique for analysis of nucleic acid sequences: Evolutionary parsimony. *Molecular Biol. Evolution* 4: 167–191.

——— 1990. Origin of the Metazoa. *Proc. Nat. Acad. Sc. USA* 87: 763–766.

Lakoff, G. 1987. *Women, Fire, and Dangerous Things*. University of Chicago Press.

Lance, G. N., and Williams, W. T. 1967. A general theory of classificatory sorting strategies. I: Hierarchical systems. *Computing Journal* 9: 373–380.

Lapage, S. P., Sneath, P. H. A., Lessel, E. F., Skerman, V. B. D., Seeliger, H. P. R., and W. A. Clark. 1975. *International Code of Nomenclature of Bacteria and Statutes of the International Committee on Systematic Bacteriology and Statutes of the Bacteriology Section of the International Association of Microbiological Societies*. (1976 revision). Washington, D.C.: American Society of Microbiological Societies.

Larson, P., and Deshpande, V. 1989. A file structure supporting traversal recursion.

Proceedings of the Association for Computing Machinery, Special Interest Group on Management of Data, International Conference on Management of Data 18: 243–252.

Lawler, E., Lenstra, J., Kan, A. R., and Shmoys, D. 1985. *The traveling salesman problem*. New York: Wiley.

Lebbe, J. 1984. *Manuel d'utilisation du logiciel XPER*. Paris: Micro Application.

——— 1986. Les champignons identifiés par ordinateur. *Science et Vie*, octobre 1986: 126–130.

Ledley, R., and Lusted, L. 1959. Reasoning foundations of medical diagnosis. *Science*, 130: 9–21.

Lee, E. 1990. User-interface development tools. *IEEE Software* 7(3): 31–36.

Lehmann, H. P. 1991. *A Bayesian computer-based approach to the physician's use of the clinical research literature*. Ph.D. Thesis, Stanford University, Stanford, Calif.

Lewis, H. R., and C. H. Papdimitrion. 1978. The efficiency of algorithms. *Sci. Amer.* 238: 96–109.

Liang, T. 1989. *Geometry guided image segmentation*. M.S. Thesis, Rutgers University, New Brunswick, N.J.

Lindberg, D., Sharp, G., Kingsland, L., Weiss, S., Hayes, S., Ueno, H., and Hazelwood, S. 1980. Computer-based rheumatology consultant. In: *Proc. Medinfo*, New York: North Holland: 1311–1315.

Lindsay, R., Buchanan, B. G., Feigenbaum, E. A., and Lederberg, J. 1980. *DENDRAL*. New York: McGraw-Hill.

Lohmann, G. P. 1983. Eigenshape analysis of microfossils: A general morphometric procedure for describing changes in shape. *Math. Geol.* 15: 659–672.

Loof P.A.A. 1978. *The genus* Pratylenchus *Filipjev, 1936 (Nematoda: Pratylenchidae): A review of its anatomy, morphology, distribution, systematics and identification*. Växtskyddsrapporter, Uppsala: Swedish University of Agricultural Sciences.

Luc, M., Baldwin, J. G., and Bell, A. H. 1986. *Pratylenchus morettoi* n.sp. (Nemata: Pratylenchidae). *Revue Nématol.* 9: 119–123.

Luger, G. F., and Stubblefield, W. A. 1989. *Artificial intelligence and the design of expert systems*. Redwood City, Calif.: Benjamin Cummings.

Lynch, K., 1964. *The image of the city*. Cambridge, Mass.: M.I.T. Press.

MacLeod, N. 1990. Digital images and automated image analysis systems. In: Rohlf, F. J., and Bookstein, F. L. (eds.), *Proceedings of the Michigan Morphometrics Workshop*. Special Publ. No. 2, Ann Arbor: Univ. Michigan Museum Zool.: 21–35.

Maddison, W. P., Donoghue, M. J., and Maddison, D. R. 1984. Outgroup analysis and parsimony. *Syst. Zool.* 33: 83–103.

Maier, D., Stein, J., Otis, A., and Purdy, A. 1986. Development of an object-oriented DBMS. *Proc. OOPSLA '86*, ACM, New York: 472–482.

Margulis, L. 1988. Systematics: The view from the origin and early evolution of life. Secession of the Protoctista from the animal and plant kingdoms. In: Hawksworth, D. L. (ed.), *Prospects in systematics*. Syst. Assoc., vol. 36. Oxford: Clarendon Press: 430–443.

Marino, O., Rechenmann, F., and Uvietta, P., 1990. Multiple perspectives and classification mechanism in object-oriented representations. *Proc. ECalifornial'90, Stockholm, Sweden, August 8–10, 1990*. London: Pitman: 425–430.

Mark, D. M., and Frank, A. 1991. Concepts of space and spatial language. In: Maguire, D., Rhind, D., and Goodchild, M. (eds.), *Handbook of GIS*. London: Longmans.

Marr, D. 1975. *Analyzing natural images: A computational theory of texture vision*. Technical Report Artificial Intelligence Memo 334, M.I.T. Artificial Intelligence Laboratory, Cambridge, Mass.

——— 1982. *Vision*. New York and San Francisco: Freeman.

Marr, D., and Hildreth, E. 1980. Theory of edge detection. *Proc. Royal Soc. Ser. B* 207: 187–280.

May, R. M. 1988. How many species are there on earth? *Science* 241: 1441–1448.

Mayr, E. 1969. *Principles of Systematic Zoology*. New York: McGraw-Hill.

——— 1988. Recent historical developments. In: Hawksworth, D. L. (ed.), *Prospects in systematics*. Syst. Assoc., vol. 36. Oxford: Clarendon Press: 31–43.

——— 1990. A natural system of organisms. *Nature* 348: 491.

Maxted, N., White, R. J., and Allkin, R. (submitted). The inference of descriptive data in a taxonomic hierarchy. *Taxon*.

McGranaghan, M. 1985. Pattern, process and a geographic language. *Ontario Geog.* 25: 15–27.

——— 1989. Context-free recursive-descent parsing of location descriptive text. *Proc. Auto-Carto 9, Baltimore, Md., April 2–7, 1989*: 580–587.

——— 1991. Schema and object matching as a basis for interpreting textual specifications of geographical locations. In: Mark, D. M., and Frank, A. U. (eds.), *Cognitive and linguistic aspects of geographic space*. Kluwer Academic Publishers.

McGranaghan, M., and Wester, L. 1988. Prototyping an herbarium collection mapping system. Technical Papers: 1988 ACSM-ASPRS Annual Convention. *GIS* 5: 232–238.

Meacham, C. A. 1981. A manual method for character compatibility analysis. *Taxon* 30: 591–600.

Meikle, R. D. 1980. *Draft index of author abbreviations* Kew, Richmond, U.K.: Royal Botanic Gardens.

Michalski, R. S., and Chilausky, R. L. 1980. Knowledge acquisition by encoding expert rules versus computer induction from examples: A case study involving soybean pathology. *Int. J. Man-Machine Studies* 12: 63–87.

Miller, R., McNeil, M., Challinor, S., Masarie, F., and Myers, J. 1986. The INTERNIST-1/Quick Medical Reference project: Status report. *Western J. Med.* 145: 816–822.

Miller, R., Pople, E., and Myers, J. 1982. INTERNIST-1: An experimental computer-based diagnostic consultant for general internal medicine. *New England J. Med.* 307: 476–486.

Minsky, M. 1975. A framework for representing knowledge. In: P. H. Winston (ed.), *The psychology of computer vision*. New York: McGraw-Hill.

Mishler, B. D., and Donoghue, M. J. 1982. Species concepts: A case for pluralism. *Syst. Zool.* 31: 491–503.

Mitchell, B. T., and Gillies, A. M. 1989. A model-based computer vision system for recognizing handwritten zip codes. *Machine Vision and Applications* 2: 231–243.

Morse, L. E. 1974. Computer programs for specimen identification, key construction and description printing. *Publications of the Museum of the Michigan State University, Biol.*, 5, 1–128.

Moss, W. W., and Power, D. M. 1975. Semi-automatic data recording. *Syst. Zool.* 24: 199–208.

Mounsey H., and Tomlinson, R. 1988. *Building databases for global science*. London: Taylor and Francis.

Mui, J. K., Fu, K. S., and Bacus, J. W. 1978. Feature selection in automated classification of blood cell neutrophils. *Proc. IEEE Pattern Recognition and Image Processing Conf.*: 486.

Naughton, J., Ramakrishnan, R., Sagiv, Y., and Ullman, J. 1989. Efficient evaluation of right-, left-, and multi-linear rules. *Proc. Assoc. Computing Machinery, Special Interest Group on Management of Data, International Conference on Management of Data*, 18: 235–242.

Nebel, B., 1985. How well does a vanilla loops into a frame? *Data & Knowledge Engineering* 1: 181–194.

Neff, N. A. 1986. A rational basis for a priori character weighting. *Syst. Zool.* 35: 110–125.

Nelson, G. 1989. Species and taxa: Systematics and evolution. In: Otte, D., and Endler, J. A. (eds.), *Speciation and its consequences*. Sunderland, Mass.: Sinauer Assoc.: 60–81.

Nevatia, R. 1982. *Machine perception*. Englewood Cliffs, N.J.: Prentice-Hall.

Newell, A. 1958. *Report on a general problem-solving program*. Technical Report P-1584, Rand Corp.

Newell, A., and Simon, H. 1963. GPS, a program that simulates human thought. In: *Computers and Thought*. New York: McGraw-Hill.

Niedenzu, F. 1890. Uber den anatomischen Bau der Laubblatter der Arbutoideae und Vaccinioideae in Bographischen Verbreitung. *Botanische Jahrbucher fur Systematik, Pflanzengeschichte und Pflanzengeographie* 11: 134–263.

Nilsson, N. J. 1980. *Principles of Artificial Intelligence*. Palo Alto, Calif.: Tioga Pub.

O'Callaghan, J., and Mark, D. M. 1984. The extraction of drainage networks from digital elevation data. *Computer Vision, Graphics and Image Processing* 28: 323–344.

Olmsted, S. 1983. *On representing and solving decision problems*. Ph.D. Thesis, Department of Engineering-Economic Systems, Stanford University, Stanford, Calif.

Ossorio, P. G. 1964. Classification space. *Multivariate Behavioral Research* 1: 479–524.

Otte, D., and Endler, J. A. (eds.), 1989. *Speciation and its consequences*. Sunderland, Mass.: Sinauer Assoc.

Paeth, A. W. 1986. *The IM Raster Toolkit design, implementation and use*. Technical Report CS-86-65, Computer Science Department, University of Waterloo, Waterloo, Ontario.

Pankhurst, R. J. 1970. A computer program for generating diagnostic keys. *Computer J.* 12: 145–151.

——— 1971. Botanical keys generated by computer. *Watsonia* 8: 357–368.

——— 1975. Identification by matching. In: Pankhurst, R. J. (ed.), *Biological identification with computers*. London and New York: Academic Press: 79–91.

——— 1978a. *Biological identification: The principles and practice of identification methods in biology*. London: Edward Arnold.

——— 1978b. The printing of taxonomic descriptions by computer. *Taxon* 27: 65–68.

——— 1983a. An improved algorithm for finding diagnostic taxonomic descriptions. *Math. Biosciences* 65: 209–218.

——— 1983b. The construction of a floristic database. *Taxon* 32: 193–202.

——— 1984. A review of herbarium catalogues. In: Allkin, R., and Bisby, F. A. (eds.), *Databases in systematics*. Syst. Assoc., vol. 26. New York: Academic Press: 155–164.

——— 1986. A package of computer programs for handling taxonomic databases. *Computer Applications in the Biosciences* 2: 33–39.

——— 1988a. Database design for monographs and floras. *Taxon* 37: 733–746.

——— 1988b. An interactive program for the construction of identification keys. *Taxon* 37: 747–755.

——— 1989. A computer program with colour graphics to identify orchids. *Orchid Review* 97: 53–55,67.

——— 1991. *Practical taxonomic computing*. Cambridge University Press.

Pankhurst, R. J., and Aitchison, R. R. 1975. An on-line identification program. In: Pankhurst, R. J. (ed.), *Biological identification with computers*. London and New York: Academic Press: 181–185.

Papadimitriou, C., and Steiglitz, K. 1982. *Combinatorial optimization: Algorithms and complexity*. Englewood Cliffs, N.J.: Prentice-Hall.

Partridge, T. R., Dallwitz, M. J., and Watson, L. 1988. *A primer for the DELTA system on MS-DOS and VMS*. 2d ed. CSIRO Aust. Div. Entomol. Rep. No. 38, 1–17.

Patrick, E. 1977. Review of pattern recognition in medicine. *IEEE Trans. Syst. Man Cybernetics* 6.

Patterson, C. 1980. Cladistics. *Biologist* 27: 234–240.

——— 1982a. Cladistics and classification. *New Scientist* 94: 303–306.

——— 1982b. Morphological characters and homology. In: Joysey, K. A., and Friday, A. E. (eds.), *Problems of phylogenetic reconstruction*. Syst. Assoc., vol. 21. London: Academic Press: 21–74.

——— 1988. The impact of evolutionary theories on systematics. In: Hawksworth, D. L. (ed.), *Prospects in systematics*. Syst. Assoc., vol. 36. Oxford: Clarendon Press: 59–91.

Pauker, S., Gorry, G., Kassirer, J., and Schwartz, W. 1976. Toward the simulation of clinical cognition: Taking a present illness by computer. *Am. J. Med.* 60: 981–995.

Pavé, A., and Rechenmann, F., 1986. Computer aides modelling in biology: An artificial intelligence approach. In: Kerckhoffs, E. J. H., Vansteenkiste, G. C., and Zeigler, B. P. (eds.), *A.I. Applied to Simulation 3*. Society for Computer Simulation, Simul. Serie 18: 52–66.

Pavlidis, T. 1982. *Algorithms for graphics and image processing*. Rockville, Md.: Computer Science Press.

——— 1986. A vectorizer and feature extractor for document recognition. *Computer Vision, Graphics and Image Processing* 35: 111–127.

Pearl, J. 1988. *Probabilistic reasoning in intelligent systems: Networks of plausible inference*. San Mateo, Calif.: Morgan Kaufmann.

Penny, D., and Hendy, M. D. 1987. TurboTree: A fast algorithm for minimal trees. *CaliforniaBIOS* 3: 183–188.

Peters, J. L. 1934–1986. *Check-list of birds of the world.* Cambridge Museum of Comparative Zoology, 16 vols.

Petrini, O., and Rusca, C. 1989. Knowledge-based expert systems in Mycology: Tolerance is a necessary attribute. *Mycologist* 3: 128–130.

Peuquet, D. 1986. The use of spatial relationships to aid database retrieval. *Proc. 2d Intern. Symp. Spatial Data Handling, Seattle, Washington.*: 459–471.

——— 1988. Toward the definition and use of complex spatial relationships. *Proc. Third Intern. Symp. Spatial Data Handling, Sydney, Australia*: 211–224.

Peuquet, D., and Zhan, C. 1987. An algorithm to determine direction between arbitrarily-shaped polygons in a plane. *Pattern recognition* 20: 65–74.

Piraud, F. 1988. *Amélioration de la convivialité de la classification dans SHIRKA par l'utilisation d'images au cours du raisonnement.* D.E.A. Thesis, Université C. Bernard, Laboratoire de Biométrie, Lyon, France.

Platnick, N. I. 1979. Philosophy and the transformation of cladistics. *Syst. Zool.* 28: 537–546.

Plowwright, R. C., and Stevens, W. P. 1973. A numerical taxonomic analysis of the evolutionary relationships of *Bombus* and *Psithyrus* (apidae: Hymenoptera). *Can. Entomol.* 105: 733–743.

Pollack, A. 1991. Transforming the decade: 10 critical technologies, software writing—from an art to a science. *New York Times*, vol. CXL, No. 48467.

Poole, D., and Goebel, R. 1988. *Representation and reasoning: A logical introduction to artificial intelligence*. Computer Science Department, University of Waterloo, Waterloo, Ontario.

Pople, H. 1982. Heuristic methods for imposing structure on ill-structured problems: The structuring of medical diagnostics. In: Szolovits, P. (ed.), *Artificial intelligence in medicine*. Boulder, Colo.: Westview Press: 119–190.

Pramanik, S., and Vineyard, D. 1990. *Query processing in partially ordered distributed database systems.* Tech. Rep. 1, Computer Science Department, Michigan State Univ.

Preparata, F., and Shamos, M., 1985. *Computational geometry: An introduction.* New York: Springer-Verlag.

Prewitt, J. M. S., and Mendelsohn, M. L. 1966. The analysis of cell images. *Ann. New York Acad. Sci.* 128: 1035.

Puvilland, P., Marcoux, E., and Rechenmann, F. 1989. *SADIG: Un système expert de classification précoce de gisements d'or*. Actes de la Convention IA 88→89 January 1989, Paris.

Quinlan, J. R. 1979. Discovering rules by induction from large collections of examples. In: Michie, D. (ed.), *Expert systems in the microelectronic age*. Edinburgh University Press: 168–186.

Radford, A. E. 1986. *Fundamentals of plant systematics*. New York: Harper and Row.

Raiffa, H., and Schlaifer, R. 1961. *Applied statistical decision theory*. Cambridge, Mass.: Harvard University Press.

Raup, D. M. 1966. Geometric analysis of shell coiling: General problems. *J. Paleontol.* 40: 1178–1190.

Rechenmann, F. 1985. SHIRKA: Mécanismes d'inférence sur une base de connaissance centrée-objet. *Cinquième Congrès Reconnaissance des Formes et Intelligence Artificielle, Grenoble, novembre 1985.*

Rechenmann, F., Fontanille, P., and Uvietta, P. 1989. *SHIRKA: Manuel d'utilisation.* Document INRIA—Lab. ARTEMIS/IMAG. Grenoble.

Reeves, A. P. 1984. Parallel computer architectures for image processing. *Computer Vision, Graphics and Image Processing* 25: 68–84.

Reynolds, G., Irwin, N., Hanson, A., and Riseman, E. 1984. Hierarchical knowledge-directed object extraction using a combined region and line representation. *Proc. IEEE Computer Soc. Conf. Computer Vision and Pattern Recognition*: 238–247.

RHODNIUS 1986. *Empress database systems, product overview.* RHODNIUS, 250 Bloor Street East, Toronto, Canada.

Riazanoff, S., Cerevelle, B., and Chorowicz, J. 1988. Ridge and valley line extraction from digital elevation models. *Intern. J. Remote Sensing* 9: 1175–1183.

Ride, W. D. L. 1988. Towards a unified system of biological nomenclature. In: Hawksworth, D. L. (ed.), *Prospects in systematics.* Syst. Assoc., vol. 36. Oxford: Clarendon Press: 332–353.

Ridley, M. 1986. *Evolution and classification: the reformation of cladism.* London and New York: Longman.

Rieppel, O. C. 1988. *Fundamentals of comparative biology.* Basel and Boston: Birkhauser Verlag.

Riskin, E. A., Daly, E. M., and Gray, R. M. 1989. Pruned tree-structured vector quantization in image coding. In: *Proc. ICASSP, May 1989, Glasgow, Scotland*: 1735–1738.

Robinson, B. L., and Greenman, J. M. 1904. Revision of the Mexican and Central American species of *Hieracium. Proc. Am. Acad. Arts* 40: 14–24.

Rogosa, M., Krichevsky, M. I., and Colwell, R. R. 1986. *Coding microbiological data for computers.* New York: Springer-Verlag.

Rohlf, F. J. 1970. Adaptive hierarchical clustering schemes. *Syst. Zool.* 18: 58–82.

——— 1982. Consensus indices for comparing classifications. *Math. Biosc.* 59: 131–144.

——— 1986. The relationships among eigenshape analysis, Fourier analysis, and the analysis of coordinates. *Math. Geol.* 18: 845–854.

——— 1990a. An overview of image processing, and analysis techniques for morphometrics. In: Rohlf, F. J., and Bookstein, F. L. (eds.), *Proceedings of the Michigan Morphometrics Workshop.* Special Publ. No. 2, Ann Arbor: Univ. Michigan Museum Zool.: 37–60.

——— 1990b. Fitting curves to outlines. In: Rohlf, F. J., and Bookstein, F. L. (eds.) *Proceedings of the Michigan Morphometrics Workshop.* Special Publ. No. 2, Ann Arbor, Mich., Univ. Michigan Museum Zool.: 167–177.

——— 1990c. Morphometrics. *Ann. Rev. Ecol. Syst.* 21: 299–316.

——— 1991. *NTSYS-pc: Numerical taxonomy and multivariate analysis system. Version 1.60.* Setauket, N.Y.: Exeter Pub.

——— 1992. The analysis of shape variation using ordinations of fitted functions. In: Sorensen, J. T. (ed.), *Ordinations in the study of morphology, evolution and systematics of insects: Applications and quantitative genetic rationales.* Amsterdam: Elsevier.

Rohlf, F. J., and Archie, J. W. 1984. A comparison of Fourier methods for the description of wing shape in mosquitoes (Diptera: Culicidae). *Syst. Zool.* 33: 302–317.

Rohlf, F. J., and Bookstein, F. L. (eds.) 1990 *Proceedings of the Michigan Morphometrics Workshop.* Special Publ. No. 2, Ann Arbor: Univ. Michigan Museum Zool.

Rohlf, F. J., Chang, W. S., Sokal, R. R., and Kim, J. 1990. Accuracy of estimated phylogenies: Effects of tree topology and evolutionary model. *Evolution* 39: 40–59.

Rohlf, F. J., Gilmartin, A. J., and Hart, G. 1982. The Kluge-Kerfoot phenomenon—a statistical artifact. *Evolution* 37: 180–202.

Rohlf, F. J., Kishpaugh, J., and Kirk, D. 1981. *NT-SYS—Numerical taxonomy system of multivariate statistical programs.* Stony Brook, N.Y.: The State University of New York.

Rohlf, F. J., and Slice, D. 1990. Extensions of the Procrustes method for the optimal superimposition of landmarks. *Syst. Zool.* 39: 40–59.

Rohlf, F. J., and Sokal, R. R. 1967. Taxonomic structure from randomly and systematically scanned biological images. *Syst. Zool.* 16: 246–260.

Rose, M.T. 1990. *The ISO development environment: User's manual.* Mountain View, Calif.: Performance Systems International.

Rosenfeld, A., and Kak, A. C. 1982. *Digital picture processing.* Vols. 1 and 2. New York: Academic Press.

Roth, L. V. 1988. The biological basis of homology. In: Humphries, C. J. (ed.), *Ontogeny and systematics.* New York: Columbia University Press: 1–26.

Rumelhart, D., Hinton, G., and Williams, R. 1986. Learning internal representations by error propagation. In: Rumelhart, D., and McClelland, J. (eds.), *Parallel distributed processing: Explorations in the microstructure of cognition. Vol. 1: Foundations.* Cambridge, Mass.: M.I.T. Press: 318–362.

Rumelhart, D., and McClelland, J. (eds.) 1986. *Parallel distributed processing: Explorations in the microstructure of cognition. Vol. 1: Foundations.* Cambridge, Mass.: M.I.T. Press.

Russ, J. C. 1990. *Computer-assisted microscopy: The measurement and analysis of images.* New York: Plenum Press.

Russell, S., and Wefald, E. H. 1989. Principles of metareasoning. In: Brachman, R. J., Levesque, H. J., and Reiter, R. (eds.), *Proc. First Intern. Conf. Principles of Knowledge Representation and Reasoning.* Toronto: Morgan Kaufmann.

Saitou, N., and Nei, M. 1987. The neighbor-joining method: A new method for reconstructing phylogenetic trees. *Molec. Biol. Evol.* 4: 406–425.

Salvini-Plawen, L. v., and Mayr, E. 1977. On the evolution of photoreceptors and eyes. In: Hecht, M. K., Steere, W. C., and Wallace, B. (eds.), *Evolutionary Biology*, vol. 10. New York: Plenum Press: 207–263.

Samadani, R. D., Mihovilovic, A., Clauer, C. R., Wiederhold, G., Craven, J., and Frank, L. 1990. Evaluation of an elastic curve technique for finding the auroral oval from satellite images. *IEEE Trans. Geosc. Remote Sensing* 28: 590–597.

Samadani, R., and Vesecky, J. F. 1990. Finding curvilinear features in speckled images. *IEEE Trans. Geosc. Remote Sensing* 28: 669–673.

Samuel, A. L. 1959. Some studies of machine learning using the game of checkers. *IBM Journal of Research and Development* 3: 211–229.

Samuel, A. L. 1967. Some studies of machine learning using the game of checkers. II:

Recent progress. *IBM Journal of Research and Development* 11(6): 601–617.
Sanderson, M. J., and Donoghue, M. J. 1989. Patterns of variation in levels of homoplasy. *Evolution* 43: 1781–1795.
Sankoff, D. D. 1975. Minimal mutation trees of sequences. *SIAM J. Appl. Math.* 28: 35–42.
Sankoff, D. D., and Cedergren, R. J. 1983. Simultaneous comparison of three or more sequences related by a tree. In: Sankoff, D. D., and Kruskal, J. B. (eds.), *Time warps, string edits, and macromolecules: The theory and practice of sequence comparison.* Reading, Mass.: Addison-Wesley: 253–263.
Sarich, V. M. 1969. Pinniped origins and the rate of evolution of carnivore albumins. *Syst. Zool.* 18: 286–295.
Saund, E. 1989. Neural networks for dimensionality reduction. *IEEE Trans. Pattern Analysis Machine Intel.* 11: 304–317.
Savage, L. 1972. *The foundations of statistics.* 2d ed. New York: Dover Pub.
Schoch, R. M. 1986. *Phylogeny reconstruction in paleontology.* New York: Van Nostrand Reinhold.
Schmidt, J. W., and Brodie, M. L. 1983. *Relational database systems: Analysis and comparison.* New York: Springer-Verlag.
Schneiderman, B. 1987. *Designing the user interface.* Reading, Mass.: Addison-Wesley.
Selfridge, P. G. 1986. Locating neuron boundaries in electron micrograph images using "primal sketch" primitives. *Computer Vis. Graphics Image Process.* 34: 156–165.
Serra, J. 1982. *Image analysis and mathematical morphology.* New York: Academic Press.
Shachter, R. 1986. Evaluating influence diagrams. *Operations Research* 34: 871–882.
Shapiro, S., and Rapaport, W. 1987. SNePS considered as a fully intensional propositional semantic network. In: Cercone, N., and McCalla, G. (eds.), *The knowledge frontier.* New York: Springer-Verlag: 262–315.
Shapiro, S., and Woodmansee, G. 1971. A net structure based relational question-answerer. *Proc. Intern. Joint Conf. AI, Washington, D.C.*: 325–346.
Shortliffe, E. 1976. *Computer-based medical consultations: MYCIN.* New York: North Holland.
Shortliffe, E., and Buchanan, B. 1975. A model of inexact reasoning in medicine. *Math. Biosc.* 23: 351–379.
Sieffer, A., 1988. *Constitution d'une base de connaissances centrée-objet en vue de l'identification des poissons des Kerguélen à l'aide du système SHIRKA.* D.E.A. Thesis, Université C. Bernard, Laboratoire de Biométrie, Lyon, France.
Siegel, A. F., and Benson, R. H. 1982. A robust comparison of biological shapes. *Biometrics* 38: 341–350.
Simon, H. 1969. *The sciences of the artificial.* Cambridge, Mass.: M.I.T. Press.
——— 1972. The theory of problem solving. *Inform. Process.* 71: 261–277.
——— 1987. Two heads are better than one: The collaboration between AI and OR. *Interfaces* 17: 8–15.
Simpson, G. G. 1945. The principles of classification and the classification of the mammals. *Bull. Amer. Mus. Nat. Hist.* 85: 1–350.
——— 1961. *Principles of animal taxonomy.* New York: Columbia University Press.

Skov, F. 1989. Hypertaxonomy—a new computer tool for revisional work. *Taxon* 38: 582–590.

Smith, G. R. 1990. Homology in morphometrics and phylogenetics. In: Rohlf, F. J., and Bookstein, F. L. (eds.), *Proceedings of the Michigan Morphometrics Workshop*. Special Publ. No. 2, Ann Arbor: Univ. Michigan Museum Zool.: 325–338.

Smith, T. R., and Peuquet, D. P. 1985. Control of spatial search for queries in a knowledge based geographical information system. *Proc. Intern. Conf. Adv. Tech. for Monitoring and Processing Global Environmental Data*, London: Remote Sensing Society: 439–452.

Smith, T. R., Peuquet, D. P., Menon, S., and Agarwal, P. 1987. KBGIS-II: A knowledge-based geographical information system. *Intern. J. Geo. Inform. Syst.* 1: 149–172.

Smyth, P. 1988. *The application of information theory to problems in decision tree design and rule-based expert systems*. Ph.D. Thesis, California Institute of Technology, Pasadena.

Sneath, P. H. A. 1967. Trend-surface analysis of transformation grids. *J. Zool.* 151(1): 65–122.

Sneath, P. H. A., and Sokal, R. R. 1973. *Numerical taxonomy.* San Francisco: Freeman.

Sober, E. 1989. *Reconstructing the past: Parsimony, evolution, and inference*. Cambridge, Mass.: M.I.T. Press.

Soper, J. H. 1975. EDP and distribution mapping. In: Brenan, J. P. M., Ross, R., and Williams, J. T. (eds.), *Computers in botanical collections.* New York: Plenum Press: 141–166.

Spetzler, C., and Stael von Holstein, C. 1975. Probability encoding in decision analysis. *Management Sci.* 22: 340–358.

Steel, M. A. 1989. *Distributions on bicoloured evolutionary trees*. Ph.D. Thesis, Massey University, Palmerston North, New Zealand.

Steel, M. A. 1990. Distributions on bicoloured evolutionary trees. (Abstract). *Bull. Australian Math. Soc.* 41: 159–160.

Stefik, M. J. 1981. Planning with constraints: Molgen, Part 1. *Artificial Intelligence* 16(2): 111–139.

Stevens, P. F. 1970. Agauria, and Agarista: An example of tropical transatlantic affinity. *Notes Royal Bot. Garden Edinburgh* 30: 341–359.

Stevensen, D. J., Smith, L., and Robinson, G. 1987. Working towards the automatic detection of blood vessels in X-ray angiograms. *Pattern Recognition Letters* 6: 107–112.

Stiassny, M. 1992. Phylogenetic analysis and the role of systematics in the biodiversity crisis. In: Eldredge, N. (ed.), *Systematics, ecology and the biodiversity crisis*. New York: Columbia University Press.

Stonebraker, M. 1986. The design of POSTGRES. *Proc. Intern. Conf. Management of Data*, ACM-SIGMOD.

Strauss, R. E., and Bookstein, F. L. 1982. The truss: Body form reconstruction in morphometrics. *Syst. Zool.* 31: 113–135.

Studier, J. A., and Keppler, K. J. 1988. A note on the neighbor-joining algorithm of Saitou and Nei. *Molec. Biol. Evol.* 5: 729–731.

Stuessy, T. 1990. *Plant taxonomy.* New York: Columbia University Press.

Suetens, P., and Oosterlinck, A. 1987. Using expert systems for image understanding. *Intern. J. Pattern Recognition Artificial Intelligence* 1: 237–250.

Suetens, P., Smets, C., Van de Werf, F., and Oosterlinck, A. 1989. Recognition of the coronary blood vessels on angiograms using hierarchical model-based iconic search. *Proc. IEEE Computer Soc. Conf. Computer Vision and Pattern Recognition:* 576–580.

Swain, P. H., and Hauska, H. 1977. The decision tree classifier: Design and potential. *IEEE Trans. Geosc. Electronics*, July 1977: 142–147.

Swofford, D. L. 1981. On the utility of the distance Wagner procedure. In: Funk, V. A., and Brooks, D. R. (eds.), *Advances in cladistics: Proceedings of the first meeting of the Willi Hennig Society,* Bronx, New York Botanical Gardens: 25–43.

——— 1984. *PAUP— Phylogenetic analysis using parsimony*. Champaign, Ill.: Natural History Survey.

——— 1985. *PAUP—Phylogenetic analysis using parsimony, version 2.4.* Champaign, Ill.: Natural History Survey.

——— 1990. *PAUP—Phylogenetic analysis using parsimony, version 3.0.* Champaign, Ill.: Natural History Survey.

Swofford, D. L., and Maddison, W. P. 1987. Reconstructing ancestral character states under Wagner parsimony. *Math. Biosc.* 87: 199–229.

Swofford, D. L., and Olsen, G. J. 1990. Phylogeny reconstruction. In: Hillis, D. M., and Moritz, C. (eds.), *Molecular systematics*. Sunderland, Mass.: Sinauer Assoc.: 411–501.

Szalay, F. S., and Bock, W. J. 1991. Evolutionary theory and systematics: Relationships between process and patterns. *Z. Zool. Syst. Evolut.-forsch.* 29: 1–39.

Szolovits, P. 1982. Artificial intelligence in medicine. In: Szolovits, P. (ed.), *Artificial intelligence in medicine*. Boulder, Colo.: Westview Press: 1–19.

Tangley, L. 1990. Cataloging Costa Rica's diversity. *Bioscience* 40: 633–636.

Tateno, Y., Nei, M., and Tajima, F. 1982. Accuracy of estimated phylogenetic trees from molecular data. I: Distantly related trees. *J. Molec. Evol.* 18: 387–404.

TDWG 1989. *Draft exchange record format standard for botanical specimen data.* International Working Group on Taxonomic Databases in Plant Sciences. TDWG 5, Las Palmas, November 8–11, 1989.

Templeton, A. R. 1989. The meaning of species and speciation: A genetic perspective. In: Otte, D., and Endler, J. A. (eds.), *Speciation and its consequences*. Sunderland, Mass.: Sinauer Assoc.: 3–27.

Thompson, D. W. 1917. *On growth and form.* London: Cambridge.

Tribus, M. 1969. *Rational descriptions, decisions, and designs*. New York: Pergamon Press.

Trobaugh, F. E., and Bacus, J. W. 1979. Design and performance of the LARC automated leucocyte classifier. *Differential Leucocyte Counting* 2: 119–133.

Tversky, A., and Kahneman, D. 1974. Judgment under uncertainty: Heuristics and biases. *Science* 185: 1124–1131.

USGS 1985. *USGS digital cartographic data standards: Digital line graphs from 1:24,000-scale maps*. U.S. Geological Survey circular 895-C. U.S.G.S.: Alexandria, Va.

——— 1987. *Geographic Names Information System. (Data Users Guide 6)*. U.S. Department of the Interior: Reston, Va.

Ullman, J. D. 1982. *Principles of database systems*. 2d ed. Rockville, Md.: Computer Science Press.

Ullman, S. 1979. *Interpretation of visual motion*. Cambridge, Mass.: M.I.T. Press.

Valdecasas, A. G., Bello, A., Reyes, J., and Becerra, J. 1989. DIRTAX: A taxonomist database. *Science Software* 5(4): 303–308.

Valdecasas, A. G., Elvira, J. R., Becerra, J. M., and Bello, E. 1990. *CONFOR, KEY, DIST, TRANSNT.* Madrid: Museo Nacional de Ciencias Naturales.

Valduriez, P. 1987. Join indices. *Assoc. Computing Machinery Trans. Database Systems* 12: 218–246.

Valduriez, P., and Boral, H. 1986. Evaluation of recursive queries using join indices. *Proc. First Intern. Conf. Expert Database Systems*. Menlo Park, Calif.: Benjamin/Cummings: 197–208.

Vesecky, J. F., Samadani, R., Daida, J. M., Smith, M. P., and Bracewell, R. N. 1988. Observation of sea ice dynamics using synthetic aperture radar images: Automated analysis. *IEEE Trans. Geosc. Remote Sensing* 26: 38–48.

Vesecky J. F., Smith, M. P., and Samadani, R. 1990. Extraction of lead and ridge characteristics from SAR images of sea ice. *IEEE Trans. Geosc. Remote Sensing* 28: 740–744.

von Neumann, J., and Morgenstern, O. 1947. *Theory of games and economic behavior*. Princeton, N.J.: Princeton University Press.

Vossen, G. 1991. Bibliography on object-oriented database management. *ACM SIGMOD Record* 20(1): 24–46.

Wagner, W. H., Jr. 1961. Problems in the classification of ferns. In: *Recent advances in botany from Lectures and Symposia Presented to the Ninth International Botanical Congress, Montreal 1959*. Vol. 1. Toronto: University of Toronto Press: 841–844.

——— 1969. The construction of a classification. In: Sibley, C. G. (ed.), *Systematic biology.* Natl. Acad. Sci. Wash. Publ. No. 1692: 67–90.

Warner, H., Toronto, A., Veasy, L., and Stephenson, R. 1961. A mathematical approach to medical diagnosis: Application to congenital heart disease. *J. Am. Med. Assoc*. 177: 177–183.

Watson, L., Aiken, S. G., Dallwitz, M. J., Lefkovitch, L. P., and Dube, M. 1986. Canadian grass genera: Keys and descriptions in English and French from an automated data bank. *Can. J. Bot.* 64: 53–70.

Watson, L., and Dallwitz, M. J. 1981. An automated data bank for grass genera. *Taxon* 30: 424–429 plus 2 microfiches.

Watson, L., Damanakis, M., and Dallwitz, M. J. 1988. *The grass genera of Greece—Descriptions, classification, keys*. (In Greek.) Heraklion: University of Crete.

Watson, L., Gibbs Russell, G. E., and Dallwitz, M. J. 1989. Grass genera of southern Africa: Interactive identification and information retrieval from an automated data bank. *S. Afr. J. Bot.* 55: 452–463.

Watson, L., and Milne, P. 1972. A flexible system for automatic generation of special-purpose dichotomous keys, and its application to Australian grass genera. *Aust. J. Bot.* 20: 331–352.

Webster, R. D., Kirkbride, J. H., and Reyna, J. V. 1989. New World genera of the Paniceae (Poaceae: Panicoideae). *SIDA* 13: 393–417.

Weiss, J., Kulikowski, C., Amarel, S., and Safir, A. 1978. A model-based method for computer-aided medical decision-making. *Artificial Intelligence* 11: 145–172.

Wellman, M. 1988. *Formulation of tradeoffs in planning under uncertainty*. Ph.D. Thesis, Department of Electrical Engineering and Computer Science, Massachusetts Institute of Technology, Cambridge.

White, R. J., and Allkin, R. 1992a. A language for the definition and exchange of biological data sets. In: Witten, M. (ed.), Mathematical models in medicine. *Mathematical and Computer Modelling* (special issue).

——— 1992b. A strategy for the evolution of database designs. In: Bisby, F. A., Russell, G. F., and Pankhurst, R. J. (eds.), *Designs for a global plant species information system*. Oxford: Oxford University Press.

Whitehead, P. 1990. Systematics: An endangered species. *Syst. Zool.* 39: 179–184.

Wieringa, R. J., and Curweil, P. H. 1986. *Final report on the farmer's aid in plant disease diagnosis*. Department of Computer Science, Agricultural University, The Netherlands.

Wiley, E. O. 1981. *Phylogenetics: The theory and practice of phylogenetic systematics*. New York: Wiley.

Willcox, W. R., and Lapage, S. P. 1975. Methods used in a program for computer-aided identification of bacteria. In: Pankhurst, R. J. (ed.), *Biological identification with computers*. London and New York: Academic Press: 103–119.

Willcox, W. R., Lapage, S. P., Bascombe, S., and Curtis, M. A. 1973. Identification of bacteria by computer: Theory and programming. *J. General Microbiol.* 77: 317–330.

Williams, P. L., and Fitch, W. M. 1989. Finding the minimal change in a given tree. In: Fernholm, B., Bremer, K., and Jornvall, H. (eds.), *The hierarchy of life*. Amsterdam: Elsevier: 453–470.

Williams, W. T. 1971. Principles of clustering. *Ann. Rev. Ecol. Systematics* 2: 303–326.

Williams, W. T., and Dale, M. B. 1965. Fundamental problems in numerical taxonomy. *Advance. Bot. Res.* 2: 35–68.

Wilson, E. O. 1988. The current state of biological diversity. In: E. O. Wilson (ed.), *Biodiversity.* Washington, D.C.: National Academy Press: 3–18.

Winograd, T., and Flores, F. 1986. *Understanding computers and cognition*. Reading, Mass.: Addison-Wesley.

Winston, P. 1975. *The psychology of computer vision*. New York: McGraw-Hill.

Winston, P. H. 1977. *Artificial intelligence*. Reading, Mass.: Addison-Wesley.

——— 1984. *Artificial intelligence*. 2d ed. Reading, Mass.: Addison-Wesley.

Woese, C. R., Kandler, O., and Wheelis, M. L. 1990. Towards a natural system of organisms: Proposal for Domains Archaea, Bacteria and Eucarya. *Proc. Nat. Acad. Sci.* 87: 4576–4579.

Wood, C. E., Jr. 1961. The genera of Ericaceae in the southeastern United States. *J. Arnold Arboretum* 42: 10–80.

Woodger, J. H. 1945. On biological transformations. In: LeGrosClark, W. E., and Medawar, P. B. (eds.), *Essays on growth and form presented to D'Arcy Wentworth Thompson*. Oxford: Clarendon Press: 94–120.

Worboys, M. F. 1991. The role of modal logics in the description of a geographical information system. In: Mark, D. M., and Frank, A. U. (eds.), *Cognitive and linguistic aspects of geographic space*. Norwell, Mass.: Kluwer Academic Publishers.

Xu Zhu, and Dallwitz, M. J. (in preparation). A study of automatic classification, identification, key making and producing descriptions for 13 species of *Elymus* in China by using CDELTA.

Yasnoff, W. A., Mui, J. K., and Bacus, J. W. 1977. Error measures for scene segmentation. *Pattern Recognition* 9: 217–231.

Yasnoff, W. A., Galbraith, W., and Bacus, J. W. 1979. Error measures for objective assessment of scene segmentation algorithms. *Analytic Quantitative Cytology J.* 1: 107–121.

Young, I. T. 1969. *Automated leukocyte recognition*. Ph.D. Thesis, Massachusetts Institute of Technology, Cambridge.

Zahn, C. T., and Roskies, R. Z. 1972. Fourier descriptors for plane closed curves. *IEEE Trans. Comp.* 21: 269–281.

Zajicek, G., and Shohat, M. 1983. On the classification of nucleated red blood cells. *Computers Biomed. Res.* 16: 553–562.

Zarucchi, J. L., Winfield, P. J., Polhill, R. M., Hollis, S. J., Bisby, F. A., and Allkin, R. 1992. The ILDIS project on the world's legume species diversity. In: Bisby, F. A., Russell, G. F., and Pankhurst, R. J. (eds.), *Designs for a global plant species information system*. Oxford: Oxford University Press.

Zerwekh R., and Jeffrey, H. J. (in preparation). Comparison of two methods for primate foot bone identification.

Index

Designed by Laury A. Egan

Composed by The Composing Room of Michigan, Inc.
in Times Roman text and Optima display

Printed on 60 lb. Glatfelter Hi Brite and bound
in Holliston Roxite by The Maple Press Company